T0305717

PRODUCTION PLANNING and INDUSTRIAL SCHEDULING

Second Edition

PRODUCTION PLANNING and INDUSTRIAL SCHEDULING

Examples, Case Studies and Applications

Second Edition

DILEEP R. SULE

CRC Press is an imprint of the
Taylor & Francis Group, an **informa** business

CRC Press
Taylor & Francis Group
6000 Broken Sound Parkway NW, Suite 300
Boca Raton, FL 33487-2742

First issued in paperback 2021

ISBN-13: 978-1-4200-4420-1 (hbk)
ISBN-13: 978-1-03-218001-4 (pbk)
DOI: 10.1201/978142004421810

Publisher's Note

The publisher has gone to great lengths to ensure the quality of this reprint but points out that some imperfections in the original copies may be apparent.

Library of Congress Cataloging-in-Publication Data

Sule, D. R. (Dileep R.)
Production planning and industrial scheduling : examples, case studies, and applications / Dileep Sule. -- 2nd ed.
p. cm.
"A CRC title."
Rev. ed. of: Industrial scheduling. 1997.
Includes bibliographical references and index.
ISBN 978-1-4200-4420-1 (alk. paper)
1. Production scheduling. I. Sule, D. R. (Dileep R.). Industrial scheduling. II. Title.

TS157.5.S85 2007
658.5'3--dc22 2007010013

Visit the Taylor & Francis Web site at
http://www.taylorandfrancis.com

and the CRC Press Web site at
http://www.crcpress.com

Table of Contents

Preface

This book is a substantially enlarged edition of the original title, *Industrial Scheduling*. It now provides a broad outlook on optimization and planning, from the initial stages of plant location and capacity determination to within-plant operations and optimization.

On occasion, production planning and control are looked upon as a collection of techniques with little relationship between the different procedures. For example, when an inventory control discussion defines an economic production quantity, it gives scant regard to the available plant capacity or the different machines through which the product must be processed. Diverse techniques are taught as independent procedures with very little coordination between how the result of one technique may influence the other. Each workstation is optimized independently without realizing its effect on the entire system.

This book integrates logistics and planning in the areas of production planning and scheduling in a broad sense. To plan for production, there must be facilities where products are made. The book starts with the development of strategies for establishing plant locations and their capacities. It discusses forecasting methods to predict demands under differing scenarios. Narrowing down to a plant level, the discussion concentrates on techniques that improve plant efficiencies in various areas. The topics include master production scheduling (MPS), material requirement planning (MRP), and inventory management.

Once decisions about when and how much to produce are made, the next step is to plan for these quantities. If the resources are not limited, the intended production can be achieved without any difficulty. However, with limited resources, scheduling of assets becomes necessary to increase overall plant efficiency, capacity utilization, as well as to reduce the required time to complete all tasks.

Major resources needed in production are plants, machines, and manpower. A manufacturing facility may be thought of as a single machine that is optimized. Proceeding one step further, operations within the plant can be examined in detail. Within a production facility, different machine configurations define various material flow patterns. Arrangements such as flow shops, job shops, open shops, and assembly lines are common. The book illustrates many of the scheduling issues associated with these patterns.

Business scheduling often requires optimization of a bottleneck machine. Some practical examples of industrial scheduling are illustrated in the later chapters, along with manpower scheduling.

For efficient scheduling, there are numerous mathematical techniques illustrated in the literature. However, these techniques are quite often cumbersome, and not easy to understand and apply. The most frequently used methods, therefore, are good heuristics. This book is primarily concerned with easy-to-follow heuristics that can be clearly understood and used.

The book is intended for diverse curricula in engineering, industrial, mechanical, and manufacturing departments as teach topics illustrated in this book, while in business schools, operations management and production departments emphasize these topics. It is projected for use in industrial and/or mechanical engineering technology programs as well. Since each procedure is completely illustrated by a solved example, the book will also be a good reference for practitioners.

Additional material is available from the CRC Web site: www.crcpress.com on *Production Planning and Industrial Scheduling* on the Web download.

Under the menu Electronic Products (located on the left side of the screen), click on Downloads & Updates. A list of books in alphabetical order with Web downloads will appear. Locate this book by a search, or scroll down to it. After clicking on the book title, a brief summary of the book will appear. Go to the bottom of this screen and click on the hyperlinked "Download" that is in a zip file.

Or you can go directly to the Web download site, which is www.crcpress.com/e_products/downloads/default.asp

Acknowledgments

In the writing of this edition, the contributions of a few graduate students need to be recognized. In particular, Manish Chavan's assistance is greatly appreciated. His constant help with whatever was needed resulted in the timely completion of this book. Kishor Joshi, Advait Damle, and James Wang also provided assistance. I thank them and others who I may have inadvertently neglected to mention.

And finally, the staff at Taylor & Francis and Newgen Imaging have been very helpful and cheerful during the entire process, especially Cindy Carelli, the senior acquisitions editor. I thank her and project editor Rachael Panthier for their support.

Author

Dileep R. Sule is the Thurman Laureate professor of industrial engineering at Louisiana Tech University. He has authored three books (one translated into Spanish) and over 70 publications in professional journals and conferences. Dr. Sule is a fellow of the Institute of Industrial Engineering. He received his B.S. in mechanical engineering (1963) from the Birla Institute of Technology, Ranchi, India, and his M.S. (1967) and Ph.D. (1969) in industrial engineering from Texas A&M University, College Station. Dr. Sule has extensive experience in industrial consulting and has worked full-time in the industry in England and Scotland.

1 Introduction to Production Planning and Scheduling

As the title of the book suggests, we discuss two important issues in manufacturing; they are: production planning and scheduling. These topics are critical because they define operational productivity. The proper application of these techniques results in reducing manufacturing cost, satisfying customer demands in a timely manner, and overall better planning and control of manufacturing operations. To compete successfully, we must be efficient and productive. Good production planning and scheduling leads to the achievement of these goals and, therefore, are integral parts of every professionally run organization.

1.1 PRODUCTION PLANNING

Manufacturing organizations exist to produce and supply products that customers need, at a price that they are willing to pay. Successful organizations achieve this aim while making profit for their shareholders. There are very few organizations that exist in monopoly, and most have to compete in the open market. The product price is therefore determined by the competition, and the only way to increase profit is to reduce production and distribution costs. This means managing and operating the organization in an efficient manner.

Production planning and scheduling are two important topics that serve to increase efficiency in manufacturing and improve effectiveness in customer service. Production planning determines what, when, and how much to produce to meet the customers' needs, without excessive inventory or back order costs. Scheduling, on the other hand, determines how to achieve the goals set in production planning when the resources are limited; and, if the goals cannot be realized, how best to set new goals that are optimum and practical with the available resources. Many aspects are involved in production planning, as discussed in the following chapters.

We start the discussion by illustrating a broad scope of planning. It involves forecasting of customer orders, determination of plant capacities, and planning for long range such as 1 year or more. The next step is that of short-range or immediate planning, narrowing the planning period down to monthly and then to weekly levels, involving topics such as production planning, inventory control, and scheduling.

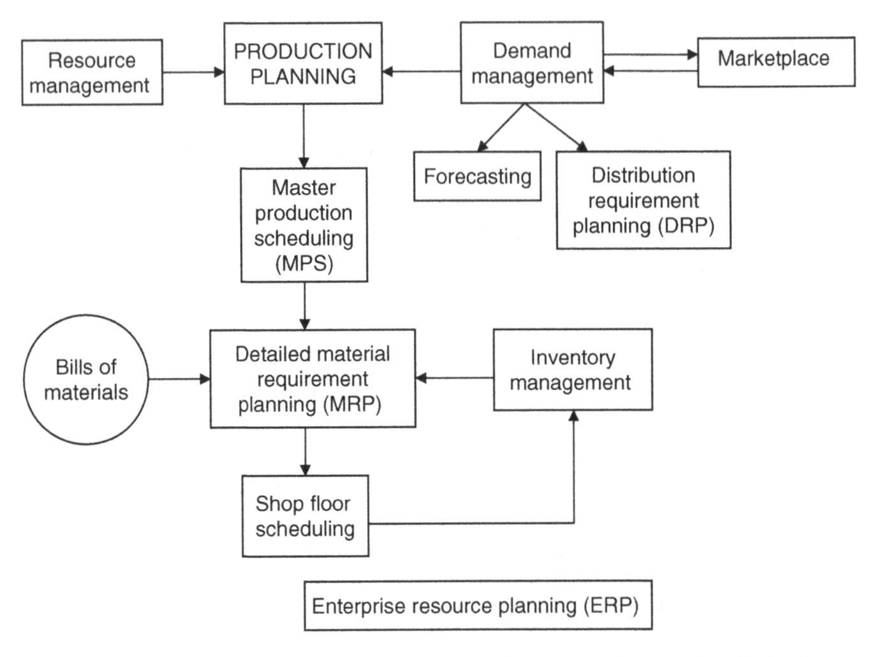

FIGURE 1.1 Diagram connecting various topics that are discussed in the book.

1.1.1 Pictorial View of a Production Planning and Control System

Figure 1.1 connects various topics that are discussed in the book. Every production planning and control system begins with the marketplace. Marketplace demand is an independent demand that relies heavily on factors such as changing consumer test, different seasons in a year, the ups and downs of national economy, and even political winds. With varying demand, it is important to match production with respect to the demand at that time. Different forecasting techniques can be used to make good demand prediction. With expected demand as an input to production and aggregate planning, all resources necessary to meet the demands and plans for the future are evaluated. These might include capital investment, labor force, full-time and/or part-time workers, overtime, and subcontracting. The outcome of an aggregate planning is a master production schedule (MPS). The next stage is material requirement planning (MRP), where a product is disintegrated into several components, and the requirement for each component is determined with help of bills of materials and inventory records. Next, a shop floor schedule for each component is developed. Production is often greater than what is required, which creates inventory, which is again looped back into the MRP phase.

Although the production planning and control system is made of these blocks, there are many important functional departments in plant operations, such as human resources, sales and marketing, operations, distribution, and finance. Enterprise resource planning (ERP) refers to software that integrates the entire system.

Next, we describe a few basic concepts in production planning.

1.1.2 Production Systems

The development of manufacturing systems and the arrangement of machines within that system lead to different plant layouts. Different features affect selection of system. They may include factors such as the variety of items produced, volume of production of each item, flexibility necessary to change the product mix, time within which customer demand must be met, and of course, the financial resources available to buy or build the capacity needed for each operation.

A large mix of different types of products demands machines that can be easily configured to produce a wide variety of products. Such machines tend to be more general purpose, perhaps slower in-production rates than otherwise possible, but allow more flexibility. Job shop and flexible manufacturing form such production systems. In a job shop, a variety of general-purpose machines are assembled, each machine type in separate departments. Thus, for example, we may have general-purpose lathes together in one department, stamping machines in another, and drilling machines in yet another. Jobs go from one department to the next as the operations demand. Machines can be set and reset to produce each type of item. Labor is more generalized, being able to perform different types of operations. The drawbacks are excessive material handling, extreme downtimes necessary for all set ups, excessive production cost, and low production rates. The quality of the product can be good, since it is possible to pay attention to each unit produced, and if necessary, inspected and reworked. Specialized, one of type item may be produced in "job shop" environment.

Material handling, set-up times, and cost of production can be reduced some what, while maintaining flexibility to produce a large variety of products by using flexible manufacturing arrangements. In flexible manufacturing, the collections of machines are more sophisticated. They mainly include numerically controlled machines, and are therefore more expensive. However, we can easily change setups on these machines to produce different types of items with very little downtime. Machines need not be placed in separate departments based on their functions as in a job shop, but are arranged in cells, based on operation flow. This arrangement is also referred to as cellular manufacturing. The type of machine in each cell can be changed, preferably physically, but if not then logically, so as to accommodate material/product flow for each product. Here, a more educated and trained labor force is needed.

Production rate can be significantly improved while reducing the material-handling cost, if all items require the same sequence of operations in their production. Different types of machines can now be arranged in a fixed sequence, as necessary in the manufacturing of products. Such an arrangement is called a flow shop. The assembly line is a special case of flow shop where very few, often as little as one product, is produced in a single flow line. Machines are built to perform very specialized operations very fast and are arranged in a specific sequence. All this adds to high production rates and very low unit cost. All excessive material handling is reduced to a minimum. Labor need not be very trained, since they repeatedly perform a limited operation to perfection. Flexibility to change the product mix of items is very limited to almost none; since most of the machines are built specifically to perform operations necessary for the present items. The following summary chart displays some of these relationships.

Labor cost	High cost		Low cost	
Product characteristics	Unique	Semi customized	Standard high volume	Continued very high volume
Jumbled flow	Job shop			
Intermittent flow		Batch		
Line flow			Repetitive	
Continuous				Oil, gas, chemical products

Production systems may also be characterized based on other factors such as those given in the chart below.

1.1.3 Characteristics of Production Systems

Labor incentive—job shops	Capital incentive—assembly lines Flexible manufacturing
Flexile—job shops, flexible manufacturing	Nonflexible—assembly lines
High-status products—job shops	Consistent products—assembly lines
Non flexible in volume of production—assembly lines	Flexible in volume of production—job shops, flexible manufacturing

No matter what the production arrangement is, production planning is essential. We must know what we can produce in a given time period, with the capacity we have, and also when we can promise the delivery of items. Limited capacity may require replanning of delivery times. Excessive capacity that remains idle is a cost that must be borne by the items that are produced. This may make the products too expensive to compete. There are ways to plan for needed capacity besides buying additional equipment. They may include inventory build-up, adding extra or part-time workers, and subcontracting, as we shall discuss in chapter "Aggregate Planning."

1.1.4 Response Time

The time within which customer demand must be met is called response time or lead time. It is the measure of time it takes to deliver the items, from the moment an order is placed till it is received. Aggregate planning helps in reducing the response time. There are other policy decisions that may also affect the response time. For example, a question may be at what stage of completeness the products should be maintained in the inventory. One extreme is to build the product when an order is received. The system working under this operating environment is called the pull system. It has the advantage that it requires no inventory of the product, thus reducing the inventory cost. But it does take a long time to build a product from the scratch, which results in a long response time. The other extreme is to keep products on hand, in inventory, and ship it as soon as demand occurs. This is called the push system. Here the response time is minimum but the inventory cost is high. Also, there are possibilities of obsolescence of products. And then there are options to build the product to a certain level of completeness and finish it when demand occurs, thus compromising on both inventory cost and response time. What level of completeness a product should be maintained is a strategic decision.

In general, products with high functional use, such as the products with basic needs found in grocery stores, low cost is the controlling factor. If a product is not available substitution is readily available. For high-status products such as computers or machine parts, response time is more important than cost, that is, customers are ready to pay slightly higher price for a better response time.

1.1.5 Supply Chain

Another means of controlling response time is to reduce the delivery time. Manufacturing all products in one central facility can be more economical because of a large quantity being produced in one plant, but may increase distribution time and cost when deliveries must be made to all customers spread across the nation. Alternately, we may have multiple production facilities and/or warehouses situated in different sections of the country, perhaps smaller in production capacities and therefore more expensive as far as unit production cost is concerned, but responding quickly to local customer needs. Building multiple facilities increases the overall cost but may satisfy the goal of maintaining certain maximum response time as the customer service criteria. Figure 1.2 shows the effect of the number of facilities on response time and cost.

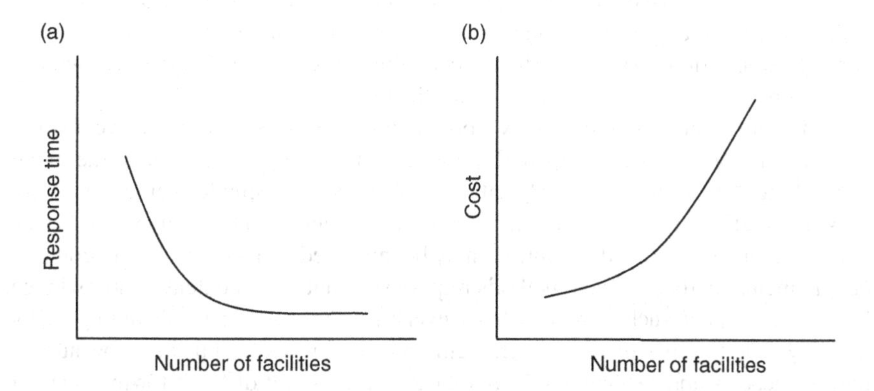

FIGURE 1.2 The effect of the number of facilities on response time and cost.

How to develop a production–distribution network to satisfy maximum response time constraint and yet minimize the total cost of production and distribution is a challenge addressed in development of supply chains.

1.1.6 Inventory

Inventory is built and used to satisfy multiple reasons. Units can be produced and stored when demand is low to satisfy demand in periods where it exceeds the production capacity. If set-up cost is high, production in large lots may reduce unit production cost as set-up cost is spread over a large number of units, but will also create inventory that must be stored. If quick response time is desired, an inventory may be necessary. Just-in-Time philosophy of a customer may mean stocking some items at our end, if we have multiple products to produce on the same facility and may not produce the parts each time immediately after demand occurs. In purchasing raw material, sometimes, due to quantity discounts, it may be advantages to purchase a large quantity from our supplier, which then must be stored till needed. Cancelled orders and poor production planning may also result in excessive inventory.

Keeping inventory is expensive. We need storage space, some amount of money is invested in the inventory, and there are carrying and maintenance costs. Good production planning optimizes issues involved in inventory policies.

1.2 SCHEDULING

Scheduling in its broadest sense is as old as mankind. Loosely defined, scheduling is an act of defining priority or arranging activities to meet certain requirements, constraints, or objectives. In olden days (and even now), time was (and still is) a major constraint. People scheduled their activities so that jobs could be accomplished within the available time. For example, time to get up, time to work, time to play, time to sleep, and so on. Time was, and still is, a limiting resource, and we need to schedule our activities, consciously or unconsciously, to utilize this limited resource in an optimum manner.

As the industrialized world develops, more and more resources are becoming critical. Machines, manpower, and facilities are now commonly thought of as resources in production and/or service activities. Scheduling these leads to increased efficiency, utilization, and ultimately, profitability for the organization.

Scheduling activities can convey a broad range of efforts. It may involve no more than working with paper and pencil, plotting charts and diagrams, or to the extent of working with sophisticated algorithms and theorems. Simple methods may not provide good results, and unless the analyzer is aware of other techniques, he/she may not even realize that the solutions may be improved. On the other hand, complex and mathematically involved methods require substantial and extensive knowledge. We cannot expect such expertise from every person who does scheduling in the industry. The number of times such techniques go unused in business, because of their intricacies and mathematical complications, makes it difficult for managers to comprehend the methods and then have the confidence to apply them and rely on the results. In this book, we shall present many scheduling techniques that give good results with minimum complexity, and which are easy to understand and apply. In fact, knowledge of basic algebra is all that is needed.

1.2.1 Scheduling Examples

The following examples illustrate the role of scheduling in modern industries.

1. In a box carton manufacturing operation, the sales personnel must go out and "beat the bushes" to find the customers. The customers vary in their orders and prices. Local manufacturers who send their products in cartons do not wish to carry a large inventory of cartons and will spread their orders over a time. Customers who are further away tend to order in larger amounts to minimize transportation costs. Sales promises a delivery date for each order. Depending on the volume of orders received, the plant may or may not be able to produce all the products on time. If there is to be a delay, which orders should be delayed to minimize the penalties? If there is similarity between the cartons by different customers so that some of them could be produced with the same setup, how should we plan the production to take advantage of this fact and yet meet the due date requirements of different customers?
2. A manufacturer of industrial lighting equipment assembles products using assembly lines controlled by operators performing their assigned tasks. Depending on the current product flowing down the assembly line, the operators have a defined set of tasks for which they are responsible. For each product, product modification, or new product, predetermined times are used to develop workstations and to make task assignments to these stations for an estimated production level and efficiency. The same techniques are used to make improvements in existing assembly lines where possible. The question is, how can assembly lines be modified for product modifications or for new products to obtain the desired production levels and efficiencies?
3. A petroleum refinery must have its continuous processes manned by operators 24 hr a day, 7 days a week. They operate with a system of three shifts, whereby operators always have at least 2 consecutive days off and every third weekend off. The schedules are developed for 3 months at a time and posted so that employees can plan for their days off and shifts worked for a horizon of 3 months. How can fair schedules be developed for all employees?
4. A machine shop has a stamping machine that stamps five different versions of a part used in compressors manufactured by one of its customers. Each of these five versions uses a different dye and requires a different time for processing. The machine shop generates weekly schedules for this machine to minimize the number of late shipments to its customers. How should these schedules be developed?
5. The quality department for an electronic printed circuit board (PCB) manufacturer is responsible for conducting tests on the PCBs produced.

For one of the tests, components are placed in a thermo-shock chamber (with limited space), so that they can be subjected to varying temperature extremes, based on the PCB types. Since the duration of tests vary and are quite lengthy (usually 24–28 hr), the greatest efficiencies are obtained when PCBs are grouped into batches. How should we form these batches and meet the customers' due dates on each PCB type?

1.2.2 Scheduling on Shop Floor

Scheduling plays an important role in shop floor planning. A schedule shows the planned time when the processing of a specific job will start on each machine that the job requires. It also indicates when the job will be completed on every machine. Thus, it is a timetable for both jobs and machines. The starting time of a job on the first machine in its sequence of operation should also be the release time for the job (assuming zero lead time). If some lead time in the shop is necessary, the release time is correspondingly adjusted. In practice, release time is the time when the job is scheduled for processing, and all purchased raw materials for the job should be available in the shop. For such a just-in-time manufacturing philosophy to work, we must know the job schedule. If the raw material is stocked before the release time, it will add to the carrying cost. On the other hand, if it is delayed beyond the release time, then not only this job but other jobs could be delayed, adding to the cost of operations.

The completion time of a job on the last machine in its sequence of operations is also the time when the job is available for shipping to a customer. If the due dates can be negotiated to match these completion times, then again just-in-time philosophy will prevail.

Capacity planning is an integral part of scheduling procedures. The term "capacity planning" is used in deciding how much of the production time of a machine (in units such as minutes, hours, or days) should be allocated to each job. Alternately, capacity planning can decide which job should be processed on which machine. If there is no capacity on the machine, a new job cannot be processed, postponing the processing and completion of the job to some future time periods.

We must know the processing time for each job on each machine that it will use in order to develop a schedule. In order to calculate the processing time for a job, we must consider both machine- and job-dependent factors such as setup time, unit processing time, machine speed and quality factors, as well as the number of units to produce to make the job order complete. If we have a choice of performing a certain operation of a job on different machines, then we may have, depending on the machine used, different processing times for the operation. Scheduling can offer choices if alternatives are available. For example, if scheduling all jobs on a single machine leads to an unacceptable cost (one reason may be overuse of the capacity), we may try a few jobs on different machines and develop schedules on each individual machine again. Alternatively, we may develop a schedule on multiple identical machines and evaluate economically the ideal number of identical machines we could use for production of the present set of jobs.

A machine schedule also displays the times when the machine is idle. This may be because no job is available for processing, or perhaps the jobs are being processed on some other machines. When a machine is idle, it is the best time to plan for preventive maintenance activities as no productive time is taken away from the machine. It will also reduce the frequencies of unplanned breakdowns. When unplanned breakdowns do occur, a new schedule could be developed after the machine is repaired with the present job conditions as the input to the scheduling models.

To blend scheduling algorithms or rules in a plant, integration of information from several sources is required. We may have data files for shop floor factors such as workers shift schedules, machines and their characteristics, maintenance data, progress on the presently scheduled jobs in the shop, and current status of the machines, whether busy, idle, broken and needing repairs, or broken and under repairs. We may also have information on customer order data such as the name of the customer, product, and quantity ordered and when the delivery is promised. The manager may assign a priority to each order in terms of quantitative weight, 1, 2, 3, 5—or perhaps in qualitative terms, "hot," "very hot," and "hottest," which is then converted into a weight with predefined rules. These priorities may change from day to day, based on the present status of the jobs. For example, a job that is hot one day may be hottest on another day as the due date nears, or a hot job may go on "hold" status if there are difficulties with the order and with the customer. Information from MRP may give the release date of the job, indicating when the raw material would be available to process the job. As mentioned earlier, ideally, MRP job release date and job scheduling date should be the same to make the system work efficiently.

We may also have a data file indicating the setup times for different types of jobs on different machines. The processing times for the jobs are calculated based on the machine- and job-related data such as setup time, machine speed and accuracy, and quantity required.

In practice, schedules are generated using either scheduling algorithms or using knowledge-based rules. Scheduling algorithms develop schedules that tend to optimize a measuring criterion, such as minimizing deviation from due dates, minimizing tardiness penalty, or minimizing the maximum delay. The rule- or knowledge-based approach tries to find a schedule that is feasible under the shop operating environment by mainly working with "If and Then" rules. For example, the rule may be: if machine A and the operator X are available, then load job type Z; if machine A and operator X and helper Y are available, then load job type P. The feasible schedule developed by using rule-based algorithms may or may not (most often not) be optimum with respect to a measuring criterion.

The scheduling information can be displayed in a variety of forms. The Gantt chart, first illustrated and used for scheduling by Henry Gantt between late 1800 and early 1900, is one of the most popular tools, even today, for displaying scheduling information. The Gantt chart is a line or block chart, where time is represented on the x-axis and other quantities of interest such as machines and/or jobs are represented on y-axis. For example, a Gantt chart may display, for each machine, the jobs that are loaded in different time periods. Similar information can also be illustrated in tables called "dispatching tables." This table may, among others, show the information on each machine as per the sequence of jobs that are scheduled on the machine, when the processing on a job is expected to start, and when it is expected to be complete.

Another chart that may be of interest is one that may display the capacities of each machine used in each time period (e.g., a day). This is called the capacity bucket chart. For example, based on the schedule developed, it may display information on machine 1, such as on day 1 machine 1 is utilized 80% of time, on day 2 it is utilized 90% of the time, and so on. Again, if the capacity used is more than what is available, then the schedule is not feasible, and alternatives must be evaluated.

Throughput charts may also be of interest to an analyzer. These show how many jobs (or units) are available and how many are processed every day. The difference between the two defines the in-process inventory, the value of which should be kept to a minimum. What is included in the scheduling chapters?

Over a number of years, many researchers have contributed to developing scheduling techniques, and some common themes and terminologies have been developed. Most of these procedures and terminologies are defined for a production environment. This is because production planning continues to be one area where these techniques dominate in their application. Also, most of us easily understand the vocabulary in industrial terms, though their use is not restricted to manufacturing alone. While machines may be viewed as a resource, jobs requiring the services of machines may be viewed as entities that need attention from the machines. Other parameters can also be easily transported from the production environment to another.

In production planning terminology, the scheduling models may be divided into the following categories.

1. *Single machine*: There is only one machine (server) available, and the arriving jobs (work) require services from this machine. Jobs are processed by the machine one at a time. Each job has a processing time and a due date and may have other characteristics such as priority. We may also have a penalty function for jobs deviating from the due date. The most common objective is to sequence jobs on the machines so as to minimize the penalty for being late, commonly called "Tardiness Penalty." Based on other objectives, there are many criteria that may serve as a basis for developing job schedules.
2. *Flow shop*: Jobs are processed on multiple machines in an identical sequence. However, the processing time of each job on each machine may be different. The objective may be to minimize the makespan, that is, the time required for the completion of all jobs.
3. *Parallel machines*: A number of identical machines are available and the jobs can be processed on any one of them. Jobs may have dependency, that is, unless the previous job in the sequence has been completely processed, the following job in the sequence may not start. The objective may be to minimize the makespan.
4. *Job shop*: This is one of the most popular generalized production systems. There are different machines in the shop, and a job may require some or all of these machines in any sequence, the only restriction being that a job cannot use the same machine more than once. The objective may be to minimize makespan or tardiness penalty.

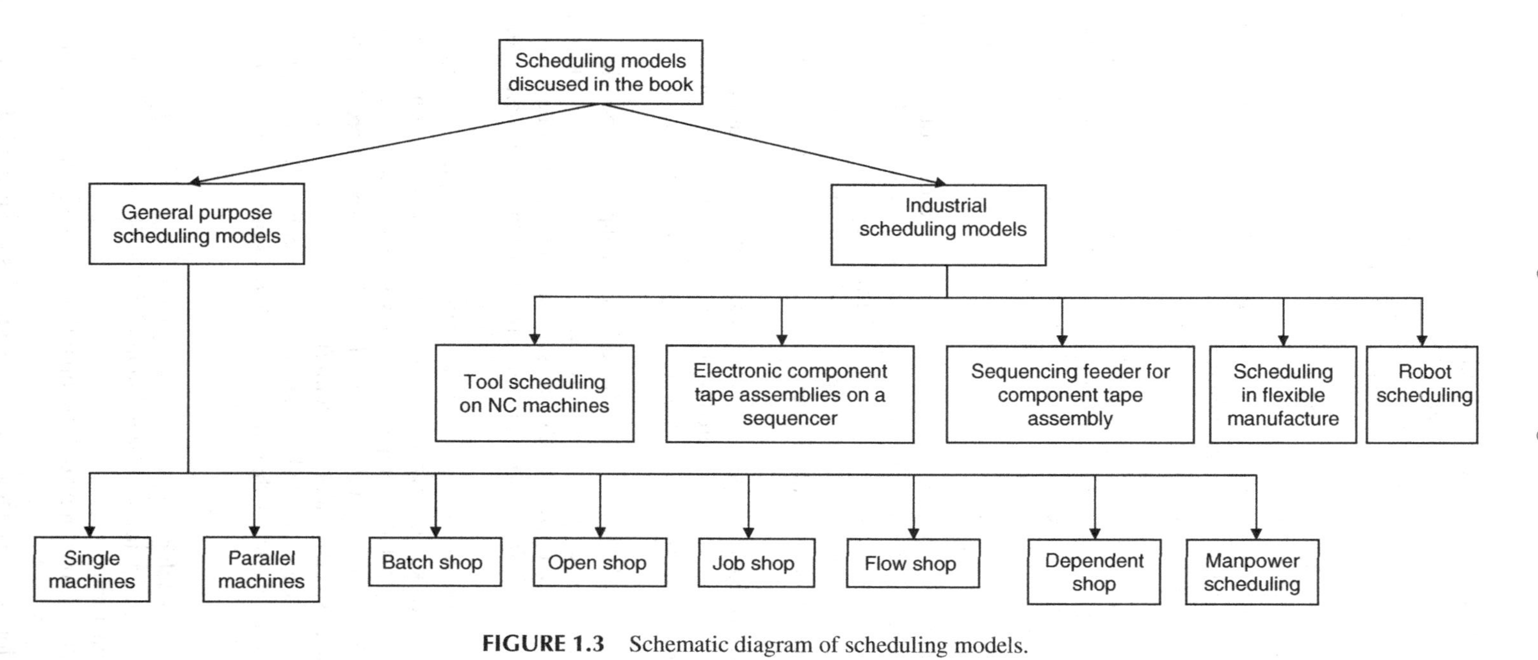

FIGURE 1.3 Schematic diagram of scheduling models.

5. *Open shop*: An open shop is similar to a job shop, except a job may be processed in any sequence on the machines that the job needs. In other words, there is no operation-dependent sequence that a job must follow. The objective is generally to minimize the makespan.
6. *Dependent shop*: In a job shop environment, if the processing order of one or more jobs depends on the processing of other jobs, it is called a dependent shop. The general objective is to minimize the makespan.
7. *Batch processing*: Jobs are processed in batches, each batch requiring a certain processing time and having a capacity limitation on how many jobs can be processed at one time. A baking oven with limited volume is one example of batch processing.
8. *Sequence-dependent setup times*: Some authors also refer to this problem as batch processing. Here, each job may belong to a type. If jobs of the same type are processed one after the other, then no additional setup is required. On the other hand, if a different type of job is processed, there is a setup cost. Each job has a due date, and we want to schedule the jobs to minimize the total penalty.
9. *Assembly line*: The job goes through a certain sequence of operations, and the objective is to define workstations and assign tasks to these stations to achieve a certain production level and efficiency.
10. *Mixed-mode assembly line*: An assembly line built to produce similar (not identical) products with different task requirements and task times.

Each topic is covered in the remaining chapters in sufficient detail. In addition, there are a number of other topics covered in the remaining chapters that do not directly relate to the standard definition of production systems. They include

11. *Manpower planning models*: To develop manpower schedules that satisfy legal and contractual requirements with a minimum workforce for a 7-day-a-week operation. One reason for concern could be safety, where employees must have enough time off, so that they stay alert on the job.

These requirements may include that each employee have, in a week: 2 consecutive days off, or "B" weekends off in a cycle of "A" weeks, or have 2 nonconsecutive days off. It is also possible that all employees start work at the same time, or an employee may start his/her shift at different times of the day to cover more than the standard 8 hr operation (also called tour scheduling).

Three chapters are devoted to industrial scheduling. These chapters illustrate different areas of scheduling that are not thought of as production systems. The topics include

a. Group technology
b. Use of group technology in minimizing tool changeovers in flexible manufacturing
c. Scheduling to minimize throughput time on numerically controlled (NC) machines
d. Development of component tape assemblies in electronic PCB production

e. Sequencing feeders for component tape assemblies in electronic component production

1.3 WHAT IS NOT INCLUDED

There are other topics that are discussed by the research community in scheduling. When all the data are known, the problem is considered to be a deterministic problem. When the data is probabilistic in nature that is, it takes random values, the problem is called a stochastic problem. Most of the scheduling problems are either deterministic or can be closely approximated by the deterministic models. The most common way to convert a stochastic problem into a deterministic one is to work with the average values. We shall restrict ourselves to only deterministic models since they illustrate many basic principles well, and are applicable to most real-life scheduling problems.

Another variation is whether preemptions are allowed. Preemption implies that we do not need to keep a job on the machine till it is completed. If another job with a higher priority arrives, it may be loaded on the machine by removing the present one. The present job may be reloaded after the completion of the priority job. We will not treat a specific model in this regard but assume that if preemption is permitted we can develop a new job sequence at any time by applying the sequencing methods again with whatever jobs are available at that time, including the job taken off of the machine with its modified (most often remaining) processing time.

We also will not discuss the phenomena of blocking. Blocking occurs if the completed job cannot be removed from the machine because of space constraints. In other words, there is no space for the completed job to move to, and it must remain on the machine. Such limitation of space is rare, and when it happens consistently, a better solution might be to investigate the reason for the space limitation and find a solution to that problem. One possible solution may be to rearrange the facilities. Yet another alternative may be to investigate material handling resources and policies. With the present emphasis on just-in-time manufacturing, excess build-up of inventories is discouraged and, consequently, the nonavailability of space is also resolved.

1.4 SUMMARY

Important topics in production planning and scheduling are really based on information and decision flow within manufacturing systems.

For planning for far and in near future, we must estimate demands by forecasting both long and short terms. Capacities for plant productions must be established. Firm customer orders and estimated immediate sales lead to short-term planning that include topics such as master production planning and MRP. To implement these plans, resources are need. Aggregate planning, rough-cut planning, and capacity requirement planning are the tools used in such planning. And, ultimately, we must distribute the products leading to distribution planning techniques.

Scheduling methods are tools that are available to allow production and other systems to run efficiently. Production planning involves activities of predicting demands, planning for production, and determining and utilizing capacity in an optimum manner. The scheduling utilizes the limited capacity in most efficient way.

The scheduling efficiency can be measured by various indexes. Two of the most popular are minimization of the time required to complete all jobs, that is, makespan, and minimization of penalty for completing jobs early or after the due dates. This chapter illustrates a few examples of scheduling in both the production and nonproduction environments. The topics listed in this chapter will be discussed in detail later on. These scheduling methods have proven to provide optimum or near-optimum results, and are based on simple algebra. Most often, the methods progress and develop solutions in successive tables.

1.5 PROBLEMS

1.1 What is production planning? Why is it important?
1.2 Describe different production systems. Why do we have so many different ways to produce a product?
1.3 What is the purpose of inventory? Is a large inventory always good? Can we reduce inventory to zero?
1.4 Why is response time important to both customer and supplier? How does inventory influence the response time? Can you relate response time to the profit and cost picture?
1.5 Discuss in your own words the importance of scheduling.
1.6 Draw a schematic diagram indicating different data files and how they may be connected to each other in scheduling within a manufacturing shop.
1.7 Give one example of a situation wherein the following scheduling methods could be applied
   a. Single machine
   b. Parallel processing
   c. Batch processing
1.8 Describe a situation where manpower scheduling would be appropriate.
1.9 What are the benefits of assembly line balancing?

REFERENCES AND SUGGESTED READINGS

Cavinato, Joseph L. 2002. "What's Your Supply Chain Type?" *Supply Chain Management Review* (May–June): 60–66.

Chopra, Sunil. 2003. "Designing Delivery Network for a Supply Chain" *Transportation Research*, Part E (39): 123–140.

Fisher Marshall, L. 1997. "What Is the Right Supply Chain for Your Product?" *Harvard Business Review* (March–April): 83–93.

Goodwin, B., M. Seegart, J. Cardillo, and E. Bergmann. 1996. "Implementing ERP in a Big Way" *APICS-The Performance Advantage*, June

Holstein, W.K. 1968. "Production Planning and Control Integrated" *Harvard Business Review*, May/June: 70–82.

Magretta, Joan. 1998. "Fast, Global and Entrepreneurial: Supply Chain Management, Hong Kong Style" *Harvard Business Review* (September–October): 102–114.

Van Dierdonck, R. and J.G. Miller. 1980. "Designing Production Planning and Control Systems" *Journal of Operations Management*, 1(1): 102–112.

2 Plant Locations and Capacity Determination

A manufacturing concern often has customers spread nationally or even internationally over wide area, and the challenge on hand may be to decide which customers should be served by which facility and what should be the capacity of each facility to minimize the costs of production and distribution. This also forms the foundation for supply chain development.

There are two situations that may be present. First, production facilities already exist, and customers have to be assigned or reassigned to improve the performance of the production/distribution system, and second, new facilities and associated capacities are to be established to develop a new network. The second alternative gives us much more flexibility since we can also choose locations for the new facilities.

2.1 EXISTING PRODUCTION FACILITIES

If we have just a single existing facility, there is no alternative but to service all customers from this facility. The capacity of the facility should ideally be able to support the total demand from all customers.

In case of multiple facilities, there are two types of problems. First, we may want to keep all existing facilities and determine optimum production and distribution systems or, secondly, we may only want to maintain production facilities that provide total optimum cost for production and distribution.

2.1.1 Distribution Network with Existing Facilities

With the existing production facilities and customers, the development of production and distribution plan requires nothing more than simple application of the transportation algorithm from linear programming (LP). If the reader is unfamiliar with the transportation algorithm, we direct him/her to any standard operations research book for a quick review. We can also solve the problem using LP formulation and using standard computer programs developed to solve LP problems, such as LINDO or Excel. (LINDO and Excel are copyrighted software from Lindo System Inc and Microsoft respectively).

Consider a company with three plants and five warehouse customers. The cost of producing a unit at a plant i and transporting it to a warehouse j, denoted as c_{ij}, is shown in the Table 2.1. For example, the cost of transporting one unit from plant 1 to warehouse 1 is 2. Also shown are the production capacities of each plant, and the requirements at each warehouse.

TABLE 2.1
Initial Data

	Warehouse					
Plant	**1**	**2**	**3**	**4**	**5**	**Supply**
1	2	4	5	3	4	30
2	5	4	6	2	5	50
3	3	3	4	4	6	80
Demand	20	30	10	15	20	

TABLE 2.2
Modified Data for Transportation Algorithm

	Warehouse						
Plant	**1**	**2**	**3**	**4**	**5**	**6**	**Supply**
1	2	4	5	3	4	0	30
2	5	4	6	2	5	0	50
3	3	3	4	4	6	0	80
Demand	20	30	10	15	20	65	160

TABLE 2.3
Final Solution

	Warehouse						
Plant	**1**	**2**	**3**	**4**	**5**	**6**	**Supply**
1	2	4	5	3	4	0	
	20				10		30
2	5	4	6	2	5	0	
				15	10	25	50
3	3	3	4	4	6	0	
		30	10			40	80
Demand	20	30	10	15	20	65	160

Since the total of supply is 160 and the demand is 95, a dummy warehouse, warehouse number 5, is introduced to make the total of demand equal to that of supply, as shown in Table 2.2.

Applying the transportation algorithm results in the following final solution of Table 2.3. For example, 20 units are transported from plant 1 to warehouse 1, 10 units from plant 1 to warehouse 5, and so on. The cost of this solution is $= 20 \times 2 + 10 \times 4 + 15 \times 2 + 10 \times 5 + 30 \times 3 + 10 \times 4 + 40 \times 0 = 290$.

Alternately, to formulate the problem in LP, we define following variables:

x_{ij} = Amount shipped from plant i to warehouse j. There are n plants and m warehouses.
c_{ij} = Cost for production and transportation of one unit from plant i to warehouse j
S_i = Maximum production or supply available at plant i
D_j = Demand from customer j

The objective is to minimize cost

$$\text{Min} \sum_{i=1}^{n} \sum_{j=1}^{m} c_{ij} x_{ij}$$

Subject to following constraints:

1. A plant cannot supply more than what is available:

$$\sum_{j=1}^{m} x_{ij} \leq S_i \quad \text{for } j = 1, 2, \ldots, n$$

2. Each customer's demand must be met:

$$\sum_{i=1}^{n} x_{ij} = D_j \quad \text{for } j = 1, 2, \ldots, m$$

For the data in Table 2.1, the LP problem results in following:

$$\text{Min } 2x_{11} + 4x_{12} + 5x_{13} + 3x_{14} + 4x_{15} + 5x_{21} + 4x_{22} + 6x_{23} + 2x_{24} + 5x_{25} + 3x_{31} + 3x_{32} + 4x_{33} + 4x_{34} + 6x_{35}$$

Subject to:

$x_{11} + x_{12} + x_{13} + x_{14} + x_{15} \leq 30$

$x_{21} + x_{22} + x_{23} + x_{24} + x_{25} \leq 50$

$x_{31} + x_{32} + x_{33} + x_{34} + x_{35} \leq 80$

$x_{11} + x_{21} + x_{31} = 20$

$x_{12} + x_{22} + x_{32} = 30$

$x_{13} + x_{23} + x_{33} = 10$

$x_{14} + x_{24} + x_{34} = 15$

$x_{15} + x_{25} + x_{35} = 20$

all $x_{ij} \geq 0$

Solving by LINDO gives an alternate solution: $x_{11} = 10$, $x_{15} = 20$, $x_{24} = 15$, $x_{31} = 10$, $x_{32} = 30$, and $x_{33} = 10$, with optimum cost of 290.

2.1.2 Response Time Consideration

One of the considerations in the development of supply chains is the response time — how quickly we can supply the demand. In the previous example, we assumed that all plants can supply to all warehouses within the response time limit set by the management. If this is not the case, only slight modifications are needed in above formulation.

In the optimum solution of the previous example, x_{11} is equal to 20. But suppose it is not possible to produce and ship from plant 1 to warehouse 1 in the required response time and, therefore, x_{11} needs to be zero. There are two ways to address this. Either assign a very high cost (cost of M for those who are used to LP), or introduce a constraint in LP formulation that says $x_{11} = 0$, and solve the problem again. In practice, we should introduce such constraints associated with variables in the initial formulation and then solve the resulting LP problem.

In our problem, the optimum value increases to 300 with following solution:

$$x_{15} = 20, \quad x_{24} = 15, \quad x_{31} = 20, \quad x_{32} = 30, \quad \text{and} \quad x_{33} = 10.$$

2.1.3 Limitation on Number of Facilities

A possible situation may exist when we are required to reduce the number of facilities from n to a number less than n, to reduce the total fixed cost associated with running n facilities. LP formulation needs slight modification to achieve this. Suppose the total number of facilities cannot exceed k. The modified formulation is as follows:

$$\text{Min} \sum_{i=1}^{n} \sum_{j=1}^{m} c_{ij} x_{ij}$$

Subject to following constraints:

1. A plant cannot supply more than what is available:

$$\sum_{j=1}^{m} x_{ij} \leq y_i S_i \quad \text{for } i = 1, 2, \ldots, n$$

2. Each customer's demand must be met:

$$\sum_{i=1}^{n} x_{ij} = D_j \quad \text{for } j = 1, 2, \ldots, m$$

3. The number of facilities is limited to k.

$$\sum_{i=1}^{n} y_i = k$$

TABLE 2.4
Variables Defined for New Problem

	Warehouse					
Plant	**1**	**2**	**3**	**4**	**5**	**Supply**
1	2	4	5	3	4	
	x_1	x_2	x_3	x_4	x_5	30
2	5	4	6	2	5	
	x_6	x_7	x_8	x_9	x_{10}	50
3	3	3	4	4	6	
	x_{11}	x_{12}	x_{13}	x_{14}	x_{15}	80
Demand	20	30	10	15	20	

4. y_i takes value of either 0 or 1. If the plant is closed, the associated $y_i = 0$, and if the plant is open, the associated $y_i = 1$.

Suppose in the previous problem, management is interested in reducing the number of open plants to 2. We solve the problem by using LINDO. The variables are modified as shown in the Table 2.4.

The LP formulation for LINDO is displayed in the following. Int instruction makes the associated variable only take the value of 0 or 1.

$$\text{Min } 2x_1 + 4x_2 + 5x_3 + 3x_4 + 4x_5 + 5x_6 + 4x_7 + 6x_8 + 2x_9 + 5x_{10} + 3x_{11}$$
$$+ 3x_{12} + 4x_{13} + 4x_{14} + 6x_{15}$$

st

$x_1 + x_2 + x_3 + x_4 + x_5 - 30y_1 \leq 0$

$x_6 + x_7 + x_8 + x_9 + x_{10} - 50y_2 \leq 0$

$x_{11} + x_{12} + x_{13} + x_{14} + x_{15} - 80y_3 \leq 0$

$x_1 + x_6 + x_{11} = 20$

$x_2 + x_7 + x_{12} = 30$

$x_3 + x_8 + x_{13} = 10$

$x_4 + x_9 + x_{14} = 15$

$x_5 + x_{10} + x_{15} = 20$

$y_1 + y_2 + y_3 = 2$

end

Int y_1

Int y_2

Int y_3

Solution is:

Objective function value

(1) 320

Variable	Value	Reduced Cost
y_1	1	−30
y_3	1	0
x_1	10	0
x_5	20	0
x_{11}	10	0
x_{12}	30	0
x_{13}	10	0
x_{14}	15	0

We have plants 1 and 3 open, and plant 2 is closed. The cost has gone up from 290 to 320.

Now, suppose the fixed cost for operations at each plant is different, and we wish to include this fact in trying to decide which plants to keep open. The only modification needed is to include the fixed cost in the objective function based on whether the plant is open or not.

In our example, suppose the fixed costs associated with plants 1, 2, and 3 are 30, 25, and 30 respectively. When all three plants are operating, the optimum cost is $290 + 30 + 25 + 30 = 375$.

The new objective function is:

$$\text{Min } 2x_1 + 4x_2 + 5x_3 + 3x_4 + 4x_5 + 5x_6 + 4x_7 + 6x_8 + 2x_9 + 5x_{10} + 3x_{11} \\ + 3x_{12} + 4x_{13} + 4x_{14} + 6x_{15} + 30y_1 + 25y_2 + 30y_3$$

Keeping the same constraint set to keep only two facilities, the solution is:

Objective function value

(1) 375

Variable	Value	Reduced Cost
y_2	1	25
y_3	1	30
x_9	15	0
x_{10}	20	0
x_{11}	20	0
x_{12}	30	0
x_{13}	10	0

This solution is to keep facilities 2 and 3 open. The solution is better than when we had plants 1 and 3 open with no fixed cost consideration. The optimum cost, if we had included the fixed cost, would have been $320 + 30 + 30 = 380$, as compared to 375 now.

2.2 NEW PLANT LOCATIONS

Plant locations are long-term decisions, and once taken cannot be easily changed. Judicious decisions can be made by first selecting a few candidate locations where plants might be located that satisfy basic production and economic requirements. Considerations in selecting the initial list of sights might include factors such as cost and availability of raw materials, as well as labor, transportation facilities and their costs, demands and sale price from different regions for the produced goods, land and building expenses, and taxes and insurance costs at the locations.

2.2.1 New Facilities with Capacity Determination

Once the acceptable prospective locations are selected, further detailed economic evaluation can be made by applying zero/one integer LP algorithm to minimize the total cost of operations.

For example, consider n possible location sites, from which we can choose any number of locations for new plants. A plant can be built with different capacities from p potential capacities. Let us define a few other variables as follows:

Indexes
For possible locations: $i = 1, 2, \ldots, n$
For customers: $j = 1, 2, \ldots, m$
For possible capacities: $k = 1, 2 \ldots, p$
C_k = Capacity of facility k
f_{ki} = Amortized annual fixed cost of placing a plant of size k at location i
m = number of different customers (distribution areas) that the facilities must serve
D_j = Unit loads of demand from customer j
c_{kij} = Cost of production and distribution of one unit load from a plant, with capacity k placed in location i to customer j
x_{kij} = Quantity (unit loads) shipped from plant with capacity k located in i to customer j
y_{ki} = 1 if plant of capacity k is placed in location i
y_{ki} = 0 otherwise

The problem of minimization of cost can then be formulated as

$$\text{Min} \sum_{k=1}^{p} \sum_{i=1}^{n} f_{ki} y_{ki} + \sum_{k=1}^{p} \sum_{i=1}^{n} \sum_{j=1}^{m} c_{kij} x_{kij}$$

TABLE 2.5
Data for Capacitated Problem

Facility Costs and Capacities

Facility	Cost	Capacity
1 (small—S)	100	50
2 (large—L)	120	80

Demand/Supply and Cost Data

		Customer			
Location	**Additional Site Cost per Year**	**1**	**2**	**3**	**Capacity**
1 for S	20	20	10	30	50
1 for L	30	18	8	27	80
2 for S	10	10	15	20	50
2 for L	30	9	13	18	80
3 for S	15	15	15	20	50
3 for L	40	14	14	18	80
Demand		50	40	70	

Subject to following constraints:

1. Demand from each customer must be met.

$$\sum_{k=1}^{p}\sum_{j=1}^{m} x_{kij} = D_j \quad \text{for each } i$$

2. Total supply from a facility in a location cannot exceed its capacity.

$$\sum_{k=1}^{p}\sum_{i=1}^{n} x_{kij} = C_j \quad \text{for each } j$$

Consider the following data from Table 2.5. We have three possible locations where either a small or a large or both facilities can be placed. There are three customers to whom products must be supplied. It cost slightly less to produce with a large facility than a small one, and therefore costs of production and distribution are different for each type.

2.2.2 Variables in Linear Programming Initial Formulation and Alternate Formulation

In Table 2.6, the original variables as defined in the formulation and alternate variables (AT) for finding the solution by computerized algorithms.

TABLE 2.6
Variable Definition

Location				Customer						
	k		**AT**	**1**	**AT**	**2**	**AT**	**3**	**AT**	**Capacity**
1 for S	1	y_{11}	y_1	x_{111}	x_1	x_{112}	x_2	x_{113}	x_3	50
1 for L	2	y_{21}	y_2	x_{211}	x_4	x_{212}	x_5	x_{213}	x_6	80
2 for S	1	y_{12}	y_3	x_{121}	x_7	x_{122}	x_8	x_{123}	x_9	50
2 for L	2	y_{22}	y_4	x_{221}	x_{10}	x_{222}	x_{11}	x_{223}	x_{12}	80
3 for S	1	y_{13}	y_5	x_{131}	x_{13}	x_{132}	x_{14}	x_{133}	x_{15}	50
3 for L	2	y_{23}	y_6	x_{231}	x_{16}	x_{232}	x_{17}	x_{233}	x_{18}	80
Demand				50		40		70		

AT: Alternate variable defination

The problem can then be stated as

$$\begin{aligned}\text{Min } & (100+20)y_{11} + (120+30)y_{21} + (100+10)y_{12} + (120+30)y_{22} \\ & + (100+15)y_{13} + (120+40)y_{23} + 20x_{111} + 10x_{112} + 30x_{113} + 18x_{211} + 8x_{212} \\ & + 27x_{213} + 10x_{121} + 15x_{122} + 20x_{123} + 9x_{221} + 13x_{222} + 18x_{223} + 15x_{131} \\ & + 15x_{132} + 20x_{132} + 14x_{231} + 14x_{232} + 18x_{233}\end{aligned}$$

Subject to

Supply constraints:

$$x_{111} + x_{112} + x_{113} \leq 50y_{11}$$

$$x_{211} + x_{212} + x_{213} \leq 80y_{21}$$

$$x_{121} + x_{122} + x_{123} \leq 50y_{12}$$

$$x_{221} + x_{222} + x_{223} \leq 80y_{22}$$

$$x_{131} + x_{132} + x_{133} \leq 50y_{13}$$

$$x_{231} + x_{232} + x_{233} \leq 80y_{23}$$

Demand constraints:

$$x_{111} + x_{211} + x_{121} + x_{221} + x_{131} + x_{231} = 50$$

$$x_{112} + x_{212} + x_{122} + x_{222} + x_{132} + x_{232} = 40$$

$$x_{113} + x_{213} + x_{123} + x_{223} + x_{133} + x_{233} = 70$$

Zero–one integer constraints

Each of the variables, $y_{11}, y_{21}, y_{12}, y_{22}, y_{13}, y_{23}$ is either 0 or 1

Alternate variables can be appropriately substituted to get an LP formulation that can be solved with available LP software such as LINDO as shown below.

$$\text{Min } 120y_1 + 150y_2 + 110y_3 + 150y_4 + 115y_5 + 160y_6 + 20x_1 + 10x_2 + 30x_3 \\ + 18x_4 + 8x_5 + 27x_6 + 10x_7 + 15x_8 + 20x_9 + 9x_{10} + 13x_{11} + 18x_{12} \\ + 15x_{13} + 15x_{14} + 20x_{15} + 14x_{16} + 14x_{17} + 18x_{18}$$

st

$x_1 + x_2 + x_3 - 50y_1 \leq 0$

$x_4 + x_5 + x_6 - 80y_2 \leq 0$

$x_7 + x_8 + x_9 - 50y_3 \leq 0$

$x_{10} + x_{11} + x_{12} - 80y_4 \leq 0$

$x_{13} + x_{14} + x_{15} - 50y_5 \leq 0$

$x_{16} + x_{17} + x_{18} - 80y_6 \leq 0$

$x_1 + x_4 + x_7 + x_{10} + x_{13} + x_{16} = 50$

$x_2 + x_5 + x_8 + x_{11} + x_{14} + x_{17} = 40$

$x_3 + x_6 + x_9 + x_{12} + x_{15} + x_{18} = 70$

end

integer 6

The "integer 6" command makes the first 6 variables in objective function to be either 0 or 1. In our case, y_1 through y_6 would have either 0 or 1 value.

The optimum solution has a cost of $2480, with following variables with values:

$$y_2 = y_3 = y_4 = 1 \quad \text{and} \quad x_5 = 40, \quad x_7 = 40, \quad x_{10} = 10, \quad x_{12} = 70$$

If only one type of plant, either large or small, can be placed in a location, we add the following constraints in the preceding formulations.

$$y_1 + y_2 \leq 1$$

$$y_3 + y_4 \leq 1$$

$$y_5 + y_6 \leq 1$$

The optimum solution, z, is slightly higher at $2490, with following values:

$$y_2 = y_4 = y_6 = 1 \quad \text{and} \quad x_5 = 40, \quad x_{10} = 50, \quad x_{18} = 70.$$

2.2.3 Spreadsheet Approach Using Solver Tool

Following are the steps for using Excel to solve an LP problem:

1. The first step is to load solver to the add-in. Go to the **Tools** menu and click **Add-ins**. In the **Add-Ins available** box, select the check box next to **Solver Add-in**, and then click **OK**. Now click **Tools** on the menu bar, and you will see the **Solver** command added to the **Tools** menu.
2. Construct an Excel sheet as shown in Figure 2.1. It consists of inputs, decision variables, constraints, and objective function. All inputs such as costs, capacities, and demands are entered from cells B_4 to H_{10}. All decision variables (X_i, Y_i) are entered from cells B_{15} to E_{20}, and all the variables are initially set to be 0. Next are the set of limitations. The constraints are of two types, one is for supply from locations, and the other is for demand from customers. Cells B_{25}–B_{30} represents supply constraints from locations and cells, and B_{33}–D_{33} represents demand constraints from all three customers. The cell B_{36} contains the objective function that minimizes the total cost.
3. Enter the formulas given below in the respective cells:

Cell	Formula	Alternative
B_{25}	$= H_4 * E_{15}$–SUM(B_{15}:D_{15})	Enter the formula in B_{25} and copy it till B_{30}
B_{26}	$= H_5 * E_{16}$–SUM(B_{16}:D_{16})	
B_{27}	$= H_6 * E_{17}$–SUM(B_{17}:D_{17})	
B_{28}	$= H_7 * E_{18}$–SUM(B_{18}:D_{18})	
B_{29}	$= H_8 * E_{19}$–SUM(B_{19}:D_{19})	
B_{30}	$= H_9 * E_{20}$–SUM(B_{20}:D_{20})	
B_{33}	$= B_{10}$–SUM(B_{15}:B_{20})	Enter the formula in B_{33} and copy it till D_{33}
C_{33}	$= C_{10}$–SUM(C_{15}:C_{20})	
D_{33}	$= D_{10}$–SUM(D_{15}:D_{20})	
B_{36}	= SUMPRODUCT(B_4:D_9,B_{15}:D_{20})+ SUMPRODUCT(G_4:G_9,E_{15}:E_{20})	—

4. Go to **Tools** and invoke the solver, as shown in Figure 2.2. In this formulation, our goal is to minimize the total cost set in the **target cell** B_{36}. Next is the addition of constraints; for this, click the **add** button and add all the constraints as shown below:

   $B_{15} : D_{20} \geq 0$ {Non negativity condition}
   $B_{25} : B_{30} \geq 0$ {Capacity constraints for 6 locations}
   B_{33}:$D_{33} = 0$ {Demand constraints of 3 customers}
   E_{15}:E_{20} = binary {Location variables are either 0 or 1}

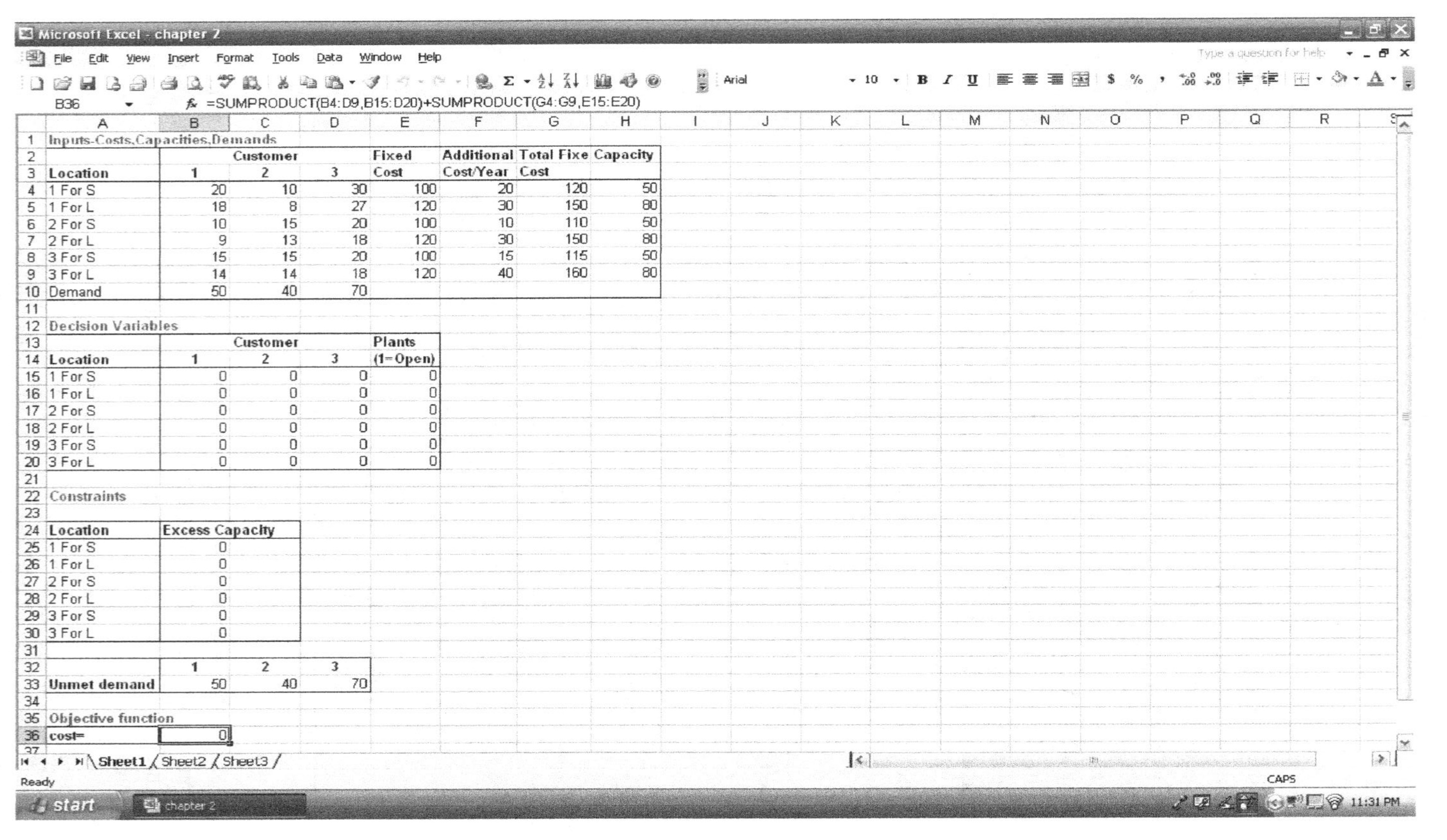

fx =SUMPRODUCT(B4:D9,B15:D20)+SUMPRODUCT(G4:G9,E15:E20)

Inputs-Costs,Capacities,Demands

Location	Customer 1	Customer 2	Customer 3	Fixed Cost	Additional Cost/Year	Total Fixe Cost	Capacity
1 For S	20	10	30	100	20	120	50
1 For L	18	8	27	120	30	150	80
2 For S	10	15	20	100	10	110	50
2 For L	9	13	18	120	30	150	80
3 For S	15	15	20	100	15	115	50
3 For L	14	14	18	120	40	160	80
Demand	50	40	70				

Decision Variables

Location	Customer 1	Customer 2	Customer 3	Plants (1=Open)
1 For S	0	0	0	0
1 For L	0	0	0	0
2 For S	0	0	0	0
2 For L	0	0	0	0
3 For S	0	0	0	0
3 For L	0	0	0	0

Constraints

Location	Excess Capacity
1 For S	0
1 For L	0
2 For S	0
2 For L	0
3 For S	0
3 For L	0

	1	2	3
Unmet demand	50	40	70

Objective function

cost= 0

FIGURE 2.1

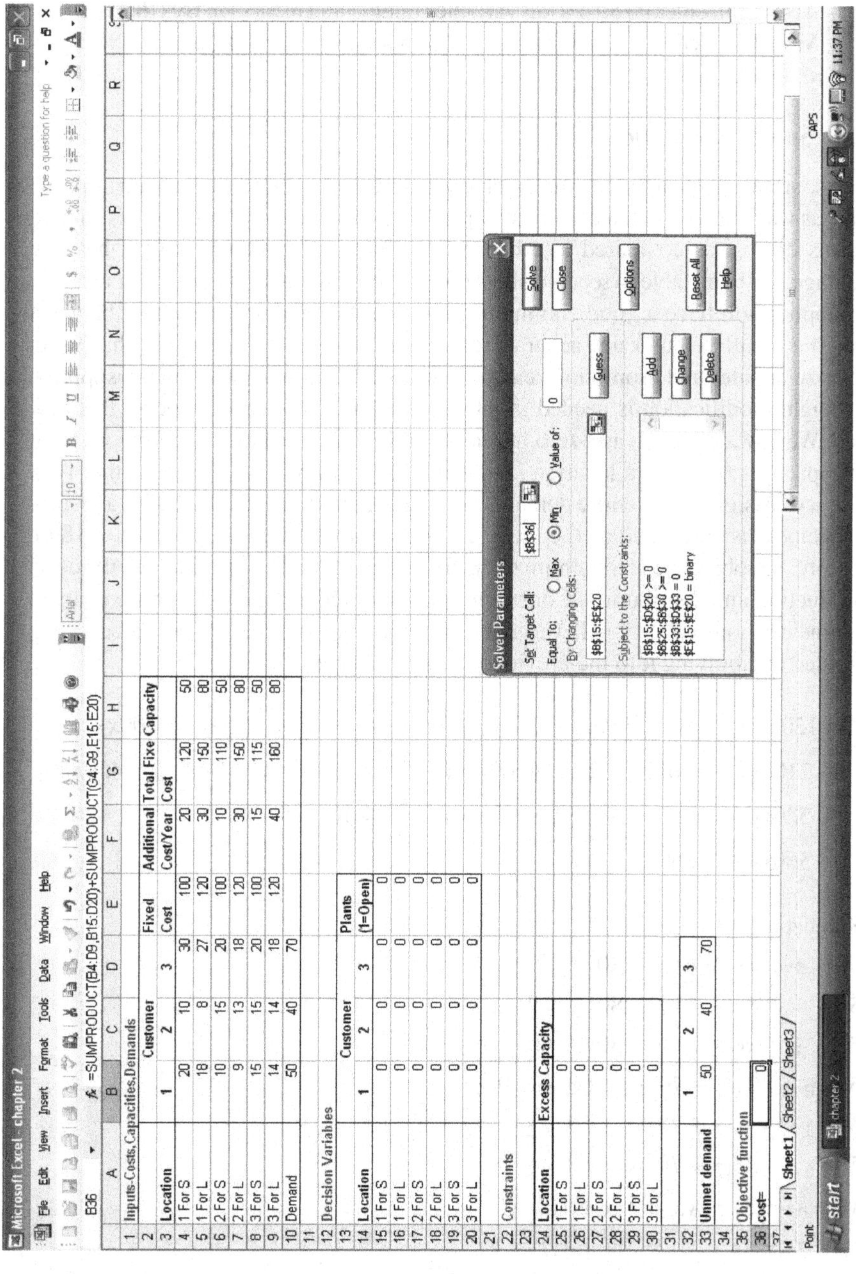

Inputs-Costs,Capacities,Demands

Location	Customer 1	2	3	Fixed Cost	Additional Cost/Year	Total Fixe Cost	Capacity
1 For S	20	10	30	100	20	120	50
1 For L	18	8	27	120	30	150	80
2 For S	10	15	20	100	10	110	50
2 For L	9	13	18	120	30	150	80
3 For S	15	15	20	100	15	115	50
3 For L	14	14	18	120	40	160	80
Demand	50	40	70				

Decision Variables

Location	Customer 1	2	3	Plants (1=Open)
1 For S	0	0	0	0
1 For L	0	0	0	0
2 For S	0	0	0	0
2 For L	0	0	0	0
3 For S	0	0	0	0
3 For L	0	0	0	0

Constraints

Location	Excess Capacity
1 For S	0
1 For L	0
2 For S	0
2 For L	0
3 For S	0
3 For L	0

	1	2	3
Unmet demand	50	40	70

Objective function

cost= 0

FIGURE 2.2

Note: At the time of adding the last constraint for binary variables, enter E_{15}:E_{20} in **By Changing Variable Cells**, then click **Add**, and enter the constraint E_{15}:E_{20} = binary.

5. At the end, enter B15:E20 (all variables integer and binary) in **By Changing Variable Cells** and click **Solve** to obtain an optimal solution of $2480, as shown in Figure 2.3.

2.2.4 Single Sourcing

Single sourcing is when a customer is served from a single plant and not from multiple plants, as was allowed in the previous section. The new policy may not be as efficient in terms of cost as compared to when order slitting was allowed, but it has one advantage. By being able to receive all its orders from one supplier, both the customer and supplier will have a good communication structure and understanding between them. This results in quick and accurate response to changes. For example, changes in order size or sales and promotions can be responded to immediately by the supplier. Only slight modification is needed in the previous model to accommodate this new policy. We now define x_{kij} as a zero or one variable. It is equal to 1 if units from plant with capacity k, located in location i, are shipped to customer j, and zero otherwise. The cost of assigning the entire demand from customer j to a location i with a facility of capacity k, is calculated as $CA_j = c_{kij}D_j$. For example, $C_{213} = 27 \times 70 = 1890$. As before, the objective is to minimize the fixed cost of facility location and the costs of production and distribution. Constraints are modified to reflect the entire demand assignment to one facility. All variables are zero–one variables. The procedure is illustrated by applying it to the previous example (Table 2.5, 2.6, 2.7).

$$\begin{aligned}\text{Min } & 120y_1 + 150y_2 + 110y_3 + 150y_4 + 115y_5 + 160y_6 + 1000x_1 + 400x_2 \\ & + 2100x_3 + 900x_4 + 320x_5 + 1890x_6 + 500x_7 + 600x_8 + 1400x_9 + 450x_{10} \\ & + 520x_{11} + 1260x_{12} + 750x_{13} + 600x_{14} + 1400x_{15} + 700x_{16} \\ & + 560x_{17} + 1260x_{18}\end{aligned}$$

Subject to:

$50x_1 + 40x_2 + 70x_3 - 50y_1 \leq 0$

$50x_4 + 40x_5 + 70x_6 - 80y_2 \leq 0$

$50x_7 + 40x_8 + 70x_9 - 50y_3 \leq 0$

$50x_{10} + 40x_{11} + 70x_{12} - 80y_4 \leq 0$

$50x_{13} + 40x_{14} + 70x_{15} - 50y_5 \leq 0$

$50x_{16} + 40x_{17} + 70x_{18} - 80y_6 \leq 0$

$x_1 + x_4 + x_7 + x_{10} + x_{13} + x_{16} = 1$

$x_2 + x_5 + x_8 + x_{11} + x_{14} + x_{17} = 1$

$x_3 + x_6 + x_9 + x_{12} + x_{15} + x_{18} = 1$

end

integer 24

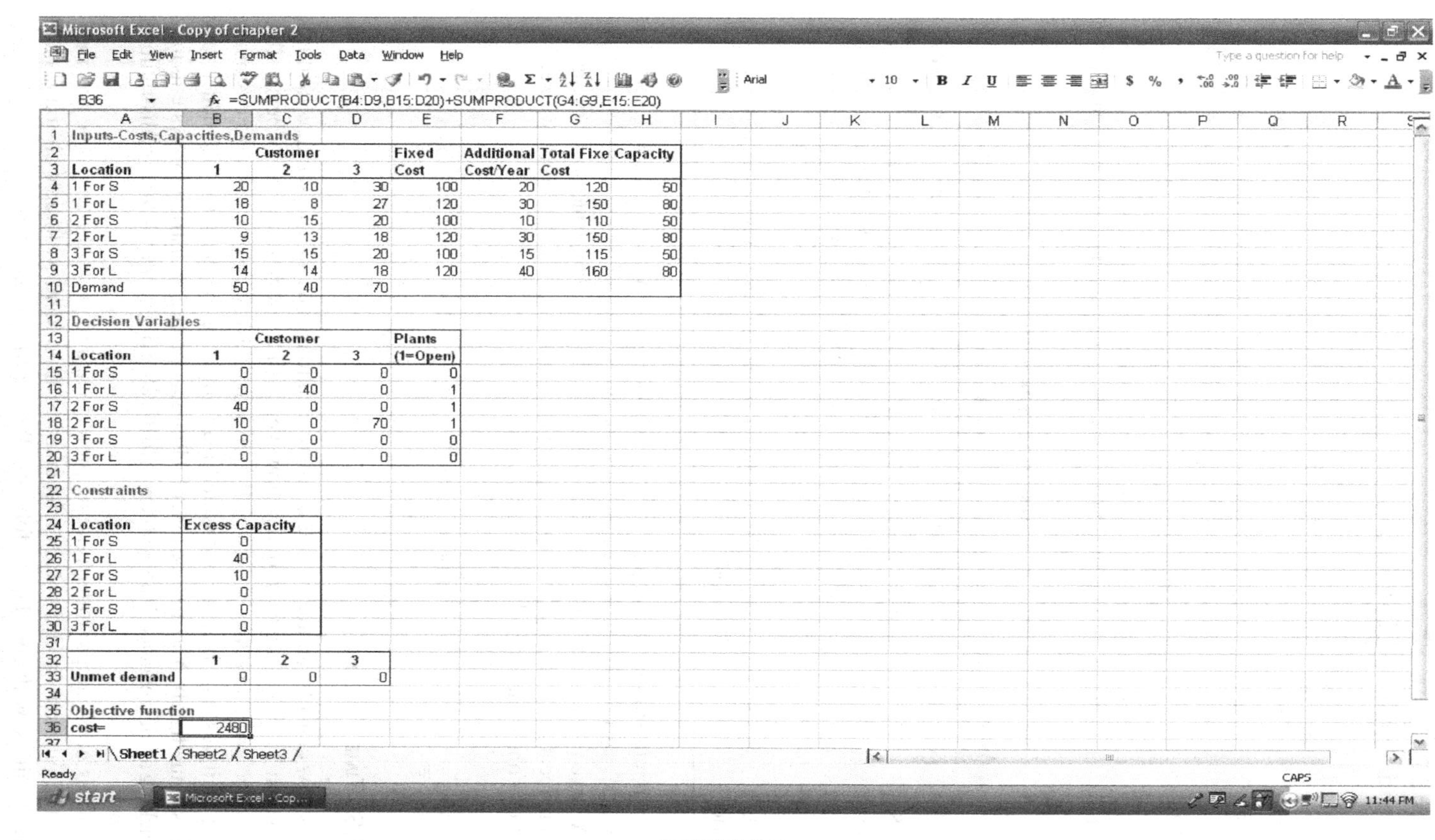

Inputs-Costs,Capacities,Demands

Location	Customer 1	Customer 2	Customer 3	Fixed Cost	Additional Cost/Year	Total Fixe Cost	Capacity
1 For S	20	10	30	100	20	120	50
1 For L	18	8	27	120	30	150	80
2 For S	10	15	20	100	10	110	50
2 For L	9	13	18	120	30	150	80
3 For S	15	15	20	100	15	115	50
3 For L	14	14	18	120	40	160	80
Demand	50	40	70				

Decision Variables

Location	Customer 1	Customer 2	Customer 3	Plants (1=Open)
1 For S	0	0	0	0
1 For L	0	40	0	1
2 For S	40	0	0	1
2 For L	10	0	70	1
3 For S	0	0	0	0
3 For L	0	0	0	0

Constraints

Location	Excess Capacity
1 For S	0
1 For L	40
2 For S	10
2 For L	0
3 For S	0
3 For L	0

	1	2	3
Unmet demand	0	0	0

Objective function

cost=	2480

FIGURE 2.3

TABLE 2.7
Demand/Supply and Cost Data

Location	Additional Site Cost per Year	Customer 1	Customer 2	Customer 3	Capacity
1 for S	20	$20 \times 50 = 1000$	$10 \times 40 = 400$	$30 \times 70 = 2100$	30
1 for L	30	$18 \times 50 = 900$	$8 \times 40 = 320$	$27 \times 70 = 1890$	40
2 for S	10	$10 \times 50 = 500$	$15 \times 40 = 600$	$20 \times 70 = 1400$	30
2 for L	30	$9 \times 50 = 450$	$13 \times 40 = 520$	$18 \times 70 = 1260$	40
3 for S	15	$15 \times 50 = 750$	$15 \times 40 = 600$	$20 \times 70 = 1400$	30
3 for L	40	$14 \times 50 = 700$	$14 \times 40 = 560$	$18 \times 70 = 1260$	40
Demand		50	40	70	

Variable Definition

Location	Additional Site Cost per Year	Customer 1 Variables	Customer 2 Variables	Customer 3 Variables	Capacity
1 for S y_1	20	$20 \times 50 = 1000\,x_1$	$10 \times 40 = 400\,x_2$	$30 \times 70 = 2100\,x_3$	30
1 for L y_2	30	$18 \times 50 = 900\,x_4$	$8 \times 40 = 320\,x_5$	$27 \times 70 = 1890\,x_6$	40
2 for S y_3	10	$10 \times 50 = 500\,x_7$	$15 \times 40 = 600\,x_8$	$20 \times 70 = 1400\,x_9$	30
2 for L y_4	30	$9 \times 50 = 450\,x_{10}$	$13 \times 40 = 520\,x_{11}$	$18 \times 70 = 1260\,x_{12}$	40
3 for S y_5	15	$15 \times 50 = 750\,x_{13}$	$15 \times 40 = 600\,x_{14}$	$20 \times 70 = 1400\,x_{15}$	30
3 for L y_6	40	$14 \times 50 = 700\,x_{16}$	$14 \times 40 = 560\,x_{17}$	$18 \times 70 = 1260\,x_{18}$	40
Demand	50	40	70		

The optimum solution is: objective $z = 2490$. $y_2 = 1$, $y_3 = 1$, $y_4 = 1$, $x_5 = 1$, $x_7 = 1$, $x_{12} = 1$.

A large plant is open is open in location 1, and both large and small plants are open in location 2. If only one type of plant is to be open in each facility, we need an addition constraint for each location, such as,

$$\text{For location } 1, y_1 + y_2 \leq 1$$

2.2.5 Time Constraints

Let us now consider another problem where locations and capacities of new facilities are known. Facilities are operational; however, at times, it may or may not be possible to produce and immediately deliver all items from every facility to every customer. Some facilities may be overloaded and may not have the capacity to produce new orders immediately. Yet, in some other facilities, appropriate distribution systems may be lacking at that instance. The problem is to develop optimum production/distribution network.

This is not a simple transportation problem and requires the zero–one programming formulation again. For example, consider a problem with three customers and three facilities, with cost data as follows (Table 2.8).

TABLE 2.8
Transportation Cost Data

	Customers		
Facilities	**1**	**2**	**3**
1	5	8	7
2	3	9	6
3	6	18	14
Demand	300	600	700

TABLE 2.9
Maximum Amount that Can be Shipped between a Facility and a Customer

	Customers		
Facilities	**1**	**2**	**3**
1	100	50	300
2	200	200	0
3	300	600	700
Demand	300	600	700

Based on the present load on each facility and the available time to produce and deliver the products, the following quantities can be delivered to each customer within the customer's time frame requirements (Table 2.9).

In addition, facility 1 can either supply 100 units to customer 1 or 50 units to customer 2, or 300 units to customer 3 but only to at the most one customer. Similarly, facility 2 can supply 200 units to customers 1 or 2, or none at all. Facility 3 can supply the entire requirement of any and all customers, but at considerably higher prices.

Let x_{ij} be the amount shipped from facility i to customer j. y_k zero–one variables are introduced to handle "either" type of constraint.

The formulation for the preceding data is as follows:

$$\text{Min } 5x_{11} + 8x_{12} + 7x_{13} + 3x_{21} + 9x_{22} + 6x_{23} + 6x_{31} + 18x_{32} + 14x_{33}$$

Subject to

$$x_{11} + x_{21} + x_{31} = 300$$

$$x_{12} + x_{22} + x_{32} = 600$$

$$x_{13} + x_{23} + x_{33} = 700$$

$$x_{11} \leq 100 - 100y_1$$
$$x_{12} \leq 50 - 50y_2$$
$$x_{13} \leq 300 - 300y_3$$
$$y_1 + y_2 + y_3 \geq 2$$
$$x_{21} \leq 200 - 200y_4$$
$$x_{22} \leq 200 - 200y_5$$
$$x_{23} = 0$$
$$y_4 + y_5 \geq 1$$
$$x_{31} \leq 300$$
$$x_{32} \leq 600$$
$$x_{33} \leq 700$$

The first set of constraints assure us that each customer's demands are met.

A little explanation about the second set of constraints—each variable has an upper limit and, from it, same constant times y_k is deducted. Since y_k can only take values of zero or one, when it is one corresponding variable x_{ij} is zero. $y_1 + y_2 + y_3 \geq 2$ assures us that at the most only one y_k can be 0, then corresponding variable can take value up to the limit to which it is defined. When all y_k are one, none of the corresponding x_{ij} can be greater than zero. These sets of constraints satisfy the requirements. A similar explanation results in the third set of constraints.

The solution is: $x_{13} = 300$, $x_{22} = 200$, $x_{31} = 300$, $x_{32} = 400$, $x_{33} = 400$, and $z = \$18{,}500$.

2.3 UNCERTAINTY IN DEMAND

It is hardly possible to predict the future with complete certainty. Prices go up and down, and so can the demand. New competitors can come into market, or product can experience obsolescence. Indeed, as we go further into the future, it becomes more difficult to predict all the factors that affect sales. The forecasting methods discussed in the later chapters provide some guidelines that may be followed; however, errors in prediction are most often present. To build a production plant/facility under such an environment is difficult. Remember, building a plant is expensive and has its influence over a long period of time.

A decision tree can be used in making decisions in this environment. Consider the following situation where we need to decide whether to build a large or small plant to meet the demand for next 2 years.

A large plant has capacity of 1000 units/year, while a small plant has a capacity of only 750 units/year. At present, the demand for the product is 700 units, and it has a sales price of \$1000/unit. Both demand and price may also change from year to year. Demand can go up or down by 10% with probabilities of 0.6 and 0.4, respectively. The price can also go up and down, based on what happens to the demand. When demand goes up, the unit price can go up by 5% with a probability of 80%, and when it goes down, the price goes down by 10% with a probability of 20%. When demand goes down, the probability price goes up is 0.3 and probability it goes down is 0.7.

Let U and D define up or down changes, and let the first letter be associated with demand and the second with price. For example, UU means demand is up and price is up in that year. Then, there are four possible states:

(UU), (UD), (DU), (DD)

The probabilities for each state are

UU = p (demand up) × p (price up) = 0.6 × 0.8 = 0.48

UD = p (demand up) × p (price down) = 0.6 × 0.2 = 0.12

DU = p (demand down) × p (price up) = 0.4 × 0.3 = 0.12

DD = p (demand down) × p (price down) = 0.4 × 0.7 = 0.28

There are four possible states at the end of first year, and for each state in the first year, there are four possible states in the second year, for a total of 16 states. They are shown in Figure 2.4. Also, shown are the demands for the associated states (details are shown in the calculation table). Since the capacity of a large plant is 1000, with a large plant we can meet all the demand. However, with a small plant, with a maximum capacity of 750, the demand can only be satisfied up to 750 units.

Year 0	Demand	Year 1	Demand at the end of year 1	Year 2	Demand at the end of year 2
	700	UU	770	UU	847
				UD	847
				DU	693
				DD	693
		UD	770	UU	847
				UD	847
				DU	693
				DD	693
		DU	630	UU	693
				UD	693
				DU	567
				DD	567
		DD	630	UU	693
				UD	693
				DU	567
				DD	567

Let us calculate the revenue at each stage for the first and second years. When demand goes up, the price goes up by 5% or 1.05 times the previous year's price,

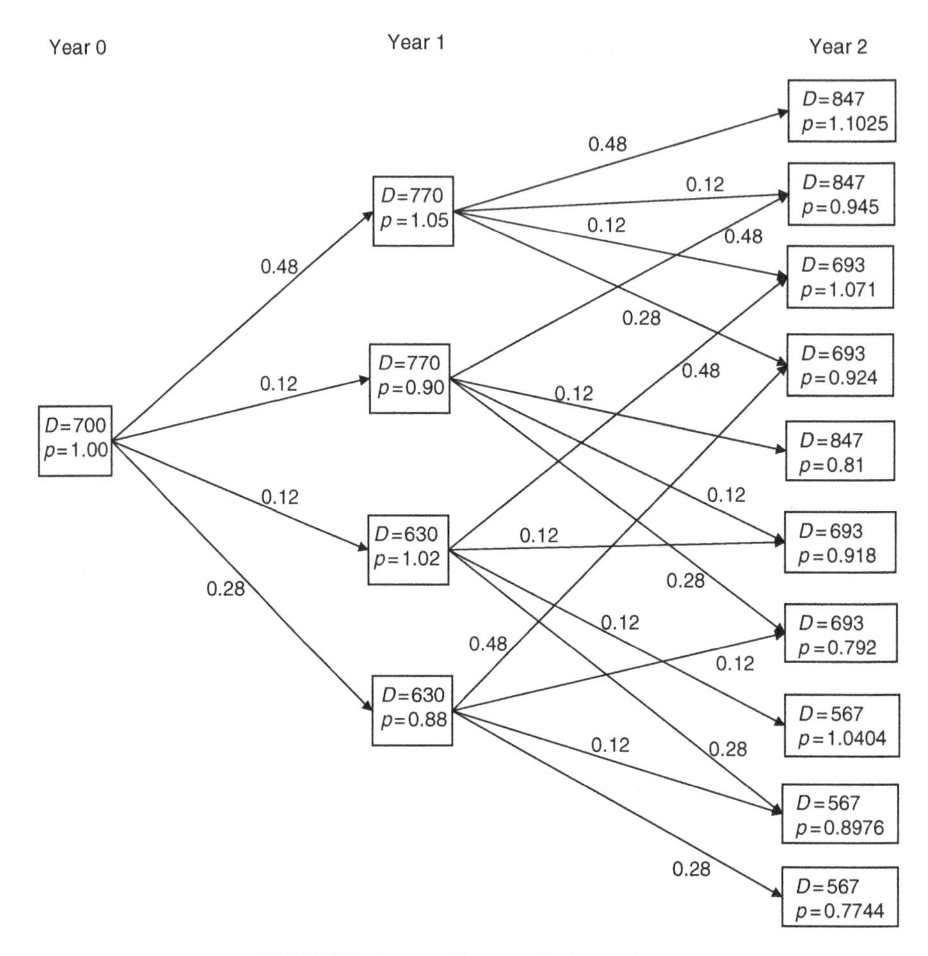

FIGURE 2.4 Decision tree for large plant.

while it goes down by 10% or 0.9 times previous year's price. When demand goes down, the price is up by 2%, and goes down by 12%. Revenues for all the states in each year are also shown.

First-Year Revenue Calculations

State	Total Demand	Revenue = Demand × Price
UU	(700 × 1.1) = 770	770 × (1,000 × 1.05) = 808,500
UD	(700 × 1.1) = 770	770 × (1,000 × 0.9) = 693,000
DU	(700 × 0.9) = 630	630 × (1,000 × 1.02) = 642,600
DD	(700 × 0.9) = 630	630 × (1,000 × 0.88) = 554,400

Second-Year Revenue Calculations

Year 1	Year 2	Total Demand	Revenue
UU	UU	$(700 \times 1.1) \times 1.1 = 847$	$847 \times (1{,}000 \times 1.05) \times 1.05 = 933{,}817.5$
UU	UD	$(700 \times 1.1) \times 1.1 = 847$	$847 \times (1{,}000 \times 1.05) \times 0.5 = 800{,}415$
UU	DU	$(700 \times 1.1) \times 0.9 = 693$	$693 \times (1{,}000 \times 1.05) \times 1.02 = 742{,}203$
UU	DD	$(700 \times 1.1) \times 0.9 = 693$	$693 \times (1{,}000 \times 1.05) \times 0.88 = 640{,}332$
UD	UU	$(700 \times 1.1) \times 1.1 = 847$	$847 \times (1{,}000 \times 0.9) \times 1.05 = 800{,}415$
UD	UD	$(700 \times 1.1) \times 1.1 = 847$	$847 \times (1{,}000 \times 0.9) \times 0.9 = 686{,}070$
UD	DU	$(700 \times 1.1) \times 0.9 = 693$	$693 \times (1{,}000 \times 0.9) \times 1.02 = 636{,}174$
UD	DD	$(700 \times 1.1) \times 0.9 = 693$	$693 \times (1{,}000 \times 0.9) \times 0.88 = 548{,}856$
DU	UU	$(700 \times 1.1) \times 0.9 = 693$	$693 \times 1{,}000 \times 1.02) \times 1.05 = 742{,}203$
DU	UD	$(700 \times 1.1) \times 0.9 = 693$	$693 \times (1{,}000 \times 1.02) \times 0.9 = 636{,}174$
DU	DU	$(700 \times 0.9) \times 0.9 = 567$	$567 \times (1{,}000 \times 1.02) \times 1.02 = 589{,}907$
DU	DD	$(700 \times 0.9) \times 0.9 = 567$	$567 \times (1{,}000 \times 1.02) \times 0.88 = 508{,}939$
DD	UU	$(700 \times 1.1) \times 0.9 = 693$	$693 \times (1{,}000 \times 0.88) \times 1.05 = 640{,}332$
DD	UD	$(700 \times 1.1) \times 0.9 = 693$	$693 \times (1{,}000 \times 0.88) \times 0.9 = 548{,}856$
DD	DU	$(700 \times 0.9) \times 0.9 = 567$	$567 \times (1{,}000 \times 0.88) \times 1.02 = 508{,}939$
DD	DD	$(700 \times 0.9) \times 0.9 = 567$	$567 \times (1{,}000 \times 0.88) \times 0.88 = 439{,}085$

2.3.1 Large Plant Evaluation

All the calculations are shown in the following table. We start by placing state probabilities for all states (as calculated in the previous table and shown in Figure 2.4) and revenues for the second year, in the table. Start from the second year and evaluate the expected revenue from the second year for each possible state of the first year. For example, the expected second-year revenue for the first year UU state is: $0.48 \times 93{,}3817 + 0.12 \times 800{,}415 + 0.12 \times 742{,}203 + 0.28 \times 640{,}332 = 812{,}639$. Assuming all revenues are end-of-year revenues, and assuming an interest rate of 8%, calculate the end of first year, that is beginning of the second year, equivalent amount by multiplying the end-of-year revenue by $1/(1 + 0.08) = 0.9259$. So, the present value at the end of the first year is $812{,}639 \times 0.9259 = 752{,}443$. This amount is added to the revenue for the first year at stage UU, which is 808,500, resulting in $808{,}500 + 752{,}443 = 1{,}560{,}943$. Similar calculations for other states results in total revenues of 1,337,951 for state UD, 1,193,522 for state DU, and 959,249 for state DD at the end of the first year. The present values at the beginning of the first year are obtained by again multiplying with 0.9259. The expected value at the beginning of year 1, year 0, is: $0.48 \times 1{,}445{,}318 + 0.12 \times 1{,}238{,}844 + 0.12 \times 1{,}105{,}113 + 0.28 \times 959{,}249 = 1{,}243{,}617$. This is the expected profit if a large plant is placed in service at the beginning of year 1, that is year 0.

Large Plant: Expected Profit Evaluation

Year 0	Year 1				Year 2				
Expected Value	Probability	Present Value	First- and Second-Year Total	State	Present Value	Expected Value	Probability	State	Revenue
1,257,746	0.48	1,445,318	1,560,943	UU	752,443	812,639	0.48	UU	933,817
							0.12	UD	800,415
							0.12	DU	742,203
							0.28	DD	640,332
	0.12	1,238,844	1,337,951	UD	644,951	812,639	0.48	UU	800,415
							0.12	UD	686,070
							0.12	DU	636,174
							0.28	DD	548,856
	0.12	1,148,746	1,240,646	DU	598,046	645,890	0.48	UU	742,203
							0.12	UD	636,174
							0.12	DU	589,906
							0.28	DD	508,939
	0.28	991,075	1,070,361	DD	515,961	557,238	0.48	UU	640,332
							0.12	UD	548,856
							0.12	DU	508,939
							0.28	DD	439,084

2.3.2 Small Plant Evaluation

Similar calculations are needed to evaluate expected profit of a small plant with 750-unit capacity, placed at at the end of year 0.

Small Plant Demands at Years 1 and 2, with Maximum Capacity of 750

Year 0	Demand	Year 1	Demand	Year 2	Demand
	700	UU	750	UU	750
				UD	750
				DU	693
				DD	693
		UD	750	UU	750
				UD	750
				DU	693
				DD	693
		DU	630	UU	693
				UD	693
				DU	567
				DD	567
		DD	630	UU	693
				UD	693
				DU	567
				DD	567

2.3.3 Revenue Evaluation

First Year

State	Total Demand	Revenue = Demand x Price
UU	(700 × 1.1) = 770	750 × (1,000 × 1.05) = 787,500
UD	(700 × 1.1) = 770	750 × (1,000 × 0.9) = 675,000
DU	(700 × 0.9) = 630	630 × (1,000 × 1.02) = 642,600
DD	(700 × 0.9) = 630	630 × (1,000 × 0.88) = 554,400

Year 1	Year 2	Total Demand	Revenue
UU	UU	(700 × 1.1) × 1.1 = 847	750 × (1,000 × 1.05) × 1.05 = 826,875
UU	UD	(700 × 1.1) × 1.1 = 847	750 × (1,000 × 1.05) × 0.9 = 708,750
UU	DU	(700 × 1.1) × 0.9 = 693	693 × (1,000 × 1.05) × 1.02 = 742,203
UU	DD	(700 × 1.1) × 0.9 = 693	693 × (1,000 × 1.05) × 0.88 = 640,332
UD	UU	(700 × 1.1) × 1.1 = 847	750 × (1,000 × 0.9) × 1.05 = 708,750
UD	UD	(700 × 1.1) × 1.1 = 847	750 × (1,000 × 0.9) × 0.9 = 607,500
UD	DU	(700 × 1.1) × 0.9 = 693	693 × (1,000 × 0.9) × 1.02 = 636,174
UD	DD	(700 × 1.1) × 0.9 = 693	693 × (1,000 × 0.9) × 0.88 = 548,856
DU	UU	(700 × 1.1) × 0.9 = 693	693 × 1,000 × 1.02) × 1.05 = 742,203
DU	UD	(700 × 1.1) × 0.9 = 693	693 × (1,000 × 1.02) × 0.9 = 636,174
DU	DU	(700 × 0.9) × 0.9 = 567	567 × (1,000 × 1.02) × 1.02 = 589,907
DU	DD	(700 × 0.9) × 0.9 = 567	567 × (1,000 × 1.02) × 0.88 = 508,939
DD	UU	(700 × 1.1) × 0.9 = 693	693 × (1,000 × 0.88) × 1.05 = 640,332
DD	UD	(700 × 1.1) × 0.9 = 693	693 × (1,000 × 0.88) × 0.9 = 548,856
DD	DU	(700 × 0.9) × 0.9 = 567	567 × (1,000 × 0.88) × 1.02 = 508,939
DD	DD	(700 × 0.9) × 0.9 = 567	567 × (1,000 × 0.88) × 0.88 = 439,085

Small Plant Expected Profit Evaluation

Year 0	Year 1				Year 2				
Expected Value	Probability	Present Value	First- and Second-Year Total	State	Present Value	Expected Value	Probability	State	Revenue
1,215,283	0.48	1,372,434	1,482,229	UU	694,729	750,307	0.48	UU	826,875
							0.12	UD	708,750
							0.12	DU	742,203
							0.28	DD	640,332
	0.12	1,176,372	1,270,482	UD	595,482	643,121	0.48	UU	708,750
							0.12	UD	607,500
							0.12	DU	636,174
							0.28	DD	548,856
	0.12	1,148,746	1,240,646	DU	598,046	645,890	0.48	UU	742,203
							0.12	UD	636,174
							0.12	DU	589,906
							0.28	DD	508,939
	0.28	991,075	1,076,644	DD	522,244	564,024	0.48	UU	640,332
							0.12	UD	548,856
							0.12	DU	508,939
							0.28	DD	439,084

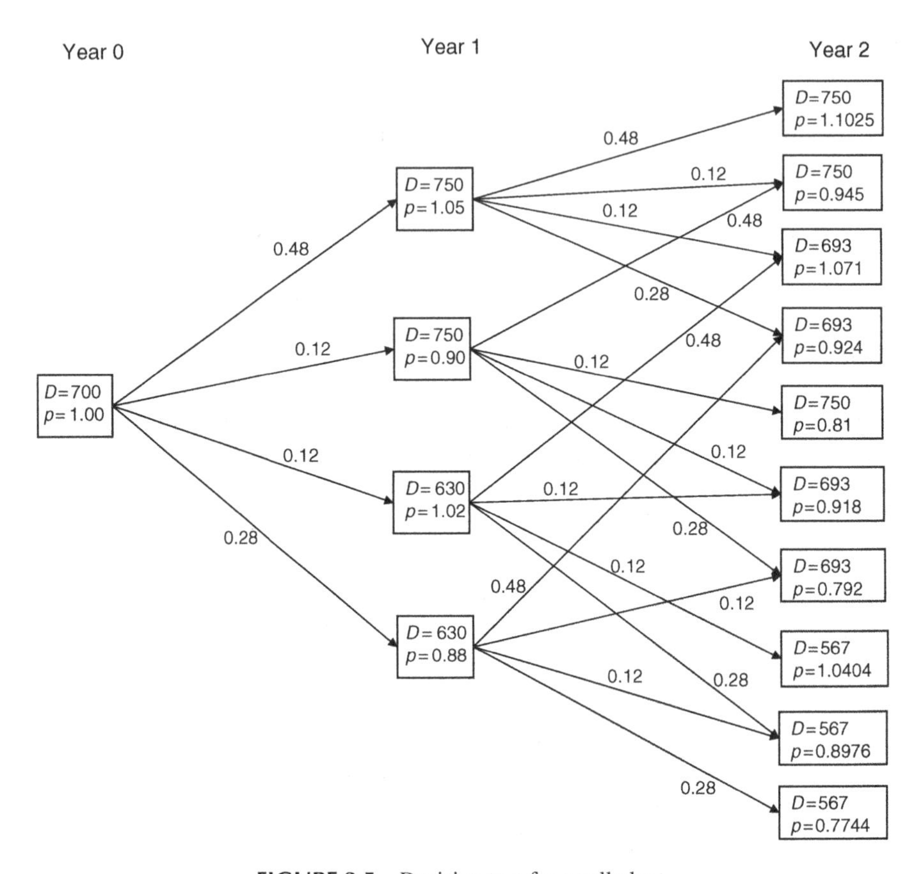

FIGURE 2.5 Decision tree for small plant.

Small plant evaluations are similar to those of a large plant. The expected income from a small plant is 1,079,063. The difference between large plant and small plant revenue is $1,257,746 – $1,215,283 = $42,463. So, as long as the cost of construction of the large plant does not exceed the small plant price by this amount, we should build a large plant (Figure 2.5).

2.3.4 Expansion in Future

Suppose we can start with a small plant and add capacity of 250 after 1 year for $50,000. Should we expand?

The answer depends on what state we are in after 1 year. Calculate the expected profit for going from small to expanded capacity, which in our case then becomes the large plant. The detailed calculations are already available in previous large and small plant expected profit evaluation tables.

State After 1 Year	Expected Profit from Small Plant	Large Plant	Difference
UU	694,729	752,443	57,714
UD	595,482	644,951	49,469
DU	598,046	598,046	0
DD	522,244	522,244	0

With a cost of expansion of $50,000, it is obvious that we should expand only if we are in state UU after 1 year.

2.4 SUMMARY

The chapter introduces models in plant location and capacity determination. The models are based on specific aim and available resources. The major objective is to develop optimum customer distribution network and to determine the associated capacity required at each plant. The last model is probabilistic, which allows decisions on plant capacity based on future prospects of demand, as judged today by the management.

2.5 EXERCISE

2.1 J & A International is a wholesaler who sells computer parts to its retailers from its three warehouses located at Los Angeles, Boston, and Chicago. It regularly supplies the parts to its retailers in different parts of the country. For this month, there is a specific demand for scanners from three retailers. The following table shows the demand from its retailers and the associated shipping costs.

	Shipping Cost per 100 Units ($) Retailer 1	Retailer 2	Retailer 3	Stock (in Units)
Los Angeles	1675	400	1160	1250
Chicago	760	1160	543	930
Boston	320	1230	440	830
Order from retailer	600	700	300	

a. Formulate the LP model.
b. Find the optimal shipping plan from each wholesaler to each retailer.

2.2 If in Problem 2.1 a complete order must be fulfilled from only one warehouse, what should be the optimum policy?

2.3 A company has three different plants producing the same products and supplying to different warehouses. The following table shows the order sizes, capacities of each plant, and the shipping costs.

	Shipping Cost Per Units ($)			
Plant	**W_1**	**W_2**	**W_3**	**Stock (in Units)**
1	400	450	375	60
2	390	330	—	40
3	—	—	370	55
order	33	21	25	

Plant 1 can distribute to all the warehouses, whereas plant 2 can distribute to warehouses 1 and 2 only, and plant 3 can distribute to warehouse 3 only. Warehouse 1 can receive any units supplied from any plant, whereas warehouses 2 and 3 can receive only 34 and 25 units, respectively.

The management wants to know how much to distribute to different warehouses from different plants, while minimizing the total shipping cost.

2.4 For the problem described in Problem 2.3, because of high operating cost, the management would like to operate only two plants to meet the demand. Keeping the shipping cost minimal, suggest from which plant it is justifiable not to produce anything.

2.5 Tricon Corporation (TC) is a manufacturer of several rubber products such as speed breakers, neoprene bearing pads, etc. Due to reconstruction work after some natural disasters and a good economic environment, the rubber industry is expecting a huge demand for the year 2007. The management is anticipating a demand of 180,000 units in the south, 150,000 in the west, 120,000 in the east, and 90,000 in the north. Considering the low labor cost and availability of natural rubber, the management has decided to build the manufacturing plants in Asia and South America. Plants can be of three types: large, medium, and small. The management has decided that no more than two plants can be built in any country. Capacities, fixed costs, and the costs of producing and shipping a unit from four locations to four marketplaces are shown in the following tables:

Capacities

	India	**Sri Lanka**	**Brazil**	**Argentina**
Large	400,000	350,000	420,000	375,000
Medium	200,000	220,000	—	205,000
Small	100,000	—	80,000	120,000

Fixed Costs

	India	**Sri Lanka**	**Brazil**	**Argentina**
Large	10 million	8 million	9 million	7 million
Medium	6 million	6.5 million	—	6.5 million
Small	4 million	—	3.75 million	3.6 million

Transportation and Shipping Costs

	India	Sri Lanka	Brazil	Argentina
East	$210	$230	$220	$235
South	$217	$220	$195	$215
West	$250	$245	$225	$233
North	$222	$236	$250	$255

So, where should TC build their factories?

2.6 From a survey, the annual demand for a product XYZ is found to be 25 units. The selling price is $1200/unit. From one year to the next, demand may go up and down by 8% with probabilities 0.5 and 0.3, respectively. When the demand goes up, the price may also go up by 2% with probability of 82%; when the demand goes up, the price may decrease by 3% with 12% probability; when the demand goes down, the price may go up by 3% with 36% probability; and finally demand and price may go down by 2% with probability 74%. Calculate the expected revenue for first and second years by using the decision tree.

2.7 A small plant has a capacity of manufacturing 900 units/year, while a large plant has a capacity of 1,500 units/year. A small plant can be built in year 0 for $100,000, and in year 1 for $120,000; similarly, a large plant can be built for $150,000 in year 0 and $180,000 in year 1. Current demand for the product is 825 units/year. The management decides to sell the product at $1,200/unit. From one year to the next, demand may go up and down by 8%, with probabilities 0.5 and 0.3, respectively. When the demand goes up, the price may go up by 2%, with a probability of 82%. When the demand goes up, the price may decrease by 3%, with 12% probability; the demand goes down and price up by 3%, with 36% probability; and demand and price go down by 2% with probability 74%. A small and large plant can be built in year 1 or year 2. Determine the optimum policy for plant placements.

2.8 Plants can be placed in any one of the 5 locations. The costs of shipping from locations to the existing warehouses are follows:

	Warehouse		
Plant	**1**	**2**	**3**
A	16	18	9
B	12	15	15
C	10	8	14
D	13	9	11
E	7	12	16
Requirements at the warehouse per year	35	50	85

Plants can be built with two capacities, of 100 or 150 units per year. The small plant costs $2000 per year to operate, while the large plant requires $3500 per year. Determine where to place plants, and what their capacity to minimize the total cost of the operation should be. Formulate the solution to the problem.

2.9 A National food company wants to build an additional new plant to meet the demand in central United States. The management is looking at four possible locations: Wichita, Omaha, St Louis, and Oklahoma City. The following table shows the associated fixed costs per year, the variable cost per unit, and the annual fixed transportation costs to the main distribution center.

Location/Cost	Wichita	Omaha	St Louis	Oklahoma City
Fixed cost/year	$500,000	$440,000	$470,000	$475,000
Variable cost/unit	$3	$4	$3	$3.30
Transportation cost	$49,000	$63,000	$60,000	$57,000

The plant can produce 40,000 units per month. Assuming that the units sell for $12 per unit, and the plant capacity being fully utilized, determine the location that will be best suited for an additional plant.

2.10 A chocolate maker produces and sells five different kinds of candies: chocolate liquor, milk chocolate, white chocolate, couverture, and ganache. The respective profits per package are $10, $12, $19, $25, and $22. All the products basically require the same process: mixing and conching, tempering, and packing. The times for these operations vary for the various items, as shown in the following table:

	Operation Time (Min)		
	Mixing and Conching	Tempering	Packaging
Chocolate liquor	7	12	3
Milk chocolate	5	9	3
White chocolate	8	14	4
Couverture	8	12	5
Ganache	12	8	6

The time available for mixing and conching, tempering, and packing are 300, 380, and 270 min, respectively. To maximize the profit, what should the manufacturer produce, and in what quantity?

REFERENCES AND SUGGESTED READINGS

Amram, Martha and Nalin Kulatilaka. 1999. *Real Options*, Cambridge, MA: Harvard Business School Press.

Ballou, Ronald H. 1999. *Business Logistics Management*, Upper Saddle River, NJ: Prentice Hall.

Daskin, Mark S. 1995. *Network and Discrete Location*, New York: John Wiley & Sons.

Harding, Charles F. 1988. "Quantifying Abstract Factors in Facility Location Decisions" *Industrial Development* (May–June): 24–28.

Korpela, Jukka, Antti Lehmusvaara, and Marku Tuominen. 2001. "Customer Service Based Design of the Supply Chain" *International Journal of Production Economics* 69: 193–204.

Luehrman, Timothy A. 1995. "Capital Projects as Real Options: An Introduction" *Harvard Business School*, case 9-295-074.

McCormack, Alan D., Lawerence J. Newman III, and Donald B. Rosenfield. 1994. "The New Dynamics of Global Manufacturing Site Location" *Sloan Management Review* (Summer): 69–79.

Note on Facility Location. 1989. *Harvard Business School*, note 9-689-059.

3 Forecasting and Aggregate Planning

3.1 FORECASTING

Objective of every manufacturing facility is to produce and sale product and make profit in doing so. They must therefore decide when to produce and how much to produce. The answer requires estimation of what the product demand would be in future periods.

Forecasting is a technique for estimating the expected demand in future. However, since it is an estimate, it seldom is exactly equal to the actual demand. There is strong likelihood of actual demand being different from the forecasted demand. Then why go through the trouble of forecasting? For one, it is better to have an estimate than to have none at all. For example, for an auto dealership, the forecasted knowledge that the sales of new cars could be 200 cars per month has a value. They can plan their activities based on this goal. If the actual demand happens to be 224 for a particular month, the planning necessary to get the additional 24 cars would be much less than if the dealer did not know the initial estimate at all ahead of time. Forecasting is more robust when it is done over an aggregated product line. For example, the cumulative error in forecasting the sale for each model of a car will be much higher than error in forecasting the total sales of the car. Low sales in some models may be compensated by increased sales in other models.

Forecasting could be over a long range with time span of years or could be over a short time frame such as weeks or even days. Errors in forecasting tend to be larger as the planning period gets longer. Uncertain factors that influence the demand may not be well estimated over a long period of time. There are many forecasting methods based on information available and market variations. There is no golden rule or model that fits all circumstances. A good forecasting method should minimize the error between the actual demand and forecasted value. We will study some of the forecasting methods in the following pages.

There are two ways of making a forecast: (1) qualitative forecasting based on informed judgment and (2) quantitative forecasting based on past data.

3.1.1 QUALITATIVE FORECASTING

Qualitative forecasting is generally subjective and is associated with a new product for which we have no historical data. Expert opinion, based on market surveys, panel discussion, or even life-cycle analogy to similar existing product lines, is used to estimate the future demand for a new product. Such forecasting may be quick but tend to have bias based on the personality of the forecaster. An optimistic person tends to overestimate, while a pessimistic person has the opposite tendency.

It is commonly believed that a group of experts can make better predictions than a single individual. It allows the experience and knowledge of multiple individuals to address the problem. They challenge and support each other's reasoning to arrive at more accurate predictions. However, it is necessary to avoid interpersonal relationships that may result in bandwagon or authority effects. One technique used for such analysis is called the Delphi technique.

Here, a group is formed where the experts are not introduced to each other and all decisions are made by polling. A questionnaire related to the question on hand is passed anonymously to the members of the team, and the results are collected and statistically analyzed. The predictions outside inter-quartile range (IQR) are challenged, and participants are asked to support or examine the assumptions and reasoning for these predictions. The new prediction, supporting or opposing arguments and other feedback, is recycled to all participants. New predictions are made, and iterations are continued till a convergence in prediction is obtained or no further improvement in prediction is possible.

Another method of forecasting is to see how future technology, economics, politics, and other related factors might influence the sales of the present products. This is called future creations. Different situations are created based on alternate scenarios associated with these factors, and the likelihood of each scenario is examined. It results in somewhat reliable sales estimates.

3.1.2 Quantitative Forecasting

Quantitative forecasting can be further broken down into three classes based on the relationship between demand and the factors that influence demand. The classifications are

1. *Casual forecasting*: Causal forecasting methods are appropriate when there is a strong relationship between demand and environmental factors. For example, housing sale is strongly influenced by interest rates. Plywood sales depend on the housing market, demand for flashlights goes up when bad weather is predicted, and so on. Regression analysis, modeling and simulation, and/or econometric modeling are the common methods used in such analyses.
2. *Time series analysis*: Forecast is based on the assumption that future demand will follow the demand pattern of the past. Past demand may have a random variation around mean demand; in addition, it may have a trend pattern and/or may even show a seasonal effect. For example, for a grocery store, demand for steady products like bread and milk are not going to change drastically from day to day. However, it may show a trend based on the overall trend of customers visiting the store over a period of time. On other hand, demand for lawn mowers may not only show a trend pattern but will also have seasonal variations. One way to understand the previous demand behavior is to plot the past data against time and observe if any pattern is present. Of course, we must have sufficient data to make such an observation. In many instances, such as in the example of

lawn mowers, seasonal variation can be predicted by knowing the product line.

Various methods such as moving average, exponential smoothing, and regression analysis are available for analysis, and they are explained later on in the chapter. Accuracy and effectiveness of these methods can be evaluated by determining the error in forecasting, that is, actual demand versus the predicted demand. We can then choose a method that gives the least forecasting error. More about analyzing forecasting errors is illustrated in Section 3.1.3. We are going to use two measures: sum of error squares and sum of absolute errors. These measures are calculated and displayed in each forecasting method shown in this section.

Time series forecasting methods can be further divided into static and dynamic techniques. In the static method, which mainly consists of regression analysis, a prediction for multiple future periods can be made based on the model developed with the present available data. The model is static and does not change with additional data available in time. In the dynamic method, on the other hand, the parameters of the model are continuously adjusted, based on the latest available information. Thus, to make forecast for the next period, and parameters are reevaluated based on the actual demand is in the present period.

To illustrate forecasting methods, let us take an example. Suppose we have the following demand data available for last 13 time periods, each period being of 3 months duration.

Period	0	1	2	3	4	5	6	7	8	9	10	11	12
Observed demand	18	30	50	60	32	50	70	45	40	60	80	60	55

Plot of the data indicates that there is a trend and may also follow a seasonal pattern, with a cycle time of four periods or 1 year, perhaps representing a season. (Figure 3.1).

3.1.3 Static Forecasting

Placing a regression equation through the data gives following relationship between period and demand. We have used Microsoft Excel software to obtain this relationship

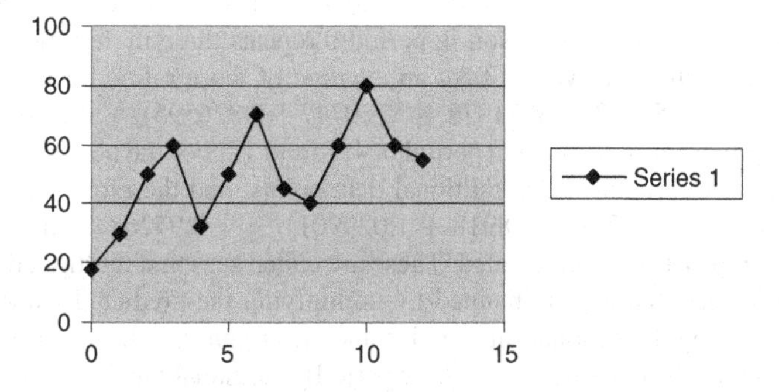

FIGURE 3.1 Plot of observed demands.

(Excel–Tools–Data Analysis–Regression), but we can use any widely used statistical analysis software for this analysis.

$$\text{Demand in period } t,\ D_t = 33.41 + 2.76\ (t)$$

ANOVA

	df	SS	MS	F	Significance F
Regression	1	1390.159	1390.159	7.254701	0.0209
Residual	11	2107.841	191.6219		
Total	12	3498			

	Coefficients	Standard Error	*t* Stat	*P*-Value	Lower 95%	Upper 95%	Lower 95.0%	Upper 95.0%
Intercept	33.41758	7.255575	4.60578	0.000758	17.44816	49.387	17.44816	49.387

Here, the intercept of line 33.41 is considered as level demand, since its value will remain constant, no matter the time period. The slope of the line, 2.76, is the trend factor. For each time period in future, the demand will increase by 2.76 units. It is an increasing trend. We can evaluate how good the linear regression is by calculating the predicted values and forecasting errors, predicted minus actual demand, for the past data based on this equation. The sum of the square of error (2107.84) from Table 3.1 gives one measure as to how good the prediction equation is, as we will see in Section 3.1.8.

3.1.4 Seasonal Correction

In the plot of the data, we have seen seasonal influence in demand. We can see that the season repeats itself every forth time period. We can probably reduce the errors and come out with better forecasted values if we can adjust the predictions for seasons. This is done by developing seasonal multipliers (Table 3.1).

For each season, calculate the ratio of actual demand to the predicted value based on regression equation. The season in period 0 repeats three more times in periods 4, 8, and 12, in different years. Take an average of these ratios for these periods, which results in $(0.538632 + 0.71978 + 0.720747 + 0.826595)/4 = 0.701439$ as the season multiplier for the season in period 0, 4, 8, and 12. For the next season starting at period 1, there are only two additional data points, and therefore, the average is calculated as $(0.829233 + 1.058918 + 1.029901)/3 = 0.972684$. Similarly, other seasons' averages also are calculated. These are called seasonal multipliers. The new prediction for each period is obtained by multiplying the predicted value from the linear equation by its seasonal multiplying factor. In general, the predicted demand value is equal to (level demand + trend x period) x seasonal factor for the trend.

TABLE 3.1
Static Forecasting with Seasonal Corrections

Period	Demand (D)	Predicted Demand (P)	Error ($D - P$)	Ratio = (D/P)	Seasonal Factors	Predicted Demand Modified by Seasonal Factor ($P1$)	Modified Error ($D - P1$)
0	18	33.418	−15.418	0.538632	0.701439	23.44067484	
1	30	36.178	−6.178	0.829233	0.972684	35.18977282	−5.4406748
2	50	38.938	11.062	1.284093	1.331932	51.8627872	−5.1897728
3	60	41.698	18.302	1.438918	1.077652	44.9359315	−1.8627872
4	32	44.458	−12.458	0.71978	0.701439	31.18455688	15.064069
5	50	47.218	2.782	1.058918	0.972684	45.92820755	0.8154431
6	70	49.978	20.022	1.400616	1.331932	66.56732186	4.0717924
7	45	52.738	− 7.738	0.853275	1.077652	56.83320915	3.4326781
8	40	55.498	−15.498	0.720747	0.701439	38.92843893	−11.833209
9	60	58.258	1.742	1.029901	0.972684	56.66664229	1.0715611
10	80	61.018	18.982	1.311089	1.331932	81.27185652	3.3333577
11	60	63.778	−3.778	0.940763	1.077652	68.73048681	−1.2718565
12	55	66.538	−11.538	0.826595	0.701439	46.67232098	8.327679
Sum of error squares			2107.85				615.43

For example, the predicted demands for periods 13, 14, and 15 are

- Predicted demand for period 13 = (33.41 + 2.76 (13) 0.972684 = 67.39
- Predicted demand for period 14 = (33.41 + 2.76 (14) 1.331932 = 95.96
- Predicted demand for period 15 = (33.41 + 2.76 (15) 1.077652 = 80.61

3.1.5 Adaptive Forecasting

Though the methods in adaptive forecasting are based on historical demand these methods are developed to respond quickly to the latest available demand information. Any sudden change in demand pattern is quickly responded to in making the new predictions.

The simplest method is termed *moving average*. It predicts the demand for the next period by simply taking the average of past n periods' demand. Each forecast, therefore, reflects the latest n periods' demand history. For example, if we select a value of $n = 4$, prediction for period 4 is an average of the previous four demands, that is, demands $(18 + 30 + 50 + 60)/4 = 39.5$. The prediction for period 5 involves the latest information on past four demands, that is, periods 1 through 4, which is $(30 + 50 + 60 + 32)/4 = 43$, and so on as shown in Table 3.2.

A variation of moving average, where we placed an equal weight on each element of the past demand data, is to give different weights for historical data based on its age. For example, we may place weight of 0.4 on the immediate past period data, 0.3 on the data that is two periods old, 0.2 on the data that is three periods old, and 0.1 for four-period-old data. Such a method is called *weighted moving average*. Based on our weights, the demand prediction for period 4 is $(.1 \times 18 + 0.2 \times 30 + 0.3 \times 50 + 0.4 \times 60)/4 = 46.8$. Similar calculations result in the predicted demands shown in Table 3.2. The table also shows the error squares and sum of error squares for each method.

3.1.6 Exponential Smoothing

It is similar to weighted moving average; however, it does have its own weighting scheme. The forecast for period t, F_t is determined as

$$F_t = F_{t-1} + \alpha(D_{t-1} - F_{t-1})$$

where α is called the exponential smoothing constant and has a value between 0 and 1. Using this equation with $\alpha = 0.2$, to determine the forecasted values, we start with period 0, assuming that the period's forecasted value was the same as the actual demand. The forecasted value for period 1 is $F_1 = F_0 + \alpha\ (D_0 - F_0) = 18 + 0.2\ (30 - 18) = 20.4$. Similarly, $F_2 = 20.4 + 0.2\ (30 - 20.4) = 22.34$.

Other values are shown in Table 3.3. Of course, other values of α will result in different forecasts and different errors. One way to choose an optimum value of α is to simulate the forecasts with different values of α and then choose for future prediction a value of α that gives the least sum of square of error.

TABLE 3.2
Moving Average and Weighted Moving Average

Period	Demand	Moving Average	Error	Weighted	Weighted Error	Moving Average Error Square	Wt. Mov. Avg. Error Square
0	18						
1	30						
2	50						
3	60						
4	32	39.5	−7.5	46.8	−14.8	56.25	219.04
5	50	43	50	43.8	6.2	2500	38.44
6	70	48	22	46.6	23.4	484	547.56
7	45	53	−8	55.4	−10.4	64	108.16
8	40	49.25	−9.25	52.2	−12.2	85.5625	148.84
9	60	51.25	8.75	48.5	11.5	76.5625	132.25
10	80	53.75	26.25	52	28	689.0625	784
11	60	56.25	3.75	62.5	−2.5	14.0625	6.25
12	55	43	12	64	−9	144	81
Sum of error squares						4113.5	2065.54

TABLE 3.3
Exponential Smoothing with $\alpha = 0.2$

Period	Observed Demand	Forecasted Demand	Error
0	18	18	0
1	30	20.40	−9.60
2	50	22.32	−27.68
3	60	27.86	−32.14
4	32	34.28	2.28
5	50	33.83	−16.17
6	70	37.06	−32.94
7	45	43.65	−1.35
8	40	43.92	3.92
9	60	43.14	16.86
10	80	46.51	33.49
11	60	53.21	6.79
12	55	54.57	0.43
Sum of error square 4712.82			

We can show that by repeated substitutions in exponential smoothing formula results in the following identity:

$$
\begin{aligned}
F_t &= F_{t-1} + \alpha(D_{t-1} - F_{t-1}) = \alpha D_{t-1} + (1-\alpha)\alpha F_{t-1} \\
&= \alpha D_{t-1} + (1-\alpha)\alpha D_{t-2} + (1-\alpha)^2 F_{t-2} = \ldots . \\
&= \alpha D_{t-1} + (1-\alpha)\alpha D_{t-2} + (1-\alpha)^2\alpha D_{t-3} + (1-\alpha)^3\alpha D_{t-4} + \cdots \\
&\quad + (1-\alpha)^{t-2}\alpha D_1 + (1-\alpha)^{t-1} F_1
\end{aligned}
$$

As can be seen in the preceding expression, the weights on prior periods' demands decline exponentially, and, therefore, this forecasting technique is called exponential smoothing.

3.1.7 Holt's and Winter's Models

Exponential smoothing can be further expanded to include trend, called *Holt's Model*, and trend and seasonal corrections called *Winter's Model.*

In both models, we start with linear regression equation through the available demand data. In our case, we have the equation

$$\text{Demand in period } t,\ D_t = 33.41 + 2.76\ (t).$$

As mentioned earlier, the intercept is considered as the initial level demand L_0, and slope 2.76 is considered as initial trend value, T_0. The seasonal factors calculated earlier are treated as the initial seasonal values for each season.

In Holt's model, level and trend factors are successively modified, based on the latest observed demand in following manner,

$$F_{t+1} = L_t + T_t$$

$$
\begin{aligned}
L_t &= \alpha D_t + (1-\alpha)(L_{t-1} + T_{t-1}) = \alpha D_t + (1-\alpha)F_t \\
T_t &= \beta(L_t - L_{t-1}) + (1-\beta)T_{t-1}
\end{aligned}
$$

In addition, with periodicity of p, the seasonal factors are modified in following manner:

$$S_t = \gamma(D_t/L_t) + (1-\gamma)S_{t-p},$$

where α, β, and γ are constants between 0 and 1.

In our example, first applying Holt's model with $\alpha = 0.2$ and $\beta = 0.3$ results in results tabulated in Table 3.4.

$L_0 = 33.41$ and $T_0 = 2.76$ give the forecast for period 1 as $F_1 = 36.17$. To calculate L_1 and T_1,

$$L_1 = 0.2D_1 + 0.8(L_0 + T_0) = 0.2D_1 + 0.8F_1 = 0.2 \times 30 + 0.8 \times 36.17 = 34.93$$

$$T_1 = \beta(L_1 - L_0) + (1 - \beta)T_0 = 0.3(L_1 - L_0) + 0.7T_0$$
$$= 0.3(34.93 - 33.41) + 0.7 \times 2.76 = 2.39.$$

And, hence, $F_1 = 34.93 + 2.39 = 37.32$.

Similarly other calculations. The forecasted demand for period 13 is 68.01, and from there on the forecast, based on the present demand, is $F_t = 66.12 + 1.89\,(t - 12)$.

To apply seasonal factors to the Winter's model, modifications require the calculation of modified seasonal factors. Remember that in our example, the periodicity $p = 4$, that is, the same season repeats every four time periods. Using successive values of demands and level factors from the Table 3.4, the new values of seasonal factors are calculated as shown in the following.

New Seasonal Factors

Period	Demand	L_t	Old Seasonal Factors	New Seasonal Factors
0	18	33.41	0.7014	0.7014
1	30	34.93	0.9726	$0.3\ (30/34.93) + 0.7 \times 0.9726 = 0.938$
2	50	39.86	1.3319	$0.3\ (50/39.86) + 0.7 \times 1.3319 = 1.307$
3	60	46.4	1.0776	$0.3\ (60/46.4) + 0.7 \times 1.0776 = 1.205$
4	32	46.85	0.7014	$0.3\ (32/46.85) + 0.7 \times 0.7014 = 0.696$
5	50	49.92	0.9726	$0.3\ (50/49.92) + 0.7 \times 0.938 = 0.957$
6	70	56.38	1.3319	$0.3\ (70/56.38) + 0.7 \times 1.307 = 1.29$
7	45	57.37	1.0776	$0.3\ (45/57.37) + 0.7 \times 1.205 = 1.08$
8	40	56.41	0.7014	$0.3\ (40/56.41) + 0.7 \times 0.696 = 0.70$
9	60	58.66	0.9726	$0.3\ (60/50.66) + 0.7 \times 0.957 = 1.02$
10	80	64.54	1.3319	$0.3\ (80/64.54) + 0.7 \times 1.29 = 1.27$
11	60	66.18	1.0776	$0.3\ (60/66.18) + 0.7 \times 1.08 = 1.03$
12	55	66.12	0.7014	$0.3\ (55/66.12) + 0.7 \times 0.70 = 0.74$

The forecasted demand is calculated as $F_{t+1} = (L_t + T_t)S_t$. The values are shown in Table 3.5.

3.1.8 Analysis of Forecasting Errors

If the forecasting method is perfect, it really would have no errors. But of course, such perfect forecasting is hardly possible. The next best thing is to have a method that produces as small an error as possible. In addition, the error should be random around the mean of zero and have variance (or standard deviation) that is as small as possible. There are number of indicators that can be useful, and they are as follows:

Variance can be estimated as being equal to (sum of square of errors)/n.

Another quick estimate of the variance of error by determining the sum of absolute mean deviation, called mean absolute deviation, MAD = (sum of absolute values of error)/n. Then, the standard deviation can be estimated as $\sigma = 1.2 \times \text{MAD}$.

TABLE 3.4
Holt Model: Forecast with Level and Trend Corrections: $\alpha = 0.2$ and $\beta = 0.3$

Period	Demand	L_t	T_t	Forecast F_{t+1}	Error	Error Square	Absolute Error
0	18	33.41	2.76	36.17			
1	30	34.93	2.39	37.32	−6.17	38.0689	6.17
2	50	39.86	3.15	43.01	12.68	160.7824	12.68
3	60	46.4	4.17	50.57	16.99	288.6601	16.99
4	32	46.85	3.05	49.9	−18.57	344.8449	18.57
5	50	49.92	3.06	52.98	0.1	0.01	0.1
6	70	56.38	4.08	60.46	17.02	289.6804	17.02
7	45	57.37	3.15	60.52	−15.46	239.0116	15.46
8	40	56.41	1.92	58.33	−20.52	421.0704	20.52
9	60	58.66	2.02	60.68	1.67	2.7889	1.67
10	80	64.54	3.18	67.72	19.32	373.2624	19.32
11	60	66.18	2.72	68.9	−7.72	59.5984	7.72
12	55	66.12	1.89	68.01	−13.9	193.21	13.9
					Sum	2410.988	150.12

TABLE 3.5
Winters Model: Forecast with Level, Trend, and Seasonal Corrections: $\alpha = 0.2$, $\beta = 0.3$, and $\gamma = 0.3$

Period	Demand	L_t	T_t	New Seasonal Factors S_t	Forecast Demand F_{t+1}	Error	Error Square	Absolute Error
0	18	33.41	2.76	0.7014	25.37			
1	30	34.93	2.39	0.938	35.01	4.63036	21.44025	4.630362
2	50	39.86	3.15	1.307	56.21	14.9938	224.8152	14.99384
3	60	46.4	4.17	1.205	60.94	3.78593	14.33327	3.78593
4	32	46.85	3.05	0.696	34.73	−28.9369	837.3413	28.93685
5	50	49.92	3.06	0.957	50.70	15.2696	233.1607	15.2696
6	70	56.38	4.08	1.29	77.99	19.2981	372.4182	19.29814
7	45	57.37	3.15	1.08	65.36	−32.9934	1088.564	32.9934
8	40	56.41	1.92	0.7	40.83	−25.3616	643.2108	25.3616
9	60	58.66	2.02	1.02	61.89	19.169	367.4506	19.169
10	80	64.54	3.18	1.27	86.00	18.1064	327.8417	18.1064
11	60	66.18	2.72	1.03	70.97	−26.0044	676.2288	26.0044
12	55	66.12	1.89	0.74	50.33	−15.967	254.9451	15.967
						Sum	5061.75	224.51652

The forecasting method should also be consistent. It should not overestimate or underestimate the demands following any pattern. To estimate the accuracy of the forecasting method, calculate the mean absolute percentage error, MAPE, which is defined as

$$\text{MAPE} = \left(\sum \text{ABS}(\text{Error}/\text{Demand}) \times 100\right) \Big/ n.$$

Also, determine if the forecasting method consistently overestimates or underestimates the actual demand by calculating bias, which is defined as:

$$\text{Bias } t = \sum_{I=1}^{t} \text{Error} \quad \text{for } I = 1 \ldots n$$

Bias should vary around 0 in a random manner.

Next determine the tracking signal, TS, defined as TS $= \text{Bias}_t/\text{MAD}_t$.

The tracking signal should be between ± 6.

In our example, we have plotted the errors for static forecasting with trend, with seasonal corrections, and for adaptive forecasting with trend and seasonal corrections. None of the errors look random (Figure 3.2).

In fact, all of them are following a cyclic pattern around the mean of zero. This indicates that may be we can take some corrective action to improve forecasting.

Let us examine one method, static with trend. Observe the errors for each season. They follow the same pattern. Errors for periods 1, 5, and 9 are very close to 0 or are negative. Errors for periods 2, 6, and 10 have large positive values and so on. Maybe we can modify the original forecasts to reduce these errors.

Let us observe the errors for all method, from Table 3.4. Plot is displayed in Figure 3.2c.

Period t	Error	Sum of Squares
1	−6.17	38.06
2	12.68	160.78
3	16.99	288.66
4	−18.57	344.84
5	0.1	0.01
6	17.02	289.68
7	−15.46	239.01
8	−20.52	421.07
9	1.67	2.78
10	19.32	373.26
11	−7.72	59.59
12	−13.9	193.21
		2410.98

Since the error is cyclic, a curve error $= a + b \sin(2\pi(t-1))/4$ may help reduce the error.

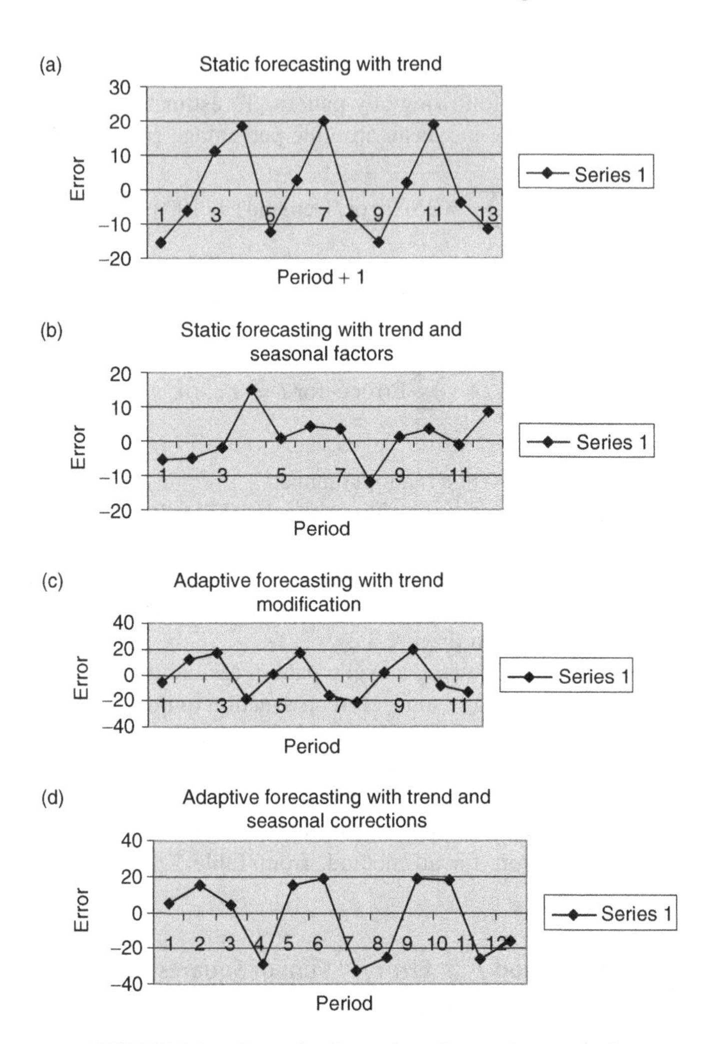

FIGURE 3.2 Error plot for various forecasting methods.

An equation through the 12 period errors leads to the following equation.

$$y = -1.213 + 17.00 \sin(2\pi(t-1))/4$$

Giving error correction for each period as

Period	Error Correction
1	−1.21
2	15.78
3	−1.21
4	−18.21

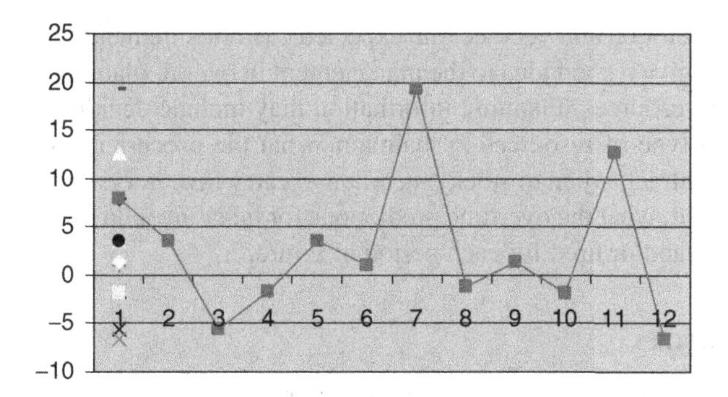

FIGURE 3.3

Therefore, the corrected forecasts and associated errors are

Period	Forecasted Demand	Correction	New Forecast	Actual Demand	New Error	Sum of Square
1	37.32	−1.21	36.11	30	6.11	37.33
2	43.01	15.78	58.79	50	8.79	77.26
3	50.57	−1.21	49.36	60	−10.64	113.20
4	49.9	−18.21	31.69	32	−0.31	0.09
5	52.98	−1.21	51.77	50	1.77	3.13
6	60.46	15.78	76.24	70	6.24	38.93
7	60.52	−1.21	59.36	45	14.36	206.20
8	58.33	−18.21	40.12	40	0.12	0.01
9	60.68	−1.21	59.47	60	−0.53	0.28
10	67.72	15.78	83.5	80	3.5	12.25
11	68.9	−1.21	67.69	60	7.69	59.13
12	68.01	−18.21	49.8	55	−5.2	27.04
						574.85

The new errors are smaller in magnitude than the original values. Sum reducing error sum of square from 241.9 (Table 3.4) to 574.85. The plot of new error also shows more random pattern (Figure 3.3).

3.2 AGGREGATE PLANNING

Once the demands for different periods are forecasted, it is time for aggregate planning. Although the exact demands are not yet known, because such planning is performed 3–18 months in advance, we must decide how to meet forecasted obligations, present planning is mainly in broad terms of product families rather than an individual product. For example, we shall plan, for each period, on how to meet the total television sales need and not how many televisions of each width to produce. We have some basic goals in aggregate planning; we should minimize the total cost of operation and yet meet the expected customer demand. Such plans, even though rough, give a good idea to the management in overall planning and in collecting necessary resources. Planning information may include deciding what and how many of each type of resources to maintain, what the process capacity should be, when to subcontract, when to stock out, when to carry inventory, items, and quantity to be outsourced, what the overtime predictions for labor are, and how many workers are to be hired and/or fired for each period in future.

3.2.1 Strategies

There are three basic strategies in aggregate planning. They are based on trade-offs between three costs: costs of production capacity; inventory; and back log. Costs of inventory and back log can be reduced if we have large production capacity, or alternatively, with minimum capacity, we might be able to meet the expected demand by producing and storing when we can and subcontracting when we cannot. These strategies are called: chase strategy, level strategy, and a combination of two, flexible strategy.

Chase strategy: In this strategy, useful production capacity is used as a driver. We have a large production capacity that can meet any forcible demand. The production capacity for each period and production rate are balanced with varying numbers of workers, and by hiring and firing the workers as needed. The result of this strategy is to maintain very low inventory, but it results in a high level of changes in capacity use and workforce. This strategy can be utilized when the cost of holding inventory is high, compared to the costs of facility and hiring/firing cost of the workers. Grocery stores, fast food facilities, and electric utility companies commonly use this strategy. In the fast food industry, for example, the production capacity of the facility is changed by hiring numbers of workers, and the number based on the expected demand in that hour. In a grocery store, the production cost is the acquisition cost, which hardly changes on daily basis and workers are required for placing units on the selves and to work as cashiers. Inventory cost is the cost of money tied down in inventory, which is proportional to the amount of grocery and time it remains on the shelf. With a number of different products stocked in a grocery store, this cost can be high when a large volume is placed on the shelves for long time. Chase strategy of storing what we can immediately (or in short time) sale, thus reducing inventory, has an advantage.

Level strategy: In this strategy, we have a limited capacity and a fixed workforce size. The production remains at a constant level, and inventory is used as the lever. When demand is lower than production, the inventory is built up, and when the situation is reversed, inventory and/or back order is used to supplement production. Thus, in this strategy, as the demand changes, inventory and stock outs will vary. The strategy is used when inventory carrying and back order costs are low. It is also used when at times capacities are hard to change. For example, in a continuous production environment, such as a chemical plant or glass and steel production, it is difficult if not impossible to change production levels in a short time. In airlines and hotel businesses, capacities are constant, and incentives such as reduced rate on weekends are used to change or level the demand pattern.

Flexible strategy: If there is enough production capacity, then utilization of the facility is used as the driver unlike chase strategy, the workforce size, that is, the number of workers, is also kept constant. However, the number of working hours each person works is changed according to demand. This results in low inventory cost since very small, if any, inventory is maintained. When both inventory carrying cost and capacity cost are relatively small, this strategy is used.

In addition, there are two options to balance the resources — internally, by changing parameters that affect capacity, and externally, by following policies that may change the demand pattern.

Internal Factors

The operational parameters that can be controlled internally are

1. *Fixed or changing production rates*: Adjusting machine capacities such that it may produce a fixed or variable number of the units in a period.
2. *Workforce*: the number of workers employed in a period.
3. *Overtime/slack times*: the number of extra hours/vexing permitted.
4. *Hire and fire*: Workers hired and fired as required in a planning period.
5. *Machine capacity level*: the number of units of machine capacity used in production.
6. *Subcontracting*: the amount of capacity or product contacted to subcontractors.
7. *Backlog*: the part of the demand that is not satisfied in the intended period and is carried forward to the next periods as a promised delivery date.
8. *Inventory on hand*: inventory that is carried over to next time periods.
9. *Lost sales*: by not meeting some or all of the demand.

External Factors

Outside plant factors that may influence demand are

1. *Pricing*: Changing the product price. Giving discounts in off seasons. Example, sale on air conditioners in winter months.
2. *Promotions*: giving incentives. Example, giving free air conditioning to a new car to a new car buyer.
3. *Package deals*: Two or more products rapped together for an attractive price. Example, vacation package that includes airfare and hotel accommodation.
4. *Advertising*: To increase brand or product awareness. Example: advertisement for department stores in news papers.
5. *Specify appointment times*: Doctors' offices use this strategy to level off demands.

The chase strategy generally influences internal factors, while the level strategy, which requires steady demand, uses external factors to maintain a constant demand rate. The following chart summarizes three strategies.

Chase strategy	• Uses production capacity as a driver • Workers are hired and fired according to need, resulting in low morale • Implemented when inventory and stock out costs are higher than hiring and firing costs
Level strategy	• Fixed workforce size • Stable working conditions, resulting in high morale • Implemented when it is difficult to change capacity and the availability of skillful workforce is limited
Flexible strategy	• Uses both production capacity and workforce as drivers • Working hours are varied according to demand • Implemented when inventory cost is relatively high and production capacity is relatively inexpensive

Linear programming is a popular tool in developing optimum aggregate plan. We shall illustrate this approach by means of an example.

3.2.2 Problem Description

Consider a manufacturing company with two product lines A and B (note that it would be mathematically more convenient if we describe products as 1 and 2, but for clarity we are noting them as products A and B). The forecasted demands for these products for next 6 months are as shown in Table 3.6. The firm has a starting inventory of 1500 for product A and 50 units for product B. The workforce at the beginning of the month is 15 full-time workers. The plant has a total of 20 working days per month, and the employees earn \$16/hr on the regular time and one and half times or \$24/hr on overtime. The employee cannot work for more than 30% of regular time on overtime in any month. The plant starts with an initial inventory of 50 units for each product and wishes to keep 50 units of inventories for each product at the end of the planning period. No initial back orders are present, and no back orders are permitted at the end of the planning period. Other information on operation of the plant is given in Tables 3.6 and 3.7. We wish to develop an optimum planning strategy that will minimize the total cost over the planning period of 6 months.

Production rate per month for product A

$$= (60 \text{ min/hr} \times 8 \text{ hr/day} \times 20 \text{ days/month}) = 480 \text{ units/month}$$

Similarly, production rate for B = 320 units/month

TABLE 3.6
Expected Product Demands

	Forecasted Demand for Product	
Month	A	B
1	1300	1000
2	1500	3000
3	2000	4000
4	1400	2000
5	2000	2000
6	1800	2200

TABLE 3.7
Associated Data for Products A and B

Cost Source	Cost for Product A in $	Cost for Product B in $
Hiring and training a new worker	200	200
Lay off of a worker	300	300
Carrying cost for an item from one month to next	2	3
Subcontracting cost per unit	35	50
Back order or stock out cost per unit	6	8
Material and production costs	10	15
Other information		
	Product A	Product B
Time required to produce one unit	0.33 hr	0.5 hr

Decision variables

We must first identify the decision variables, the values of which we can decide to achieve our objective. For $t = 1, 2, 3 \dots 6$, the variables are

DA_t : Demand for product A in period t
DB_t : Demand for product B in period t
W_t : Total number of workers working in month t
WA_t : Number of workers working on product A in month t
WB_t : Number of workers working on product B in month t
H_t : Number of workers hired at the beginning of the month t
L_t : Number of employees laid off at the beginning of the month t
PA_t : Number of units of A produced at the end of the month t
PB_t : Number of units of B produced at the end of the month t
IA_t : Inventory of product A at the end of the month t

IB_t : Inventory of product B at the end of the month t
SA_t : Number of units of product A stocked out/backlogged at the end of the month t
SB_t : Number of units of product B stocked out/backlogged at the end of the month t
CA_t : Number of units of product A subcontracting for month t
CB_t : Number of units of product B sub contracting for month t
O_t : Number of overtime hours worked in month t
OA_t : Number of overtime hours worked on product A in month t
OB_t : Number of overtime hours worked on product B in month t

Analysis

Objective function is to minimize cost. The components of the objective functions are

1. Full-time workers cost per month $= \Sigma$ (20 days/month $\times$ 8 hr/day $\times$ \$16/hr) $\times W_t = \Sigma 2560\ W_t$
2. Overtime cost per month—$\Sigma\ 24\ O_t$
3. Cost of hiring and layoff—$\Sigma\ 200\ H_t + \Sigma\ 300\ L_t$
4. Cost of the inventory and stocking out
   Product A: $\Sigma\ 2\ IA_t + \Sigma\ 6\ SA_t$
   Product B: $\Sigma\ 3\ IB_t + \Sigma\ 8\ SB_t$
5. Cost of materials and subcontracting
   Product A: $\Sigma\ 10\ PA_t + \Sigma\ 35\ CA_t$
   Product B: $\Sigma\ 15\ PB_t + \Sigma\ 50\ CB_t$

With these cost data, the objective function is

$$\text{Min } \Sigma\ 2560\ W_t + \Sigma\ 24\ O_t + \Sigma\ 200\ H_t + \Sigma\ 300\ L_t + \Sigma\ 2\ IA_t + \Sigma\ 6\ SA_t$$
$$+ \Sigma\ 3\ IB_t + \Sigma\ 8\ SB_t + \Sigma\ 3\ IB_t + \Sigma\ 8\ SB_t + \Sigma\ 10\ PA_t + \Sigma\ 35\ CA_t$$
$$+ \Sigma\ 15\ PB_t + \Sigma\ 50\ CB_t$$

Constraints

1. *Workforce, hiring and layoff constraints*: Workers available in period t equal to workers available in period $t-1$ plus new hires in period t minus layoffs in period t. Again, the linear programming will prevent both hiring and layoffs in the same period, simultaneously as they both incur costs.

$$W_t = W_{t-1} + H_t - L_t$$

and

$$W_t = WA_t + WB_t$$

2. *Capacity constraints*: Total production capacity of each product is what regular workers can produce in a month plus what they produce in overtime hours during the month. Thus,

$$PA_t <= 480\ WA_t + \frac{OA_t}{0.33}$$

$$PB_t <= 320\ WB_t + \frac{OB_t}{0.50}$$

3. *Overtime constraint*: Overtime cannot exceed 30% of regular time.

$$OA_t <= 0.3(8 \times 20 \times WA_t) = 48\ WA_t$$

and

$$OB_t <= 0.3(8 \times 20 \times WB_t) = 48\ WB_t.$$

Therefore

$$O_t = OA_t + OB_t$$

4. *Inventory balance*: This equation balances what is or can be available with what is required or can be carried forward for the next period.

   Net available is inventory from the previous month plus what is produced this month plus what is subcontracted for this month. Net requirement is demand for this month plus back order from previous month that is due this month plus remaining inventory for this month minus any back orders to be fullfilled next month. Note, as before in optimization, if we have an inventory in this period then we will not have back order in this period, and *vice versa*.

$$IA_{t-1} + PA_t + CA_t = DA_t + SA_{t-1} + IA_t - SA_t$$
$$IB_{t-1} + PB_t + CB_t = DB_t + SB_{t-1} + IB_t - SB_t$$

5. Initial conditions

$$IA_0 = 50,\quad DA_1 = 1300,\quad DA_2 = 1500,\quad DA_3 = 2000,\quad DA_4 = 1400,$$
$$DA_5 = 2000,\quad DA_6 = 1800,\quad IB_0 = 50,\quad DB_1 = 1000,\quad DB_2 = 3000,$$
$$DB_3 = 4000,\quad DB_4 = 2000,\quad DB_5 = 2000,\quad DB_6 = 2200$$

$$W_0 = 15,\quad SA_0 = 0,\quad SB_0 = 0$$

TABLE 3.8
Linear Programming Solution

Month	1	2	3	4	5	6
Number of workers, W_t	5.54	12.5	16.66	9.79	9.79	7.13
Overtime hours, O_t						
New hire, H_t		6.95	4.16			
Workers laid off, L_t	9.45			6.87		2.65
Inventory of product A, IA_t				300		50
Inventory of product B, IB_t						50
Production of A, PA_t	1250	1500	2000	1700	1700	1850
Production of B, PB_t	940	3000	4000	2000	2000	1050
Workers on product A, WA_t	2.60	3.12	4.16	3.54	3.54	3.85
Workers on product B, WB_t	2.93	9.37	12.50	6.25	6.25	3.28
Overtime for product A, OA_t						
Overtime for product B, OB_t						

6. Final requirements

$$IA_6 = 50, \quad IB_6 = 50, \quad SA_6 = 0, \quad SB_6 = 0$$

The results of linear programming solution, solved by software LINDO, is given in Table 3.8. It is assumed in all solutions displayed that internal workers can be switched between products A and B as needed.

The minimum cost solution follows a flexible strategy. It has all options available and selects the alternatives that results in a minimum cost of \$460,875. As seen in Table 3.8, it has fraction workers working during various months. This is possible only if workers, can be vexed as needed.

Linear Programming Results
Minimum cost \$460,875.00

If workers cannot be vexed, then the number of full-time workers, new hires, and the workers laid off must have integer values. Placing those conditions in our formulation leads to the results in Table 3.9. Now, the cost has increased, but it still follows a flexible strategy. It allows for hiring, firing, build up of inventory and stock outs as needed, and the solution does take advantage of the available options.

Integer solution \$462,127.30

Next, we examine level and chase strategies. In level strategy, the number of workers are kept constant, $W = 15$, in our case, for all time periods. No layoffs or

TABLE 3.9
Integer Solution

Month	1	2	3	4	5	6
Number of workers, W_t	6	12	16	10	10	7
Overtime hours, O_t		7	62			
New hire, H_t		6	4			
Workers laid off, L_t	9			6		3
Inventory of product A, IA_t						50
Inventory of product B, IB_t	146.6	0.66		174.66	43.33	50
Production of A, PA_t	1250	1500	2000	1400	2000	1850
Production of B, PB_t	1086.6	2854	3910.66	2265.30	1866.66	1006.66
Workers on product A, WA_t	2.60	3.08	3.78	2.92	4.16	3.85
Workers on product B, WB_t	3.39	8.91	12.22	7.08	5.83	3.14
Overtime for product A, OA_t		7	62			
Stock out of product B, SB_t			88.66			

TABLE 3.10
Level Workforce Strategy

Month	1	2	3	4	5	6
Number of workers, W_t	15	15	15	15	15	15
Overtime hours, O_t						
New hire, H_t						
Workers laid off, L_t						
Inventory of product A, IA_t		800				50
Inventory of product B, IB_t						50
Production of A, PA_t	1250	2300	1200	1400	2000	1850
Production of B, PB_t	940	3000	4000	2000	2000	1050
Workers on product A, WA_t	2.60	5.63	2.5	8.75	8.75	11.72
Workers on product B, WB_t	12.39	9.37	12.50	6.25	6.25	3.28
Overtime for product A, OA_t						
Overtime for product B, OB_t						

new hires are permitted by making H_t and L_t equal to zero for all time periods. The cost of solution has now gone up to $527,100 (Table 3.10).

In chase strategy, no inventories or stock outs are allowed for any period. The resulting integer solution is $464,314, a little more than flexible strategy (Table 3.11).

Chase strategy: No inventory or stock out
Minimum cost $464,314

TABLE 3.11
Chase Strategy Results

Month	1	2	3	4	5	6
Number of workers, W_t	6	12	16	9	10	7
Overtime hours, O_t		80	107	27	92	5
New hire, H_t		6	4		1	
Workers laid off, L_t	9			7		3
Inventory of product A, IA_t						50
Inventory of product B, IB_t						50
Production of A, PA_t	1250	1500	2000	1400	2000	1800
Production of B, PB_t	940	3000	4000	2000	2050	1050
Workers on product A, WA_t	2.60	3.12	3.50	2.92	4.16	3.75
Workers on product B, WB_t	3.39	8.87	12.50	6.08	5.83	3.25
Overtime for product A, OA_t		107				
Overtime for product B, OB_t		80		27	92	5

3.2.3 Other Factors

3.2.3.1 Human Factors

Flexible strategy with its ability to choose from all possible alternatives normally does provide optimum aggregate planning; however, it does have some drawbacks. Layoffs are never easy and does incur human cost, even on the people who are not laid off. Seeing co-workers, relatives, and friends getting fired creates a feeling of fear and guilt. Loyalty to the company can hardly be demanded in these circumstances. A worker will try to find alternate employment as soon as the opportunity arises. Labor laws and contracts may also restrict the firing of workers. Hiring also presents a problem. New workers need to be trained, and it takes to bring them to the speed. These people also affect the performance of the present workers. Use of temporary workers is an acceptable alternative that some companies follow.

Large storages space is necessary if unlimited storage is allowed. Also, building of a large inventory has problems associated with pilferage, breakage, and risk of sudden change in customer preference. Back orders are not always accepted by customers, and they eventually may look for an alternate supplier who is more reliable and consistent.

3.2.3.2 Changing Demand

Aggregate planning can avoid the costly alternatives of hiring/firing, storage, and back orders by adjusting the forecasted demands. The demands could suit the available capacities and known operating policies. But external factors used to influence demands that are mentioned earlier also have some limitations. The customer, for example, may get used to, and indeed expect, discounts and promotions as a normal way of doing business. Labor contracts, government regulations, and competition from other suppliers may also limit the flexibility in applying promotional policies.

3.2.3.3 Spreadsheet Approach

To develop an aggregate plan, we have used lincar programming in this chapter. This approach allows selection of alternatives from among the largest possible options. It is also possible to develop the aggregate plan using the spreadsheet approach. But it requires some pre-analysis. Let us illustrate this approach as applied to the level strategy.

In our example, recall that it requires 0.33 hr for product A and 0.5 hr for product B. Each worker, therefore, can produce 480 units of product A or 320 units of product B. The worker is paid \$16/hr on regular time and \$24/hr on overtime. No more than 30% of overtime is allowed. Carrying cost for A is \$2/month and for B is \$3/month. Material cost for A is \$10 and for B is \$15. Based on these data, we can draw some conclusions.

1. Total cost of production for A is $16 \times 0.33 + 10 = \$15.28$ in regular time and $24 \times 0.33 + 10 = \$17.92$ in overtime. However, in level production strategy, a worker will not be fired and, therefore, is available all the time. Hence, labor cost is constant and does not change with the production quantity; therefore, only the material and production costs can be considered as the price of products. Thus, price for A and B are only \$10 and \$15, respectively. With subcontract cost of \$35 for A, we can add as much as $35 - 10 = \$25$, in inventory carrying or back order cost. This allows for as much as $25/2 = 12.5$ months of carrying a unit in inventory or $25/6 = 4.16$ months of back order. Similar analysis for product B reveals 11.66 months of carrying and 4.37 months of back orders. In other words, over the planning period of 6 months, it is always profitable to produce A and/or B for future use whenever there are idle workers available than to give a subcontract to an outside company.
2. To compare carrying and back order for product A with a carrying cost of \$2/month and back order cost of \$6.00, it is cheaper to carry inventory for as much as $6/2 = 3$ months before back ordering. In other words, if there is not sufficient workforce to build product A to meet a demand for a period, go back as many as 3 months to see if the inventory can be built before back ordering. Similarly, for product B, the period is $8/3 = 2.33$ months.
3. It is cheaper to carry product A in the inventory than product B.

To proceed with the analysis, let us fist determine net requirement for each product. An initial inventory of 50 units for each product and final (period 6) requirements of 50 units for each product lead to the net monthly requirement as displayed. Dividing each requirement by the production rate of each worker gives monthly manpower requirement for products A and B. Adding the requirements for each product gives the total manpower requirement. With a level workforce of 15 workers, we can meet the manpower requirements of all periods except for period 3. We need to transfer $16.33 - 15 = 1.33$ units manpower capacity in previous months, if possible. In month 2, the requirement is 12.49, giving an idle capacity of $15 - 12.49 = 2.51$ units. We can, therefore, turn out a product in month 2. Now the question is, which product to

TABLE 3.12
Spread Sheet Approach

Month	1	2	3	4	5	6
Demand for product A	1300	1500	2000	1400	2000	1800
Demand for product B	1000	3000	4000	2000	2000	2200
Net requirement for A	1250	1500	2000	1400	2000	1850
Net requirement for B	950	3000	4000	2000	2000	2250
Workers needed, product A	2.60	3.12	4.16	2.91	4.16	3.84
Workers needed, product B	2.9	9.37	12.50	6.25	6.25	7.03
Total	5.5	12.49	16.66	9.16	10.41	10.83
Shift worker, product A		1.66	−1.66			
Product B						
New assignments	5.5	14.15	15	9.16	10.41	10.83
Workers available	15	15	15	15	15	15
Inventory for A			796			

produce. Since product A has lower carrying cost than product B, we will produce all of B in period 3 and 1.33 man months of product A in period 2. That gives inventory of $1.33 \times 480 = 796$ units for A in period 2 (Table 3.12).

The approach is logical and simple once some basic rules are established.

3.3 SUMMARY

The chapter presents two important concepts. How to determine the expected demand for the future and how to plan for it. Commonly used methods and models are presented and are illustrated with examples.

3.4 EXERCISE

3.4.1 Discussion and Review Questions

1. What advantages and disadvantages does the quantitative forecasting have over qualitative forecasting as a forecasting tool?
2. Compare and contrast the use of exponential smoothing and moving averages in evaluating forecasts.
3. Discuss how the flexibility in production systems relates to the forecast horizon and forecast accuracy.
4. What are the main advantages and limitations of exponential smoothing over moving averages?
5. Discuss how the number of periods in a moving average affects the responsiveness of the forecast?

3.1 Two different techniques (F1 and F2) were used to forecast the demand of hamburgers in a university cafeteria. The actual demand and the two sets of forecasts are as follows:

		Forecast	
Period	**Demand**	**F1**	**F2**
1	108	106	106
2	115	108	108
3	110	112	110
4	114	111	112
5	109	112	114
6	112	110	116
7	120	111	118
8	118	114	120

a. Compute MAD and MSE for each forecast. Justify which forecasting appears to be more accurate.
b. Compute the tracking signal for the eighth period of each forecast. What does it show?

3.2 Weekly sales of a vehicle for a dealer for the past 12 weeks are

Week	Sales
1	20
2	24
3	18
4	14
5	25
6	20
7	18
8	21
9	24
10	22
11	28
12	32

a. The manager would like to predict the sales for the coming week. Use a 3-week moving average.
b. Also, find out whether or not a 2-week moving average is better than a 3-week moving average or moving average of longer periods. (Hint: Use mean absolute deviation.)
c. Compare the graph (week versus sales) for actual data, 2-week moving average, and 3-week moving average.

3.3 For the data in Problem 3.1, compute the forecast for the next week using the 3-week weighted average with weights 0.5, 0.3, and 0.2.

3.4 The monthly accidents in a certain county is given as

Month	Number of Accidents
July	22
August	15
September	20
October	18
November	23
December	20
January	25

a. Compute the forecast for the month of February using exponential smoothing with F1 = 22 and $\alpha = 0.5$.
b. Using linear regression, forecast for the next month.

3.5 The demand and forecast of an item are given for the respective years. Find the missing demand.

Year	Demand	Forecast
1982	620	—
1983	520	—
1984	360	—
1985	730	500
1986	???	537
1987	460	580
1988	530	613
1989	570	547
1990	680	520

3.6 Determine the forecast demand for the next four periods, using linear regression. Also determine if the error is random; or additional nonlinear term is needed in the regression equation.

Time Period	Demand
1	1600
2	2100
3	2500
4	3100
5	2050
6	2800
7	3300
8	3600
9	2200
10	2600
11	2900
12	3400

Continued

Time Period	Demand
13	3200
14	3600
15	4250
16	4500

3.7 Jane Martin is a senior staff in the purchasing and marketing department at Whipple Electronics Products. She has been working to develop a forecasting system for the monthly price of a particular type of insulated wire. She has accumulated the historical data for the past 12 months.

Month	Price/feet ($)	Month	Price/feet ($)
1	1.15	7	1.20
2	0.98	8	0.96
3	1.25	9	1.40
4	0.87	10	1.25
5	1.03	11	1.25
6	1.15	12	1.15

a. Forecast for all the months with $\alpha = 0.1$, $\alpha = 0.3$, $\alpha = 0.5$. Use exponential smoothing for the forecast.
b. Calculate the least mean absolute deviation for all the above three values of α over the 12-month period.
c. Using the best value of α from above, calculate the forecast for the next period (i.e., month 13).
d. Compare the above forecast with the three-point moving average method.

3.8 You have been hired as an industrial engineer in the purchase and sales department of an electronics product company. The planning director aims to forecast for the next 12 months from the following historical data:

Month	2004	2005	2006
January	575	578	608
February	527	507	597
March	540	562	612
April	502	533	603
May	508	516	658
June	573	580	605
July	508	537	627
August	498	440	578
September	485	511	585
October	526	480	581
November	552	499	632
December	587	542	656

a. Forecast the monthly demand for the year 2007 using moving average, simple exponential smoothing, Holt's model, and Winter's model methods.
b. Make an analysis for the above part, and suggest the best method to the planning director. (Include bias, tracking signal, MAPE, MSE, and MAD in your analysis.)

3.9 ABC Instruments manufactures various electronic equipments for students. Of these, it sells calculators to K&P Supplies, with orders to deliver 650 for winter quarter (starting January 1), 720 for spring quarter (starting April 1), 400 copies for summer quarter (starting July 1), and 900 copies for fall quarter (starting October 1). The production cost (material + labor) is $18 per piece but, starting summer quarter, the material and labor cost is going to increase by 10%. Also, any order below 700 costs $1.2 per piece for shipping and handling; otherwise, it costs $0.9 per piece.

Because of the cost factor, the management wants to keep the production steady to avoid overtime and layoff. It has been decided that production in any quarter cannot vary by more than 25% of the previous quarter. The previous year fall quarter production was 600. It costs $1.0 per piece per quarter for the inventory.

The following table shows the production and delivery for each quarter and the variables.

Production (in)	**Delivery (in)**	**Variable**
Winter	Winter	P_{11}
Winter	Spring	P_{12}
Winter	Summer	P_{13}
Winter	Fall	P_{14}
Spring	Spring	P_{22}
Spring	Summer	P_{23}
Spring	Fall	P_{24}
Summer	Summer	P_{33}
Summer	Fall	P_{34}
Fall	Fall	P_{44}

where, P_{ij} denotes the production in quarter i and delivery in quarter j.

Assuming no starting inventory, develop a production schedule for ABC Instruments in each of the three quarters for the coming year, in order to minimize the total cost.

3.10 The following table shows the production and demand of a table manufacturer for a local company.

Month	Demand
May	200
June	500
July	300
August	550
September	600
October	700

The company has gathered the following data.

Costs

Holding cost	$8/table/month
Subcontracting	$20/table
Regular labor	$10/hr
Overtime labor	$16/hr above 8 hr/worker/day
Hiring cost	$40/worker
Layoff cost	$80/worker

Other data

Current workforce	12
Labor hr/table	5
Work days/month	20
Beginning inventory	120

What will be the cost of the two following strategies?

a. Vary the workforce to have exact production to meet the forecast demand.

b. Vary overtime, and use the constant (given) workforce.

3.11 Plan production for a 4-month period: February through May. For February and March, you should produce to exact demand forecast. For April and May, you should use overtime and inventory with a stable workforce; stable means the number of workers needed for March will be held constant through May. However, government constraints have put a maximum of 5000 hr of overtime labor per month in April and May (zero overtime in February and March). If demand exceeds supply, then back orders will occur. There are 100 workers on January 1. You are given the following demand forecast:

February	80,000
March	64,000
April	100,000
May	40,000

Productivity is four units per worker hr/8 hr per day/20 days per month.

Assume zero inventory on February 1. Costs are

Hiring	=	$50 per new worker
Layoff	=	$70 per worker laid off
Inventory holding	=	$10 per unit month
Straight time labor	=	$10 per hr
Overtime labor	=	$15 per hr
Back order	=	$20 per unit.

Find the total cost of this plan.

REFERENCES AND SUGGESTED READINGS

Armstrong, J.S. 1986. "The Ombudsman: Research on Forecasting: A Quarter Century Review, 1960–1984" *Interfaces*, January–February.

Box, G.E.P. and G.M. Jenkins. 1970. *Time Series Analysis: Forecasting and Control*, New-York: Holden-Day.

Chambers, J.C., S.K. Mullick, and D.D. Smith. 1971. "How to Choose the Right Forecasting Technique" *Harvard Business School*, July–August: 45–74.

Dougherty, John R. "Getting Started with Production Planning" Readings in Production and Inventory Control and Planning, *APICS 27th Annual Conference*, 1984, pp. 176–79.

Gupta, S. and P.C. Wilton. March 1987. "Combinations of Forecasts: An Extension" *Management Science*, 33(3).

Hodgeson, T.J., R.E. King, and C.U. King. 1990. "Development of Production Planning System: A Case History" *Production and Inventory Management* 31(4): 18–24.

Holt, C.C., F. Modigliani, J.F. Muth, and H.A. Simon. 1960. *Planning Production, Inventories and Workforce*, New York: Prentice Hall.

Konijnendijk, P.A. 1982. *Coordination of Production and Sales*, Eindhoven, the Netherlands: Technical Eindhoven University.

Lawrence, M.J., R.H. Edmundson, and M.J. O'Connor. December 1986. "Accuracy of Combining Judgmental and Statistical Forecasts" *Management Science*, 32(12).

Ling, R.C. and W.E. Goddard. 1995. *Orchestrating Success: Improved Control of Business with Sales and Operations Planning*, New York: John Wiley & Sons.

Winters, P.R. April 1960. "Forecasting Sales by Exponentially Weighted Moving Averages" *Management Science*, 6: 324–342.

4 Master Production Scheduling and Material Requirement Planning

As the title suggests, this chapter presents two topics in planning. Both use a spreadsheet approach and use similar logic in planning and are therefore grouped together in one chapter.

4.1 MASTER PRODUCTION SCHEDULE

In aggregate planning, we had established overall policies, determined the resources needed, and made other necessary plans to meet the forecasted demands for different product families. The time frame was fairly long, in terms of months, if not years. The next step is to develop a detailed planning for individual products. The short planning horizon is in terms of weeks rather than months. The purpose of these plans is to transfer each product requirement into time-phased production and purchasing plans. They also serve to develop capacity and resource requirement in production and distribution. In some cases, planning information is directly based on actual customer orders for immediate future and not based on forecasted values, displaying real-time resource needs. Most of these plans use simple spreadsheets approach to present information. We shall demonstrate some commonly used planning tools in this chapter.

Master production schedule (MPS) determines how many units are to be produced in a specific period. Generally, the period is small, such as a week. Two major factors are considered: first, for which products we should develop the MPS, and second, what information we should take into account. This information may include factors such as customer order or replenishment needs.

Ideally, we like to develop detailed plans for each independent product. However, the number of products can soon become unmanageable. For example, should we consider automobile with different colors or different options such as radios and air conditioners as different product lines? If we do, then the combinations and therefore the number of products could be very large indeed. A simple solution is to consider a basic product as an independent product and additional options as add-ons.

Production environment may also influence what should be considered as a "product" in MPS. Indeed, the production environment may even influence the earlier stage of aggregate planning in deciding what the parent product is.

Products are made in three ways. In the make-to-stock (MTS) environment, suppliers produce products to their specifications and customers have the option to either purchase it or not. Here, the customers have no influence on the design of the product or any other details. Products in a department store, such as toasters, hot pots, and coffee makers, follow this mode. We either purchase the product as displayed or look for another alternative. Typically, in MTS, the numbers of different items produced by a company are limited and are produced in an assembly line environment. The final products are stored, and customers are supplied from an inventory. MPS is used to determine when to build the stock level or inventory for each product. They are not used to satisfy individual orders, as these orders are filled from the built-up inventory.

In the assemble-to-order (ATO) environment, different options may be used to assemble a product that a customer wants, for example, building a computer system on customer specifications, or adding options to a basic product based on customer preference, such as in an automobile. Here, although the number of basic products is limited, the number of final items can be large. MPS is developed for each basic product.

In the make-to-order (MTO) environment, a product is made to customer specification. Here, a large number of final products can be made using the basic skills and expertise of the company. The MPS is made for raw material or basic resources that are used to develop the final products.

Production Environment	Master Production Schedule (MPS)
Make to stock (MTS)	Final product
Assemble to order (ATO)	Basic components
Make to order (MTO)	Raw material and basic resources

The development of MPS is a simple procedure and requires sequential evaluations. However, some basic principles must be followed. The sum of individual product quantities in MPS must add to those with the production requirements for the parent product of this family in aggregate planning. Each individual product must be given appropriate resources such as storage space, labor, raw materials, and machining capacities in each period to satisfy the production needs. The production lot size should be determined by using appropriate inventory models, which are discussed in the next chapter, which includes detailed cost analysis.

In aggregate, imagine we had two parent products, A and B. Suppose product A has three subproducts. For example, product A is table and subproducts A1 is a table with side supports, A2 is a table with central support, and A3 is a large table. In another situation, where the parent product A is wooden chair and A1 is a high-back chair, A2 is an armchair, and A3 is a cushioned chair. Assume, in our example, that the rough distribution for product A is as given in Table 4.1.

Roughly, 30% of A is A1, 40% is A2, and 30% is A3. Then, the monthly requirements of products A1, A2, and A3 when distributed on weekly requirements may be as follows.

By the way, the time per unit, used in aggregate planning for the parent product A, is the weighted average of production times for the family, namely $0.3 \times 30 + 0.4 \times 20 + 0.3 \times 10 = 20$ min or 0.33 h.

TABLE 4.1
Data on Product A Distribution

Parent Product A

Subproducts	A1	A2	A3
Distribution %	30	40	30
Production time/unit in min	30	20	10

TABLE 4.2
Percent Distribution of Main Product A into Subproducts

Month	Week	A1	A2	A3	Product A Requirements in Aggregate Plan
1	1	200	100	50	1085
	2		300	100	
	3	125		50	
	4		35	125	
2	1	200		50	750
	2	25	100		
	3		100	100	
	4		100	75	

Continuing with A1 for illustration, weekly gross requirements are listed in Table 4.2. Initially, we have 50 units available and it takes 1 week between the time an order is placed (released) and the time it is received, that is, a lead time is 1 week. We have one order of 120 units, scheduled to be received in week 1 from previous planning. Incidentally, we will use the same information for all MPS plans illustrated in this chapter.

The basic principle used in MPS development is that in any given week, we should have sufficient, but not excessive, inventory on hand to satisfy the demand for that week. This is achieved by planning receipts, and therefore order releases, in an optimum manner. We can determine inventory on hand by applying the following equation:

$$\begin{pmatrix}\text{Inventory on}\\ \text{hand at the}\\ \text{end of period } t\end{pmatrix} = \begin{pmatrix}\text{Inventory on}\\ \text{hand at the}\\ \text{end of period } t-1\end{pmatrix} + \begin{pmatrix}\text{Planned}\\ \text{Receipts}\\ \text{in period } t\end{pmatrix} - \begin{pmatrix}\text{Gross}\\ \text{requirements}\\ \text{in period } t\end{pmatrix}$$

4.1.1 Fixed Order Quantity (FOQ)

Suppose the optimum production quantity for A1 is 120 (Table 4.3) and is fixed at that level.

In our example, in week 1, we expect to receive 120 units from the previous order and have 50 units on hand, making total of 170. However, demand is for 200 units and,

TABLE 4.3
Replenishment Policy for A1 Using FOQ

	Weeks							
	1	2	3	4	5	6	7	8
Gross requirements or demand	200		125		200	25		
Scheduled receipts from earlier orders	120							
On-hand inventory 50	−30	90	85	85	5	100	100	100
Planned receipts		120	120		120			
Planned order release	120	120		120	120			

after satisfying the demand, we will have −30 units of inventory on hand or will have 30 units of back order in period 1. This does not follow the principle that we should have sufficient inventory on hand to satisfy the demand (no back order as far as possible), but here we did not have any choice since we had started our planning with only 50 units on hand, and the maximum quantity we can receive is 120.

To satisfy this back order, which, in reality, is the demand for week 2, we must have an order arriving in period 2. Since the optimum order quantity is 120 units with the lead time of 1 week, replenishment order for 120 units must be released in week 1. In week 2, we receive 120 units. The total demand is $0 + 30$ units of back order $=30$, giving on hand inventory of 90 units. Looking forward to period 3, we see there is a demand for 125 units, which will require one order receipt. Therefore, an order for 120 units must be released in period 2. With a total inventory of $90 + 120 = 210$ units, 125 will be used to satisfy the demand in product 3, leaving 85 of on-hand inventory. There is no demand in period 4; however, 200 units are on demand in period 5, which means we must receive one order in period 5. In period 5, the remaining on-hand inventory is $85 + 120 - 200 = 5$. To satisfy the demand of 25 units in period 6, we must receive a new order. Thus, an order is placed in period 5. Net on-hand inventory in period 6 is $5 + 120 - 25 = 100$. Since there are no demands in periods 7 and 8, this inventory remains in those periods.

The MPS, developed for product A1, was based on ordering and receiving a fixed quantity of 120 units per order. This policy is called fixed order quantity (FOQ) rule. This quantity may be decided based on economic analysis called economic production quantity (explained in Inventory Control Chapter 5) or may be decided based on some physical constraint, such as the amount that can be accommodated in a truck or minimum quantity that must be purchased, called FOQ.

However, if the product requirements are not more or less uniform from period to period, then an FOQ cannot be ordered based on economic order quantity (EOQ) principle. Indeed, if the requirement variations are very large from period to period, FOQ policy may lead to excessive inventory in some periods and excessive backorders or stock-outs in others. Increase or decrease in FOQ may also lead to corresponding increases in the costs of shipping and purchase along with inventory and stock-out costs. There are other alternative order policies that may be appropriate when requirements are very irregular. They are discussed next.

TABLE 4.4
Replenishment Policy Using POQ

	Weeks							
	1	2	3	4	5	6	7	8
Gross requirements or demand	200		125		200	25		
Scheduled receipts from earlier orders	120							
On-hand inventory 50	−30	125	0		25			
Planned receipts		155			225			
Planned order release	155			225				

4.1.2 Periodic Order Quantity (POQ)

In this policy, the order quantity per order varies and is based on the quantity needed during a predetermined fixed time periods between orders. In this policy, since periods when orders are placed are preplanned, the corresponding order processing cost is expected to decrease.

In our example, suppose we fix time between orders to 3 weeks (Table 4.4).

During the first 3 weeks, the total demand is $200 + 0 + 125 = 325$. On-hand inventory in period 1 consists of initial inventory of 50 units and scheduled receipts of 120 units. So, we must order $325 - (50 + 120) = 155$. Since the lead time is 1 week, orders placed in week 1 will be received in week 2. After satisfying 30 units of back order from week 1, we have 125 units of on-hand inventory, which takes care of demand of period 3. The total demand for next 3 weeks, i.e., weeks 4, 5, and 6, is $0 + 200 + 25 = 225$. Since there is no demand in period 4, we should receive the order in period 5 and therefore place the order in period 4. There is no demand in periods 7 and 8; hence, no new order is placed. Note, strictly following POQ, we should have placed order in period 3 to receive it in period 4, but that involves carrying additional inventory cost for period 4. This could be avoided by obvious modification to the firm POQ policy. Such corrections are not unusual in POQ.

4.1.3 Lot for Lot (L4L)

The policy states to order just enough to meet the immediate requirement. This policy should result in no inventory in any period.

The application of lot for lot (L4L) policy to our example is shown in Table 4.5 and hardly needs any explanation. There are multiple orders, increasing the total order cost, but no inventory carrying cost, since there is no inventory in any period.

4.1.4 Least Total Cost

Here, lot size and order interval are adjusted so that order (or set up) cost per order and carrying cost per unit of time, generally a week, for order quantity are about the same. Thus, we have, order cost/order = carrying cost/period × quantity ordered for the period.

TABLE 4.5
Replenishment Policy Using L4L

	Weeks							
	1	2	3	4	5	6	7	8
Gross requirements or demand	200		125		200	25		
Scheduled receipts from earlier orders	120							
On-hand inventory 50	−30							
Planned receipts		30	125		200	25		
Planned order release	30	125		200	25			

Suppose in our example, order cost is $30 per order and carrying cost per week is $0.15. Then Q for least total cost (LTC) is $30/0.15 = 200$. Then, policy calls for ordering the exact quantity to meet number of periods of demand that is as close to 200 as possible.

In our example, the net requirements are as follows.

Week	1	2	3	4	5	6	7	8
Net requirement	30		125		200	25		

Let us determine the first order. Cumulative demands are:

Week	1	2	3	4	5	6	7	8
Cumulative net	30	30	155	155	355	380	380	

It is obvious that, in this case, the first order should be for 155 units, the closest number to 200 units LTC value. It covers periods 1 through 3. The remaining cumulative net requirements are:

Week	1	2	3	4	5	6	7	8
Cumulative net					200	225	225	225

The second order ideally should be for 200 units, and a third order will have to be placed just for 25 units of demand in period 6.

The preferred alternative for the last two orders would have been to order 225 units in the second order and not placing the third order, reducing the order cost while slightly increasing the carrying cost.

4.1.5 Incremental Cost Analysis (ICA)

A slight variation to LTC procedure is to calculate the incremental carrying cost of adding a demand week to the previous order. If the incremental cost is greater than the order cost, we place a new order. If it is less, we include this demand period in the previous order. The policy provides lower cost than LTC, since it incorporates multiple periods in carrying cost determination.

Week	1	2	3	4	5	6	7	8
Net requirement	30		125		200	25		

An order is necessary to acquire 30 units for week 1. If we include week 2, there is no additional cost, since there is no requirement for period 2. If we include the next period, week 3, the additional carrying cost is 125 units received in period 1 to be used in period 3 or 125 units $\times$ 2 periods $\times$ \$0.15/period = \$37.5. Since additional carrying cost is more than the order cost of \$30, restrict the first order to first week's demand. The remaining requirements are:

Week	1	2	3	4	5	6	7	8
Net requirement			125		200	25		

The second order for 125 units will be placed to receive 125 units in period 3. Check if addition of demand from month 5 to this order is economical.

Additional carrying cost for period 5 demand when demand is added to period 3 order $= 200 \times 2 \times 0.15 =$ \$60.

Again the carrying cost is greater than order cost, and therefore do not include month 5 in month 3 order.

The third order is for month 5. To check if month 6 should be included or not, follow the same procedure.

The additional carrying cost for period 6 when demand is added to period 5 order $= 25 \times 1 \times 0.15 =$ \$3.75. Hence, add requirements for month 6 in order for period 5. The final assignments, with 1 week of lead time, are listed in Table 4.6.

TABLE 4.6
Replenishment Policy Using ICA

	Weeks							
	1	**2**	**3**	**4**	**5**	**6**	**7**	**8**
Gross requirements or demand	200		125		200	25		
Scheduled receipts from earlier orders	120							
On-hand inventory 50	−30							
Planned receipts		30	125		200	25		
Planned order release	30	125		225	25			

TABLE 4.7
ATP Calculations

	Weeks							
	1	2	3	4	5	6	7	8
Gross requirements or demand (forecasted)	200		125		200	25		
Actual customer orders	100	50	100	60	70	30	5	15
Scheduled receipts from earlier orders	120							
Planned receipts		120	120		120			
Projected inventory	20	90	110	50	100	70	65	50
Available to promise (ATP)	20	90	50	50	50	50	50	50

4.1.6 Available-to-Promise Chart (ATP)

The MPS clearly shows, based on forecasted demands, when orders should be released, when they are received, and how much inventory on hand we have in each period. The information on planned receipts can be used further to develop another chart, called available-to-promise (ATP) chart. This chart indicates how many items we can promise to a new demand, based on actual customer demands that are on our books right now, and not on the forecasted values of the demands.

For example, suppose we are following the FOQ policy with $Q = 120$ and have actual demands on record for our product A1 during the next 8 weeks of the planning period as shown in Table 4.7.

The ATP is calculated as the available inventory till next planned receipts after satisfying customers' orders for all periods from present till one period before next planned receipt. For example, in period 1, scheduled receipt from the earlier order is 120, next planned receipt is in period 2, and hence, ATP is $120 - 100 = 20$. Similarly for period 2, ATP $= 120 - 50 = 90$. ATP for period 3 is ATP available in inventory from period 2 + planned receipts − amount promised from now till the next planned receipts, which is period 5. So, ATP $= 90 + 120 - (100 + 60) = 50$. In period 4, projected inventories goes down from 110 in period 3, by actual demand in period 4 to 50, while ATP stays the same as in period 3. ATP for period 5 is available inventory in period 4 + receipt in period 5 − all actual orders till next planned receipt. In our case, we have no new planned receipted in the present planning period. Hence, ATP for period 5 is $50 + 120 - (70 + 30 + 5 + 15) = 50$. For the remaining periods, ATP remains the same, while projected inventory decreases based on actual customer orders.

4.1.7 Concluding Remarks on MPS

Generally, in MPS, receipts are planned so that all the gross requirements are met and there are no back orders. In our example, in all MPS policies, we observe back order in period 1. Really, we do not have any control on that since they are the results of the past decisions.

Each policy plans when and by how many units to order. We can evaluate which policy is better by calculating some indicators and costs such as order cost and inventory carrying cost, which are discussed in the Chapter 5. However, some general observations can be made here.

Fixed order quantity policy normally carries high level of inventory, because the order level generally does not match the weekly requirements and there is some inventory left over from week to week. Many times, an order may have to be placed to satisfy a demand even though there is some but not enough inventory on hand for that period. Periodic order quantity (POQ) reduces average inventory on hand, as it matches order quantity to the required amount for the period. However, based on the period initial inventory or number of orders could be large. L4L policy has no inventory but large number of orders. However, these policies also need to carry some safety stock to respond to any unpredictability of demands. The incremental cost analysis (ICA) technique requires some calculations, but reduces the total operating cost.

4.2 MATERIAL REQUIREMENT PLANNING AND OTHER TECHNIQUES

Material requirement planning (MRP) translates or "explodes" demand requirements of parent items from MPS to the requirements of all their components. These components may include manufactured items, purchased items, subassemblies, and even the indirect products that are required for assemblies. The purpose is simple. Determine the exact time and quantity when each item is required so that the inventory within the system is minimized and production and purchases are properly planned.

To develop MRP, we need some basic information. Some information is independent of time, while others are time dependent, based on the time over which MRP is developed. The information includes:

1. MPS chart(s) for parent item(s). This is a time-dependent chart(s) and, to develop MRP for a predetermined period, select MPS for product(s) over the same period.
2. Bill of material (BOM) for a product: BOM is developed in levels. It shows product structure or relationship between parent product at level zero with all its components and subassemblies at successive lower levels.
3. Lead times for all products whether purchased, manufactured, or assembled.
4. Initial inventories of all products.

The MRP for each level follows a logic similar to one used in development of MPS. We shall illustrate development of MRP by an example.

Let us continue with the example of tables, which we have divided into three subproducts A1 (with side supports), A2 (with central support), and A3 (circular table). Now consider a case in which subproducts A1, A2, and A3 require components A,

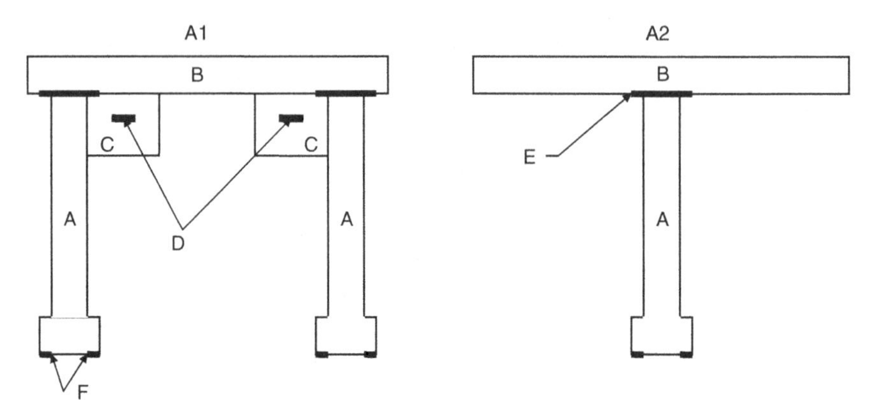

FIGURE 4.1 Tables (A1 and A2) and their components.

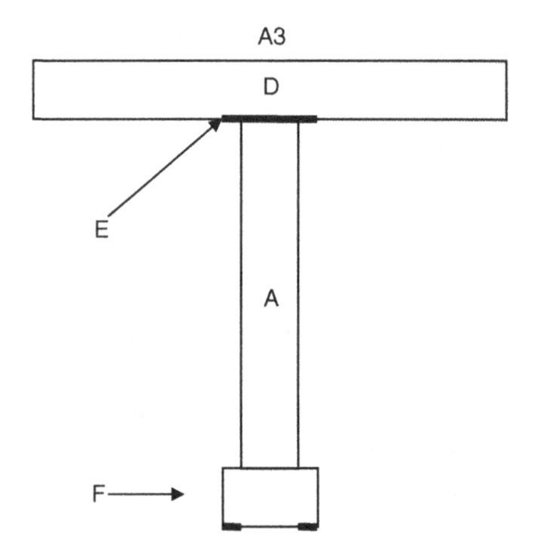

FIGURE 4.2 Table (A3) and its components.

B, C, D, and E in quantities and in order as shown in Figures 4.1 and 4.2. This is a popular graphical way of representing BOM as shown in Figure 4.3.

BOM is developed in levels, showing at each level immediate components needed to make a unit at that level. For example, at level 0 is product A1, which requires components at level 1, with components A, B, and C in quantities indicated in parenthesis. To make components in level 1, we need components in level 2, again in quantities needed shown in parenthesis.

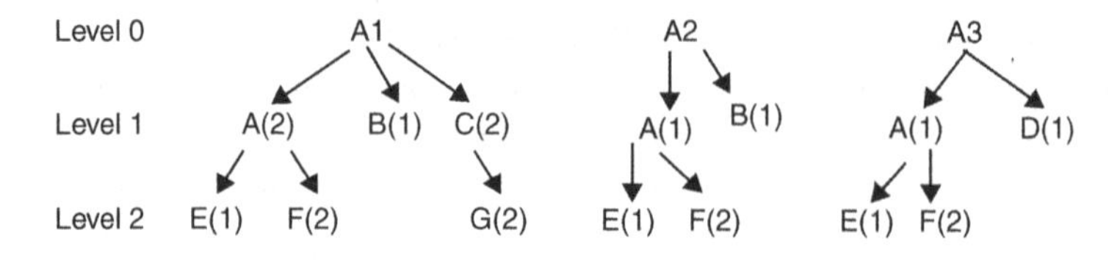

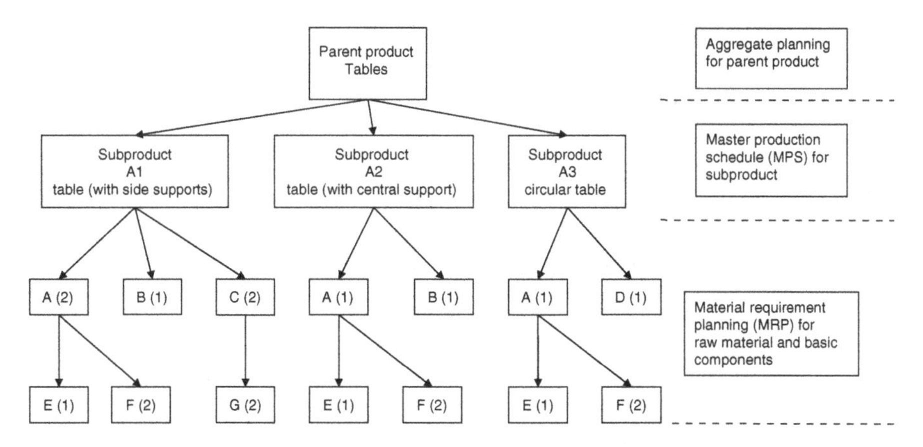

FIGURE 4.3 Flow diagram of product A (table).

TABLE 4.8
Component Data

Product/ Component	Initial Inventory on Hand	Replenishment Policy	Lead Time in Weeks
A1	50	FOQ of 120 units	2
A2	120	L4L	1
A3	100	L4L	1
A	600	L4L	1
B	600	L4L	1
C	300	FOQ of 300 units	1
D	350	FOQ of 200 units	2
G	650	L4L	2
E	400	FOQ of 400	0
F	600	L4L	1

In addition, we have information on the products as shown in Table 4.8.

MPS for subproducts A1, A2, and A3 for the next 8 weeks are as given in Tables 4.9, 4.10, and 4.11.

Now we must understand timings. Planned order release time for a lower-level item, (say item A1 at level 0), is when the next higher-level components, (items A and B at level 1 in our example), needed for that item must be delivered. While, the lead time within an item, (item A1 with 1 week lead time), is the time required to process (and deliver) the incoming components, (components A and B in our example), to develop the final product, (A1).

Based on order release time for products A1 and A2, requirements for the next higher-level components, items A and B, are as displayed in Tables 4.12 and 4.13.

TABLE 4.9
Requirements for Subproduct A1

	Weeks							
FOQ = 120/Lead Time = 2	**1**	**2**	**3**	**4**	**5**	**6**	**7**	**8**
Gross requirements or demand	200		125		200	25		
Scheduled receipts from earlier orders	120							
On-hand inventory 50	−30	90	85	85	5	100	100	100
Planned receipts		120	120		120			
Planned order release	120	120		120	120			

TABLE 4.10
Requirements for Subproduct A2

	Weeks							
L4L/Lead Time = 1	**1**	**2**	**3**	**4**	**5**	**6**	**7**	**8**
Gross requirements or demand	100	300		35		100	100	100
Scheduled receipts from earlier orders								
On-hand inventory 120	20							
Planned receipts		280		35		100	100	100
Planned order release	280		35		100	100	100	

TABLE 4.11
Requirements for Subproduct A3

	Weeks							
L4L/Lead Time = 1	**1**	**2**	**3**	**4**	**5**	**6**	**7**	**8**
Gross requirements or demand	50	100	50	125	50		100	75
Scheduled receipts from earlier orders								
On-hand inventory 100	50							
Planned receipts		50	50	125	50		100	75
Planned order release	50	50	125	50		100	75	

Applying appropriate replenishment policies for each item results in production/purchase policies as shown in Tables 4.14, 4.15, and 4.16.

Similarly, continuing the analysis for the next level components E and F results in Tables 4.17 and 4.18.

TABLE 4.12
Total Requirements for Component A

	Weeks							
	1	**2**	**3**	**4**	**5**	**6**	**7**	**8**
Gross requirements for A1	240	240		240	240			
Gross requirements for A2	280		35		100	100	100	
Gross requirements for A3	50	50	125	50		100	75	
Total requirements	570	290	160	290	340	200	175	

TABLE 4.13
Total Requirements for Component B

	Weeks							
	1	**2**	**3**	**4**	**5**	**6**	**7**	**8**
Gross requirements for A1	120	120		120	120			
Gross requirements for A2	280		35		100	100	100	
Total requirements	400	120	35	120	220	100	100	

TABLE 4.14
Production/Purchase Planning for Product A

	Weeks							
L4L/Lead Time = 1	**1**	**2**	**3**	**4**	**5**	**6**	**7**	**8**
Gross requirements or demand	570	290	160	290	340	200	175	
Scheduled receipts from earlier orders								
On-hand inventory 600	30							
Planned receipts		260	160	290	340	200	175	
Planned Order release	260	160	290	340	200	175		

Similarly, production/purchase plans for items C and G are developed in Tables 4.19 and 4.20.

4.2.1 Least Unit Cost Purchase Policy

We have seen in the previous example that all replenishment policies discussed in MPS can be used to replenish in MRP analysis. In all these policies, it was assumed that the unit price of an item remains unchanged, no matter what the order quantity is.

TABLE 4.15
Production/Purchase Planning for Product B

	Weeks							
L4L/Lead Time = 1	**1**	**2**	**3**	**4**	**5**	**6**	**7**	**8**
Gross requirements or demand	400	120	35	120	220	100	100	
Scheduled receipts from earlier orders								
On-hand inventory 600	200	80	45					
Planned receipts				75	220	100	100	
Planned order release			75	220	100	100		

TABLE 4.16
Production/Purchase Planning for Product D

	Weeks							
FOQ = 200/Lead Time = 2	**1**	**2**	**3**	**4**	**5**	**6**	**7**	**8**
Gross requirements or demand	50	50	125	50		100	75	
Scheduled receipts from earlier orders								
On-hand inventory 350	300	250	125	75	75	175	100	
Planned receipts								
Planned order release				200				

TABLE 4.17
Production/Purchase Planning for E

	Weeks							
FOQ = 400/Lead Time = 0	**1**	**2**	**3**	**4**	**5**	**6**	**7**	**8**
Gross requirements or demand	260	160	290	340	200	175		
Scheduled receipts from earlier orders								
On-hand inventory 400	140	240	350	10	210	35		
Planned receipts		400	400		400			
Planned order release		400	400		400			

However, when a quantity discount exists, that is, when the unit price changes based on the order quantity, it may be profitable to order a larger quantity than what is immediately required to reduce the total cost of operation. Least unit cost analysis can be used to determine economic purchase quantities that satisfy planned order releases. We illustrate least unit cost (LUC) policy by an example.

TABLE 4.18
Production/Purchase Planning for Product F

	Weeks							
L4L/Lead Time = 1	**1**	**2**	**3**	**4**	**5**	**6**	**7**	**8**
Gross requirements or demand	520	320	580	680	400	350		
Scheduled receipts from earlier orders								
On-hand inventory 600	80							
Planned receipts		240	580	680	400	350		
Planned order release	240	580	680	400	350			

TABLE 4.19
Production/Purchase Planning for Product C

	Weeks							
FOQ = 300/Lead Time = 1	**1**	**2**	**3**	**4**	**5**	**6**	**7**	**8**
Gross requirements or demand	240	240		240	240			
Scheduled receipts from earlier orders								
On-hand inventory 300	60	120	120	180	240			
Planned receipts		300		300	300			
Planned order release	300		300	300				

TABLE 4.20
Production/Purchase Planning for Product G

	Weeks							
L4L/Lead Time = 2	**1**	**2**	**3**	**4**	**5**	**6**	**7**	**8**
Gross requirements or demand	600		600	600				
Scheduled receipts from earlier orders								
On-hand inventory 650	50	50	0	0				
Planned receipts			550	600				
Planned order release	550	600						

Suppose item E is a purchase item with following purchase price structure.

Quantity	Purchase Price/Unit
1–400	$3.00
401–800	$2.75
801 to up	$2.50

The order cost is $100/order, and the carrying cost is $0.50/unit/week. Based on this information, we need to decide the optimum quantity to purchase for the requirements determined in Section 4.2.

The calculations are displayed in Tables 4.21, 4.22, and 4.23. In each table, we are trying to decide the order quantity that gives the least cost per unit. In Table 4.21, demands are tabulated for all weeks. The next column is cumulative demand. It is never optimum to place an order for less than a full week's demand. The order cost is fixed to one order, which is $100. We are seeking order quantity for the first order.

First order is placed at the beginning of the first week. Carrying cost is based on the order quantity. For example, if the order is for 360, it will be consumed in the first week, and the carrying cost is zero. If the order is for the first and second weeks, that is, the next cumulative quantity of 520 units, then 160 units, the demand for

TABLE 4.21
Calculations for the First Order Quantity

Week	Demand	Cumulative Demand	Order Cost	Carrying Cost	Unit Purchase Price	Total Cost	Cost Per Unit of Demand
1	360	360	100	0	3	1180	3.277
2	160	520	100	80	2.75	1610	**3.096**
3	400	920	100	480	2.5	2880	3.130
4	40	960	100	540	2.5	3040	3.166
5	140	1100	100	820	2.5	3670	3.336
6	60	1160	100	970	2.5	3970	3.422
7	0	1160	100	970	2.5	3970	3.422
8	0	1160	100	970	2.5	3970	3.422

TABLE 4.22
Calculations for the Second Order Quantity

Week	Demand	Cumulative Demand	Order Cost	Carrying Cost	Unit Purchase Price	Total Cost	Cost Per Unit of Demand
1							
2							
3	400	400	100	0	3.0	1300	3.250
4	40	440	100	20	2.75	1330	**3.022**
5	140	580	100	160	2.75	1855	3.198
6	60	640	100	250	2.75	2110	3.296
7	0	640	100	250	2.75	2110	3.296
8	0	640	100	250	2.75	2110	3.296

TABLE 4.23
Calculations for the Third Order Quantity

Week	Demand	Cumulative Demand	Order Cost	Carrying Cost	Unit Purchase Price	Total Cost	Cost Per Unit of Demand
1							
2							
3							
4							
5	140	140	100	0	3.0	520	3.714
6	60	200	100	30	3.0	730	**3.650**
7	0	200	100	30	3.0	730	3.650
8	0	200	100	30	3.0	730	3.650

the second week, will be carried in inventory in week 1, for 1 week, and the cost is $360 \times 0 + 160 \times 1 \times 0.5 = 80$. Similarly, carrying cost for ordering the first 3 weeks of demand of 920 units is $360 \times 0 + 160 \times 1 \times 0.5 + 400 \times 2 \times 0.5 = 480$ and so on. The unit purchase cost is based on order quantity. The total cost is, purchase cost per unit × quantity ordered + order cost + carrying cost. The total cost divided by the order quantity gives the cost per unit, the last column data.

The least cost is associated with the order quantity of 520. Therefore, the first order is for 520 units. Perform the analysis again with the remaining demands. The details are shown in the Table 4.22.

The second order is for 440 units. To determine the third order, perform the analysis again for remaining demands. The details are in Table 4.23

The third order is for 200 units.

If the order quantity must remain constant, then similar analysis can be made using fixed quantities. More about quantity discount analysis is given in Inventory and Capacity Planning chapter.

4.2.2 General Considerations

Some observations associated with planning factors in MRP are as follows:

4.2.2.1 Planning Period

What is the minimum planning horizon? Ideally, it should be at least as long as the cumulative lead times for the entire product structure. This allows planning on the parent product propagation through to the highest-level component. In our example in Table 4.8, for product A_1, the longest total lead time through the structure tree is lead times 2 for A_1 + 1 for A + 1 for $F = 4$. Similarly, for product A_2, the minimum planning period is 1 for A_2 + 1 for A + 1 for $F = 3$ (or 1 for A_2 + 2 for $B = 3$). Similarly for A_3 planning period is 3. Hence, the minimum planning period should be four periods.

4.2.2.2 Product Structure

When many products are planned simultaneously, as we had with products A_1, A_2, and A_3 there might be some common components among the products. These components may appear at different levels in each product tree. It is important that when planning for the component, we collect requirements from all levels and from all parent products. We had performed this analysis for component A, B, and E in our example, even though they appear as the same levels in all products.

4.2.2.3 Manufacturing Resource Planning

Development of MRP assumes that we can receive the input components/material when required. We can then use the lead time in that stage to process and produce the number of components as the MRP plan calls for. This is possible if we have sufficient manpower and processing capacities, so that there are no bottlenecks and delays. However, if not, we must consider the limitations imposed by the production system and incorporate those in our planning. Manufacturing Resource Planning, also known as MRP II, allows us to modify MRP plans to respond to system limitations. MRP II is a computerized software that modifies the production quantities, order releases, and receiving dates based on the current available resources. MRP II is an extended mode of planning that incorporates the available recourses and capacities in MRP planning.

4.2.2.4 Enterprise Resource Planning

The next phase of integration is to include most of the supporting departments in the planning picture. Enterprise resource planning (ERP) is such a software. It includes, in planning, the departments of sales and marketing, engineering, finance, human resources, distribution, and operations.

There are a number of reasons for using an ERP software.

1. ***Assimilate customer order information***: All customer order information, from the time orders are received till they are delivered, is maintained in the system, making it easier to keep up with the orders at all stages of production and distribution. All departments concerned with the status and progress of an order have access to the information and can act on it.
2. ***Standardize manufacturing***: The system can be standardized by all departments, following the same rules with integrated information management system.
3. ***Inventory reduction***: With systems like MRP and MRP II integrated within ERP allows efficient and smooth manufacturing without excessive inventories at any point in the system.
4. ***Human resource planning***: ERP has means of following every employee's allocation and efforts. It improves human resource utilization.
5. ***Combine financial information***: ERP creates a single financial measure across all departments, such as finance, sales, and operations, which

everyone understands and follows. Financial information is presented in a manner that allows the accounting department to coordinate payments and collections while facilitating the finance department to plan for working capital and resource planning, and so on.

With all explicit advantages of ERP, there are also some difficulties. ERP software can be very expensive and difficult to implement completely throughout the system. The need for key personnel to buy into the system is essential, and it is not always successful.

Detailed discussion of both MRP II and ERP is beyond the scope of this book. However, we will discuss capacity evaluations in a later chapter.

4.2.3 Distribution Requirement Planning

Instead of supplying products directly to customers, many large companies distribute their products through the warehouses that are set in different parts of the country. Customer demands are collected at retail warehouses and then transferred to the central warehouses and from there transferred back to the plants. In each stage, order quantity is adjusted for the best economics. In this case, most of the uncertainty in demand is absorbed by the inventories available at different points in the distribution system. These uncertainties may arise in response to marketing strategies, advertising, changing customer preferences, or excessive competition.

Distribution requirement planning (DRP) integrates the distribution system with the objective of meeting customers' demands, which are independent demands without incurring excessive inventory. It influences demand management at the production source plant, and hence influence MPS. It follows the same basic principles as MRP in developing distribution policies. Replenishments to distribution centers are, in general, by truckload quantities, which help in reducing transportation cost. However, it may also create a large quantity in replenishments. DRP encourages logistics savings through better planning of aggregate transportation capacity needs and order dispatching policies.

An example of DRP is shown below. It hardly needs any explanations, since it follows the same logic as MRP even though the notation may be different. Safety stock is maintained to respond to any unexpected demand highs. It is used in emergencies as needed and must be replenished as soon as possible, without increasing inventory unnecessarily.

4.2.3.1 DRP Example

Consider a company having two field warehouses and a central one. Table 4.24 represents a typical DRP record for a particular field warehouse. The table consists of the following information:

Requirements: Predicted or expected demand.
In transit: Quantity or order that is available.
Lead time: Time between placement of order and receiving it.

TABLE 4.24
Data for DRP Example

		Warehouse 1[a]						
Time period		1	2	3	4	5	6	7
Marketplace forecast		15	50	5	15	15	35	15
In transit (order to receive)			40	40			40	40
Available balance	22	7		37	22	7	12	37
Planned shipments (order release)		40	40			40	40	
Safety stock		5	2	5	5	5	5	5
		Warehouse 2[b]						
		1	2	3	4	5	6	7
Marketplace forecast		20	20	70	20	30	30	30
In transit (order receive)				60	60		60	
Available balance	45	25	5		35	5	35	5
Planned shipments (order release)		60	60		60			
Safety stock		10	10	5	10	10	10	10

[a] Safety stock = 5; shipping quantity = 40; lead time = 1.
[b] Safety stock = 10; shipping quantity = 60; lead time = 2.

Available balance: The amount of inventory available. Balance before the first period shows the beginning inventory.
Planned shipments: Released order quantity.
Safety stock: Quantity maintained in reserve to respond to unexpected demand spike.

Gross requirements for the central warehouse are addition of planned shipments for warehouse 1 and warehouse 2. Gross requirements for the central warehouse are dependent on the demand of the field warehouse (Table 4.25).

Firm order is the same as planned order release, unless there are some reasons to change them. These reasons may include available transportation capacities and optimum shipping quantities that may depend on truckload size. Firm orders are then used in master production system or ERS planning. A safety stock of 20 units is used to supplement on-hand inventory when demand is greater than stock on hand, and then it is replenished as soon as possible.

DRP shows when the central warehouse should receive its orders. These periods in turn are incorporated in MRP and MPS as the demand points for further planning. DRP also allows us to develop an efficient transportation/distribution system, knowing when and how many units are to be shipped from one point to another.

4.3 SUMMARY

The chapter presents a number of planning techniques that follow a similar spreadsheet approach. It is important to know when to produce, how many units

TABLE 4.25
Gross Requirements at Central Warehouse

Warehouse 1								Warehouse 2						
1	2	3	4	5	6	7	+	1	2	3	4	5	6	7
40	40			40	40			60	60		60			

Central warehouse						
1	2	3	4	5	6	7
100	100		60	40	40	

		Central warehouse						
		1	2	3	4	5	6	7
Gross requirement		100	100		60	40	40	
Scheduled receipts			100		100		100	
Available balance	90				30		50	50
Planned order release			100		100		100	
Firm planned order (MPS)			100		100		100	
Safety stock		10	10	10	20	10	20	

Safety stock = 20; shipping quantity = 100; lead time = 0.

to produce, and when to deliver. Different replenishment policies may result in different costs, and we should try to apply one that minimizes the cost. ERP is inclusive software that interlinks operations from different departments in the plant, allowing common information, without any distortions, being made available to all.

These techniques assume that sufficient resources are available when needed. If this is not the case, we need to schedule the available resources in an optimum manner. How to achieve this objective is illustrated in scheduling chapters.

4.4 EXERCISE

4.1 The gross requirement for a material from an MRP schedule is:

Period (week)	1	2	3	4	5	6	7	8	9
Gross requirements	25	30	30	70	25	15	70	50	30

Units ending as inventory at the end of a period t must be carried over as beginning inventory for the next period, with \$1.50 per unit as holding cost per week. The set up cost is \$120, and the lead time is 1 week. The beginning inventory is 35 units.

a. Develop a L4L 9-week schedule and calculate the total relevant costs.

b. Develop POQ solution if time between the orders is fixed to 2 weeks. Calculate the total relevant costs, assuming that the stock out cost is $10 per unit.
c. Resolve the above problem with zero lead time.
d. In the problem, if actual demands for period 1, 2, and 3 were 15, 20, and 35, respectively, find the APT for these periods.

4.2 The MRP gross requirements for item A for next 8 weeks are shown in the following table.

Period (week)	1	2	3	4	5	6	7	8
Gross requirements	20	15	15	45	30	70	100	40

The beginning inventory is 65 units. The lead time for the item is 2 weeks, the setup cost is $15, and the carrying cost is $0.050 per unit per day. Using least total cost method, determine the time and the magnitude of the first order to be placed.

4.3 In Figure 4.1, two tables A1 and A2 are illustrated. Suppose that the requirements for each type for next 4 weeks are as follows:

Week	1	2	3	4
A1	20	25	30	15
A2	12	8	9	10

Develop the requirements for all its components.

4.4 The annual demand of item X is 10,400 units over a 52-week-per-year schedule. It costs $300 for the setup. When one unit of this item must be carried in inventory from 1 week to another, it incurs $0.60 per unit.

The net requirements for the item from an MRP schedule are:

Period (week)	1	2	3	4	5	6	7	8	9	10
Gross requirements	75	150	400	200	150	140	300	200	150	500

Determine which of the following lot-sizing methods results in the least carrying and ordering costs for the 10-week schedule:

a. L4L
b. POQ with a period of 3 weeks
c. FOQ of 200 units

4.5 In example 4.4.3, if the units are bought from an outside vendor with following purchase price:

Quantity	Price/Unit ($)
1–150	2.00
151–300	1.80
301 and more	1.75

Determine the least cost purchase policy.

4.6 A product's demand varies during 8 weeks for warehouse 1 and is given as follows:

Weeks	1	2	3	4	5	6	7	8
Demand	150	75	90	—	165	50	200	—

a. Determine the DRP schedule if the shipping quantity is 120 units, lead time is 1 week, and the required safety stock is for 50 units. If it costs $200/trip and $5/period for storage, what is the total cost of the policy?

b. We have the choice to purchase a larger truck with the cost of $275/trip. It can carry 180 units. With a storage cost of $5/period/unit, which truck we should utilize—the one with capacity of 120 or the new one with a capacity of 180?

4.7 Repeat problem 4.4.6 if in addition to the warehouse 1 we have warehouse 2 with demands as follows:

Weeks	1	2	3	4	5	6	7	8
Demand	100	—	190	50	105	50	20	80

In addition, central warehouse is also supplied using one of the two types of trucks of capacity 120 or of capacity 180. All other data also remains the same for warehouse 2 and the central warehouse.

REFERENCES AND SUGGESTED READINGS

Berry, W.L., T.E. Vollmann, and D.C. Whybark. 1979. *Master Production Scheduling: Principles and Practice*, Falls Church, VA: American Production and Inventory Control Society.

Davis, E.W. 1979. *Case Studies in Materials Requirement Planning*, Falls Church, VA: American Production and Inventory Control Society.

Dougherty, J.R. and J.F. Proud. "From Master Schedules to Finishing Schedules in the 1990s" *American Production and Inventory Control Society, 1990 Annual Conference Proceedings*, 1990, pp. 368–370.

Ford, Q. "Distribution Requirement Planning and MRP" *APICS 24th Annual Conference Proceedings*, 1981, pp. 275–278.

Funk, P.N. "The Master Schedulers Job Revisited" *American Production and Inventory Control Society, 1990 Annual Conference Proceedings*, pp. 374–377.

Harvard Business Review, September–October, 1975.

Kinsey John, W. "Master Production Planning: The Key to Successful Master Scheduling" *APICS 24th Annual Conference Proceedings*, 1981, pp. 81–85.

Orlicky, J. 1975. *Materials Requirement Planning*, New York: McGraw-Hill.

Proud John, F. "Controlling the Master Schedule" *APICS 23rd Annual Conference Proceedings*, 1980, pp. 413–416.

Smith, B. July 1985. "DRP improves Productivity, Profit and Service Levels" *Modern Materials Handling*, 63–65.

5 Inventory and Capacity Planning

Chapter presents two topics, inventory planning and capacity planning. In inventory control, we determine the optimum inventory levels, ordering and restocking policies, and factors that may influence inventory cost. In capacity planning, we decide how to determine the capacity required at each resource center and how to plan or manage the available capacity to meet the requirements.

5.1 INVENTORY PLANNING

There are a number of reasons to have inventories. Inventories are built as a result of production or purchase of quantities that cannot be utilized immediately. We want to buy or build an economic quantity to minimize a unit cost of production and storage. This may require purchase or building of larger lots. Inventories can also be a result of taking advantage of quantity discounts offered by suppliers. Here, a large quantity is purchased to reduce the unit purchase price. Inventories are used between two production stations to decouple two operations if they have imbalanced production rates. Inventories are used as safety stocks to respond to unexpected demands. They can be used to level production capacities, producing more when demand is low to compensate for the periods when demand is higher than the production capacity. Inventories are also needed to display the available items for sale in retail stores.

With all these reasons to have inventory, why not have an unlimited inventory? The answer is obvious. There is cost associated with inventories, cost of storage, cost of spoilage, cost of taxes and insurance, and even cost of obsolescence. As the size of inventory and time of storage goes up, so do these costs. So, we must find a good balance between cost and benefits.

5.1.1 Economic Order Quantity

There are two basic costs in inventory management—first, order cost if the item is purchased, (or setup cost, if the units are produced in our own facilities), and second, carrying cost, which is a time-dependent cost. The longer an item is in storage, the more it costs to keep it. This cost might include factors such as cost of storage, taxes, insurance, spoilage, and obsolescence. For a demand that is uniform over a time period, as in Figure 5.1, we may have different alternatives in satisfying the demand.

For example, we might just place one order and order a large quantity at one time, incurring the minimum order cost but a large carrying cost, or we may order in very small quantities frequently, reducing the carrying cost but increasing the order costs. The objective is to determine the optimum order quantity.

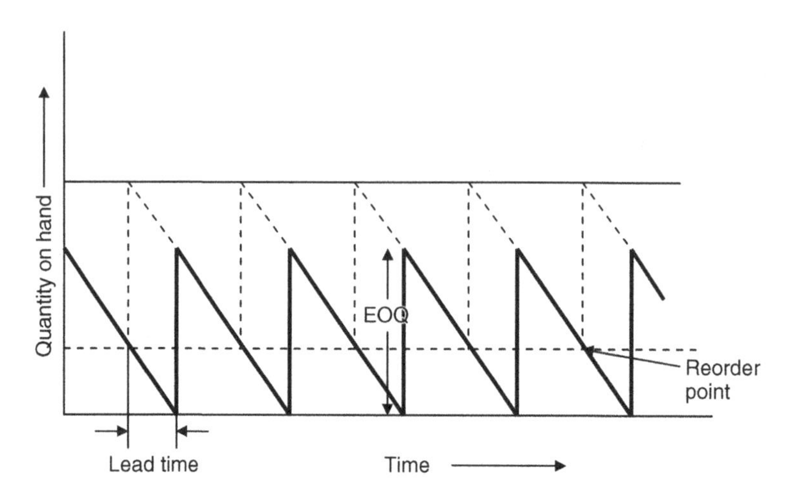

FIGURE 5.1 Economic order quantity with uniform demand.

Let us define following notation:

Q Quantity to order (or produce) per order
D Demand per unit time, normally a year
C Cost of a unit
S Setup or order cost/order
H Holding or carrying cost per unit time. Holding cost may be defined as \$/unit/unit time multiplied by the cost of a unit or \$/\$ invested/unit time

Then, the total cost, TC, is

$$\text{TC} = CD + \frac{Q}{2} \times H + \left(\frac{D}{Q}\right) S \quad (5.1)$$

The first term reflects the total unit purchase cost per year. The second term represents inventory carrying cost. It is equal to inventory carrying cost per unit multiplied by the average inventory over the year. The average inventory is calculated with the following logic. The best time to receive a new order is when the inventory on hand is zero. In a situation where demand rate is known and constant, there is no need to keep any safety stock. If we have a policy that receives the order when there is some inventory present, that inventory will remain in the system all year without ever being utilized. The total inventory at this point when the new shipment is received is then $0 + Q = Q$. Thus, average inventory in a cycle is $(Q + 0)/2 = Q/2$. Since the demand, D, is at a uniform rate, this quantity is depleted to zero in Q/D time period, or the cycle time is Q/D. The same cycle repeats throughout the year and hence the average stock in the system over the year is also $Q/2$. The third term represents order cost per year. With Q units ordered per order, there are D/Q number of orders that must be placed to receive the total of D units in a year. With S dollar cost per order, order cost is $(D/Q)S$ (Figure 5.2).

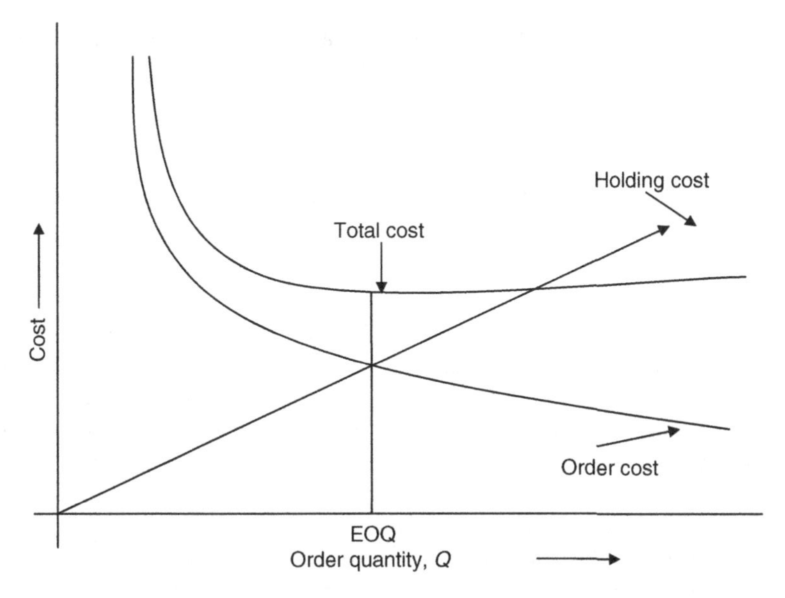

FIGURE 5.2 Order quantity, Q.

To minimize the total cost, take derivative of (5.1) with respect to Q, equate it to zero, and solve for Q, which results in:

$$Q = \text{SQRT}\left(\frac{2DS}{H}\right),$$

where SQRT stands for square root.

This quantity s called economic order quantity (EOQ).

Example

A company uses 600 units per month. It incurs an order cost of \$200/order consisting of placing order, transportation cost and receiving the product. Each unit costs \$40/unit and holding cost is 25% per year per dollar. Determine the optimum order quantity.

$$Q = \text{SQRT}\left(\frac{2DS}{H}\right)$$

Here, H is the carrying cost of the unit per year, which is $0.25 \times 40 = \$10$/unit/year. Demand is $600 \times 12 = 7200$ units per year. Substituting appropriate values leads to an EOQ of

$$\text{Q} = \left(\frac{2 \times 7200 \times 200}{10}\right)^{1/2} = 536.6 \text{ or } 537$$

Total cost is $= 7{,}200 \times 40 + (536.6/2) \times 10 + (7{,}200/536.7) \times 200 = 288{,}000 + 2{,}683 + 2{,}683 = \$293{,}366$.

It is interesting to note that the carrying cost and order cost are equal. This is one of the properties of EOQ. At EOQ, the carrying cost and order cost are always equal.

Suppose we order 500 units at time rather than EOQ of 537. Now, the total cost per year is,

$$7{,}200 \times 40 + \left(\frac{500}{2}\right) \times 10 + \left(\frac{7{,}200}{500}\right) \times 200 = 288{,}000 + 2{,}500 + 2{,}880$$
$$= \$293{,}380$$

or an increase of $14 from EOQ quantity. This is hardly a significant increase in cost for deviating from economic order quantity. In fact, the second property of EOQ is that the total of carrying cost and order cost are fairly flat (remains relatively same) around EOQ. Therefore, it is advantages to order what is convenient to order near EOQ quantity rather than exact EOQ. This property also allows for small errors in estimating exact costs for carrying or order and even some variation in the basic assumption that the demand is at a constant rate.

To receive an order when inventory goes to zero, it may be necessary to reorder before inventory actually reaches zero. Time between placement of order and receiving it is called replenishment lead time. The demand during this time t (in weeks) is, assuming demand D is for a year (52 weeks), $t \times D/52$. This quantity is called reorder point. Whenever the inventory level on hand (stock level) reaches the reorder point, a new order for Q units is placed. This rule is often used in practice since it is not difficult with modern computers to keep the record of inventory on hand. It can also absorb small variations from week-to-week demands since only inventory on hand is the trigger mechanism for placing a new order and not a fixed time period.

5.2 SAFETY STOCK DETERMINATION

Safety stock is needed when demand though fairly uniform over a period, is not necessarily constant from day to day and we may have some unexpected spicks. It does not matter what daily demands are till we reach reorder point R since a quantity of items are there to take care of the demands (see Figure 5.3). However, to make sure we have sufficient inventory on hand to meet expected demand during the replenishment time, we need to adjust both reorder point, R, and the order quantity compare to the EOQ model. The analysis and procedure is shown with the following example.

Assume demand distribution during lead time is normal with mean μ, (which is also the reorder point) and standard deviation of σ. This is fairly good assumption when demand has random variation about the mean value. Determination of μ and σ are based on historical data of the demand during the lead time. A decision must be made as to what is the most percent of time, $\alpha\%$, we can be without stock and thus lose the customer. Then, $1 - \alpha\%$ is called level of customer service or customer level of service. If $\alpha = 0$, then we must have 100% level of service, that is, we must satisfy all customer demand during the lead time. In this case, safety stock would be very large, theoretically close to infinity.

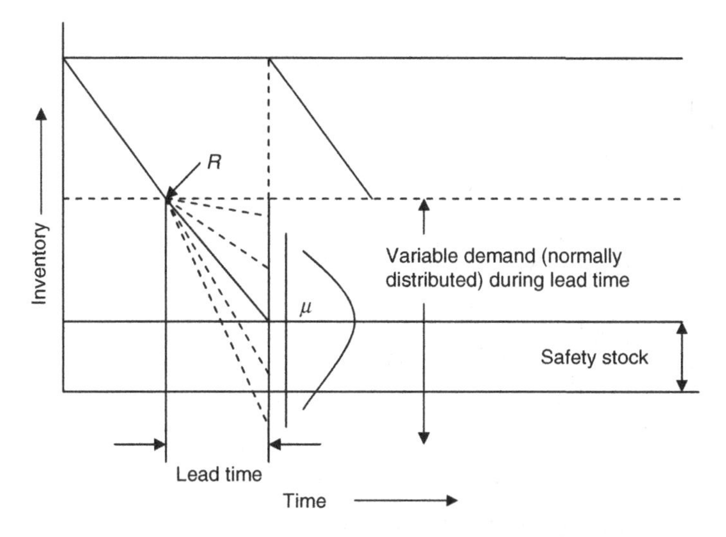

FIGURE 5.3 Safety stock requirements.

Using standard Z analysis, determine the quantity, U, the new reorder point, as follows:

$$Z = \left(\frac{U - \mu}{\sigma}\right) \quad \text{or} \quad U = \sigma Z + \mu,$$

where

$$\begin{aligned} Z &= 1.645 \text{ for 95\% customer level of service} \\ &= 2.33 \text{ for 99\% customer level of service.} \end{aligned}$$

In our example, suppose the lead time is 1 week and standard deviation of demand during this time is 20. Reorder point, without safety stock, is $1 \times (7200/52) = 138$, which is also the mean value of reorder point μ. Then, to provide 99% customer service,

$$U = 20 \times 2.33 + 138 = 46.6 + 138 = 184.6 \text{ or } 185$$

Quantity σZ is called safety stock. In our case, safety stock is 46.6 or 47.

Sometimes, it is easier to estimate mean and variance of demand during the lead time by estimating mean daily (or weekly) demand and its variation. Knowing the lead time duration in days, the use of central limit theorem from statistics allows us to estimate mean and variance of the demand during lead time. For example, suppose the average daily demand is 27.6, with standard deviation of 5 and lead time of 1 week or 5 days. Then, mean and standard deviation of demand during the lead time is

$$\text{Mean} = 5 \times 27.6 = 138 \text{ and variance of } 5 \times 5^2 = 125.$$

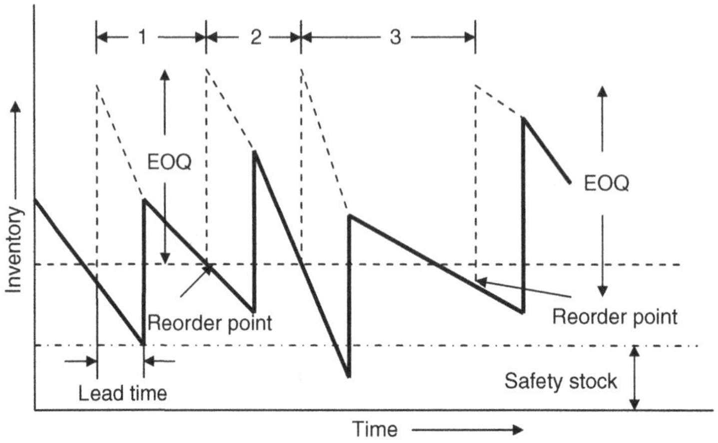

FIGURE 5.4 Fixed order quantity-varying cycle system.

Or standard deviation of 11.18. That is,

$$\text{Mean demand during lead time} = \text{Lead time in days} \times \frac{\text{Mean demand}}{\text{Day}}$$

and

$$\text{Standard deviation during lead time} = (\text{Lead time in days})^{1/2} \times \left(\frac{\text{Standard deviation of demand}}{\text{Day}}\right).$$

5.2.1 Fixed Quantity-Varying Cycle System

With safety stock, the real reorder point is 185. In fixed order quantity-varying cycle length policy, any time the inventory level goes to U or below, we order EOQ units. In our example with reorder point set at 185, any time the inventory level falls to or below 185, we should order 537 units. With the order of 537 units and if there had not been an abnormal demand, the inventory on hand after ordering would have been $537 + 185 = 722$ units. Figure 5.4 shows varying cycle lengths for each cycle because of variations in daily demands. Again, inventory on hand is also not depleted at the same constant rate, but may vary somewhat from day to day, based on actual demand.

5.2.2 Periodic Review System

Rather than monitoring inventory continuously, it may be more convenient to review it every fixed interval of time. At that time, we check on-hand inventory and place an order so that the inventory reaches a predetermined level. To match with EOQ

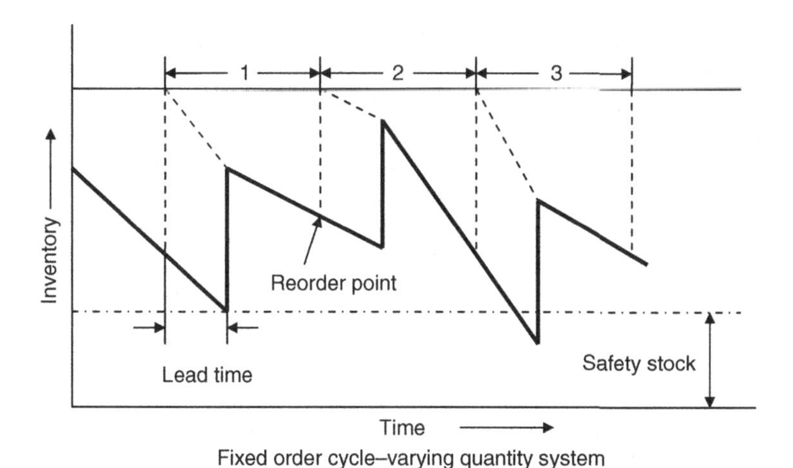

FIGURE 5.5 Fixed order cycle-varying quantity system.

calculations, the review period could be as close to cycle time in EOQ as possible. Recall that the cycle time in EOQ is Q/D. In our example (Section 5.1.1), EOQ is 537, with $D = 7200$ units/year. Cycle time is $(537/7200) \times 52 = 3.87$, or say 4 weeks. Using our safety stock analysis (Section 5.2.1), we will review the inventory every 4 weeks and order sufficient quantity to bring the level back to target value of 722. For example, if the inventory level on Monday morning is 200, and there is demand of 54 during Monday, the net available inventory on Tuesday is $200 - 54 = 146$. Since this is less than the reorder point of 185, by $185 - 146 = 39$, we have used this amount from safety stock. So, we should place an order for $537 + 39 = 576$ units, or simply for $722 - 146 = 576$ units.

Average demand during week 4 is $(4 \times 7200)/52 = 553.84$. Safety stock of 185 remains in the system throughout the year, and is used when needed. If the demand was very predictive and constant, we could avoid the use of safety stock, and order only the 4 weeks requirement of 553.84 or 554 units (Figure 5.5).

Assuming safety stock is maintained, cost of the inventory system per year is

$$
\begin{aligned}
7{,}200 \times 40 + \left(\frac{553.84}{2}\right) \times 10 + 185 \times 10 + 200\left(\frac{52}{4}\right) &= 288{,}000 + 2{,}769.2 \\
&\quad + 1{,}850 + 2{,}600 \\
&= \$295{,}219.2
\end{aligned}
$$

5.2.3 Safety Stock with Substitute Products

Suppose we have two or more similar products in stock, it may be possible to substitute one product by another if demand arises and the demanded item is out of stock. For example, in a grocery store if a milk carton of one brand is not available, the customer generally picks up another brand without any complaints. Under these circumstances, it is possible to reduce total safety stock for each item and still provide the desired level of service.

Consider three brands A, B, and C, each having demand that is normally distributed with a mean of 30 per day and a standard deviation of 6 per day. If one brand of product is not available, demand can be substituted by another brand. Lead time is 2 days. We wish to provide service level of 95%.

Demand during lead time for each product has normal distribution with mean of $2 \times 30 = 60$ and standard deviation of $\sqrt{2} \times 6 = 8.48$.

For 95% service level, $z = 1.645$ and hence safety stock $S = \sigma Z$, is

$$8.48 \times 1.645 = 13.94 \text{ or } 14.$$

If substitution of products is not allowed, we would require 14 units of safety stock for each brand for the total of $3 \times 14 = 42$ units in safety stocks.

But since substitution is allowed, the total demand for all products is determined by applying the central limit theorem as follows:

$$\text{Mean demand} = \sum \text{Demand for each product in lead time,}$$

and

$$\text{Variance of demand} = \sum \text{Variance of demand in lead time for each product}$$

Or standard deviation, if the variance of each product is same $= \sqrt{(\text{Number of products})} \times$ Standard deviation for a product in lead time.

Or, in our case, the mean of total demand during lead time is $3 \times 60 = 180$, and the standard deviation of the total demand is $\sqrt{3} \times 8.48 = 14.68$.

Therefore, total safety stock S_t is,

$$1.645 \times 14.68 = 24.14, \quad \text{or} \quad S_t = 25$$

As expected, safety stock when the products could be substituted is less than the combined safety if the products cannot be substituted. That is,

$$S_t < 3 \times S.$$

5.3 QUANTITY DISCOUNTS

At times a seller offers to reduce the cost of a unit if larger quantities are purchased as an incentive to make consumer buy bigger quantities at one time. Sellers can distribute their fixed cost over a larger quantity and reduce their unit cost of production. They are thus able to pass some of the savings to customers in terms of quantity discounts.

As buyers, how should we respond to this offer? Our objective should be same as before: reduce the total cost of inventory, including the total purchase cost. We should determine what is the optimum quantity we should purchase to minimize this cost.

There are two types of quantity discounts that are offered in practice. They are all unit quantity discount and marginal unit quantity discount.

In all unit quantity discount, the unit price for a unit is constant, based on the purchase quantity per order. In marginal unit quantity discount, the unit price is based on quantity ordered in each break point. The average price of a unit will therefore vary based on the total units bought.

5.3.1 All Unit Quantity Discount

Here, all units are charged the same amount based on the quantity order in a single order.

Consider our problem again, except the unit price of the product now depends on the quantity purchased. Recall that the company uses 600 units per month. It incurs an order cost of $200/order, consisting of placing the order, transportation cost, and receiving the product. Holding cost, *h*, (carrying cost) is 25% per year per dollar. Unit cost is as follows:

Quantity	Unit Price
1–699	$40/unit
700–999	$38/unit
1000 and up	$37/unit

Our objective is to determine the optimum order quantity (Figure 5.6).

We can start with applying EOQ formula, except we have a problem of deciding what is the carrying cost here. The carrying cost per unit is calculated as unit cost multiplied by the holding cost. So, we have different carrying costs based on the quantity ordered.

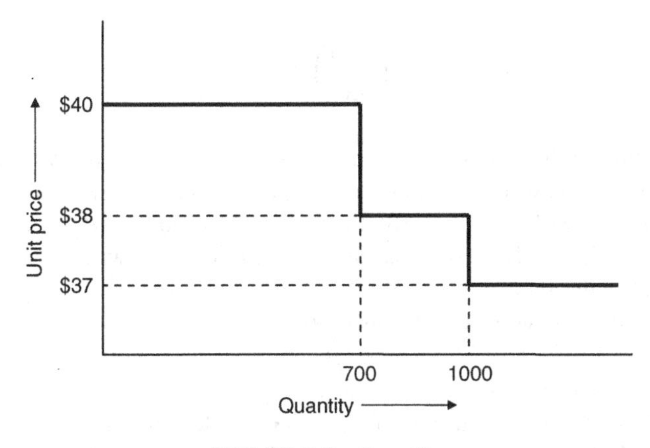

FIGURE 5.6 Quantity.

The procedure can start with the least unit cost and by calculating the EOQ. If it is feasible, that is, order quantity is greater than 1000, this would be the least-cost solution, and there is no need to proceed further.

$$\text{For } Q \geq 1000, \quad C = 37$$

and

$$H = C \times h = 37 \times 0.25 = 9.25$$

Hence

$$\text{EOQ} = \text{SQRT}\left(\frac{2SD}{H}\right) = \text{SQRT}\left(2 \times 200 \times \frac{7200}{9.25}\right) = 557.9$$

This EOQ quantity cannot be purchased for \$37 per unit. The minimum quantity we will have to purchase to get the unit price of \$37 is 1000 units. Calculate the total yearly cost with this quantity to see what the cost/year would be if 1000 units are ordered.

$$\text{Total cost/year} = 7{,}200 \times 37 + \left(\frac{1{,}000}{2}\right) 9.25 + \left(\frac{7{,}200}{1{,}000}\right) \times 200 = \$272{,}465.$$

Let us now check the next least cost bracket, when the unit price is \$38. The holding cost is $38 \times 0.25 = 9.5$. The EOQ with this holding cost is

$$\text{EOQ} = \text{SQRT}\left(2 \times 200 \times \frac{7200}{9.5}\right) = 550.59$$

Again, this quantity cannot be bought for \$38 unit cost. Again, check what the annual cost would be if we place the minimum order, $Q = 700$, with the present unit cost.

$$\text{Total cost/year} = 7{,}200 \times 38 + \left(\frac{700}{20}\right) 9.5 + \left(\frac{7{,}200}{700}\right) \times 200 = \$275{,}989$$

This cost is higher than the one with unit cost of \$37. There is no need to check any further bracket(s) where the unit cost is higher than the one we had just checked since the total cost will only go up. By the way, had we calculated the total cost with $C = 40$, it is \$293,380. This cost was calculated earlier when we solved the problem without quantity discount.

The procedure can now be stated as follows:

1. Start with the lowest unit cost bracket.
2. Determine EOQ for this cost. If EOQ is feasible, determine the associated cost. Mark this as present minimum, PM, and go to step 4.

3. If the EOQ is not feasible, calculate the cost associated with purchasing the smallest possible quantity with the present unit cost. Mark this as PM.
4. Go to next higher price bracket.
5. Determine EOQ for this cost. If the EOQ is feasible, determine the associated cost. Mark this as New Minimum, NM, and go to step 7.
6. If the EOQ is not feasible, calculate the cost associated with purchasing the smallest possible quantity with the present unit cost. Mark this as NM.
7. Compare PM with NM. If PM is less than NM, stop. Quantity associated with PM is the optimum quantity to order. If not go to step 8.
8. Make PM = NM and go to step 4.

5.3.2 Marginal Unit Quantity Discount

In the marginal quantity discount, the unit price is bracket dependent. In the following chart, for example, the first 699 units are charged at $40.00 per unit, next 300 units cost at $38/unit, and the remaining order, if greater than 1000, costs $37/unit,

Bracket	Quantity	Unit Price
0	1–699	$40/unit
1	700–999	$38/unit
2	1000 and up	$37/unit

The carrying cost depends on the unit price, which in turn depends on the quantity ordered. Let us define break points by q_i. So, $q_0 = 1$, $q_1 = 700$, and $q_2 = 1000$.

Let R_i define the cost of ordering q_i units. Total cost at the end of bracket i, is,

$$R_i = C_0(q_1 - q_0) + C_1(q_2 - q_1) + \cdots + C_{i-1}(q_i - q_i - 1)$$

In our case, $R_0 = 0, R_1 = (700 - 1) \times 40 = 27{,}960$, and $R_2 = 27{,}960 + 38(1000 - 700) = 39{,}360$.

The average cost per unit for ordering Q which is in bracket i is

$$= \left(\frac{R_i + (Q - q_i)C_i}{Q}\right)$$

Therefore, the holding cost is: average inventory × unit price × holding cost in $/$ invested

$$= \left(\frac{Q}{2}\right) \times \left(\frac{R_i + (Q - q_i)C_i}{Q}\right) \times h = \left(\frac{R_i + (Q - q_i)C_i}{2}\right) \times h \tag{5.2}$$

And the purchase cost is

$$D \times \frac{\text{Average cost}}{\text{Unit}} = D \times \left(\frac{R_i + (Q - q_i)C_i}{Q}\right) \tag{5.3}$$

Order cost is

$$\frac{\text{Number of orders}}{\text{Year}} \times \text{order cost per order} = (D/Q) \times S \tag{5.4}$$

Total cost/year is the sum of (5.2), (5.3), and (5.4).

To minimize the total cost, apply basic calculus. Take derivative with respect to Q, equate it to zero, and solve for Q results in

Optimum quantity for $i = 0$

$$Q_0 = \text{SQRT}\left(\frac{2DS}{hC_0}\right)$$

and

Optimum Q in cost bracket $i = \text{SQRT}(2D \times (S + R_i - q_iC_i))/(h \times C_i)$, where $i = 1, 2\ldots$

The procedure to solve a problem is as follows:

1. Calculate EOQ for each cost bracket. If feasible, calculate the total cost for the associated Q.
2. If Q calculated in step 1 is not feasible, calculate the total cost with quantity at each break point.
3. Select the least total cost alternative.

In our example:

For bracket $i = 0$

$$Q_0 = \text{SQRT}\left(2 \times 7200 \times \frac{200}{0.25 \times 40}\right) = 536.6$$

which is a feasible quantity.

With total cost of

$$7{,}200 \times 40 + \left(\frac{536.6}{2}\right) \times 10 + \left(\frac{7{,}200}{536.7}\right) \times 200 = 288{,}000 + 2{,}683 + 2{,}683$$
$$= \$293{,}366$$

For bracket $i = 1$

$$R_1 = 40 \times 699 = 27{,}960 \quad \text{and} \quad q_1 = 700$$

$$Q_1 = \text{SQRT}\left(2 \times 7{,}200 \times \frac{(200 + 27{,}900 - 700 \times 38)}{0.25 \times 38}\right) = 1{,}507.8$$

This quantity is outside the range of bracket 1. Calculate the total cost at the break point that is, $Q = 999$.

$$\text{Average cost/unit is } \frac{\text{R}_2}{999} = \frac{39{,}360}{999} = \$39.4$$

$$\text{Total cost} = 39.4 \times 7{,}200 + \left(\frac{999}{2}\right) \times 39.4 \times 0.25 + \left(\frac{7{,}200}{999}\right) \times 200 = \$290{,}041$$

The total cost has decreased from bracket 0.

For bracket $i = 2$

$$R_2 = 3{,}9360 \quad \text{and} \quad q_2 = 1{,}000$$

$$Q_2 = \text{SQRT}\left(2 \times 7{,}200 \times \frac{(200 + 39{,}360 - 1{,}000 \times 37)}{0.25 \times 37}\right) = 1{,}996.3 = 1{,}996$$

This is a feasible quantity in bracket 2.

$$\text{Average unit price} = \frac{(R_2 + (1996 - 1000) \times 38)}{1996} = \$38.68$$

$$\text{Total cost} = 7{,}200 \times 38.68 + \left(\frac{1{,}996}{2}\right) \times 38.68 \times 0.25 + \left(\frac{7{,}200}{1{,}996}\right) \times 200$$

$$= 264{,}096 + 9{,}650.6 + 721.44 = \$274{,}468$$

This is the least total cost of all the total costs we have examined. Hence, optimum order policy is to order $Q = 1996$ units.

Note that, in case of quantity discount, no matter whether all unit or marginal unit case, in optimum policy, the annual order cost and annual carrying costs may not be equal as was the case when we did not have quantity discount in the EOQ model from Section 5.1.1.

5.3.3 One-Time Unit Price Discount

At times, the supplier offers a one-time deal. Order more units now (or in a short time span) and receive certain percentage of unit purchase price off. Consider the earlier problem as modified in the following.

The company uses 600 units per month. It incurs an order cost of $200/order, consisting of placing the order, transportation cost, and receiving the product. Holding cost, h, (carrying cost) is 25% per year per dollar. Unit cost is $40/unit.

Now, we have received a promotional advertisement. We can receive $p\%$, or in our case 5%, off if we buy more than the regular number of EOQ units. The question is, how many units should we order now?

Assuming our demand is not going to be affected by any additional purchases, we can purchase more than the EOQ quantity as long as there is net savings. If we order n multiples of EOQ at the same time to take advantage of this offer, we would save $(n - 1)$ future orders. In addition, savings in purchase cost savings would be $C \times p \times n \times \text{EOQ}$. The additional cost would be for carrying excess inventory. We need to calculate this savings as follows:

Had we ordered Q units at a time for n cycles, the inventory carrying cost would be

Average inventory × cycle time in years) × (unit cost) × (unit carrying cost/$/year) × number of cycles

$$= \left(\frac{Q}{2}\right) \times \left(\frac{Q}{D}\right) C \times h \times n \tag{5.5}$$

On other hand, inventory carrying cost for ordering nQ units with discount price, at one time, is

$$= \left(\frac{nQ}{2}\right) \times \left(\frac{nQ}{D}\right) (1 - p) \times C \times h \tag{5.6}$$

The difference between (5.6) and (5.5) is the additional carrying cost by ordering nQ units to take the advantage of the discount.

Determine the maximum value of n for which the savings are more than the cost.

In our example, we have

$$H = C \times h = 40 \times 0.25 = 10$$

The optimum EOQ is

$$Q = \text{SQRT}\left(2 \times 7200 \times \frac{200}{10}\right) = 536.6 \text{ or } 537$$

Savings by ordering $n \times Q$ units,

$$\text{Order cost saving} = (n - 1) \times 200$$

$$\text{Purchase saving} = 0.05 \times 40 \times n \times 537 = 1074\, n$$

$$\begin{aligned}\text{Additional carrying cost} &= \left[\left(\frac{n \times 537}{2}\right) \times \left(\frac{n \times 537}{7200}\right) \times (1 - 0.05) \times 40 \times 0.25 \times n\right] \\ &\quad - \left[\left(\frac{537}{2}\right) \times \left(\frac{537}{7200}\right) \times 40 \times 0.25 \times n\right] \\ &= 190.24\, n^2 - 198.69\, n\end{aligned}$$

To determine optimum 'n', check effects on savings and cost for different values of n.

The results of simulating on n are given in Table 5.1. We see that cost is more than savings at $n = 8$. Select $n = 7$ and order $Q = 537 \times 7 = 3759$ units at the given discount of 5%.

TABLE 5.1
Simulation on *n*

n	Order Cost Savings	Purchase Cost Saving	Total Savings	Additional Carrying Cost
2	200	2,148	2,348	363.56
3	400	3,222	3,622	1,118.16
4	600	4,296	4,896	2,249.08
5	800	5,370	6,170	3,762.55
6	1,000	6,444	7,444	5,656.5
7	1,200	7,518	8,718	7,930.93
8	1,400	8,592	9,992	10585.84

5.3.4 Multiple Products Order Jointly

At times, we may have one supplier shipping more than one product. The advantage of delivering multiple products at the same time is reduced total shipping cost. For example, we may have one incoming delivery truck on which all the items could be loaded. However, each product may have different demand and, therefore, it may not be economical to ship every product in each shipment.

Suppose we have n items. Let S be the fixed order cost and s_i the additional order cost of placing item i in that order. Let other costs be the same as before, except defined for each item i as follows:

- Q_i Quantity to order for item i per order
- D_i Demand per unit time, normally a year, for item i
- C_i Cost of a unit of item i
- S Order cost/order
- s_i Additional order cost associated with item i
- h_i Holding or carrying cost per unit time for item i in \$/\$ invested/year
- H_I Holding cost for item i in \$/unit/unit $= C_i \times h_i$

We define one more variable, m_i, as the yearly frequency of order for item i. The most frequently ordered item will have some of the other items included with its orders, based on the values of m_i.

The problem is to find optimum order quantity for each product, so that shipping can be coordinated, which requires determination of the optimum values of m_i.

Let us consider following data for three products.

Items	1	2	3	
Demand/year	6000	3000	1000	
Cost/unit	50	80	100	
Holding cost/unit/year, H_i	12.5	20	25	
Item order cost, s_i	50	150	200	
Order cost/order				800

Steps of the solution procedure are as follows:

1. Determine the optimum order quantity and, from that, the optimum order frequency for, each product.

   When each product i is ordered separately, the order cost is $S + s_i$, and therefore the associated EOQ is

$$Q_i = \text{SQRT}\left(\frac{2D(S + s_i)}{H}\right)$$

   In our case

$$Q_1 = \text{SQRT}\left(2 \times 6000 \times \frac{(800 + 50)}{12.5}\right) = 903.3 \; Q_2 = 553.8 \; Q_3 = 282.8$$

   And the associated order frequencies are

$$m_1 = \frac{D_1}{Q_1} = \frac{6000}{903.3} = 6.64, \quad m_2 = 5.4 \quad m_3 = 3.53$$

2. Determine the item associated with maximum m_i. This is the item that would be ordered most frequently. All other items' order frequencies will be adjusted so that their ordering time coincides with the ordering time of item i. Recalculate m_i.

   In our case, the maximum $m_i = m = \max(6.64, 5.4, 3.53) = 6.64$. So, the most frequently ordered item is item 1.
3. Calculate the relative frequency of order r_i for each product by taking the ratio of m_i/m. If the ratio is not an integer, round to an integer value. Rounding up or down is based on if the decimal portion is greater than or less than 0.5.

   In our case, $m = \max(6.64, 5.4, 3.53) = 6.64$.

$$r_1 = 1, \quad r_2 = 6.64/5.4 = 1.22 \text{ or } 1, \quad \text{and} \quad r_3 = 6.64/3.53 = 1.88 \text{ or } 2$$

   This means, for every order, items 1 and 2 are included, and for every second order, items 1, 2, and 3 are included.
4. Reevaluate order quantity and relative frequency of the most frequently ordered item with modified fixed portion of order cost, S, so that it is distributed in appropriate proportion to each item.

   The frequency calculations reveal relative frequencies when each item would be in order. In our example, items 1 and 2 are in every order while item 3 is in every other order. Reevaluate the fixed setup cost S allocation to distribute amongst items ordered in each order. In our case it is as follows:

Order	1	2	3	4
Items ordered	1, 2	1, 2, 3	1, 2	1, 2, 3
Percentage fixed cost distribution per item in the order set	50%	33%	50%	33%

Since same ordering pattern repeats itself after second order, we can denote two orders as the cycle time. Average percentage of fixed cost carried by each item is then as follows:

	Order 1 (%)	Order 2 (%)	Average or Per-Order Cost Distribution
Item 1	50	33	$(0.5 + 0.33)/2 = 0.415$
Item 2	50	33	$(0.5 + 0.33)/2 = 0.415$
Item 3	0	33	$(0.0 + 0.33)/2 = 0.165$

5. Distribute fixed cost in the proportion calculated in step 4 and reevaluate the order quantities and order frequencies.

   In our case

$$Q_1 = \text{SQRT}\left(2 \times 6000\left(\frac{0.415 \times 800 + 50}{12.5}\right)\right) = 605.5$$

$$m_1 = \frac{6000}{605.5} = 9.9$$

$$Q_2 = \text{SQRT}\left(2 \times 3000\left(\frac{0.415 \times 800 + 150}{20}\right)\right) = 380.2$$

$$m_2 = \frac{3000}{380.2} = 7.89$$

$$Q_3 = \text{SQRT}\left(2 \times 1000 \times \left(\frac{0.165 \times 800 + 200}{25}\right)\right) = 162.97$$

6. Repeat steps 3, 4, and 5 till r_i from two successive iterations are the same.

   In our case, $m = \max(9.9, 7.89, 5.95) = 9.9$.

$$r_1 = 1, \quad r_2 = 9.9/7.89 = 1.25 \text{ or } 1, \quad \text{and} \quad r_3 = 9.9/5.89 = 1.68 \text{ or } 2.$$

   Since relative frequencies in two consecutive iterations are the same, the procedure is stopped.

   Order frequencies for each product per year are: Item 1 = 9.9, Item 2 = 9.9, and Item 3 = 9.9/2.

7. Calculate the cost.

$$Q_1 = \frac{6000}{9.9} = 606.06 \quad Q_2 = \frac{3000}{9.9} = 303.03 \quad \text{and}$$

$$Q_3 = \frac{1000}{(9.9/2)} = 202.02$$

Determine cost. In our case,

$$\text{Holding cost/year} = \left(\frac{Q_1H_1 + Q_2H_2 + Q_3H_3}{2}\right)$$
$$= \left(\frac{606.06 \times 12.5 + 303.03 \times 20 + 202.02 \times 25}{2}\right)$$
$$= 9343.42$$

$$\text{Order cost/year} = m \times S + \left(\frac{m}{r_1}\right) \times s_1 + \left(\frac{m}{r_2}\right) \times s_2 + \left(\frac{m}{r_3}\right) \times s_3$$
$$= 9.9 \times 800 + \left(\frac{9.9}{1}\right) \times 50 + \left(\frac{9.9}{1}\right) \times 150 + \left(\frac{9.9}{2}\right) \times 200$$
$$= 10{,}890$$

$$\text{Total purchase cost/year} = D_1 \times C_1 + D_2 \times C_2 + D_3 \times C_3$$
$$= 6{,}000 \times 50 + 3{,}000 \times 80 + 1{,}000 \times 100$$
$$= 640{,}000$$

$$\text{Total cost of inventory} = 640{,}000 + 9{,}343.42 + 10{,}890 = \$660{,}233.42$$

At times, a better answer may be obtained by applying steps 8 and 9.

8. Take the average of order frequencies for items associated with m, in previous two cycles, and fix that as the optimum order frequency. Call it p. The reason for this is, choosing an optimum frequency of only one item and adjusting all others to suit it may not, and most often does not, form the optimum order frequency for the entire system. The average value of two trials tries to make the necessary compensation.

   In our case, m is associated with item 1. Initial frequency in first iteration for item 1 is 6.64 and in the second iteration it is 9.9. Average is $(9.9 + 6.64)/2 = 8.27$ call it p.
9. Determine the order quantity for each product which is D_i/p, and then calculate the total cost

   The most frequently used item, item 1 is ordered 8.27 times a year. Item 2 is ordered every time item 1 is ordered, and item 3 is ordered every other time. Hence,

$$Q_1 = \frac{6000}{8.27} = 725.5 \quad Q_2 = \frac{3000}{8.27} = 362.75 \quad \text{and}$$
$$Q_3 = \frac{1000}{(8.27/2)} = 120.9$$

Determine the optimum cost. In our case,

$$\text{Holding cost/year} = \left(\frac{Q_1H_1 + Q_2H_2 + Q_3H_3}{2}\right)$$
$$= \left(\frac{725.5 \times 12.5 + 362.75 \times 20 + 120.9 \times 25}{2}\right)$$
$$= 9673$$

$$\text{Order cost/year} = m \times S + \left(\frac{m}{r_1}\right) \times s_1 + \left(\frac{m}{r_2}\right) \times s_2 + \left(\frac{m}{r_3}\right) \times s_3$$
$$= 8.27 \times 800 + \left(\frac{8.27}{1}\right) \times 50 + \left(\frac{8.27}{1}\right) \times 150$$
$$+ \left(\frac{8.27}{2}\right) \times 200$$
$$= \$9097$$

$$\text{Total purchase cost/year} = D_1 \times C_1 + D_2 \times C_2 + D_3 \times C_3$$
$$= 6{,}000 \times 50 \times 3{,}000 \times 80 + 1{,}000 \times 100$$
$$= 640{,}000$$

$$\text{Total cost of inventory} = 640{,}000 + 9{,}673 + 9{,}097 = \$658{,}770$$

In our case, the application of steps 8 and 9 results in a better solution.

5.4 SINGLE PERIOD PLANNING

Inventory management of seasonal goods presents a special challenge. The items must be stocked before the season begins, and if not sold during the season, the same item cannot be sold for full price in off-season. Christmas trees are good examples. The trees must be stocked before the season starts, and once the Christmas is over, they have no value. Breads and other perishable items in a grocery store present similar problem. Changing fashion presents adequate stocking of fashionable clothings a similar challenge. A newsboy must purchase newspapers early in the morning and must sell them during the day. At the end of day, the remaining papers have only salvage value.

The problem is typically referred to in literature as a "newsboy problem" or "Christmas tree problem."

5.4.1 Discrete Order Quantity

When demand is small and can be approximated by discrete quantities, and the order quantity is also same as one of these values, the following procedure, illustrated by an example, can be applied to evaluate the optimum order quantity.

TABLE 5.2
Profit from Purchasing Q Units with Demand of D Units

	D (Demand)				
***Q* (Purchase)**	**10**	**20**	**30**	**40**	**50**
10	200	200	200	200	200
20	100	400	400	400	400
30	0	300	600	600	600
40	−100	200	500	800	800
50	−200	100	400	700	1000

Consider the following example. A clothing store must decide how many new shirts to stock at the beginning of fall season. Demand is probabilistic and, based on the past experience, is estimated to follow the distribution given in the following table. Other pertinent data is also given. Determine the optimum order quantity, if order can only be placed at the beginning of the fall season.

Demand	10	20	30	40	50
Probability $f(x)$	0.1	0.2	0.4	0.2	0.1

Purchase price, c, is \$15/unit. The sale price, p, is \$35/unit. If a shirt is not sold in the season, it can be sold at the end of the season for a salvage value, s, of \$5/shirt. Determine how many units we should order.

Unlike in the previous models, here the order cost does not play any part in the determination of optimum Q. This is because we must place the order at the beginning, and it does not matter what the order cost is, only one order is placed. The carrying cost is also ignored since inventory holding time is very short and is limited to only one cycle or period.

Let us develop a table that displays profit if Q units are purchased and D is the demand for the units.

$$\text{Profit} = D \times p - Q \times c + (Q - D) \times s \quad \text{for } Q > D$$
$$= (p - c)\, Q \quad \text{for } Q \le D$$

Calculate the expected profit for each possible value of Q based on demand distribution. Assuming we can purchase in multiples of 10 units, the profit is shown in Table 5.2.

Since demand is not known, but has probability distribution, the expected profit over the entire range of demand is found by calculating the expected value (Table 5.3).

We know the expected value, $E(x) = \sum xf(x)$.

TABLE 5.3
Expected Profit

Q	10	20	30	40	50
Expected profit	200	370	480	470	400

For example, in our case,

$$\begin{aligned}\text{Expected value of profit for } Q = 40 \text{ is, } & -100 \times 0.1 + 200 \times 0.2 \\ & + 500 \times 0.4 + 800 \times 0.2 + 800 \times 0.1 \\ & = 470\end{aligned}$$

Maximum expected profit is associated with $Q = 30$, and therefore we should place the order for 30 units.

5.4.2 Single Period Ordering with Continuous Demand

With large demands and possible order quantity values, it is more convenient to consider demand as a continuous variable. In general, such demand has normal distribution with a mean of μ and a standard deviation of σ.

Let us define the following terms:

Cost of understocking $C_u = p - c$
Cost of overstocking $C_o = c - s$
$Q^* =$ optimum order size
Optimum service level (OSL) = probability that demand $\leq O^*$
The probability that demand is $\leq Q^*$ is OSL, and the probability that it is higher than Q^* is (1 – OSL).

If we purchase one unit more than optimum Q^*, the probability that we will sell that unit is (1 – OSL), and we will have a profit of C_u, or the expected profit is $(1 - \text{OSL}) \times C_u$.

If the demand does not exceed Q^*, the probability of that is OSL, then by storing one additional unit will incur a loss due to overstocking. The expected cost of the loss is $\text{OSL} \times C_o$.

The optimum order quantity is such that any unit stocked more than optimum quantity will incur a net loss, or

$$\text{Expected profit} - \text{Expected loss} \leq 0$$

The minimum possible loss by adding one more unit than optimum Q, that is, Q^*, is zero. From the preceding expression, the condition is

$$\text{Expected profit} = \text{expected loss}$$

or

$$(1 - \text{OSL}) \times C_u = \text{OSL} \times C_o$$

or

$$\text{Optimum OSL} = \frac{C_u}{(C_u + C_o)}$$

To illustrate, consider the clothing store example again, with few modifications. We now have a large demand for the shirts, which can be approximated by a normal distribution, with a mean of 300 and a standard deviation of 66.6. The purchase price, c, is \$20/unit. The sale price, p, is \$55/unit. If a shirt is not sold in the season, it can be sold at the end of the season for a salvage value, s, of \$15/shirt. Determine the optimum order quantity. Here,

$$C_u = p - c = 55 - 20 = \$35$$

$$C_o = c - s = 20 - 15 = \$5$$

$$\text{OSL} = \frac{35}{35 + 5} = 0.875$$

In standard normal distribution, determine the value of z below which the area is 0.875. From a standard normal table, $z^* = 1.15$.

$$z^* = \frac{Q^* - \mu}{\sigma}$$

$$1.15 = \frac{Q^* - 300}{66.6}$$

$$Q^* = 300 + 66.6 \times 1.15 = 376.59$$

Note that the optimum quantity is greater than the mean value of the demand. This is because cost of understocking is more than the cost of overstocking, which suggests we should meet most of the demand rather than just the mean demand.

When demand is normally distributed, the expected profit of stocking Q^* is given by

$$[(p - s) \times \mu \times \text{OSL}] - [(p - s) \times \sigma \times vz^*] - [Q^* \times (c - s) \times \text{OSL}] + [Q^* \times (p - c) \times (1 - \text{OSL})]$$

Where $v(z^*)$ is Probability mass function (value of z function at z^*). This value can be obtained by evaluating the following expression.

$$v(z) = f(z^*) = \frac{1}{\sqrt{2\pi}} e^{-(z/2)*2}$$

Or evaluating NORMDIST (z^*, 0, 1, 0) from Excel.
In our case,

$$(p - s) = 55 - 15 = 40 \quad (c - s) = 20 - 15 = 5 \quad (p - c) = (55 - 20) = 35$$
$$\text{OSL} = 0.875 \quad z^* = 1.15 \quad \mu = 300 \quad \sigma = 66.6 \quad Q^* = 376.59$$

OSL = 0.875 and $v(z^*)$ = NORMDIST (1.15, 0,1, 0) = 0.2059 (from Excel).

$$\begin{aligned}\text{Profit} &= (40 \times 300 \times 0.875) - (40 \times 66.6 \times 0.2059) - (376.59 \times 20 \times .875) \\ &\quad + (376.59 \times 35 \times (1 - 0.875) \\ &= \$5008.76\end{aligned}$$

5.4.3 Buyback Policy

One interesting observation in formulas developed before, namely,

$$C_\mathrm{u} = p - c \quad C_\mathrm{o} = c - s$$

and

$$\text{OSL} = \frac{C_\mathrm{u}}{C_\mathrm{u} + C_\mathrm{o}}$$

If the supplier follows a buyback policy, that is, buying back any units not sold at the end of the season for full purchase price, then $c = s$. In this case, $C_\mathrm{o} = 0$ and OSL = 1, indicating an optimum service level of 100% and the associated $z = \infty$. The corresponding optimum Q is also ∞, indicating that we should stock as many units as physically possible.

If the buyback policy is $s = f \times c$, where f is a certain percentage value, optimum Q can be easily determined for different possible percentage values by applying the previously developed procedure.

5.4.4 Effect of Reduction in σ

Standard deviation represents uncertainties in demand. If we knew the demand with complete certainty, σ should have been zero. It is possible to reduce σ by various means such as collecting more information on expected demand or placing an order closer to the start of the season, so that the demand distribution can be more accurately estimated or by accumulating multiple similar products as a group for placing an order and then customizing each item based on its demand. For example, in a clothing store that sells customizes logo blazers, we might order all the basic blazers together and then place different logos when demand arises rather than ordering each logo blazer separately.

Some such alternative might increase the purchase price of a unit but may still improve the overall profit. For in the previous example, supposing ordering closer to the start of fall season reduces σ from 66.6 to 33 but suppose purchase price c also increases from 20 to 22.

The new values are

$$C_u = p - c = 55 - 22 = \$33$$
$$C_o = c - s = 22 - 15 = \$7$$

Or OSL $= C_u/(C_u + C_o) = 33/(33 + 7) = 0.825$; the associated z^* is 0.935.

$$z^* = \frac{Q^* - \mu}{\sigma}$$
$$0.935 = \frac{Q^* - 300}{33}$$

therefore,

$$Q^* = 330.855$$
$$v(z^*) = \text{NORMDIST}\ (0.825, 0, 1, 0) = 0.2576$$

$$\begin{aligned}
\text{Profit} &= [(p - s) \times \mu \times \text{OSL}] - [(p - s) \times \sigma \times v(z^*)] \\
&\quad - [Q^* \times (c - s) \times \text{OSL}] + [Q^* \times (p - c) \times (1 - \text{OSL})] \\
&= [(55 - 15) \times 300 \times 0.825] - [(55 - 15) \times 33 \times 0.2576] \\
&\quad - [330.855 \times (22 - 15) \times 0.825] + [330.855 \times (55 - 22) \times 0.175] \\
&= \$9560
\end{aligned}$$

Hence, reducing uncertainty in demand distribution does increase profit. The profit might improve even if the unit price goes up, if σ can be reduced sufficiently.

Now, consider anther problem where we sell three different embalmed novelty souvenirs for a football game. It is estimated that each item has a demand that is normally distributed with a mean of 300 and a standard deviation of 60. Units can be bought with three distinctive emblems already affixed. Or, alternatively, all souvenirs can be bought as generic souvenirs, and emblems are attached when demand arises. The second alternative does create additional cost. All impertinent information is as follows:

$P = \$35$, $c = \$10$, $s = \$0$, Additional cost for second alternative \$1.

1. For first alternative: $C_u = p - c = 35 - 10 = 25$. $C_o = c - s = 10 - 0 = 10$.

$$\text{OSL} = \frac{25}{25 + 10} = 0.714$$

Associated z* = 0.568

$$z^* = \frac{Q^* - \mu}{\sigma}$$

$$0.568 = \frac{Q^* - 300}{60}$$

$$Q^* = 300 + 60 \times 0.568 = 334.08$$

$$\text{Profit} = [(p - s) \times \mu \times \text{OSL}] - [(p - s) \times \sigma \times v(z^*)]$$
$$- [Q^* \times (c - s) \times \text{OSL}] + [Q^* \times (p - c) \times (1 - \text{OSL})]$$

$$\text{Profit} = [(35 - 0) \times 300 \times 0.714] - [(35 - 0) \times 60 \times 0.339)]$$
$$- [334.08 \times (10 - 0) \times 0.714] + [334.08 \times (35 - 0) \times (1 - 0.714)]$$
$$= \$7743.91 \text{ for each product.}$$

Total profit for all three products = 3 × 7,743.91 = $23,231.73

2. For the second alternative: $C_u = p - (c + 1) = 35 - 11 = 24$. $C_o = (c + 1) - s = 11 - 0 = 11$

$$\text{OSL} = \frac{24}{24 + 11} = 0.685$$

The associated $z^* = 0.482$.

$$z^* = \frac{Q^* - \mu}{\sigma}$$

New $\mu = 3 \times 300 = 900$, and new $\sigma = \sqrt{3} \times 60 = 103.92$.

$$0.482 = \frac{Q^* - 900}{103.92}$$

$$Q^* = 900 + 103.92 \times 0.482 = 950.08$$

$$\text{Profit} = [(35 - 0) \times 900 \times 0.685] - [(35 - 0) \times 103.92 \times 0.355]$$
$$- [950.08 \times (11 - 0) \times 0.685] + [950.08 \times (35 - 11) \times (1 - 0.685)]$$
$$= \$20,310.04$$

In this example, ordering independent products preamble is a better alternative than combining the products, even though σ is affected. The additional cost involved in embalming does not make the proposition economical.

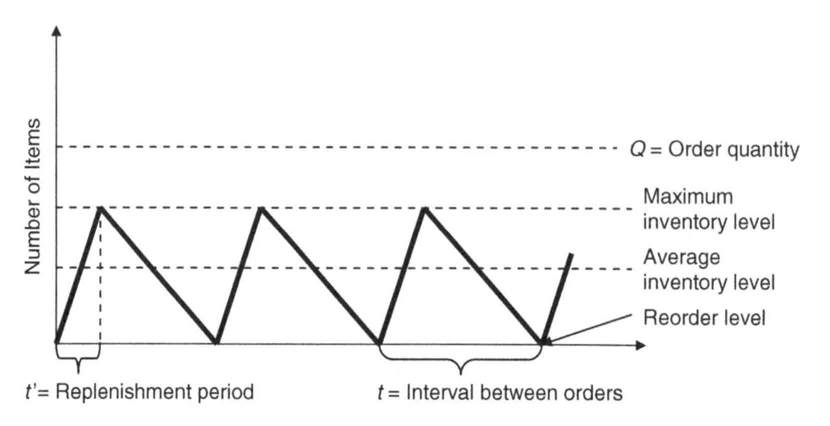

FIGURE 5.7 Consumption during production.

5.5 CONSUMPTION DURING PRODUCTION

When a product is produced in our plant while it is being continuously shipped to customers or continuously consumed in production of some other products in the plant, then it is continuously depleted at the rate of its demand even while it is being produced. If the production rate is P units/year and demand rate is D units/year, then inventory is accumulated at a rate of $(P - D)$ while units are produced. For inventory to build, production rate has to be greater than the demand rate. The time required to produce Q units at a rate of P units is Q/P. Maximum inventory built during the production time, I_{max}, is the net production rate multiplied by the time of production, that is,

$$I_{max} = (P - D) \times \frac{Q}{P}$$

$$\text{Average cycle inventory} = \frac{\text{Maximum cycle inventory}}{2} = \frac{I_{max}}{2}$$

$$\text{Total annual cost} = D \times c + \left(\frac{I_{max}}{2}\right) \times H + \left(\frac{D}{Q}\right) \times S$$

Taking derivative with respect to Q and equating it to zero gives

$$Q^* = \left[\text{SQRT}\left(2 \times D \times \frac{S}{H}\right)\right] \times \left[\text{SQRT}\left(\frac{P}{P - D}\right)\right]$$

This quantity is called economic production lot size or simply economic lot size (ELS) (Figure 5.7).

5.6 JIT INVENTORY SYSTEM

Just-in-Time (JIT) is a philosophy in which raw material or products are scheduled to arrive at the production facility just in time for processing. It eliminates keeping a large inventory at the production facility, and at the same time allows a small variation in the daily production rate. JIT in its expanded format also optimizes manufacturing processes by eliminating all waste, including wasted steps, wasted material, and excess inventory. A lean manufacturing system depends on JIT inventory systems. A large part of the JIT system depends on logistics, which include transportation, warehousing, and several strategies for handling the potential supply chain uncertainties. JIT is easy to grasp conceptually; everything happens "just-in-time." Conceptually, there is no problem about this; however, achieving it in practice is likely to be difficult!

A JIT system minimizes investment in inventory. Since materials arrive at the time they are needed, the company minimizes inventory investment by having only work-in-process (WIP) inventory. This eliminates the need for safety stocks, and reduces inventory on hand. However, there must be good coordination between the company, the supplier, and the shipping company to meet schedules for the production line. It has been found that a well-balanced JIT system results in increased quality, a decrease in product costs, reduction in investments, reduction in requirements, and minimum total manufacturing, distribution, and inventory costs.

However, a JIT system does have a few drawbacks. JIT is not possible without short distances or reliable and prompt transportation system between the customer and supplier. JIT requires good, consistent quality, so that throughput is unaffected.

5.6.1 Distribution Strategy

Since distribution plays so important apart in JIT, let us look at some of the typical delivery/distribution systems used in a typical supply chain. They can be broadly classified into three types:

1. *Direct*: Trucks travel directly from supplier to a plant. Often a large one type of product shipment is made in this manner.
2. *Milk-run (peddling)*: Trucks pick up products at one or more suppliers and deliver them to one or several plants. Different products are collected, perhaps one type from each supplier and bought to the customer.
3. *Cross-dock*: This system is used when we have multiple suppliers and customers. Products are delivered from suppliers to a cross dock facility, and then are shipped from the cross dock to each customer. Different products are collected at the cross dock facility, and only the ones that are required by a given customer are shipped to that customer.

The three distributions strategies are illustrated in the following diagram (Figure 5.8).

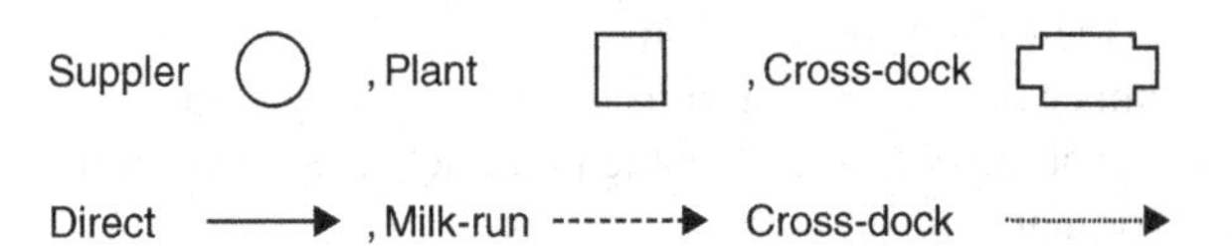

The different distribution strategies have dissimilar transportation costs and times. For example, direct delivery has the shortest transportation distance and, therefore,

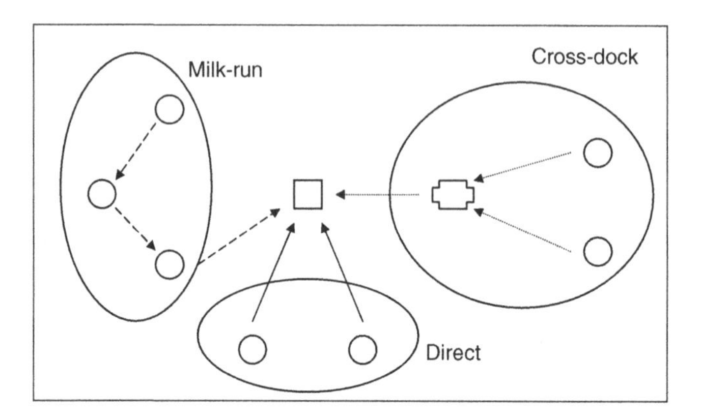

FIGURE 5.8 Types of distribution strategies.

the lowest transportation cost if full load (full truck) can be delivered every time. The delivery times can also be very short, and frequent deliveries can be made. A delivery through a cross-dock is used when the distances are long and therefore the direct transportation cost is high and the quantity for direct shipment may not fill the entire truck. Delivery times might be long compare to direct run. Milk run falls somewhere in between.

There are various rules for collection and distribution that can be developed to minimize procurement and inventory cost in a JIT system. We illustrate basic concepts as an illustration, by applying two different rules in a milk-run procurement strategy. One can develop many additional logical rules based on the particular situation on hand. In our example, we have multiple sources (suppliers), each shipping a single product to a single destination (customer), with the following facts.

1. Daily demands for all products are known, and can vary from day to day.
2. Batch size of each product may vary from day to day based on the policy followed. Here, the "batch size" is the number of units picked up to fulfill the "daily demand" of a particular product. Only units associated with full day's demand are collected in a pick-up trip since that is the only way we can avoid a trip to the associated supplier.
3. There is one truck with known capacity available for daily transportation. The truck capacity is large enough to accommodate one milk run through all suppliers.
4. There is sufficient inventory at each supplier to ship the required number of units as a pick-up policy may demand.
5. Cost of procurement for each product is known and constant and is incured each time the truck visits a supplier.
6. If the number of units required to fulfill the "daily demand" of a product is available at the customer site, then no truck needs to visit the associated suppler for that day.

The objective is to develop a pick-up policy so as to minimize the total cost of pick up and inventory for a finite time period.

The procedure is to perform daily evaluation of how to use excess capacity, if one is available. Starting with the first day, apply the following steps to each day of planning

1. Calculate the total required pick up for the day. Subtract it from the capacity of the truck to get available excess capacity (AEC).
2. To utilize excess capacity, we must determine which product quantities (if any) that are not immediately needed should be picked up. Additional pick up is only possible from a supplier whom we are presently visiting. Picking up extra units may avoid a trip to that supplier but will create an additional inventory and associated cost.

For each supplier that is visited, calculate the possible savings as equal to the cost of procurement minus cost of inventory. It is assumed that if additional units are picked from a supplier, only full demand quantities for a day would be picked. This policy assures that no visit for that day would be required and thus incurring savings in procurement cost. Since the maximum additional quantity that can be picked is equal to the excess capacity for that day, the analysis is only performed only for future demands that are less than or equal to the available excess capacity.

Select the possible pick up, the product with maximum net savings. If there is a tie between products for selecting maximum net savings, then initially, select the one with minimum quantity. This allows maximum possible excess capacity still made available for additional pick ups.

Subtract this quantity from the available excess or AEC, to determine the new value for the excess capacity. Continue selection with next maximum savings, as long as it is in continuous fashion, that is, either from the next cell in the row where the previous pick up was made or the same day demands where present visit is being made. If, for example, next cell's demand exceeds the excess capacity, the associated pick up cannot be made. Do not go to following cells (days) for the same product, even though the savings in those cells are higher than any other product that can be picked up presently. This is because we will have the opportunity to obtain the additional savings due to decrease in carrying costs, as we visit the prior infeasible cell in the following trips.

The objective is to fill the truck as much as possible. If under tie rule we had selected a minimum quantity and even with additional pick ups, there still some capacity left in the truck that cannot be filled, try selecting the larger demand product from the tied pair at this time. This is because by picking up a larger quantity now, we may have released some capacity for the following trips (example problem illustrates these concepts).

3. Repeat step 2 until AEC becomes zero or there are no possible pick-up policies for the excess capacity.

Backward pass: Check if any site visit can be avoided by performing the analysis from the last day of planning and moving towards day 1. The trip can be avoided to a site if the quantity required for a day can be picked up earlier, even in broken down amounts. This is possible if the associated carrying cost is less than the site visit cost.

TABLE 5.4
Data for Milk Run Example

		Master Schedule						
Pick-Up Cost	**Product/Day**	**1**	**2**	**3**	**4**	**5**	**6**	**7**
20	1	3	5	7	3	5	7	5
18	2	13	11	9	11	13	9	8
16	3	7	9	7	10	7	10	9
14	4	5	7	9	5	7	9	5
12	5	12	10	8	12	8	10	8
Total		40	42	40	41	40	45	38

Example
There are five products, each from different suppliers, that are picked up every day for use in our plant. Expected daily requirements for next 7 days are displayed in Table 5.4. Every day, a truck with capacity of 50 units can be used to pick up these products in a milk run. Procurement or pick-up cost inures every time the truck visits a supplier to pick up a product. This is in addition to travel cost, which does not change since the truck makes the round trip every time. Inventory carrying cost is \$1/unit/day for all products.

5.6.1.1 Day 1 Evaluation

On day 1, the total pick up is for 40 units, giving AEC $= 50 - 40 = 10$ units. There are a number of products on day 2, for which demand is less than 10. There are also demands on day 3, 4 . . . that are less than 10. Savings are calculated for each feasible demand as follows:

$$
\begin{aligned}
\text{Savings} &= \text{pick-up cost for the product} - \text{inventory carrying cost} \\
&= \text{pick-up cost for the product} - (\text{number of units picked up} \\
&\quad \times \text{days picked up earlier} \times \text{inventory carrying cost/day})
\end{aligned}
$$

For example, savings for product 1 on day 2, savings $= 20 - 5 \times 1 \times 1 = 15$. For the same product on day 4, savings is $20 - 3 \times 3 \times 1 = 11$. The savings are noted. When savings are zero or negative, they are marked by "–" in associated cells.

The maximum savings is 15 with product 1 on day 2, so pick up demand of 5 units, leaving $10 - 5 = 5$ units of excess capacity. Next maximum savings is for day 4 of product 1, but since demand for product 1 has not been picked yet (pick ups in a row are not continuous), examine other products first (go in the column for day 1).

Products 2, 3, and 4 all have savings of 7 for day 2, but associated demands of 11, 9, and 7 respectively are greater than presently available excess capacity of 5, and

TABLE 5.5
Day 1 Evaluation

Day 1: Requirements and Calculations

Product	Pick-Up Cost	Days	1	2	3	4	5	6	7
1	20	Daily demand	3	5	7	3	5	7	5
		Savings		15	6	11	–	–	–
		Additional pick up		**5**	x	**3**			
2	18	Daily demand	13	11	9	11	13	9	8
		Savings		7	–	–	–	–	–
		Additional pick up		x					
3	16	Daily demand	7	9	7	10	7	10	9
		Savings		7	2				
		Additional pick up		x					
4	14	Daily demand	5	7	9	5	7	9	5
		Savings		7					
		Additional pick up		x					
5	12	Daily demand	12	10	8	12	8	10	8
		Savings		2					
		Additional pick up		x					
Total pick up			40	45		**48**			

x: cell examined but units cannot be picked
Bold entries: units picked

hence, none can be picked up. Next maximum savings is 6 associated with 1 day 2, but again the available excess capacity (AEC) does not allow us a pick up. Product 5 for day 2 cannot be picked up because of AEC. Since there is no immediate feasible demand cell (demand cell that can be reached in a row without a non-visit cell before it) available with savings, now go to product 1, day 4 with savings of 11 and pick up 3 units, which is feasible (Table 5.5).

No other pick up is possible, and our total load for the truck is 48 units.

$$\text{Total savings for day1} = 15 + 11 = 26$$

5.6.1.2 Day 2 Evaluation

Table 5.6 shows day 2 calculations. Total required pick up for day 2 is for products 2, 3, 4, and 5, resulting in 37 units. Since demand requirement for product 1 in day 2 was picked up in day 1 trip, there is no visit to product 1 site in day 2. AEC $= 50 - 37 = 13$. Since no site visit is made for product 1, no additional units can be picked up for that product and therefore no savings calculations are made. Other saving calculations are shown in the Table 5.6.

$$\text{Total savings for day 2} = 9 + 4 = 13$$

TABLE 5.6
Day 2 Evaluation

Day 2: Requirements and Calculations

Product	Pick-Up Cost	Days	1	2	3	4	5	6
1	20	Daily demand		7		5	7	5
		Savings		No visit				
		Additional pick up						
2	18	Daily demand	11	9	11	13	9	8
		Savings		9	–	–	–	–
		Additional pick up						
3	16	Daily demand	9	7	10	7	10	9
		Savings		9	–			
		Additional pick up		**7**				
4	14	Daily demand	7	9	5	7	9	5
		Savings		5	4			
		Additional pick up			**5**			
5	12	Daily demand	10	8	12	8	10	8
		Savings		4		–		
		Additional pick up						
Total pick up			37	44	**49**			

Maximum savings of 9 is associated with products 2 and 3, Demands are 9 and 7, respectively. Initially, choose product 3 with smaller demand to see if any additional pick ups can be made. AEC is $13 - 7 = 6$. We can pick up product 4 day 4 demand of 5 units. So, pick up those units. The truck is filled to capacity 49.

5.6.1.3 Day 3 Evaluation

Table 5.7 shows the remaining pick up requirements. It also shows the days and products cells where the quantities that are already picked up and therefore no visits are to be made. Savings are shown in Table 5.7. Again no savings are calculated for the product 3, since on day 3 no visit is made to pick up product 3.

$$\text{Total savings for day } 3 = 10 + 7 = 17$$

Based on savings 5 units for product 1 in day 5 are picked up first and then 11 units for product 2 for day 4 are selected next.

5.6.1.4 Day 4 Evaluation

Again, savings are only calculated for products where visit is made for day 4, namely products 3 and 5 (Table 5.8).

TABLE 5.7
Day 3 Evaluation

Day 3: Requirements and Calculations

Product	Pick-Up Cost	Days	1	2	3	4	5
1	20	Daily demand	7	No visit	5	7	5
		Savings			10	–	
		Additional pick up			**5**		
2	18	Daily demand	9	11	13	9	8
		Savings		7	–	–	
		Additional pick up		**11**			
3	16	Daily demand	No visit	10	7	10	9
		Savings					
		Additional pick up					
4	14	Daily demand	9	No visit	7	9	5
		Savings			2	–	
		Additional pick up					
5	12	Daily demand	8	12	8	10	8
		Savings		–			
		Additional pick up					
Total pick up			33	**49**	38		

TABLE 5.8
Day 4 Evaluation

Day 4: Requirements and Calculations

Product	Pick-Up Cost	Days	1	2	3	4
1	20	Daily demand	No visit	No visit	7	5
		Savings				
		Additional pick up				
2	18	Daily demand	No visit	13	9	8
		Savings				
		Additional pick up				
3	16	Daily demand	10	7	10	9
		Savings		9	–	–
		Additional pick up		**7**		
4	14		No visit	7	9	5
		Savings				
		Additional pick up				
5	12	Daily demand	12	8	10	8
		Savings		4	–	–
		Additional pick up		**8**		
Total pick up			22	**37**		

TABLE 5.9
Day 5 Evaluation

Day 5: Requirements and Calculations

Product	Pick-Up Cost	Days	5	6	7
1	20	Daily demand	No visit	7	5
		Savings			
		Additional pick up			
2	18	Daily demand	13	9	8
		Savings		9	0
		Additional pick up		**9**	
3	16	Daily demand	No visit	10	9
		Savings			
		Additional pick up			
4	14	Daily demand	7	9	5
		Savings		5	4
		Additional pick up		**9**	**5**
5	12	Daily demand	No visit	10	8
		Savings			
		Additional pick up			
Total pick up			20	38	**43**

Only requirements for day 5 for products 3 and 5 can be selected at this time.

$$\text{Total savings for day } 4 = 9 + 4 = 13$$

5.6.1.5 Day 5 Evaluation

The calculations are shown in Table 5.9. In addition to remaining day 5 pick up, additional pick ups are product 2 day 6 and product 4 days 6 and 7.

$$\text{Total savings for day } 5 = 9 + 5 + 4 = 18$$

5.6.1.6 Day 6 Evaluation

The details are in Table 5.10. Visits to products 1, 3, and 5 are made, and a total of 49 units are picked up.

$$\text{Savings for day } 6 = 15 + 7 + 4 = 26$$

5.6.1.7 Day 7 Evaluation

Table 5.11 shows the details. Only 8 units for product 2 need to be picked.

$$\text{Savings for day } 7 = 0.$$

TABLE 5.10
Day 6 Evaluation

Day 6: Requirements and Calculations				
Product	**Pick-Up Cost**	**Days**	**6**	**7**
1	20	Daily demand	7	5
		Savings		15
		Additional pick up		**5**
2	18	Daily demand	No visit	8
		Savings		
		Additional pick up		
3	16	Daily demand	10	9
		Savings		7
		Additional pick up		**9**
4	14	Daily demand	No visit	No visit
		Savings		
		Additional pick up		
5	12	Daily demand	10	8
		Savings		4
		Additional pick up		8
Total pick up			27	**49**

TABLE 5.11
Day 7 Evaluation

Day 7: Requirements and Calculations			
Product	**Pick-Up Cost**	**Days**	**7**
1	20	Daily demand	No visit
		Savings	
		Additional pick up	
2	18	Daily demand	8
		Savings	
		Additional pick up	
3	16	Daily demand	No visit
		Savings	
		Additional pick up	
4	14	Daily demand	No visit
		Savings	
		Additional pick up	
5	12	Daily demand	No visit
		Savings	
		Additional pick up	
Total pick up			**8**

$$\begin{aligned}\text{Overall cost of operation} &= \left[\left(\frac{\text{Cost}}{\text{Day to visit all product sites}}\right)\right.\\ &\qquad \left.\times \text{ Number of days in planning}\right]\\ &\quad - [\text{Sum of daily savings in planning}]\\ &= [(20 + 18 + 16 + 14 + 12) \times 7]\\ &\quad - [26 + 13 + 17 + 13 + 18 + 26]\\ &= 560 - 113\\ &= \$447\end{aligned}$$

5.6.2 Backward Check

As a final step, a backward check is made. There are only 8 units to be picked up on day 7 from site 2. If we can be pick up these units when the truck is visiting that site in previous planning, that is, in days 1, 2, 3, and 5, we may be able to save the site procurement cost of $18. Presently used and available truck capacity in these days is

Days	1	2	3	5
Truck capacity used	48	49	49	43
Capacity available	2	1	1	7

Since the total available is 11 units, it is possible to accommodate 8 units from site 2 and avoid a trip on day 7. To reduce additional carrying cost that this policy will incur, we should collect the excess products as close to day 7 as possible. So, if we pick up 7 units on day 5 and 1 unit on day 3, the additional carrying cost will be units $\times$ days picked up earlier $\times$ carrying cost $= 7 \times (7-5) \times 1 + 1 \times (7-3) \times 1 = \$18$. There is savings of $18 - 18 = 0$, or there is no net savings by following this policy; however, it may be good to give the driver off on the seventh day since it is not costing us any additional dollars.

Total cost of operation is 447.

No other backward pass is possible due to truck capacity limitation. The present solution is the optimum solution.

One variation of the problem is to include cost of travel. Suppose travel cost for the truck is $50.00 per round trip, on day 7, the savings are not only for not visiting site 2 on that day, but also due to not making the entire trip together. Thus, the savings are $50 + 18 = 68$. This value should be compared with additional carrying cost incures, 18 in our case. The final pick-up table is displayed in Table 5.12 with the cost of $447 without considering travel cost and $\$447 + 6 \times 50 = \$747$ by considering travel cost.

TABLE 5.12
Final daily Pick Up Schedule

	Days						
Product	**1**	**2**	**3**	**4**	**5**	**6**	**7**
1	3 + 5 + 3 = 11	–	7 + 5 = 12	–	–	7 + 5 = 12	–
2	13	11 = 11	9 + 11 + 1 = 21	–	13 + 9 + 7 = 29	–	–
3	7	9 + 7 = 16	–	10 + 7 = 17	–	10 + 9 = 19	–
4	5	7 + 5 = 12	9	–	7 + 9 + 5 = 21	–	–
5	12	10	8	12 + 8 = 20	–	10 + 8 = 18	–
Truck capacity utilized	48	49	50	37	50	49	–

5.7 RECOURSE CENTER CAPACITY PLANNING

All production plants are comprised of resource centers. They might include work centers with multiple machines and workers and/or with single production facility or a worker. Resource centers is where detailed everyday production takes place. They must know what items to produce and how many units of each item to produce on a daily basis. If capacity is available, such a schedule can be met; if not, we must do alternate planning. This generally involves scheduling of resources.

We discuss here some of the topics in resource center planning.

5.7.1 Rough Cut Planning

This is a quick and dirty way of calculating roughly how much capacity of each center has been promised based on the present schedule, whether such a plan is feasible, and if additional products can be added in these periods. The estimates are based on information that is readily available on product demand, labor and/or machine standards, and historic utilization of each work center.

For example, suppose the MRP schedule for products A and B for the next 5 weeks is as shown in Table 5.13.

The time standard for each product is: product A 0.4 hr/unit and for product B 0.3 hr/unit. Products A and B use work centers W10, W11, and W12. Historically, utilization of each work center is 50%, 40%, and 10% to produce these products. That is, of the total time required for products roughly the time distribution for each center is 50% in W10, 40% in W11, and 10% in W12.

The total time required for each week for each product is obtained by multiplying the required quantity and its time standard. For example, for product A in week 1, the time required is $50 \times 0.4 = 20$ hr (Table 5.14).

With the given utilization of each work center, the total time distributed in proportion by which products require work centers are given below. For example, for week 1, capacity from WC 10 $= 29 \times 0.5 = 14.5$, for WC 11 $= 29 \times 0.4 = 11.6$, and

TABLE 5.13
Demand for Products for Each Week

Week	1	2	3	4	5
Product A	50	0	50	0	50
Product B	30	30	30	30	30

TABLE 5.14
Weekly Time Requirements

Week	1	2	3	4	5
Product A	20	0	20	0	20
Product B	9	9	9	9	9
Total	29	9	29	9	29

TABLE 5.15
Time Distribution to Each Work Center

Week	1	2	3	4	5
WC 10	14.5	4.5	14.5	4.5	14.5
WC 11	11.6	3.6	11.6	3.6	11.6
WC 12	2.9	0.9	2.9	0.9	2.9

A		B	
X(2)	Y(1)	X(1)	Y(1)

FIGURE 5.9 Bill of material for products A and B.

for WC 12 = $29 \times 0.1 = 2.9$ (Table 5.15). We can also draw bar charts to visualize this information.

Many times, in a stable production, this information is sufficient to make any adjustments, such as rescheduling demands and increasing capacities.

5.7.2 Capacity Bills

More detailed and therefore a more accurate capacity determination technique depends on additional information such as what is available in bill of materials and routing information.

Bill of materials give information structure and subcomponents of the product. For example, suppose bill of materials for products A and B are shown in Figure 5.9.

TABLE 5.16
Routing Information Part X

Operation		Machine	Aux Equipment	Setup Time	Hr/pc	Production/hr
1 of 2	Molding	10-Injection		0.05	0.02	50
2 of 2	Cutting	20-Saw	Auto Jaw	0.1	0.10	10

TABLE 5.17
Expanded Routing Sheet Information

Operation Sequence		Lot Size	Machine	Aux Equipment	Set Up Time	Prod Run Time per Piece	Total Hours per Unit	Time per Lot in Hours
1 of 2	Molding	50	10-Injection		0.05	0.02	0.021	1.05
2 of 2	Cutting	100	20-Saw	Auto Saw	0.1	0.10	0.11	10.1

TABLE 5.18
Routing Sheet

Product/ Part	Operation	Lot Size	Machine/ Work center	Setup Time in Hours	Production Run Time/Piece	Total Hr/Unit	Time for Lot
A	1 of 1	30	W10	0.6	0.18	0.2	6
B	1 of 1	50	W10	2.5	0.1	0.15	7.5
X	1 of 2	100	W11	1	0.02	0.03	5
X	2 of 2	100	W12	1	0.01	0.02	2
Y	1 of 1	100	W11	2	0.08	0.1	10

It shows that product A is made with two subcomponents, X and Y, It needs 2 units of X and 1 unit of Y. Similarly, B is made of 1 unit of X and 1 unit of Y.

Routing sheet gives even further detailed information on the production of an item. There are some variations on how exhaustive information it can have, but two versions are shown in Tables 5.16 and 5.17.

If the number of units to produce is large and the setup time is negligible compare to total production time, we might have routing sheet similar to the following one (Table 5.16).

When lot size is predetermined, we might be able to distribute set time to each piece produced. For example, we might transfer the above routing information sheet to Table 5.17.

In our example, suppose the routing information is condensed in Table 5.18.

From here, we can determine the time taken to produce each product A and B in each work center.

TABLE 5.19
Work Center Utilization

Week	1	2	3	4	5
WC 10	$50 \times 0.2 + 30 \times 0.15 = 14.5$	4.5	14.5	4.5	14.5
WC 11	$50 \times 2 \times 0.03 + 50 \times 0.1 + 30 \times 0.03 + 30 \times 0.1 = 11.9$	3.9	11.9	3.9	11.9
WC 12	$50 \times 0.02 + 30 \times 0.02 = 1.6$	0.6	1.6	0.6	1.6

To produce product A, we need 2 units of X and 1 unit of Y. The time requirements in each center are displayed in a routing chart.

$$\begin{aligned}\text{Time for product A} &= \text{Time for A in W10} + 2 \times \text{Time for X in W11}\\ &\quad + 2 \times \text{Time for X in W12} + \text{Time for Y in W11}\\ &= 0.2 + 2 \times 0.03 + 2 \times 0.02 + 0.1 = 0.4\end{aligned}$$

Similarly Time for B is $= 0.15 + 0.03 + 0.02 + 0.1 = 0.3$

For weekly demand displayed below:

Week	1	2	3	4	5
Product A	50	0	50	0	50
Product B	30	30	30	30	30

Utilization of each work center would be as shown in Table 5.19.

These utilization hours are different from the ones where we had based our allocation on historical data alone. In a historical prospective, all products manufactured in a center are used to develop average utilization data. When we use bill of materials and routing data, we are much more specific to the products on hand. And that also illustrates the difference between rough cut and capacity bill planning.

5.7.3 Number of Machines Needed

Once we have determined work load at each work center, next step would be to establish resources needed to perform the work. For example, if the plant is operating at 95% efficiency, realistically we only have $480 \times 0.95 = 456$ min of working time. Collected production data as shown above must be modified to account for shrinkage allowances, personal times and for factors such as material handling efficiencies, tools availability, and other factors that may stop production. For example, consider WC 10; if shrinkage allowance is 3% for product A and 2% for product B and personal allowance is 8%, then the data can be modified as follows:

$$\text{Number of units of A production should start} = \text{Number of units needed} \times (1 + \text{Shrinkage allowance})$$

$$\text{Standard time for production} = \text{Actual production time} \times (1+\text{Personal allowances})$$

$$\frac{\text{Available production time}}{\text{Machine}} = \text{Daily production time} \times \text{Plant efficiency Available}$$

In our case, for week 1

$$\text{Number of A required} = 50 \times (1 + 0.03) = 51.5 \text{ or } 52$$

$$\text{Number of B required} = 30 \times (1 + 0.02) = 30.6 \text{ or } 31$$

$$\text{Standard time per unit for A} = 0.2 \times (1 + .08) = 0.216$$

$$\text{Standard time per unit for B} = 0.15(1 + 0.08) = 0.162$$

$$\text{Total load} = 52 \times 0.216 + 31 \times 0.162 = 16.254$$

$$\text{Machines needed} = \frac{16.254}{(8 \times 0.96)} = 2.116 \text{ or } 3.$$

5.8 THEORY OF CONSTRAINTS

The operation of a manufacturing plant is a process that combines many work centers together. In fact, if we look at the entire production operation, it involves many work centers including purchasing of raw material, manufacturing of products, distribution of the finished products, and even includes departments such as finance, sales, and human resources. Again, within each department, there may be many individual work centers.

Not all departments and work centers perform at the same pace nor do they have equal resources to do so.

Consider a simple example, consisting of three work centers in series. If production rate of first station is 60/hr, second work center 45/hr, and third 30/hr, then maximum production rate of this system is 30/hr.

WS 1	WS 2	WS 3
60/hr	45/hr	30/hr

The only way to increase the production rate is to increase the rate at WS3. Thus, WS3 is the bottleneck work station or constraint in this flow. It is not productive to spend additional resources to increase flow rates in nonbottleneck work stations.

The theory of constraints has some basic rules. Make sure that constraint work center is working to the fullest extent possible ‘Drumings’ all times. Take out all inefficiencies from that work center. Only well-trained workers should work at the constrain station. Ensure that machines do not break down; that means having a good preventive maintenance program. Ensure that the bottleneck machine is not starved or chocked, and that is units are available when needed and moved immediately out of machine when processing is complete. This may mean altering scheduling at other work stations so that units are available at the bottleneck machine when needed, by providing 'buffer' for inventories and 'rope' to pull units from the bottleneck machine. This is called Drum, Buffer, Rope system.

Increasing the capacity of the bottleneck station by adding additional shifts, overtime, and additional parallel machines and work centers are other means of increasing capacity. We need to be careful though, as it may not increase throughput capacity as much as increase of capacity in the bottleneck machine as some other station may now become the bottleneck. For example, the addition of one more WS 3 increases the total capacity of WS 3–60, but now WS 2 is the bottleneck, with throughput of 45/hr.

It was fairly easy in our example to identify the bottleneck machine. But this may not be the case when the systems are complex. And, even if we can identify the bottleneck machine, how do you make sure that continuous flow of units can be maintained? Many products are manufactured, perhaps using different machines in different sequences with different times on each machine. It is very difficult to sequence the jobs without some help.

Scheduling techniques discussed in the next chapters are the ways to optimize the entire system, including the bottleneck machine. Different methods are applied, based on the system configurations and objectives, as we shall discuss in scheduling chapters.

5.9 SUMMARY

All the topics covered so far, such as forecasting, material requirement planning, and inventory control, lead to decisions on what to produce, when to produce, and how much to produce to make the system run efficiently and economically. Even when to ship and how much to ship are decisions made with distribution requirement planning.

All the planning is feasible if sufficient production capacity is available when needed. Aggregate planning tries to resolve capacity issue to a certain extend, by subcontracting, increasing capacity by over times, and/or by adding work shifts and by changing the workforce. Even with such planning, we ultimately need capacity to produce.

In the next chapters, we go to each production center within the plant, where day-to-day capacity planning must be done. This planning directly lead us to scheduling, the next important topic in production planning.

5.10 EXERCISE

5.1 An item has a demand of 650 units for a year. Cost of each unit is $55, reorder cost is $100, and holding cost is 15% of value of the unit cost per year. No shortages are allowed.

a. Describe the economic order quantity.

b. Determine the total minimum cost per year and order and carrying cost per year.
c. If only 100 units can be ordered at a time, calculate the costs.

5.2 A university bookstore sells the book "*Things You Learn at University.*" The annual demand of the book is 1500 units. The book can be ordered any time, and the ordering cost is $34 per order. The holding cost for the book is $0.50 per unit per month. The store is looking to reduce its inventory cost by determining the optimal number of books to obtain per order.

Calculate
a. Optimal order quantity
b. Expected number of orders and expected time between orders
c. Annual order cost
d. Total annual cost for the store

Also, verify that to reduce the optimal lot size by a factor k, the fixed cost of order cost has to be reduced by a factor of k^2.

5.3 For the Problem 5.2, for the next 2 years it has been predicted that the demand, which presently is 1500, will increase by 50% in the first year and then reduce by 25% in year 2. Recalculate the optimal lot size(s) and the number of orders to be placed.

5.4 A company that operates 52 weeks a year is concerned about its inventory of plastic tubes. There is a demand of 7500 m a week, which costs $10 a meter. The replenishment cost of these tubes is $42 for administration and $58 for delivery, while holding costs are determined as 25% of value held a year. No shortages are allowed.
a. What can be the inventory policy for these tubes?
b. What is the gross profit if the company sells tubes for $15 a meter?
c. If the lead time is 1 week, what is the reorder point?

5.5 In Problem 5.4, if a periodic review policy is used, determine the optimum review period.

5.6 A product has an average demand of 800 units per year. The unit cost of the product is $50. The carrying cost is $20/unit/year, and the order cost is $300. Assume that the preferred safety stock is for 20 units. For past 10 weeks, the demand has been as follows:

Week	1	2	3	4	5	6	7	8	9	10
Demand	50	69	45	32	54	71	33	48	59	70

Determine the actual order quantities per order over the past 10 weeks based on the periodic review policy.

5.7 KP Electronics needs to determine the order quantities and reorder points for computer parts that it sells. The following data refers to the computer monitor:

Cost to place an order	$49
Holding cost	15% of the product cost per year
Cost of computer monitor	$150 each
Annual demand	700
Standard deviation during lead time	15
Lead time	7 days

The demand for the monitor occurs 365 days a year.

Calculate the economic order quantity. If the firm wants to provide 95% of customer level of service, what reorder point, R, should be used?

5.8 The monthly demand of hamburger patties in a university cafeteria is 3000 units. The ordering cost for the meat patties is $17/order, and the holding cost is $0.25 per unit per year. For quality of food, the management does not want to keep the patties more than 15 days in its freezer.

Calculate the reorder point and size. Also, find the average inventory level.

5.9 There are four products in the store. Demand for each product are as follows:

Product	Demand/Day
1	10 ± 2
2	12 ± 3
3	15 ± 3
4	11 ± 2

If the lead time for each product is 5 days, determine the safety stock to serve 95% of the time if:

a. Each product is independent and hence cannot be substituted for one another.
b. If demand for any product can be substituted by other available product if the demanded product is not available.

5.10 ABC manufacturers needs ball bearings to manufacture it various products. The supplier of ball bearing has offered quantity discounts if the company purchases more than the present order quantities. The new volume and prices are:

Order Quantities	Cost per Bearing (in $)
0–499	3.20
500–999	3.00
1000 and more	2.90

Further, the ordering cost is $32 per order, the annual demand is 6000 units, and the cost of inventory is $0.6 per year per bearing.

a. Determine the optimum policy if all unit quantity discount is allowed.
b. Determine the optimum policy if marginal quantity discount is allowed.

5.11 The demand of a product is 520 per month. To maintain a customer level satisfaction of 95% with a lead time of 1 week, find the safety stock and reorder point.

5.12 There are four similar products of different brands, each having demand of 75 units per day and standard deviation of 6 per day. If one product is not available, the customers are satisfied with the other brands. With a lead time of 4 days to provide 99% customer service determines the safety stock and when to reorder the products.

5.13 Freeman Motor Company has a normally distributed demand for a new car model during its reorder period. The average demand during this period is 250, and the standard deviation is 10 cars. The manager wants to know how much safety stock it should maintain if they maintain stock outs only 5% of the time.

5.14 A company uses 1200 units per year. It cost $300 per order, and the carrying cost is $12 per unit per year. The unit cost is $100. A new proposition manufacturer has offered one-time 7% off if we buy more than EOQ units. How many units should we buy?

b. Another manufacturer has agreed to sell the unit for $95 if the quantity ordered is greater than 300. Should we go with this alternative?

5.15 In the following example, either joint or individual order can be placed.

Item	1	2	3	
Demand/year	4,000	3,000	10,000	
Cost/unit in $	20	30	40	
Holding cost/unit/year in $	10	15	13	
Item cost in $	100	150	225	
Order cost/item in $	40	50	40	
Order cost in $				1,000

5.16 A local toy manufacturer uses computer chips in one of its products. The annual usage of these computer chips is 25,000 units and maybe assumed to be uniform over the year. The carrying cost for any inventory in the plant is taken as 20% for every dollar invested for a year. The current supply comes from four different distributors, and the maximum quantity a supplier can supply and some other data are shown in the following table.

Distributor	Quantity	Cost per Unit ($)	Order Cost
Ruston	8,000	4	$20
Monroe	10,000	4.5	$30
Shreveport (A)	11,000	5.0	$25
Shreveport (B)	5,000	6.0	$10

Develop a strategy of purchasing to minimize the annual cost.

5.17 An item has an annual demand of 2500 units, each order costs $13 to place, and the yearly holding cost is 35% of unit cost. The unit cost depends on the quantity ordered as given:

a. Ordered quantity less than 999—unit cost is $1.5
b. Quantities between 999 and 1499—unit cost is $$1.1
c. Quantities 1500 and more—unit cost is $.80

What is the optimal ordering policy?

5.18 An order must be placed for the coming New Year celebration. Past experience has shown that the demand for firework boxes have the following distribution.

Demand	150	200	225	250	300
Probability	0.1	0.2	0.3	0.2	0.2

The purchase price is $30/box, and the sales price is $50/box. Based on the county ordinance, any fireworks left over after the week has no value since it cannot be stored in the city limits. Determine the order quantity.

5.19 Suppose in Problem 5.18 that the demand is large enough to be approximated by normal distribution with a mean of 230 and a standard deviation of 25. Determine how many units to order.

5.20 In Problem 5.18, suppose the manufacturer will buy back any leftover firework boxes for $20/box. Determine how many units of order should be placed.

5.21 The expected demand for two products for the next 6 weeks is as follows:

Week	1	2	3	4	5	6
Product X	50	30	20	10	40	20
Product Y	10	20	15	15	20	15

Time required to produce X is 13 hr/unit and for Y is 1 hr/unit. Product X uses work centers W1, W2, and W3, while Y uses W2 and W3. The work center efficiencies are: W1 80%, W2 90%, and W3 70%.

Perform rough cut planning for three work centers.

5.22 Two products are produced in the plant P1 and P2. Demands for the products are 500 units/week and 1000 units/week, respectively. The plant works 40 hr/week. The following parts go into making each product.

P1		
A1 (2 units)	A2	C1
P2		
A1	A2	C2

The routing sheets for different products are as follows:

Operation	Machine	Setup Time/hr	Production/hr
Product P1			
1 of 2	M1	1.1	50
2 of 2	M2	2.3	15
A1			
1 of 2	M3	0.2	10/hr
2 of 2	M2	1.4	10/hr
A2			
1 of 1	M3	1.5	15/hr

C1 and C2 are purchased from outside.

Determine the number of machines of each type that are needed.

5.23 In Problem 5.22, product P1 sells for $50 and P2 for $30. Each hour of working on machines M1, M2, and M3 cost $6, $7, and $8, respectively. If we have two machines of each type, determine the number of units of P1 and P2 to produce to optimize the profit.

5.24 Develop an efficient truck pick-up schedule for a milk run strategy to pick up the following. The inventory carrying cost is $0.8/day/unit

Master Schedule

Pick-up Cost	Product/Day	1	2	3	4	5	6	7
30	1	7	5	8	3	5	7	5
18	2	13	11	9	15	13	9	11
22	3	7	7	7	10	7	12	9
14	4	5	7	9	5	7	9	5
16	5	12	10	8	12	8	10	8

5.25 If a trip travel cost in Problem 5.24 is $75, determine the pick-up schedule and the associated cost.

REFERENCES AND SUGGESTED READINGS

Baker, K.R., M.J. Magazine, and H.L.W. Nuttle. August 1986. "The Effect of Commonality on Safety Stock in a Simple Inventory Model" *Management Science*, 32(8).

Berry, W.L., T. Schmitt, and T.E. Vollmann. November 1982. "Capacity Planning Techniques for Manufacturing Control Systems: Information Requirements and Operational Features" *Journal of Operations Management*, 3(1).

Blackstone, J.H. Jr. 1989. *Capacity Management*, Cincinnati, OH: South-Western Publishing.

Brown, R.G. 1967. *Decision Rules for Inventory Management*, New York: Holt, Rinehart & Winston.

Burlingame, L.J. "Extended Capacity Planning" *APICS Annual Conference Proceedings*, 1974, pp. 83–91.

Constable, G.C. and D.C. Whybark. October 1978. "The Interaction of Transportation and Inventory Decisions" *Decision Sciences*, 9(4): 688–699.

Crowther J. 1964. "Rationale for Quantity Discounts" *Harvard Business Review*, (March–April): 121–127.

Dolan Robert, J. 1987. "Quantity Discounts: Managerial Issues and Research Opportunities" *Marketing Science*, 6: 1–24.

Lankford, Ray. "Short Term Planning of Manufacturing Capacity." *APICS 21st Annual Conference Proceedings*, 1978, pp. 37–68.

Lee Hau L. and Corey Billington. 1992. "Managing Supply Chain Inventory" *Sloan Management Review*, Spring: 65–73.

6 Single Machine Scheduling

The broad subject of scheduling is introduced next in Part II. There are various topics and problems in scheduling, and they are discussed in the subsequent chapters. We shall start with the broadest application and narrow it down in the subsequent chapters to detail shop floor planning.

In this chapter, we introduce the basic problem of single machine sequencing. The basic problem can be stated as follows (See also Figure 6.1): there are a number of jobs (works, demands, etc.) requiring the services of a single machine (a facility, a person, etc.), and we want to schedule these jobs in the best possible manner to optimize some objectives. There are, in practice, many different objectives, and one may be more dominant over another in a specific instance. There are many variations of the problem and solution procedures. In some instances, for example, it may be more important to minimize the penalty for completing jobs late (called tardiness penalty), while in other instances, it may be paramount that we meet certain specified due dates without exception. In yet another problem, a job may be delivered within a certain range of due dates, and the penalty occurs only if we are earlier than the earliest specified due date or later than the latest specified due date for the job. There are many other objectives; however, we will study the basic problem in this chapter, while its variations are introduced in Chapter 7. Fortunately, most of the variations require slight modifications to the heuristic procedure developed in Section 6.3.

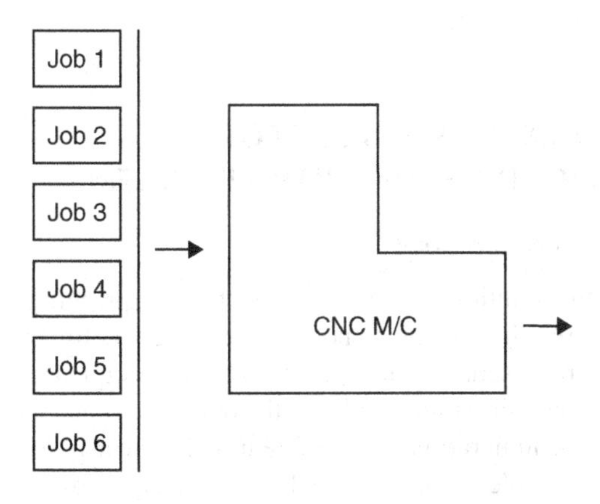

FIGURE 6.1 Single machine scheduling setup.

6.1 TARDINESS PROBLEM

The definition of the basic single machine tardiness problem is as follows.

We have n jobs to schedule on a single machine. For each job, the processing time P_i, the due date D_i, and the cost per unit time of tardiness L_i are known. The tardiness cost for a job is assumed to be linear, that is, if a job i is finished t time units after the due date, a penalty of $L_i \times t$ is incurred. If a job is completed before its due date, there is no penalty. The objective is to minimize the total tardiness penalty, which is defined as the sum of the tardiness cost of each job. This problem is commonly referred to as the single-machine scheduling problem with tardiness penalty.

Implied in the problem statement are some assumptions: All jobs are available for processing at time zero. Setup times for the jobs are sequence independent and may be included in the processing times. Once the job is taken up for processing, it is continuously worked on without interruptions, and there is no machine failure; therefore, the machine is available for work all the time.

The practical and industrial applications of the problem are considerable. For instance, in scheduling n jobs on a single machine such as a lathe, the processing times for each job may vary based on the number of units and the amount of work needed for the job. The due dates and penalty values are established due to idleness or availability of the succeeding resources such as next required machines or transportation facilities. In scheduling n jobs in a job shop environment, for example, the entire manufacturing facility may be treated as a "single machine." Here, the jobs are customer specified, and the due dates and penalty functions are negotiated in advance. In an office environment, a manager might schedule his or her meeting appointments, based on time required for the meeting and the importance of the meeting measured as consequences or penalties that may result from not making on time the decisions that may result from the meetings. A bottleneck facility may also be viewed as a single machine. Bottlenecks occur because of capacity limitations, and jobs requiring services of this facility must be scheduled to satisfy one or more of the objectives discussed in Chapters 6 and 7.

6.2 SURVEY OF EXACT METHODS FOR SINGLE-MACHINE SCHEDULING PROBLEM

6.2.1 EXHAUSTIVE ENUMERATION

Computationally, the solution to the single-machine sequencing problem could be very demanding. We could develop all possible sequences defining the order in which the jobs could be processed, find the penalty for each sequence, and then select a sequence that gives the minimum penalty as the optimum sequence. This approach is called an exhaustive enumeration. The problem with exhaustive enumeration is that examination of all possible sequences would entail analysis of $n!$ combinations. For a large value of n, this is an impossible task. Mathematicians call such a problem as NP-complete or NP-hard, meaning the problem cannot be solved in polynomial time as n gets large. Quite often, the time required to solve such a problem (by hand or on computer) increases exponentially with respect to n. This makes it impractical to apply exhaustive enumeration in almost all cases except in a few problems where "n" is relatively small, generally less than 10.

6.2.2 Branch-and-Bound Algorithm

An approach to reduce the number of sequences that must be examined from $n!$ to a somewhat manageable level is to apply a branch-and-bound method. This method can be used to solve any optimization problem that has a finite number of feasible solutions, and is called *combinatorial optimization*.

The steps of the branch-and-bound method as they are applied to a sequencing problem are as follows:

Step 1: Determine the job to be processed last. In any sequence, there is always a job that is processed last, and if there are no restrictions, any job can be processed last. Draw a tree diagram indicating a node for each possible job to be processed last.

Step 2: Calculate the lower bound on the total penalty (W) associated with the node. The total penalty is the due date of the job subtracted from the sum of the processing time of the remaining jobs (jobs that are not yet assigned to nodes in the branch, including the job under consideration for the present node) multiplied by the penalty weight (L) for the job.

Step 3: Select the node with the lowest lower bound penalty value (parent node) for branching. Each of the remaining jobs can be processed in the next position, so construct nodes for each of these jobs. Develop branches connecting these nodes from the parent node.

Step 4: Calculate the late penalty for each of the jobs on the nodes developed in step 3. The penalty is calculated as in step 2 and added to the lower bound of the parent node to obtain the lower bound of the present node.

Step 5: Repeat steps 3 and 4 until the first sequence is determined. This penalty value is the initial minimum penalty value.

Step 6: Search the tree diagram, eliminating all nodes that have lower bounds above the current minimum value.

Step 7: Begin branching on each of the nodes that remain, eliminating nodes where lower bounds are above the current value. Replace the current minimum lower bound with the new minimum value if one is found.

Step 8: Once all other nodes have been eliminated, the current lower bound will be the total penalty for the associated sequence of jobs.

6.2.2.1 Illustrative Example 1

Let us illustrate these steps by applying them to the data in Table 6.1.

Applying the branch-and-bound method to the data in Table 6.1 yields the tree diagram in Figure 6.2.

In step 1, nodes 1, 2, 3, and 4 are constructed, indicating that each of the four jobs could be processed in the last position of the sequence.

TABLE 6.1
Data for an Example Problem 1

Job Number	Processing Time P_i	Due Date D_i	Weight L_i
1	37	49	1
2	27	36	5
3	1	1	1
4	28	37	5

In step 2, we get the lower bounds of the penalties at this stage. For example, for node 1, the total processing time for all remaining jobs, including the job to be scheduled at this point (presently all four jobs), is $37+27+1+28 = 93$. Subtracting the due date of job 1 yields $93 - 49 = 44$. Multiplying 44 by the penalty weight for job 1, yields $W > 44$ (44×1), where W indicates the total penalty for the sequence. Similarly, the lower bound on the penalty values for the remaining jobs is $W > 285$ (node 2), $W > 92$ (node 3), and $W > 280$ (node 4).

Step 3 selects job 1 as the branching node since 44 is the lowest penalty value of the four nodes. Since job 1 is now assigned to position 4, step 3 also constructs nodes branching from job 1 for jobs 2, 3, and 4.

Step 4 calculates the lower bound for these jobs. For example, consider node 6. The total processing times for jobs that are not yet scheduled, that is, jobs 2, 3, and 4 is $27 + 1 + 28 = 56$. The due date for job 3, which is being considered for this node, is 1. Hence, the penalty for job 3 is $(56 - 1) \times 1 = 55$ added to the penalty of node 1 from where the branching has started, gives the lower bound on the penalty as $W > 55 + 44 = 99$. Similarity, $W > 144$ for node 5 and $W > 139$ for node 7.

Of all the nodes that are open (from which further branches can be developed) node 3 has the lowest cost of 92. Develop nodes 8, 9, and 10 by placing jobs 1, 2, and 4 respectively in Position 3 of the sequence.

From amongst the nodes 2, 4, 5, 6, 7, 8, 9, and 10, node 6 has the least cost. Proceed from that node to develop nodes 11 and 12 by placing Jobs 2 and 4 in the second position.

Of nodes that are open, node 8 has the least cost. Develop nodes 13 and 14.

Continuing in the same manner node 17 gives the minimum cost of 139. It is also a completed sequence with jobs assigned in each position. No other open node has a cost less than 139 and therefore no further analysis is needed.

The optimum sequence is 3-2-4-1 with a minimum penalty of 139.

6.3 COMMONLY USED HEURISTIC RULES

The branch-and-bound method is computationally more efficient than exhaustive enumeration, but for a large value of n, it still requires high computational time and effort. Most of the real-life problems, therefore, are solved by some heuristic methods. Eight different rules have been prominently used in literature to solve the single

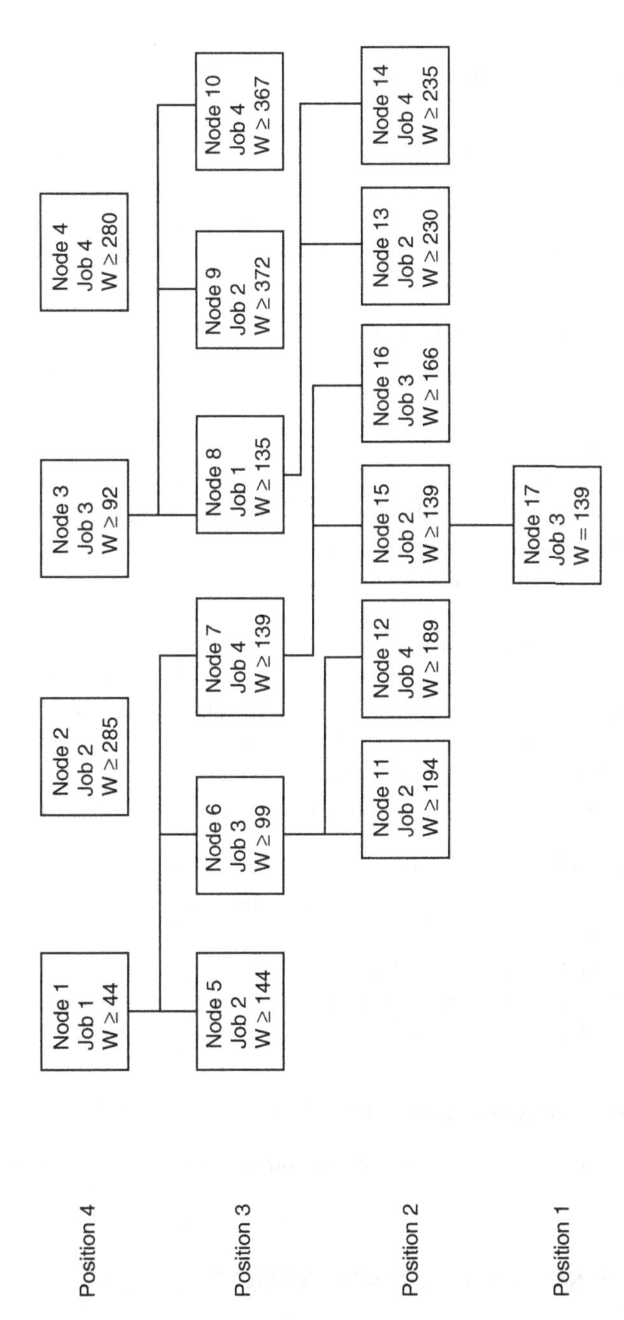

FIGURE 6.2 Branch and bound tree for single machine scheduling example.

machine problem (Ahmed, 1992; Baker, 1974; Conway, 1967; Johnson, 1974). They are briefly described, then illustrated by an example.

6.3.1 Earliest Due Date Rule

A schedule is developed considering the due dates of the jobs. Start by placing the job with the earliest due date (EDD) in the first position. The next job scheduled is the job that has the EDD from among the jobs that are yet to be scheduled. Continue this process until all the jobs are scheduled. The resulting sequence is the same if the jobs are arranged and processed in ascending order of their due dates.

6.3.2 Cost Over Time (COVERT) Rule

Few additional notations are defined to explain this rule.

TT: Sum of all processing times.
RT: Sum of processing times of the jobs that are yet to be scheduled.
ST: Starting time for the next scheduled job. It is zero for the first job.
CF: Coefficient that is calculated as explained below.
PR: Priority that is calculated as explained below.

The COVERT rule has three steps.

Step I. Calculate PR. For all jobs that are yet to be scheduled, PR is calculated as follows:
Case 1: If $D_i < (\text{ST} + P_i)$, then $\text{PR} = 1$
Case 2: If $(D_i > (\text{ST} + P_i))$ and $D_i < \text{TT}$, then $\text{PR} = ((\text{TT} - D_i)/(\text{RT} - P_i))$
Case 3: If $\text{TT} \leq D_i$, then $\text{PR} = 0$.
Step II. Calculate CF for a job i, yet to be scheduled as:
$\text{CF}_i = \text{PR} \times L_i/P_i$
Step III. Job with the maximum CF is scheduled next.
ST and RT are recalculated with the remaining jobs, and the steps are repeated.

6.3.3 Shortest Processing Time (SPT) Rule

Arrange the jobs in the ascending order of their processing times and schedule them in the same order.

6.3.4 Largest Penalty per Unit Length (LPUL) Rule

For each job, calculate a ratio, $U_i = L_i/P_i$. Schedule the jobs in the descending order of U_i. In case of a tie, from among the tied jobs, the job with the smallest processing time is selected next.

6.3.5 SHORTEST PROCESSING TIME AND LPUL RULE

Use SPT rule to schedule the jobs. In case of a tie, break it using U_i from LPUL rule.

6.3.6 SHORTEST WEIGHTED PROCESSING TIME (SWPT) RULE

Calculate ratio $S_i = P_i/L_i$ for each job. Jobs are scheduled using the ascending order of the ratios.

6.3.7 LARGEST WEIGHT (WT) AND LPUL RULE

Arrange the jobs in descending order of their weights W_i and schedule them in the same order. Break a tie by using LPUL rule.

6.3.8 CRITICAL RATIO (CR) RULE

Calculate the value for T, which is the sum of processing times for all the jobs that have been scheduled. Calculate the ratio CR_i for each unscheduled job i, which is equal to $(D_i - T)/P_i$. The job with the smallest value of CR is scheduled next. Repeat the process until all the jobs are scheduled.

6.3.8.1 Illustrative Example 1

Next, we apply the rules to the data in Table 6.1.

6.3.8.1.1 EDD Rule

According to EDD rule, the job having the EDD is processed first, the job with the next EDD is processed second, and so on. In our example, the job processing sequence would be 3-2-4-1. The evaluation of this sequence is shown in four lines. The first line is the job sequence line. It shows the starting time, processing time, and completion time for a job. The notation is: the number to the left of parenthesis is the starting time for the job, within parenthesis are the job number and the job's processing time, and the number to the right of parenthesis is the completion time of the job, which also is the starting time of the next job. The next line is the due date line, which indicates a job number in the parenthesis and its due date to the right. The third line is the deviation line. It compares the completion time and due date for each job and determines if the job is early (E), late (L), or on time (W), and by how many days. In this line, we note the number of days of deviation from the due date and mark it with the associated symbol (E, L, or W). The penalty for the sequence is calculated in the fourth line.

Job sequence:	0	(3/1)	1	(2/27)	28	(4/28)	56	(1/37)	93.
Due dates:		(3)	1	(2)	36	(4)	37	(1)	49
Deviation:			0/W		8/E		19/L		44/L
Penalty:				$(5 \times 19) + (1 \times 44) = 139$					

6.3.8.1.2 COVERT

Calculations associated with the application of the COVERT rule to the example problem are shown next.

Iteration 1. TT = 93; RT = 93; ST = 0.

The unscheduled job list consists of all jobs, namely job 1, 2, 3, and 4. The calculations for PR and CF for each unscheduled job are

Job 1: $49 > (0 + 37)$ hence $PR_1 = (93 - 49)/(93 - 37) = 0.785$.
$CF_1 = 0.785 \times 1/37 = 0.0212$

Job 2: $36 > (0 + 27)$ hence $PR_2 = (93 - 36)/(93 - 27) = 0.864$.
$CF_2 = 0.864 \times 5/27 = 0.16$

Job 3: $1 = (0 + 1)$ hence $PR_3 = 1$. $CF_3 = 1 \times 1/1 = 1$

Job 4: $37 > (0 + 28)$ hence $PR_4 = (93 - 37)/(93 - 28) = 0.862$.
$CF_4 = 0.862 \times 5/28 = 0.153$

Since job 3 has the largest CF value, it is selected and scheduled in the first place. Job 3 is also taken off from the unscheduled job list.

Iteration 2. TT = 93 RT = (93 − processing time for job 3) = 93 − 1 = 92. ST = 1

Unscheduled jobs are 1, 2, and 4. The calculations for PR and CF are

Job 1: $49 > (1 + 37)$ hence $PR_1 = (93 - 49)/(92 - 37) = 0.800$;
$CF_1 = 0.8 \times 1/37 = 0.022$

Job 2: $36 > (1 + 27)$ hence $PR_2 = (93 - 36)/(92 - 27) = 0.877$;
$CF_2 = 0.877 \times 5/27 = 0.162$

Job 4: $37 > (1 + 28)$ hence $PR_4 = (93 - 37)/(92 - 28) = 0.875$;
$CF_4 = 0.875 \times 5/28 = 0.156$

Since job 2 has the largest CF, it is selected for the second position.
Iteration 3. TT = 93 RT = (92 − 27) = 65; ST = 1 + 27 = 28.
Unscheduled jobs are 1 and 4. The calculations for CF and CR are:

Job 1: $49 < (28 + 37)$ hence $PR_1 = 1$; $CF_1 = 1 \times 1/37 = 0.027$
Job 4: $37 < (28 + 28)$ hence $PR_4 = 1$; $CF_4 = 1 \times 5/28 = 0.178$

Job 4 is scheduled in the third position, and the remaining job, job 1, is scheduled in the last position. The final sequence is 3-2-4-1.

This is the same sequence as obtained with EDD rule and hence the penalty for the sequence is the same, that is, 139.

6.3.8.1.3 SPT Rule

Ranking jobs based on shortest processing time leads to the sequence 3-2-4-1. This sequence is the same as we have obtained using the EDD and COVERT rules.

6.3.8.1.4 LPUL Rule

The calculations for LPUL rule are shown below.

Job 1: $U_1 = 1/37 = 0.027$
Job 2: $U_2 = 5/27 = 0.185$
Job 3: $U_3 = 1/1 = 1.0$
Job 4: $U_4 = 5/28 = 0.178$

Scheduling jobs in descending order of U results in the sequence: 3-2-4-1, the same as other methods.

6.3.8.1.5 SPT and LPUL Rule

Since the application of LPUL resulted in no ties, the SPT/LPUL rule leads to the same schedule.

6.3.8.1.6 SWPT Rule

The calculations for S_i are as follows:

$S_1 = 37/1 = 37$
$S_2 = 27/5 = 5.4$
$S_3 = 1/1 = 1.0$
$S_4 = 28/4 = 5.6$

Arranging jobs in ascending order of S_i leads to the schedule 3-2-4-1, the same as before.

6.3.8.1.7 Largest WT and LPUL

Jobs 2 and 4 have the same weight of 5. Since $U_2 = 5/27 = 0.185$ is greater than $U_4 = 5/28 = 0.178$, job 2 is placed in position 1 and job 4 in position 2 of the sequence. Jobs 1 and 3 again have the same weights, namely 1. The corresponding U values are 0.027 and 1.0, respectively. Hence, job 3 is placed in position 3, and job 1 is placed in position 4. The final sequence is 2-4-3-1.

Job sequence:	0	(2/27)	27	(4/28)	55	(3/1)	56	(1/37)	93.
Due dates:		(2)	36	(4)	37	(3)	1	(1)	49
Deviations:			9/E		18/L		55/L		44/L
Penalties:			$(18 \times 5) + (55 \times 1) + (44 \times 1 = 171)$						

6.3.8.1.8 Critical Ratio Rule

Iteration 1: T = 0. All jobs are unscheduled.

$\text{CR}_1 = (49 - 0)/37 = 1.324$
$\text{CR}_2 = (36 - 0)/27 = 1.333$

$$CR_3 = (1 - 0)/1 = 1.000$$
$$CR_4 = (37 - 0)/28 = 1.321$$

Least ratio is associated with job 3. Schedule job 3 in the first position.

Iteration 2: T = 1

$$CR_1 = (49 - 1)/37 = 1.297$$
$$CR_2 = (36 - 1)/27 = 1.296$$
$$CR_4 = (37 - 1)/28 = 1.285$$

Job 4 has the smallest ratio, and therefore job 2 is placed in the second position. The sequence so far is 3-4.

Iteration 3: T = 1 + 28 = 29

$$CR_1 = (49 - 29)/37 = 0.540$$
$$CR_2 = (36 - 29)/27 = 0.259$$

Job 2 is placed in the third position, and the remaining job 2 is placed in the last position, giving the sequence of 3-4-2-1.

Job sequence:	0	(3/1)	1	(4/28)	29	(2/27)	56	(1/37)	93.
Due date:		(3)	1	(4)	37	(2)	36	(1)	49
Deviations:			0/W		8/E		20/L		44/L
Penalties:				(20 × 5) + (44 × 1) = 144.					

The best sequence among all the sequences obtained using the various rules is 3-2-4-1, giving a penalty of 139. A number of procedures gave this sequence, as shown in Figure 6.3.

6.4 DESCRIPTION OF AN EFFICIENT HEURISTIC

The eight rules described in the previous section, though simple, do not produce optimum or even near optimum results consistently. In fact, a few rules do not even use all the available information. For example, the SPT rule only considers the processing times, the EDD rule only considers the due dates. Thus, more often than not, the resulting solutions are nowhere near the optimum values for the objective of minimization of the total tardiness penalty. A heuristic procedure is presented here, which is simple to apply and has proven to be very effective. The procedure is developed in two phases; the backward phase is applied first and then the forward phase is implemented. The details of the two phases are described next.

6.4.1 Backward Phase

In this phase, the initial job sequence is developed. The sequential job assignments start from the last position and proceed backwards toward the first position. The

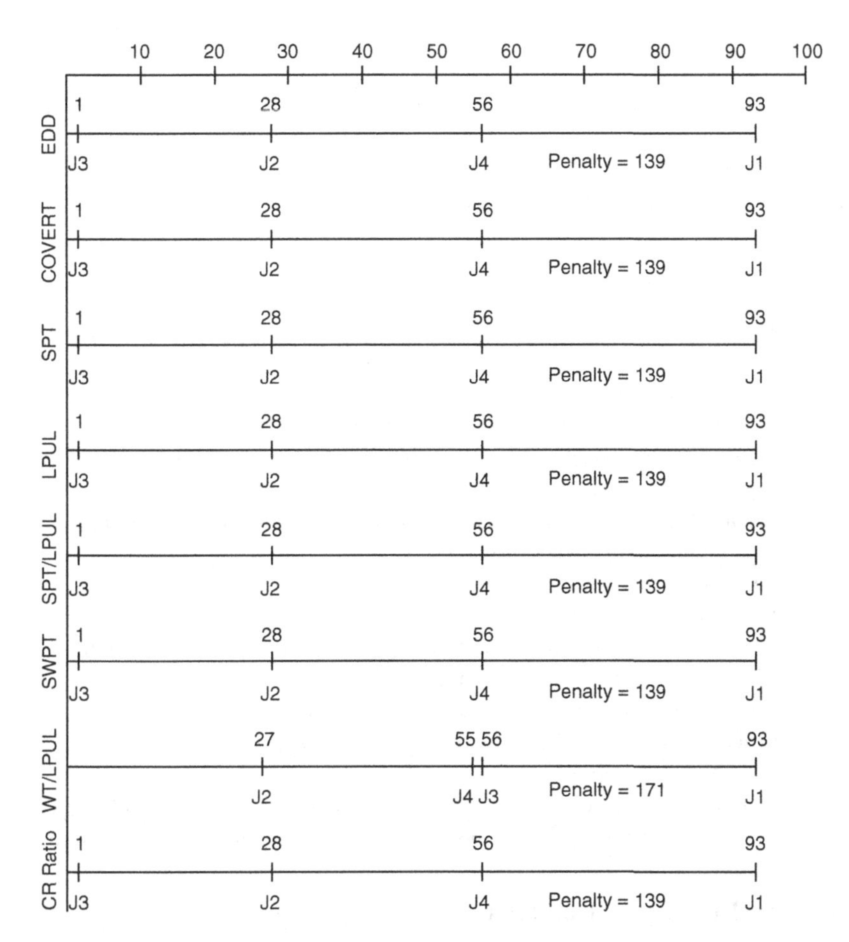

FIGURE 6.3 Sequence obtained using different methods and their associated penalties.

assignments are complete when the first position is assigned a job. The process consists of the following steps (for clarity the number of jobs n is now indicated as N):

1. Note the position in the sequence (the value of the position counter) in which the next job is to be assigned. The sequence is developed starting from position N and continuing backwards to position 1. The initial value of the position counter is N.
2. Calculate T, which is the sum of processing times for all unscheduled jobs.
3. Calculate the penalty for each unscheduled job I as $(T - D_i) \times L_i$.
4. The next job to be scheduled in the designated position (the value of the position counter) is the one having the minimum penalty from step 3. In case of a tie, choose the job with the largest processing time.
5. Reduce the position counter by 1.

Repeat steps 1 through 5 until all the jobs are scheduled.

6.4.2 Forward Phase

Perform the forward pass on the job sequence found in the backward phase, which is the "best" sequence at this stage. The forward pass progresses from the job in position 1 towards the job in position N. Let k define the lag between two jobs in the sequence that are exchanged. The steps of the forward phase are as follows:

1. Set $k = N - 1$
2. Set $j = k + 1$.
3. Determine the savings (or cost) by exchanging two jobs in the best sequence with a lag of k. The job scheduled in position j is exchanged with the job scheduled in position $j-k$ (if $j-k$ is zero or negative, go to step 6). Calculate the penalty after exchange, and compare it to the "best" sequence penalty.
4. If there is either positive or zero savings in step 3, go to step 5; otherwise, there is cost associated with this exchange and the exchange is rejected. Increase the value of j by one. If j is equal to or less than N, go to step 3. If j is greater than N, go to step 6.
5. If the total penalty has decreased, the exchange is acceptable. Perform the exchange. The new sequence is now the "best" sequence, go to step 1. Even if the savings is zero, make the exchange and go to step 1, except if the set of jobs associated in this exchange had been checked and exchanged in an earlier application of the forward phase. In that case, no exchange is made at this time. Increase the value of j by one. If j is less than N, go to step 3. If $j = N$, go to step 6.
6. Decrease the value of k by one. If $k > 0$, go to step 2. If $k = 0$, go to step 7.
7. The resulting sequence is the best sequence generated by this procedure.

6.4.3 An Illustrative Example Using the Heuristic Algorithm

The procedure is illustrated by applying it to the problem given in Table 6.1. For convenience, the data is reproduced below.

Job Number	Processing Time P_i	Due Date D_i	Weight L_i
1	37	49	1
2	27	36	5
3	1	1	1
4	28	37	5

Start the procedure with the backward phase.

Since all the jobs are unscheduled at this point, the total processing time of all unscheduled jobs is the sum of processing times for all jobs, namely

$$T = 37 + 27 + 1 + 28 = 93$$

Calculate the penalty for each job if it was to be completed at $T = 93$.

Job	Penalty
1	$(93 - 49) \times 1 = 44$
2	$(93 - 36) \times 5 = 285$
3	$(93 - 1) \times 1 = 92$
4	$(93 - 37) \times 5 = 280$

Since job 1 has the least cost, it is scheduled in the last position, that is, position 4.
Since job 1 now has been scheduled, the new value of $T = (93 - \text{processing time of job 1}) = (93 - 37) = 56$.
The new penalty values for the remaining jobs are as follows:

Job	Penalty
2	$(56 - 36) \times 5 = 100$
3	$(56 - 1) \times 1 = 55$
4	$(56 - 37) \times 5 = 95$

The least penalty is with job 3, and hence it is placed in position 3.
The next value of $T = (56 - \text{processing time for job 3}) = (56 - 1) = 55$.
The new penalty values for jobs 2 and 4 are as follows:

Job	Penalty
2	$(55 - 36) \times 5 = 95$
4	$(55 - 37) \times 5 = 90$

Job 4 is placed in position 2 and the remaining job, job 2, is placed in position 1. This results in the job sequence 2-4-3-1. Thus, we have:

Job sequence:	0	(2/27)	27	(4/28)	55	(3/1)	56	(1/37)	93
Due dates:		(2)	36	(4)	37	(3)	1	(1)	49
Deviations:			9/E		18/L		55/L		44/L
Penalty:			$(18 \times 5) + (55 \times 1) + (44 \times 1) = 189$						

The total penalty for the above schedule is 189.

The next step is to apply the forward phase. We should note that exchanging jobs in positions j and $j + 1$ does not change the completion times for the jobs scheduled before position j or the starting times of the jobs scheduled after position $j + 1$. For example, switching jobs 4 and 3 will not change the completion time of job 2, that is, 27 and the starting time of job 1, which is 56. Therefore, we only need to calculate the increase or decrease of penalty associated with completion times of the switched jobs to measure the change in penalty of the entire sequence. However, in a small problem like the one that we are illustrating, it may be just as quick to recalculate the penalty for the entire sequence after an exchange is made.

Setting $k = N - 1$ results in $k = 3$. Switching jobs with lag of 3, that is, job 2 and 1, results in the sequence 1-4-3-2. The associated penalty is 420. Since the penalty is higher than the original sequence penalty, no switch is made. The best sequence so far is still 2-4-3-1.

There are no other jobs with lag of 3; therefore, the value of k is changed to 2.

With $k = 2$, exchanging jobs 2 and 3 results in sequence 3-4-2-1, with the penalty of 144. Since the new penalty is less than that obtained for the original sequence of 2-4-3-1, the new sequence 3-4-2-1 becomes the best sequence.

The evaluation continues, starting from the first job in the new best sequence and k set back to 3. Exchanging jobs 3 and 1 results in sequence 1-4-2-3 with the penalty of 644, and hence the exchange is rejected. The best sequence is still 3-4-2-1.

Setting $k = 2$ and exchanging 3 and 2 results in 2-4-3-1 with the penalty of 189, and the exchange is rejected. Switching 4 and 1 results in sequence 3-1-2-4 with the penalty higher than 144, and the switch is again rejected. Since no more exchange with $k = 2$ is possible, the k is set to the new value of 1. The best sequence so far is still 3-4-2-1.

Exchanging 3 and 4 results in 4-3-2-1 as the nonpromising sequence with a penalty of 172. The next exchange, namely 3-2-4-1, has a penalty of only 139. Hence, the sequence 3-2-4-1 becomes the best sequence.

Performing all the exchanges again on the current best sequence leads to no further improvement; hence, the sequence 3-2-4-1 also becomes the optimum sequence (see Figure 6.4). An exhaustive enumeration also displayed 3-2-4-1 as the optimum sequence.

Incidentally, though the procedure seems long and tedious in this instance, this example was specifically chosen to display the importance of the backward phase and to show how to make the exchanges. Experience with the method has shown that in most instances the best sequence is obtained either immediately after the application of the forward phase or with very few additional iterations of the backward phase.

6.4.4 Validity of the Heuristic and Conclusions

A computer program was developed (included in the appendix) that computes many of the heuristic algorithms. Also, a program that exhaustively enumerates the optimum solution by checking $N!$ combinations was developed. The tardiness factor index T, first introduced by Srinivasan (1971), and also illustrated by Ow and Morton (1989), was used to generate 100 data sets for each file of 4, 5, 50, and 100 jobs. T is defined as $T = 1 - D/NP$, where D is the average due date, P is the average processing time, and N is the number of jobs. The value of T can vary from 0 to 1. A data set with T close to zero has large positive slack (slack = due date − processing time) or very loose schedule. The slack becomes smaller or the schedule becomes tighter as we examine data sets with increasing value of T. In fact, with T close to 1, we may have jobs in the data set such that their processing times are greater than the times when the jobs are due.

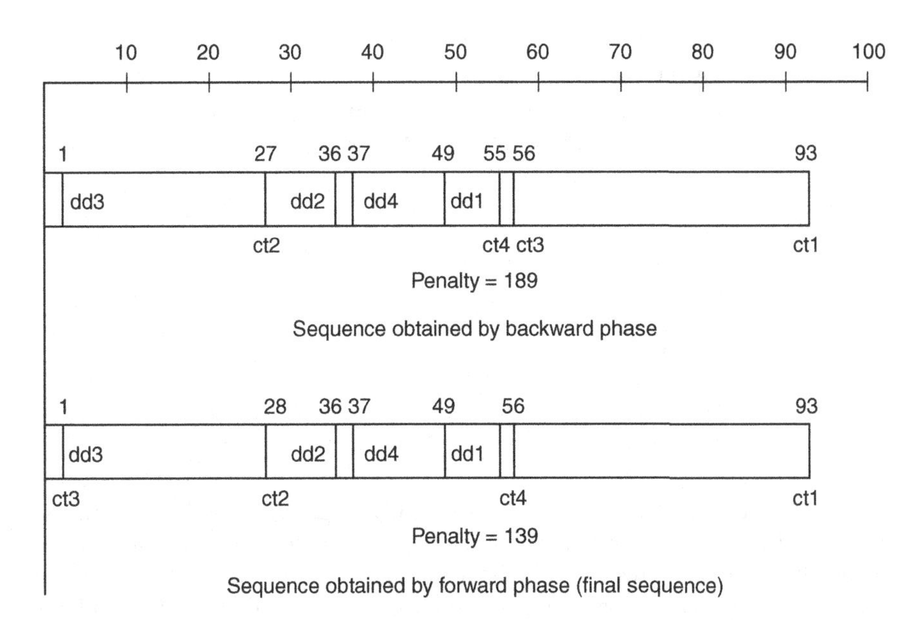

FIGURE 6.4 Sequence obtained applying the heuristic of section 2.4 (note ct = completion time; dd = due date).

TABLE 6.2
Comparisons Statistics: Minimum Tardiness Penalty in 100 Trials—4 Jobs/5 Jobs

Method	$T = 0.1$	$T = 0.5$	$T = 0.7$
EDD	100/100	13/44	16/15
COVERT	57/59	42/31	59/41
SWPT	83/77	54/40	92/76
SPT/LPUL	100/100	48/48	23/15
WT/LPUL	32/30	15/11	26/12
CRITICAL RATIO	100/100	13/29	16/9
New heuristic	100/100	98/97	99/98
Exhaustive search	100/100	100/100	100/100

We developed the data sets of 100 problems each for $T = 0.1$, 0.5, and 0.7. For each job, the processing time was chosen randomly between 1 and 50, and the corresponding due date was established by $D_i = P_i \times N \times (1 - T)$. This approach of assigning due dates is more consistent and in line with what a manufacturer may wish. Very few manufacturers can survive a trauma of assigning random due dates, especially when there are penalties associated with not meeting the due dates. The tardiness penalty between 1 and 5 was assigned randomly to each job. For 4 and 5 job case, each problem was also solved with exhaustive enumeration. The relative performances of different methods are tabulated in Table 6.2. The table indicates the

TABLE 6.3
Comparisons Statistics: Minimum Tardiness Penalty in 100 Trials—50 Jobs/100 Jobs

Method	$T = 0.1$	$T = 0.5$	$T = 0.7$
EDD	100/100	39/38	0/0
COVERT	100/69	7/11	0/0
SWPT	43/27	0/0	0/0
SPT/LPUL	43/27	0/0	0/0
WT/LPUL	100/0	0/0	0/0
CRITICAL RATIO	100/51	17/25	0/0
New heuristic	100/100	100/100	100/100

number of times, in 100 test problems, each method gave the minimum tardiness sequence, the first number representing the results from the 4 job data set and the second representing the results from the 5 job data set. The average computational times on the Sun workstation for the methods ranged between 0.3 and 0.5, with the heuristic procedure taking the longest time.

Because of the long computational time for the problems with 50 and 100 jobs, the exhaustive search was not possible. In these cases, we compared the solutions from the heuristic with the solutions from all other methods. The heuristic procedure gave the least penalty sequence in each of 100 problems. Table 6.3 shows the number of times the other methods had the same tardiness penalty as the heuristic. The computational times for the other methods varied from 0.1 to 0.4 s for a 50-job problem and from 0.1 to 0.9 s for a 100-job problem. The average computational time for the new heuristic was 7.6 s for a 50-job problem and 11.99 s for a 100-job problem.

The accuracy and dependability of the eight heuristic rules decreased rapidly as the value of T increased, making the schedule tighter. The heuristic procedure consistently performed better. Where it did differ from the optimum, the best answers were within an average of 3% of the optimum.

6.5 SINGLE MACHINE PROBLEM WITH EARLY AND LATE PENALTIES

In this section, we extend the basic problem to include early penalties if a job is completed earlier than the due date, D_i. Let us assume that for job I, the cost per unit time of being early E_i is known. The earliness and tardiness costs for a job are assumed to be linear, that is, if a job I is finished t time units after the scheduled due date, a penalty of $L_i \times t$ is incurred, and if it is completed t time units before the due date, a penalty of $E_i \times t$ is incurred. The objective is to minimize the total penalty, which is defined as the sum of the penalty cost of each job. This problem is commonly referred to as the single machine scheduling problem with early and tardiness penalties.

The motivation for early and tardy penalties comes from the just-in-time (JIT) philosophy, to produce goods only when they are necessary. In general, the earliness cost can be considered as holding cost for finished goods, deterioration of perishable goods, and opportunity costs. The tardiness cost may be the backlogging cost, which includes performance penalties, lost sales, and lost goodwill.

The procedure described in Section 6.4 is slightly modified to account for earliness penalties. The step 3 of backward phase now requires calculations of both early penalties and late penalties. A job is selected in step 4 using the minimum penalty rule, which now includes both early and late penalties. Thus, the modifications needed are as shown below.

6.5.1 Backward Phase

In step 3, calculate the penalty for each unscheduled job I as $(T - D_i) \times L_i$ if $T > D_i$ or $(D_i - T) \times E_i$ if $T < D_i$.

6.5.1.1 Illustrative Example 2

The procedure is illustrated by applying it to another 4-job problem with the data in Table 6.4 that includes early penalties.

Start the procedure with the backward phase. Since all the jobs are unscheduled at this point, the total of processing times of all unscheduled jobs is the sum of processing times for all jobs, namely:

$$T = 1 + 49 + 10 + 27 = 87$$

Calculate the penalty for each job if it was to be completed at $T = 87$.

Job	Penalty
1	$(87 - 3) \times 3 = 252$
2	$(176 - 87) \times 0 = 0$
3	$(87 - 35) \times 5 = 260$
4	$(97 - 87) \times 2 = 20$

TABLE 6.4
Data for an Example Problem 2

Job Number	Processing Time P_i	Due Date D_i	Early Penalty E_i	Late Penalty L_i
1	1	3	2	3
2	49	176	0	9
3	10	35	3	5
4	27	97	2	3

Since job 2 has the least cost, it is scheduled in the last position, that is, position 4. Since job 1 now has been scheduled, the new value of $T = 87$ – processing time of job $2 = 87 - 49 = 38$.
The new penalty values for the remaining jobs are as follows:

Job	Penalty
1	$(38 - 3) \times 3 = 105$
3	$(38 - 35) \times 5 = 15$
4	$(97 - 38) \times 2 = 118$

The least penalty is with job 3, and hence it is placed in position 3.
The next value of $T = 38$ – processing time for job 3, that is, 10. Thus, $T = 28$.
The new penalty values for jobs 1 and 4 are as follows:

Job	Penalty
1	$(28 - 3) \times 3 = 75$
4	$(97 - 28) \times 2 = 138$

Job 1 is placed in position 2, and the remaining job, job 4, is placed in position 1. This results in the job sequence of 4-1-3-2 with following start and completion times.

0 - (4/27) - 27 - (1/1) - 28 - (3/10) - 38 - (2/49) - 87.

The total penalty for the above schedule is 230.

Application of the Forward Phase results in the optimum sequence of:

1 3 2 4

With penalty of 96.

6.5.1.2 Modified Backward Phase When Early Penalties are Present

A slight modification to the backward phase presented earlier leads to a solution that is very close to, if not, the optimum. Notice as we move from the last position to the first position in the sequence, the completion times of the jobs are decreasing. When examining a job for the present position, if it has an early penalty now and we do not schedule it, the penalty will only go up when scheduled in an earlier position. For job that has late penalty now, this penalty will decrease if we schedule it in an earlier position. So the new rule is as follows:

In backward phase, select a job with maximum early penalty and schedule it in the position under examination. If no job with penalty is available, select a job with minimum late penalty and place it in that position.

Applying this rule to the previous example leads to following:
If a job is completed at T = 87

Job	Late/Early	Penalty
1	L	
2	E	$(171 - 87) \times 0 = 0$
3	L	
4	E	$(97 - 87) \times 2 = 20$

Maximum E penalty is 20 associated with job 4. Place job 4 in last position.

– – – 4

New T = 87 – Processing Time of Job 4 = 87 – 27 = 60
If Job is completed at $T = 60$

Job	Late/Early	Penalty
1	L	
2	E	$(171 - 60) \times 0 = 0$
3	L	

Select Job 2 with early penalty for position 3.

– – 2 4

New T = 60 – Processing Time of Job 2 = 60 – 49 = 11
If a Job is completed at T = 11

Job	Late/Early
1	L
3	E

Select Job 3 since it is the only job with early penalty.

– 3 2 4

Since job 1 is the only for position 1, place it in that position. The final sequence from Backward Phase is:

1 3 2 4

Which is also the optimum sequence when applied the forward phase.

The procedure leads to 1-3-2-4 as the final sequence with the cost of 96. Incidentally, exhaustive enumeration also displayed 1-3-2-4 as the optimum sequence.

6.5.2 Validity of the Heuristic

Similar to the validation of the heuristic for tardiness penalty alone, a computer program was developed that evaluates many of the existing heuristic methods along with the heuristic for early and late penalties. Also, a program that exhaustively enumerates the optimum solution, by checking $N!$ combinations, was developed. The tardiness factor T was used to generate 100 data sets for each file of 4, 5, 50, and 100 jobs. T is defined as $T = 1 - D/NP$, where D is the average due date, P is the average processing time, and N is the number of jobs.

We developed the data sets of 100 problems each for $T = 0.1$, 0.5, and 0.7. For each job, the processing time was chosen randomly between 1 and 50, and the corresponding due date was established by $D_i = P_i \times N \times (1 \times T)$. This approach of assigning due dates is more consistent and in line with what a manufacturer may wish. Very few manufacturers can survive the trauma of assigning random due dates, especially when there are penalties associated with not meeting the due dates. The tardiness penalty between 1 and 5 was assigned randomly to each job. For 4 and 5 job case, each problem was also solved with exhaustive enumerations. The relative performances of different methods are tabulated in Table 6.5. The table indicates the number of times, in 100 test problems, each method gave the minimum tardiness sequence, the first number representing the results from the 4-job data set and the second representing the results from the 5-job data set. The average computational times on the Sun workstation for the methods ranged between 0.3 and 0.5, with the heuristic procedure taking the longer time.

Because of the long computational time for problems with 50 and 100 jobs, an exhaustive search was not possible, and we compared the solutions from the heuristic with the solutions from all other methods. The heuristic procedure gave the least penalty sequence in each of the 100 problems. Table 6.6 shows the number of

TABLE 6.5
Comparisons Statistics: Minimum Tardiness Penalty in 100 Trials—4 Jobs/5 Jobs

Method	$T = 0.1$	$T = 0.5$	$T = 0.7$
EDD	100/100	46/44	22/15
COVERT	57/59	42/31	59/41
SWPT	83/77	54/40	92/76
SPT/LPUL	100/100	48/48	23/15
WT/LPUL	32/30	15/11	26/12
CRITICAL RATIO	100/100	13/29	16/9
New heuristic	100/100	98/97	99/98
Exhaustive search	100/100	100/100	100/100

TABLE 6.6
Comparisons Statistics: Minimum Tardiness Penalty in 100 Trial—50 Jobs/100 Jobs

Method	$T = 0.1$	$T = 0.5$	$T = 0.7$
EDD	100/100	37/37	0/0
COVERT	100/67	7/11	0/0
SWPT	41/27	0/0	0/0
SPT/LPUL	100/98	43/37	0/0
WT/LPUL	100/0	0/0	0/0
CRITICAL RATIO	100/51	17/25	0/0
New heuristic	100/100	100/100	100/100

times the other methods had the same tardiness penalty as the heuristic. The computational times for the other methods varied from 0.1 to 0.4 s for a 50-job problem and from 0.1 to 0.9 s for a 100-job problem. The average computational time for the heuristic procedure was 7.6 s for a 50-job problem and 11.99 s for a 100-job problem.

Again, we see that the new heuristic consistently performed well. If it did differ from the optimum on average, the answers were within 2% of the optimum.

6.6 SOME WELL-KNOWN THEOREMS

The validity checks in Sections 6.5 and 6.6 showed that the eight heuristic rules do not necessarily perform well when the objective is to minimize total tardiness penalty or the early-tardy penalty. The question that arises now is, if the heuristic rules are not performing well, why are they so popular? The answer is: Some of them do give optimal sequences with other important objectives. These objectives measure different criteria in a schedule. A few well-known theorems in single machine scheduling literature associated with these criteria are presented here without their proofs. The reader interested in the proofs is directed to other excellent scheduling books (Baker, 1974; Pinedo, 1995) listed in the references.

1. One measure of scheduling efficiency is to minimize the total weighted completion time or weighted flow time. It is defined as $3W_iC_i$, where W_i is the weight proportional to the value invested (perhaps indicating the importance of the job) and C_i is completion time for job I. This measure gives an indication of the holding or inventory cost incurred by the schedule.

Theorem 6.1: Weighted flow time is minimized if jobs are sequenced based on LPUL rule (here $L_i = W_i$) or the SWPT rule. It should be noted that both LPUL and SWPT rules will produce the same sequence.

Lemma 6.1: If all jobs are equally important (i.e., W_i for all jobs is same), then the SPT rule minimizes both the total flow time and the mean flow time.

2. Let us measure the lateness of job I as $LT_i = C_i - D_i$, where C_i is the completion time for job I. LT_i can be both negative and positive, depending on whether the job is completed earlier or later than the due date. This is unlike tardiness $T_i = C_i - D_i$, which can only take positive values since $T_i = 0$ if $C_i - D_i < 0$. Lateness, thus, measures the conformity of the schedule to known due dates. A schedule that has lower mean value of lateness is completing jobs closer to the due dates than the schedule that has higher mean lateness value.

Theorem 6.2: The mean lateness is minimized by SPT rule sequencing.

Theorem 6.3: The maximum job lateness, that is, the maximum value among LT_i, is minimized by EDD rule sequencing.

Theorem 6.4: If all jobs have a common due date, then SPT rule sequencing minimizes the mean tardiness.

Theorem 6.5: The maximum tardiness is minimized by EDD rule sequencing.

6.7 SUMMARY

This chapter examined the basic single machine scheduling problem, where jobs have due dates and there are penalties associated with their being tardy. The objective of scheduling jobs is to minimize the total tardiness penalty. Exhaustive enumeration, although it gives an optimum sequence, is only feasible for small-size problems, that is, where the numbers of jobs are less than ten. The branch-and-bound method reduces the computations, but is still overwhelming. The heuristic rules, although they work efficiently for other objectives, do not provide a good solution for the objective of minimizing total early/tardy cost. A heuristic for tardy jobs and its modification for early and tardy jobs is suggested, which has proven to be very effective in minimizing the total penalty cost. Some theorems have been stated without proofs to illustrate the importance of the heuristic rules if the scheduling efficiency is measured using other criteria.

6.8 PROBLEMS

6.1 Using the BF heuristic presented in this chapter, develop a schedule for the following data to minimize the lateness penalty. Also, solve the problem using the branch-and-bound method, and compare the results.

Job #	Processing Time	Due Date	Weight
1	24	38	2
2	13	22	6
3	32	61	6
4	9	47	2

6.2 For the data given in Problem 2.1, develop a schedule using each of the following heuristics and compare each result to that obtained in Problem 2.1.

a. EDD
b. COVERT
c. SPT
d. LPUL
e. SWPT
f. CR

6.3 Develop a schedule for the following data using the BF heuristic presented in this chapter.

Job #	Processing Time	Due Date	Weight
1	24	38	2
2	13	22	6
3	32	61	6
4	9	47	2
5	3	8	6

6.4 For the data given in Problem 2.3, develop a schedule using each of the following heuristics and compare each result to that obtained in Problem 2.3.

a. EDD
b. COVERT
c. SPT
d. LPUL
e. SWPT
f. CR

6.5 Rework Problem 2.1 with the addition of the following early and late penalties.

Job #	1	2	3	4
Early penalty	0	2	1	2
Late penalty	2	6	6	2

6.6 Rework Problem 2.3 with the addition of the following early and late penalties.

Job #	1	2	3	4	5
Early penalty	0	2	1	2	2
Late penalty	2	6	6	2	2

REFERENCES AND SUGGESTED READINGS

Ahmed, I. and W.W. Fisher. 1992. "Due Date Assignment, Job Order Release and Sequencing" *Decision Sciences*, 23: 633–644.

Baker, K.R. 1974. *Introduction to Sequencing and Scheduling*, New York: Wiley.

Baker, K.R. and G.D. Scudder. December 1991. "Sequencing with Earliness and Tardiness Penalties: Part I" *Computers and Operations Research*, 38: 22–36.

Cambadi, B.V. December 1991. "One Machine Scheduling to Minimize Expected Mean Tardiness: Part I" *Computers and Operations Research*, 18(9): 787–796.

Conway, R.W., Maxwell, W.L., and L.W. Miller. 1967. *Theory of Scheduling*, Reading, MA: Addison-Wesley.

Fathi, Y. and H.W.L. Nuttle. September 1990. "Heuristics for the common Due Date Weighted Tardiness Problem" *IIE Transactions*, 22: 215–225.

Hall, N.G., Kubiak, W., and S. Sethi. September–October 1991. "Weighted Deviation of Completion Times about a Restrictive Common Due Date" *Operations Research*, 39: 847–856.

Hall, N.G. and M.E. Posner. September–October 1991. "Weighted Deviation of Completion Times about a Common Due Date" *Operations Research*, 39: 836–846.

Johnson, L.A. and D.C. Montgomery. 1974. *Operations Research in Production Planning, Scheduling and Inventory Control*, New York: Wiley.

Lee, C.Y., Danusaputro, S.L., and C.S. Lin. 1991. "Minimizing Weighted Number of Tardy Jobs and Weighted Earliness-Tardiness Penalties about a Common Due Date" *Computers and Operations Research*, 379–389.

Ow, P.S. and T.E. Morton. February 1989. "The Single Machine Early/Tardy Problem" *Management Science*, 35: 177–187.

Pinedo, M. 1995. *Scheduling Theory, Algorithms and Systems*, Englewood Cliffs, NJ: Prentice Hall.

Srinivasan, V. 1971. "A Hybrid Algorithm for the One Machine Sequencing Problem to Minimize Total Tardiness" *Naval Research Logistics Quarterly*, 18: 317–127.

Sung, C.S. and U.G. Joo. November 1992. "A Single Machine Scheduling Problem with Earliness, Tardiness and Starting Time Penalties under a Common Due Date" *Computers and Operations Research*, 19(8): 757–766.

7 Other Objectives in Single-Machine Scheduling

In this chapter, we shall study numerous alternative objectives commonly used to schedule jobs on a single machine. One or more of these goals may be more appropriate in a particular work surroundings than merely to minimize early/tardy penalties discussed in Chapter 6. The basic assumptions stated in Chapter 6 are still applicable, unless specifically modified to accommodate a goal. Most of these problems are generally solved by a heuristic, and many researchers have developed different heuristic procedures to address these problems. The heuristic methods described in this chapter are, to a large part, extensions of the methods studied in Chapter 6. All of the procedures are well tested and found to give optimum or near-optimum solutions; therefore, independent validly test results for each goal are not shown.

7.1 COMMON DUE DATE

Scheduling jobs on a single machine or facility, when all jobs must be shipped at the same time, is a typical example of a common due date problem. Such instances occur, for example, when a customer orders different variations of a product, all of them being made on the same facility. For example, a paper manufacturer may supply paper rolls with different widths (24″, 30″, and 36″) to a customer who may process them further in bags. Because of the transportation cost, the customer may ask all rolls to be delivered in the same truck, if possible. If not, there would be additional cost of inconvenience charged as the lateness cost per unit of time. There may also be an early penalty indicating storage and carrying cost.

In such a case, one might think that the optimum due date would be one that is equal to or greater than the total processing time of all jobs. However, this may not be the case. Since the penalties are associated with both early and late completions, an optimum due date should balance these penalties.

If the common due date specified by the customer is less than the summation of the processing times of all the jobs, the jobs are shipped in two shipments, one on the specified due date and second on the completion time of all the jobs. This proposed heuristic tries to find the minimum penalty without considering any specific common due date. So, if the customer-specified common due date is somewhere near the common due date obtained by this heuristic, one can bargain with the customer, and reduce the loss incurred, by reducing the penalty.

Assume that the number of jobs to be performed and their respective processing times, early penalty and late penalty are known. The objective is to find the minimum penalty. The end result will give the common due date, the sequence to be followed and the total penalty incurred, so that the penalty is minimum most. The steps for the procedure are as follows:

1. Check the early and late penalties for all the jobs yet to be sequenced and find the minimum.
2. If the minimum is associated with the early penalty, place the corresponding job in the earliest possible position in the sequence. If the minimum is associated with the late penalty, place the corresponding job in the latest possible position in the sequence. If there is a tie in the early penalties, the job with the largest processing time is scheduled at the earliest possible position. If there is a tie in the late penalties, the job with the largest processing time is scheduled at the latest possible position in the sequence. If the tie remains, it is solved randomly. If the early and the late penalties are the same for a job, two sequences are obtained by first placing that job at the earliest possible position, and second, by placing that job at the latest possible position in the sequence.
3. Mark the job as being sequenced and scratch the associated early and late penalties.
4. If all the jobs are placed in the sequence, go to step 5; otherwise go to step 1.
   - The sequences obtained are called the initial sequences, or the best sequence obtained so far. By considering the completion time of each job in the initial sequence as the common due date, the penalty is calculated. The job completion time where the penalty is minimum is the best common due date for the initial sequence.
5. Apply only the backward phase of B/F method to the initial sequence and select the one with minimum cost.
6. Find 6%, a percentage figure chosen somewhat randomly but which has proved to be very good, of the total processing time. Call the obtained integer value as deviation. This deviation is subtracted and added to the common due date obtained so far. Check the penalties by applying the backward phase of B/F method to initial sequence with each value as the common due date. The sequence and associated common due date that gives minimum cost is the best policy.

7.1.1 Illustrative Example 7.1

A problem given by Feldmann and Biskup [19] is used to illustrate the method. The data are as follows in Table 7.1.

Following this procedure will help solve the preceding problem.

TABLE 7.1
Data for Problem

i	1	2	3	4	5	6	7	8	9	10
p_i	20	6	13	13	12	12	12	3	12	13
Early penalty	4	1	5	2	7	9	5	6	6	10
Late penalty	5	15	13	13	6	8	15	1	8	1

Step 1: The early and late penalties of the corresponding jobs are given in the third and forth rows, respectively. The initial sequence obtained by following steps 2 through 4 is as follows.

2	4	1	3	7	9	6	5	8	10

The completion times of the jobs can be given as follows:

0 (2/6) 6 (4/13) 19 (1/20) 39 (3/13) 52 (7/12) 64 (9/12) 76 (6/12) 88 (5/12) 100 (8/3) 103 (10/13)116.

By considering the job completion time as the common due date, the penalty is calculated (Table 7.2).

In the preceding table, 819 is the lowest penalty obtained for the same initial sequence or the best sequence obtained so far with the common due date as 76. Six percent of the total processing time is 6.96. The common due date obtained so far is 76. The time span obtained by subtracting and adding the integer value of 6

TABLE 7.2
Penalties Calculated with Each Due Date

Due Date	Penalty
116	$(110 \times 1) + (97 \times 2) + (77 \times 4) + (64 \times 5) + (52 \times 5) + (40 \times 6) + (28 \times 9) + (16 \times 7) + (13 \times 6) + (0 \times 1) = 1874$
103	$(97 \times 1) + (84 \times 2) + (64 \times 4) + (51 \times 5) + (39 \times 5) + (27 \times 6) + (15 \times 9) + (3 \times 7) + (0 \times 1) + (13 \times 1) = 1302$
100	$(94 \times 1) + (81 \times 2) + (61 \times 4) + (48 \times 5) + (36 \times 5) + (24 \times 6) + (12 \times 9) + (0 \times 6) + (3 \times 1) + (16 \times 1) = 1191$
88	$(82 \times 1) + (69 \times 2) + (49 \times 4) + (36 \times 5) + (24 \times 5) + (12 \times 6) + (0 \times 8) + (12 \times 6) + (15 \times 1) + (28 \times 1) = 903$
76	$(70 \times 1) + (57 \times 2) + (37 \times 4) + (24 \times 5) + (12 \times 5) + (0 \times 8) + (12 \times 8) + (24 \times 6) + (27 \times 1) + (40 \times 1) = 819$
64	$(58 \times 1) + (35 \times 2) + (15 \times 4) + (12 \times 5) + (0 \times 8) + (12 \times 8) + (24 \times 8) + (36 \times 6) + (39 \times 1) + (52 \times 1) = 843$

TABLE 7.3
Penalties over the Range of Due Dates

Due Date	70	71	72	73	74	75	76	77	78	79	80	81	82
Penalty	860	853	846	839	832	825	818	825	832	839	846	853	860

to this common due date is 70–82. Start applying the complete backward-forward heuristic with the common due date as 70. Increment the common due date by 1 until we reach 82. The results are given in Table 7.3.

The minimum value obtained is 818, associated with the due date 76.

The associated sequence using only the backward phase of the B/F method on the initial sequence is: 4-2-1-3-7-9-6-5-8-10. The details of this sequence are as follows:

0 (4/13) 13 (2/6) 19 (1/20) 39 (3/13) 52 (7/12) 64 (9/12) 76 (6/12) 88 (5/12) 100 (8/3) 103 (10/13) 116.

7.2 COMMON DUE DATE SPECIFIED BY A CUSTOMER

Often, a common due date is specified by a customer. All the jobs are to be delivered on the specified due date. If the entire set of jobs are not produced by then, the customer may still accept the jobs that are completed but will associate penalties for the jobs that are not delivered on the due date. The penalties for the jobs may vary based on the importance of the jobs, perhaps the customer's way of specifying which jobs he would like to get done first and which may be later.

Both early and late penalties might apply for each job. The late penalties are customer defined, while early penalties may indicate the cost of storing and inventorying of the finished jobs until they are delivered.

7.3 EARLY AND LATE DUE DATES

On occasions we have customers who may specify a range of due dates. An early due date is a date before which the order is not desired and will be assigned an early penalty, and a late due date is a date after which the order is considered late and carries a late penalty. The penalties are proportional to the time units of deviations from early due date, DE_i or the late due date, DL_i. We are required to develop the production schedule to minimize the penalty costs.

Again, a minor modification to the heuristic is all that is needed to solve this problem. When a job is scheduled, determine if the job completion time C_i is earlier than the early due date DE_i, within due dates (on time indicated by W) or later than the late due date, DL_i. The penalties for application of step 3 in Section 6.4 are assessed as follows:

1. If $C_i < ED_i$ — Penalty $= (ED_i - C_i)\, E_i$
2. If $ED_i < C_i < LD_i$ (i.e., $C_i = W$) — Penalty $= 0$
3. If $C_i > LD_i$ — Penalty $= (C_i - LD_i)\, L_i$

TABLE 7.4
Data for Early and Late Due Dates: Example 7.2

Job	P_i	ED_i	LD_i	E_i	L_i
1	10	20	25	2	3
2	15	25	33	1	4
3	5	30	35	1	3
4	20	40	50	3	5

One more modification in the backward phase; if there are a number of jobs tied for placement in a sequence position because of the equal penalties, select the job with the longest processing time from the jobs that are tied, and place it in the position under consideration. The remaining steps of the heuristic remain unchanged.

7.3.1 Illustrative Example 7.2

Consider a four job problem with the following data (Table 7.4).

Application of the backward phase:

Iteration 1: $T = 10 + 15 + 5 + 20 = 50$. (Note, T indicates the possible completion time, C_i, for a job, and the penalties are calculated based on the question: What if the job is completed at time T?)

Determine the job to schedule in position 4:

Job	Status	Penalty
1	L	$(50 - 25) \times 3 = 75$
2	L	$(50 - 33) \times 4 = 68$
3	L	$(50 - 35) \times 3 = 45$
4	W	0

The minimum penalty is associated with job 2, place the job in position 4.

Iteration 2: $T = 50 - 20 = 30$

Job	Status	Penalty
1	L	$(30 - 25) \times 3 = 45$
2	W	0
3	W	0

Jobs 2 and 3 are tied with the least penalty. Schedule the job with longest processing time, job 2 and place it in position 3. The partial sequence so far is –, –, 2-4.

Iteration 3: $T = 30 - 15 = 15$

Job	Status	Penalty
1	E	$(20 - 15) \times 2 = 10$
3	E	$(30 - 15) \times 1 = 15$

Job 2 with minimum penalty is scheduled in position 2, and the remaining job, job 2, goes in position 1.

The best sequence is 3-1-2-4.

Schedule:	0	(3/5)	5	(1/10)	15	(2/15)	30	(4/20)	50
Due dates:		(3)	30/E	(1)	20/E	(2)	W	(4)	W
Delays:			25/E		5/E				
Penalty:			$25 \times 1 + 5 \times 2 = 35$						

Application of the forward phase.

Cycle 1. Iteration 1. Best sequence 3-1-2-4. $K = 3$. Exchange 3-4. Sequence to examine: 4-1-2-3

Schedule:	0	(4/20)	20	(1/10)	30	(2/15)	45	(3/5)	50
Due dates:		(4)	40/E	(1)	25/L	(2)	33/L	(3)	35/L
Delays:			20/E		5/L		12/L		15/L
Penalty:			$20 \times 3 + \ldots\ldots > 35$, no exchange.						

Cycle 1. Iteration 2. Best sequence 3-1-2-4. $K = 2$. Exchange 3-2. Sequence to examine: 2-1-3-4.

Schedule:	0	(2/15)	15	(1/10)	25	(3/5)	30	(4/20)	50
Due dates:		(2)	25	(1)	W	(3)	W	(4)	W
Delay:			10/E						
Penalty:			$10 \times 1 = 10$, <35 Exchange.						

The new best sequence is 2-1-3-4.

Cycle 2: Iteration 1: Best sequence 2-1-3-4. $K = 3$. Exchange 2-4. Sequence to examine: 4-1-3-2.

Schedule:	0	(4/20)	20	(1/10)	30	(3/5)	35	(2/15)	50
Due dates:		(4)	W	(1)	25/L	(3)	W	(2)	33/L
Delays:					5/L			17/L	
Penalty:		$5 \times 3 + 17 \times 4 = 83 > 10$, no exchange.							

Cycle 2: Iteration 2. Best sequence 2-1-3-4. $K = 2$. Exchange 2-3. Sequence to examine: 3-1-2-4.

The sequence was already examined and has a penalty of 35. No exchange is made.

Cycle 2: Iteration 3. Best sequence 2-1-3-4. $K = 2$. Exchange 1-4. Sequence to examine 2-4-3-1.

Schedule:	0	(2/15)	15	(4/20)	35	(3/5)	40	(1/10)	50
Due dates:		(2)	25/E	(4)	40/E	(3)	35/L	(1)	25/L
Delays:			10/E		5/E		5/L		25/L
Penalty:	$10 \times 1 + 5 \times 3 + 5 \times 3 + 25 \times 3 = 115 > 10$, no exchange.								

Cycle 2: Iteration 4. Best sequence 2-1-3-4. $K = 1$. Exchange 2-1. Sequence to examine 1-2-3-4.

Schedule:	0	(1/10)	10	(2/15)	25	(3/5)	30	(4/20)	50
Due dates:		(1)	20/E	(2)	W	(3)	W	(4)	W
Delays:			10/E						
Penalty:			$10 \times 2 = 20 > 10$, no exchange.						

Cycle 2: Iteration 5. Best sequence 2-1-3-4. $K = 1$, Exchange 1-3. Sequence to examine 2-3-1-4.

Schedule:	0	(2/15)	15	(3/5)	20	(1/10)	30	(4/20)	50.
Due dates:		(2)	25/E	(3)	30/E	(1)	20/L	(4)	50
Delays:			10/E		10/E		10/L		W
Penalty:		$10 \times 1 + 5 \times 3 + 10 \times 3 = 55 > 10$, no exchange.							

Cycle 2: Iteration 6: Best sequence 2-1-3-4. $K = 1$, Exchange 3-4. Sequence to examine 2-1-4-3.

Schedule:	0	(2/15)	15	(1/10)	25	(4/20)	45	(3/5)	50
Due dates:		(2)	25/E	(1)	W	(4)	W	(3)	35/L
Delays:			10/E					15/L	
Penalty:			$10 \times 1 + 15 \times 3 = 55 > 10$ no exchange.						

Since all the exchanges have been examined, the current best sequence 2-1-3-4 is the overall best sequence, with the penalty of 10 units.

7.4 QUADRATIC OR NONLINEAR PENALTY FUNCTION

So far, we have been considering both early and late penalties as linear functions of the difference between job completion times and the due dates. Often, the thought is that as a job completion time deviates further and further away from the due date, it should be assessed at a much higher penalty and not just be charged a penalty that is linearly proportional to the deviation. Quadratic penalty function reflects this sentiment. The cost for early completion and/or late completion may be stated as E_iX^2 or L_iX^2, respectively, where X is the deviation between completion time and due date for the job. As X increases, the cost increases in a quadratic manner. In some instances, not all jobs have quadratic penalty functions, reflecting perhaps the importance that is attached toward completion of a job on or very close to the due date. For example, some job cost may still be linearly proportional to X while others may be a step function such as cost $= A + BX$.

To develop the best schedule requires a simple modification to the procedure applied in Section 6.4. The penalties are calculated based on the individual job's cost function. All other steps of the procedure remain the same.

7.4.1 Illustrative Example 7.3

The cost functions for four jobs in Section 6.2 are now modified to reflect acceptable early and late completion dates. The data are shown in Table 7.5.

Here, the user is placing more emphasis on completion of job 2 on time (not too early or too late), while job 2 may be early but not too late. The penalty values are calculated using appropriate functions for jobs 2 and 3. For example, at $T = 50$, the backward process leads to penalty values as follows:

Job	Status	Penalty
1	L	$(50 - 25)^2 \times 3 = 1875$
2	L	$(50 - 33) \times 4 = 68$
3	L	$(50 - 35)^2 \times 3 = 675$
4	W	0

The remaining steps being the same as shown in Section 6.4, we leave it to the reader to complete the calculations and develop the best sequence.

7.5 MINIMIZATION OF THE AVERAGE DELAY

Another objective that is frequently preferred by schedulers is the minimization of the average delay. Here, each job is equally important and has an associated due date. The preference is of course to deliver all jobs on time; however, if there are going

TABLE 7.5
Data for Quadratic Penalty: Example Problem 7.3

Job	P_i	ED_i	LD_i	E_i	L_i
1	10	20	25	$2X^2$	$3X^2$
2	15	25	33	1	4
3	5	30	35	1	$3X^2$
4	20	40	50	3	4

to be some delays then we want to develop a schedule that will minimize the average of all delays.

This objective can be accommodated by modifying the procedure described in Section 6.4. Since we are only concerned with the delays, only the late penalties are of interest. Remember that the procedure in Section 6.4 develops a sequence that minimizes the total penalty, which is defined as the sum of delay times weight for each job. Here, all jobs are equally important and all have equal weights (e.g., one), and hence, the application of the Section 6.4 procedure, which leads to minimization of the total penalty, also leads to minimization of the total delay. With the number of jobs being constant, this is also equivalent to minimization of the average delay.

7.6 MINIMIZATION OF THE MAXIMUM DELAY

Another objective may be to minimize the maximum delay associated with any job. Scheduling jobs based on earliest due dates (EDD) rule leads to a schedule that gives the least possible value for the maximum delay (Theorem 2.5).

7.7 MINIMIZE THE NUMBER OF JOBS THAT ARE DELAYED

It is not the magnitude of the total delay or the delay associated with an individual job that may be of importance but rather the number of jobs that are delayed may be critical. Here, the objective is to maximize the number of jobs that are performed by their due dates. A simple procedure to achieve this objective has the following steps:

1. Arrange the jobs in ascending order of the due dates. Let T denote the completion time of all jobs that are in the "select set." Initially, there are no jobs in the select set, and therefore, the value of T is zero.
2. If all the jobs have been examined go to step 6, if not, select the first unexamined job, call it job i, in the order established in step 1. Add the processing time of the job i to the current value of T and denote it by T temp. This indicates the completion time of the job i if it is selected to process next. Compare it with its due date. If T temp is less than or equal to the due date for job i, go to step 3; if it exceeds the due date, go to step 4.
3. Add job i to the select set. Make T equal to T temp. Go to step 2.
4. Select the job with maximum processing time from the select set, call it job j, and compare its processing time with job i. If the processing time of the job i is greater or equal to the processing time of job j, job i is not selected, go back to step 2. If the processing time of job i is less than that of job j, go to step 5.
5. Check if by removing the largest processing job from select set, job j, and replacing it with the job under consideration, job i, we will be able to finish the job i by its due date. If we can, perform this operation and modify the value of T by making it equal Ttemp and subtracting from it the difference between the processing times of job j and I. Go back to step 2.

TABLE 7.6
Data for Minimization of Number of Jobs Delay: Example 7.4

Jobs	1	2	3	4	5
Processing times	10	15	8	12	22
Due dates	15	25	27	32	40

If we cannot complete the job i by its due date, job i is not selected and the value of T remains the same. Go back to step 2.

6. All the jobs have been examined. The select set gives the maximum number and sequence of jobs that can be done by their due dates.

7.7.1 Illustrative Example 7.4

Consider the data given in Table 7.6 where the objective is to minimize the number of jobs that are delayed.

The jobs are arranged in the ascending order of their due dates. Initially, all the jobs are unexamined and there are no jobs in the select set. Thus, $T = 0$. Select the first job. $T\text{temp} = T + 10 = 10$.

Since Ttemp is less than the due date for the job 2, it is selected and included in the select set. T is now made equal to Ttemp and hence has a value of 10.

The next job to examine is job 2. $T\text{temp} = 10 + 15 = 25$. Since this value is equal to the due date of job 2, the job is selected and added to the select set. The select set now has jobs 2 and 2. $T = T\text{temp}$, which is equal to 25.

For job 2, $T\text{temp} = 25 + 8 = 33$, which exceeds the due date for job 2. According to step 4, the job with the maximum process time in the select set is job 2, with the processing time of 15. This is greater than the processing time for job 2. Removing job 2 from the select set will make $T = 25 - 15 = 10$. Now $T\text{temp} = 10 + 8 = 18$, which is lower than the due date for job 2. Hence, according to step 5, job 2 is removed from the select set and job 2 added. The new value of T is 18.

The next job is job 2. $T\text{temp} = 18 + 12 = 30$ which is less than 32, the due date for the job and hence it is selected and added to the select set, which now has jobs 2, 3, 4 with $T = 30$.

When the processing time of job 2 is added in T, Ttemp becomes 52, which is greater than its due date. Applying step 4, the job with the largest processing time in the select set is job 2. Its processing time of 12 is smaller than the processing time for job 2, and hence job 2 is not selected.

All the jobs are now examined. The maximum number of jobs that can be processed on time is three, in sequence of 1-3-4 with the following results:

Schedule:	0	(1/10)	10	(3/8)	18	(4/12)	30
Due dates		(1)	15	(3)	27	(4)	32
Deviation			5/E		8/E		2/E

Note, none of the three accepted jobs are delayed.

7.8 MAXIMIZE THE NUMBER OF JOBS PROCESSED WHEN THE AVAILABLE TIME IS LESS THAN TOTAL PROCESSING TIME

Suppose we have a set of jobs that cannot be performed in the available time, that is, the sum of processing times is greater than the time available on the facility. Also, each job has a weight that is proportional to the importance of the job to the customer and also has late and early due dates and associated penalties. Which jobs should be completed and which ones should be discarded? The following steps suggest a procedure.

1. Calculate for each job the ratio $M_i = L_i/P_i$. Since the late penalty indicates the importance of the job (e.g., higher the late penalty, the more important the job is to the customer), the ratio indicates the importance of the job per unit time.
2. Check if the present total processing time required is equal to or less than the available process time on the facility. If yes, go to step 3. If not, eliminate the job with the smallest value of M_i. If there is a tie, apply the following rules:
   a. Choose a job with the largest processing time, if (present time required – processing time of the job) is greater than the total time available for processing on the facility.
   b. Choose a job such that (the present time required – processing time) is as close to the time available. Perform step 2 until we can go to step 3.
3. Apply the procedure from Section 7.3 on the remaining jobs to obtain the best sequence.

7.8.1 ILLUSTRATIVE EXAMPLE 7.5

Consider the job data given in Table 7.7. Suppose the facility is available for only 50 time units, and we want to maximize the number of jobs processed.

The total of all processing times is 77. Since only 50 time units are available on the facility, we cannot process all jobs. The first job to drop is job 2 with the lowest value of M. The processing time still required is $77 - 13 = 64$ and hence at least one more job needs to be dropped. The next smallest value of M is 0.20 and jobs 2, 7, and 8 are tied. Based on step 2a, the largest processing time among these is 16, which is associated with job 2. $(64 - 16) = 48$, which is less than 50. Hence, we apply step 2b. Based on step 2b, trying jobs 2, 7, and 8, elimination of job 2 brings us closest to the available time of 50 units, and hence job 2 is dropped.

Applying step 3, the best sequence obtained for the remaining job is 4-5-6-7-3-1-9.

TABLE 7.7
Data to Maximize the Number of Jobs Processed: Example 7.5

Job	Processing Time	Early Due Date	Late Due Date	Early Penalty	Late Penalty	*M*
1	10	15	17	1	2	0.20
2	13	20	24	1	1	.077
3	5	25	27	2	3	0.6
4	8	10	11	2	5	0.62
5	7	15	16	2	6	0.85
6	3	17	19	1	2	0.66
7	5	13	17	0	1	0.20
8	16	30	32	1	3.2	0.20
9	12	50	51	2	5	0.41

TABLE 7.8
Data for Sequence Dependent Jobs: Example 7.6

Job Number	Processing Time P_i	Early Due Date ED_i	Late Due Date LD_i	Early Penalty E_i	Late Penalty L_i
1	15	20	25	1	3
2	20	25	30	0	3
3	30	30	40	2	3
4	15	20	25	2	2
5	30	40	50	1	2

7.9 SEQUENCE-DEPENDENT JOBS

At times, especially in the chemical industry, the same facility is used to produce different products. For example, to make chemical X may require adding chemicals Y and Z, which are both produced on the same facility. To keep the setup costs down, it may be required or strongly preferred to produce chemical Y before Z. How can we develop a sequence that accommodates such precedence relationships?

Only a minor change is needed in the procedure Section 6.4. When calculating the penalties in the backward phase, assign a very large penalty for a job whose succeeding job has not yet been scheduled. Also, any job exchanges in the forward phase must satisfy the precedence relationships or that sequence is invalid and has an infinite cost.

7.9.1 Illustrative Example 7.6

Consider the data given in Table 7.8 in which a constraint is that job 3 cannot be processed until both jobs 1 and 2 are done.

Applying backward phase

$$T = 15 + 20 + 30 + 15 + 30 = 110$$

Job	Penalty	Remarks
1	Infinity	Job 2 cannot be scheduled before 3 in the backward phase.
2	Infinity	Job 2 cannot be scheduled before 3 in the backward phase
3	$(110 - 40) \times 3 = 210$	
4	$(110 - 25) \times 2 = 170$	
5	$(110 - 50) \times 2 = 120$	

The minimum penalty job is job 2; therefore, it is placed in the fifth position.

$$T = 110 - 30 = 80$$

Job	Penalty	Remarks
1	Infinity	Job 1 cannot be scheduled before job 3 in the backward phase
2	Infinity	Job 2 cannot be scheduled before job 3 in the backward phase
3	$(80 - 40) \times 3 = 120$	
4	$(80 - 25) \times 2 = 110$	

Place job 2 in sequence position 4. The partial sequence so far is –, – ,– , 4, 5

$$T = 80 - 15 = 65$$

Job	Penalty	Remarks
1	Infinity	Job 1 cannot be scheduled before job 3 in the backward phase
2	Infinity	Job 2 cannot be scheduled before job 3 in the backward phase
3	$(65 - 40) \times 3 = 75$	

Place job 2 in position 3. The partial job sequence so far is –, –, 3, 4, 5

$$T = 65 - 30 = 35$$

Job	Penalty	Remarks
1	$(35 - 25) \times 3 = 30$	
2	$(35 - 30) \times 3 = 15$	

Place job 2 in position 2 and therefore job 1 in position 1. The sequence is 1-2-3-4-5. The cost of the sequence is 325.

7.9.2 Forward Phase

A summary of forward phase results are displayed in the following:

Cycle 1: Best sequence. 1-2-3-4-5. Cost = 325

$K = 4$	5-2-3-4-1	Not feasible	
$K = 3$	4-2-3-1-5	Not feasible	
	1-5-3-4-2	Not feasible	
$K = 2$	3-2-1-4-5	Not feasible	
	1-4-3-2-5	Not feasible	
	1-2-5-4-3	Cost = 370	Do not exchange
$K = 1$	2-1-3-4-5	Cost = 335	Do not exchange
	1-3-2-4-5	Not feasible	
	1-2-4-3-5	Cost = 310	Exchange

Cycle 2: Best sequence: 1-2-4-3-5 Cost = 310

$K = 4$	5-2-4-3-1	Not feasible	
$K = 3$	3-2-4-1-5	Not feasible	
	1-5-4-3-2	Not feasible	
$K = 2$	4-2-1-3-5	Cost = 340	Do not exchange
	1-3-4-2-5	Not feasible	
	1-2-5-3-4	Cost = 385	Do not exchange
$K = 1$	2-1-4-3-5	Cost = 320	Do not exchange
	1-4-2-3-5	Cost = 315	Do not exchange
	1-2-3-4-5	Cost = 325	Do not exchange
	1-2-4-5-3	Cost = 340	Do not exchange

The optimum sequence is 1-2-4-3-5 with the penalty of 310.

7.10 SEQUENCE-DEPENDENT JOBS WITH MINIMUM/MAXIMUM SEPARATIONS

A rapid deterioration of chemical Y may dictate how quickly chemical Z must be produced and then mixed with each other to make a good quality of chemical X. Thus, the maximum time within which the two chemicals can be produced may have a limit. In other instances, there might be minimum separation that is required between the production of two products (jobs). Such separation may allow a sufficient lead time to purchase outside components (or chemicals) before the two jobs produced on our facility can be assembled. The due dates in each case specify the target date of production. Early penalties may signify the storage charges, while the late penalties may be due to additional charges incurred because of the deviations from the target dates.

Changes required in the procedure from Section 7.9 are as follows: In the backward phase, a job cannot be assigned unless all its prerequisites are met. These include the precedence relationship and separation requirement. Similarly, in the forward phase the job prerequisites are checked to see if the sequence is feasible. Only on a feasible sequence are the penalties calculated and further comparisons made.

7.10.1 Illustrative Example 7.7

Consider the problem illustrated in Section 7.9. Now, let us impose an additional constraint that does not allow the lag between job 3 and job 1 to be more than 15 days. The allocation in the backward phase changes slightly. When $T = 35$ and the sequence developed so far is –, –, 3, 4, 5, job 2 cannot be placed in position 2 because the processing time of job 2 is 20 units. This means if job 2 is placed in position 2, there would be time lag of 20 units between job 1 and job 3, thus violating the constraint. Indeed, any job with a processing time of greater than 15 would be ineligible in the second position. Thus, the feasible sequence is 2-1-3-4-5, which has the penalty of 335. Incidentally, this sequence remains optimum in the forward phase.

7.11 MINIMIZE VARIATION OF FLOW TIME

In a service industry, it may be desirable to provide each customer (job) with approximately the same degree of service by making him/her spend roughly the same time in the system. Since all the jobs are available initially, one way to achieve this objective is by specifying a common due date. If the due date is greater than the sum of processing times of all jobs, then all customers have received exactly the same degree of service. In achieving such goal, however, we will also carry a large penalty cost since the completed jobs will have to be held in inventory until all the jobs are processed.

Kanet (1981) has suggested an interesting way to develop a job sequence if we can deliver the jobs whenever they are processed. Here, each job is equally important, and there are no due dates or penalties defined by the customer. The time a job spends in the system includes not only its own processing time but also waiting for other jobs ahead in the sequence to be processed. The completion times for all jobs should be as close together as possible to satisfy our objective. Alternatively, this objective may be to minimize the total absolute difference in completion (TADC) times or the variance of completion times.

The procedure starts by arranging jobs in descending order of processing times. The job sequence is built from both directions, forward and backward. The first job in the arrangement is assigned to the first position in the job sequence, the second in the

TABLE 7.9
Data for Minimization of Variation in Flow Time: Example 7.8

Job #	1	2	3	4	5
Processing time	14	9	18	15	10

last or nth position, the third in the second position, the fourth in the $n-1$ position, and so on. This is called a V-shape arrangement. The resulting job sequence minimizes the TADC times given by the following expression.

$$\text{TADC} = \sum_{I=1}^{n} \sum_{j=1}^{n} |C_i - C_j|$$

7.11.1 Illustrative Example 7.8

Let us illustrate this procedure by applying it to the following example. The objective is to develop schedule to minimize the variation in flow time (Table 7.9).

The job arrangement in the descending order of processing times is: 3-4-1-5-2. The corresponding job sequence is developed by placing job 2 in the first position, job 2 in the fifth position, job 1 in the second position, job 2 in the forth position and finally job 2 in the third position. The final sequence is 3-1-2-5-4. The corresponding completion dates are

0 (3/18) 18 (1/14) 32 (2/9) 41 (5/10) 51 (4/15) 66.

The objective value is $(32-18)+(41-18)+(51-18)+(66-18)+(41-32)+(51-32)+(66-32)+(51-41)+(66-41)+(66-51)=230$.

7.12 SEQUENCE-DEPENDENT SETUP TIMES

So far, we have assumed that all jobs are independent of each other and each has its own due date and penalty values. The setup times for the jobs were implicitly included in the total processing times. However, if the setup times are sequence dependent and substantial, the objective may be to develop a schedule that would minimize the total setup time.

7.12.1 Illustrative Example 7.10

Table 7.10 shows the setup times when job j is processed after job i. We can solve the problem by considering all possible combinations, that is, by applying exhaustive enumerations. For small problems, this is possible, but for large n the method becomes impractical. We can also treat the problem as a traveling salesman problem (Taha, 1987) and apply a heuristic such as the "next best (NB) rule." According to the rule, if

TABLE 7.10
Sequence-Dependent Setup Times: Example 7.10

		Follower Job j					
		1	2	3	4	5	6
Predecessor Job i	1	0	5	8	3	6	4
	2	8	0	6	7	5	9
	3	6	9	0	2	4	9
	4	1	4	3	0	6	4
	5	3	10	12	10	0	8
	6	2	9	14	8	9	0

job i is being processed, select the next unprocessed job j that requires the minimum setup time".

Suppose we start with job 2; the next job selected is job 2, since it has the lowest setup cost following job 2. After job 2, job 2 is selected, since it has the lowest setup cost of all unprocessed jobs following 4. Continuing the process leads to the sequence 1-4-3-5-6-2, with total setup time of 27. We might try different starting jobs and determine the best sequence and associated cost and then select the minimum cost sequence as the operational sequence. In our example, the sequences and associated costs are as follows:

$$2-5-1-4-3-6=23, \quad 3-4-1-6-2-5=23,$$
$$3-4-1-6-5-2=23, \quad 4-1-6-2-5-3=31,$$
$$5-1-4-3-2-6=27, \quad 6-1-4-3-5-2=22$$

7.13 DUAL CRITERIA

Each procedure discussed so far develops a schedule that performs the best for the intended criterion. What would happen if we have two or more objectives that we want to optimize simultaneously? Is it possible to develop a schedule that will be performed best for multiple objectives? The answer is it may or may not be. The optimum sequence for one criterion may not be optimum for another, and we may have to seek a compromise solution. The procedure is demonstrated by the following example.

7.13.1 Illustrative Example 7.11

For example, consider the data from Table 7.11

Supposing we have two objectives: (1) to minimize penalty (P) and (2) to minimize the number of tardy jobs (NT).

It is not possible to apply both criterions simultaneously within a solution procedure. We might apply the procedure for minimization of penalty and determine

TABLE 7.11
Data for Multiobjective Scheduling: Example 7.11

Job number	Processing Time	Due Date	Early Penalty	Late Penalty
1	5	45	0	1
2	5	20	1	1
3	5	15	0	1
4	10	20	1	2

the value of the second criteria for every sequence as it develops and then select a sequence that best serves both criteria. The solution is shown in the following:

Backward Phase

2-3-4-1 $P = 15$ NT $= 0$

Forward Phase

Cycle 1	$K = 3$	1-3-4-2	$P = 5$	NT $= 1$	Accept the exchange
Cycle 2	$K = 3$	2-3-4-1	$P = 15$	NT $= 0$	
	$K = 2$	4-3-1-2	$P = 15$	NT $= 0$	
	$K = 2$	1-2-4-3	$P = 20$	NT $= 1$	
	$K = 1$	3-1-4-2	$P = 5$	NT $= 1$	Accept the exchange
Cycle 3	$K = 3$	2-1-4-3	$P = 25$	NT $= 1$	
	$K = 2$	4-1-3-2	$P = 20$	NT $= 2$	
	$K = 2$	3-2-4-1	$P = 10$	NT $= 0$	
	$K = 1$	1-3-4-2	$P = 5$	NT $= 1$	
	$K = 1$	3-4-1-2	$P = 10$	NT $= 1$	
	$K = 1$	3-1-2-4	$P = 15$	NT $= 1$	

The calculations show some dominating sequences, and we could select from these based on our inclination towards each of the objectives. For example, we could select sequence 3-2-4-1 with a penalty cost of 10 and with no tardy jobs or sequence 1-3-4-2 with the penalty of 5 but one tardy job.

7.14 DELAY OF EARLY COMPLETING JOBS

Let us examine the concept of early penalty more critically. Early cost is incurred when the job is completed before its due date. If, however, the processing time available on the facility is *larger* than the total processing times required by all jobs, then it may be possible to delay the start of one or more jobs (introduce slack in the facility) so that the completion times are as close to the due dates as possible. The delay or slack introduction may factually mean keeping the facility idle or subcontracting the facility for special order jobs. Such non use or sub use of the facility may involve a penalty. The question is: where and how to introduce slack in order to minimize the total cost which includes the job related penalties and idle facility related cost?

The algorithm is presented in two phases (1) slack introduction and (2) optimal sequence search.

We start the procedure by first obtaining the initial best sequence using the Section 6.4 procedure. To that sequence, apply the following steps of slack introduction:

7.14.1 Phase I. Slack Introduction

1. Calculate the completion times of all the jobs in the sequence. This is done to determine the due date slack. If no job is completed early this method cannot be applied and the sequence is the best. If one or more jobs can be completed early, it might be possible to reduce the cost by introduction of slack in the facility. Determine the cost of idle facility "s" and precede to step 2.
2. Arrange the jobs in the first column of the savings table, Table 7.12, in the *reverse* sequence. For example, if 1-2-3 is the initial sequence, then job 3 is entered in the first row and jobs 2 and 1 are entered in rows 2 and 3 in rows, respectively.
3. Calculate the *due date slack* $Y_i = (D_i - C_i)$ where $i = 1, 2, \ldots, n$, and tabulate them in column 4 of the savings table. The slack in each job is calculated to determine the maximum amount of slack that can be added in the sequence at this time. If for all jobs $Y_i < 0$, then no additional slack can be added, go to step 10. If at least one value of $Y_i > 0$, go to step 4.
4. The value of slack that can be introduced in this iteration "S" is the minimum *positive value* of Y_i. It is note in column 5. For jobs with $Y_i > 0$, the addition of slack may give savings. On the other hand, the addition of minimum slack ensures minimal increase in the late penalties for jobs with $Y_i < 0$.
5. Calculate the value of $S \times L_i$ for jobs with $Y_i L0$ and the product $S \times E_i$ for jobs with $Y_i > 0$ and enter them in columns 5 and 6, respectively. Each value represents a savings in early penalty or an additional cost in late penalty for each job, as appropriate.
6. Compute the slack cost (idle machine cost) as the product of the S and slack cost per unit time, s. Note, introduction of slack renders the machine/facility idle, and hence the slack cost of the facility is determined to find the net savings.
7. Calculate cumulative savings and tabulate in column 10. The cumulative savings are calculated to see the amount of savings that are realized by the addition of slack S before the corresponding job. Note the cumulative savings of a job is the result of sum of the savings of the subsequent jobs in the actual sequence shown in the reverse order in the savings table.
8. Calculate the net profit for all the jobs as the difference between cumulative savings and slack cost. Enter the net profit in column 11, and proceed to

TABLE 7.12
Savings Table

Jobs Reverse Sequence	Due Date D_i	Adjustable Completion Time C_i	$Y_i = D_i - C_i$	Min $(+Y_i)$ S	L_i	E_i	Cost (If $Y_i <= 0$) $S \times L_i$	Savings (If $Y_i > 0$) $S \times E_i$	Cumulative Savings	Net Savings $(CS - S \times s)$
(1)	(2)	(3)	(4)	(5)	(6)	(7)	(8)	(9)	(10)	(11)

Net savings = Point of introduction of slack is

step 9. Slack S is added immediately before the job for which the maximum cumulative positive savings are observed. If there are no positive savings, go to step 10.

9. Add slack value S computed in step 4 to the completion times of the jobs that follow the job that has maximum savings and go to step 2. This is done because the addition of slack would delay the completion of all succeeding jobs by that amount. Since *only* the minimum positive amount of Y_i (due date slack) is added every time, the next iteration (steps 2 through 9) is performed to find if more savings can be realized by the addition of some more slack in the sequence.
10. Add the values of the savings from all the savings tables and subtract it from the cost of the initial best sequence obtained in step 1. This represents the cost of the sequence due to the addition of slack(s). Determine the start and completion times of all the jobs.

7.14.2 Phase II. Optimal Sequence Search

The above solution does not always represent the optimal sequence, even though the initially selected sequence represents the best sequence with the least cost. It may happen that addition of slack to a different sequence may lower the cost even further. Forward phase of Section 6.4 is performed on the best sequence obtained at the end of introduction of slack phase, and phase I of slack introduction is repeated, till there is no further job switch.

7.14.3 Illustrative Example 7.12

To illustrate the above methodology, the following example is used. Also, consider the cost of idle facility as 5 unit/time unit (Table 7.13).

The initial job sequence using the procedure from Section 6.4 is 3-4-2-5-1.

Step 1: calculation the completion times of jobs in sequence.

Sequence:	0	(3/8)	8	(4/6)	14	(2/35)	49	(5/43)	92	(1/45)	137
Delay:		(3)	33/E	(4)	25/E	(2)	146/E	(5)	180/E	(1)	188/E

Penalty: $25 \times 1 + 11 \times 3 + 97 \times 1 + 88 \times 2 + 51 \times 3 = 484$.

TABLE 7.13
Data for Example 7.12

Job Number	Processing Time (P_i)	Due Date (D_i)	Late Penalty (L_i)	Early Penalty (E_i)
1	45	188	11	3
2	35	146	2	1
3	8	33	7	1
4	6	25	8	3
5	43	180	4	2

Most of the jobs (in this case in fact all the jobs) are completed early and therefore delaying some jobs may prove profitable. Phase I of the slack introduction method is now applied.

Iteration 1: Sequence 3-4-2-5-1

Step 2: Construction of a savings in Table 7.14 as follows:

Step 3: The difference between the due dates and the job finish times are noted in column 4.

Step 4: The minimum positive value of Y_i is 11 (i.e., $S = 11$) for job number 4.This is the slack value S (col 5) in this iteration.

Step 5: The product of the slack value and early penalties (since slack here represents early slack) for the respective jobs is calculated and entered in column 9.

Step 6: Cumulative values of col 9 are entered in col 10.

Step 7: Since the slack cost is 5 / unit time, the slack cost in this case is $5 \times 11 = 55$.

Step 8: The cumulative savings are calculated and entered in column 11. They are calculated as (cumulative savings in column 10 - slack cost from step 7).

Step 9: The maximum positive savings of $55 is observed for job 2. A profit of $55 can be realized by adding a slack value of 11 at the start of job 2, that is, at the start of the sequence. The start and completion times of the jobs after iteration 1 are given in the following. Note that the introduction of slack is shown as +11.

$$0 + 11 \quad (3/8) \quad 19 \quad (4/6) \quad 25 \quad (2/35) \quad 60 \quad (5/43) \quad 103 \quad (1/45) \quad 148$$

Iteration 2: Sequence 3-4-2-5-1

Repeat steps just illustrated (step 2 through 9) on the same sequence, 3-4-2-5-1, with the new starting and completion times that result from the introduction of slack. The calculations are shown in Table 7.15.

In iteration 2, the slack value S is 14 (col 5), and the maximum savings realized is 14 (col 11). Here, the slack should be added at the beginning of job 2, the job for which this savings is realized. The starting and completion times for the jobs after iteration 2 are

$$0 + 11 \quad (3/8) \quad 19 \quad (4/6) \quad 25 + 14 = 39 \quad (2/35) \quad 74 \quad (5/43) \quad 117 \quad (1/45) \quad 162$$

Again, note the slack times are indicated by +11 and +14, giving starting times for job 2 as 11 and for job 2 as 39.

Iteration 3: Sequence 3-4-2-5-1

Continue with the starting and completion times for the sequence in iteration 2. The resulting calculations are shown in Table 7.16.

TABLE 7.14
Savings Table for Initial Best Sequence (Iteration 1)

Jobs Reverse Sequence (1)	D_i (2)	C_i (3)	$Y_i = D_i - C_i$ (4)	Min $(+Y_i)$ S (5)	L_i (6)	E_i (7)	Cost If $Y_i <= 0$ $S \times L_i$ (8)	Savings If $Y_i > 0$ $S \times E_i$ (9)	Cumulative Savings (CS) (10)	Net Savings $(CS - S \times s) =$ $(CS - 11 \times 5)$
1	188	137	188 − 137 = 51		11	3	—	33	33	−22
5	180	92	180 − 92 = 88		4	2	—	22	55	0
2	146	49	146 − 49 = 97		2	1	—	11	66	11
x4	25	14	25 − 14 = 11	11	8	3	—	33	99	44
3	33	8	33 − 8 = 25		7	1	—	11	110	55*

Net savings = 55 Point of addition of slack is before job 2

TABLE 7.15
Savings Table for Initial Best Sequence (Iteration 2)

Jobs Reverse Sequence (1)	D_i (2)	C_i (3)	$Y_i = D_i - C_i$ (4)	Min $(+Y_i)$ S (5)	L_i (6)	E_i (7)	Cost If $Y_i <= 0$ $S \times L_i$ (8)	Savings If $Y_i > 0$ $S \times E_i$ (9)	Cumulative Savings (CS) (10)	Net Savings $(CS - S \times s) =$ $(CS - 14 \times 5)$
1	188	148	188 − 148 = 40		11	3	—	42	42	−28
5	180	103	180 − 103 = 77		4	2	—	28	70	0
2	146	60	146 − 60 = 86		2	1	—	14	84	14*
4	25	25	25 − 25 = 0		8	3	-112	—	−28	−98
3	33	19	33 − 19 = 14	14	7	1	—	14	−14	−84

Net savings = 14 Point of addition of slack is before job 2

The maximum positive savings of 14 is associated with job 2. The value of slack is 14 (minimum positive value of Y_i), and it should be added at the beginning of job 2. This is in addition to slack introduced in the previous iteration. The following are the start and finish times of the jobs after iteration 3:

$$0+11 \quad (3/8) \quad 19 \quad (4/6) \quad 25+14+14=53 \quad (2/35) \quad 88 \quad (5/43) \quad 131 \quad (1/45) \quad 176$$

Iteration 4: Sequence 3-4-2-5-1(Table 7.17)

Iteration 4 has shown positive savings and hence 12 units of slack is added in front of job 2. The next sequence is $0+11=11(3/8)19(4/6)25+14+14+12=77(2/35)112(5/43)155(1/45)200$ (Table 7.18).

There is no further savings and hence the previous solution is best with cost of

$$0+11=11(3/8)19(4/6)25+14+14+12=77(2/35)112(5/43)155(1/45)200.$$

Due date	33	25	146	180	180
	14/E	0/E	34/E	25/E	20/L
Cost	$14\times1+0\times3+34\times1+25\times2+20\times11=318$				

7.14.4 Phase II: Optimal Sequence Search

The best sequence examined so far is 3-4-2-5-1. Proceeding with the forward phase, the next sequence to examine is one check with $K=4$, that is, sequence 1-4-2-5-3. The slack introduction phase is applied to this sequence in the following iterations.

Iteration 1: Best sequence 3-4-2-5-1. K = 4. **Sequence to examine: 1-4-2-5-3**

The calculations are summarized in Table 7.19.

There is no reduction in the cost, and therefore the best sequence remains 3-4-2-5-1.

Iteration 2: Best sequence: 3-4-2-5-1: $K=3$. **Sequence to examine 5-4-2-3-1**

Since there are no other jobs with a lag of 4, the value of the lag, K, is changed to 3. The new sequence with lag 3 to examine is 5-4-2-3-1. The initial cost of this sequence is $1094. Since there is no reduction of cost after slack introduction, the sequence 3-4-2-5-1 remains the best sequence.

Iteration 3: Best sequence 3-4-2-5-1: $K=3$. **Sequence to examine 3-1-2-5-4**

With $K=3$, exchanging jobs 2 and 4 resulted in the sequence 3-1-2-5-4 which had a higher cost and did not improve on the best sequence.

Iterations 3, 4, and 5. Best sequence 3-4-2-5-1: $K=2$

None of the sequences with $K=2$ exchanges showed any cost reductions over the best sequence though sequences 3-4-1-5-2 did reduce in cost with slack introduction.

TABLE 7.16
Savings Table for Initial Best Sequence (Iteration 3)

Jobs Reverse Sequence (1)	D_i (2)	C_i (3)	$Y_i = D_i - C_i$ (4)	Min $(+Y_i)$ S (5)	L_i (6)	E_i (7)	Cost If $Y_i <= 0$ $S \times L_i$ (8)	Savings If $Y_i > 0$ $S \times E_i$ (9)	Cumulative Savings (CS) (10)	Net Savings $(CS - S \times s) =$ $(CS - 14 \times 5)$
1	188	162	188 – 162 = 26		11	3	—	42	42	– – 28
5	180	117	180 – 117 = 63		4	2	—	28	70	0
2	146	74	146 – 74 = 62		2	1	—	14	84	14*
4*	25	25	25 – 25 = 0		8	3	–112		–28	–98
3*	33	19	33 – 19 = 14	14	7	1		14	-14	–84

Net savings = 26 Point of addition of slack is before job 2

TABLE 7.17
Savings Table for Initial Optimal Sequence (Iteration 4)

Jobs Reverse Sequence (1)	D_i (2)	C_i (3)	$Y_i = D_i - C_i$ (4)	Min $(+Y_i)$ S (5)	L_i (6)	E_i (7)	Cost If $Y_i <= 0$ $S \times L_i$ (8)	Savings If $Y_i > 0$ $S \times E_i$ (9)	Cumulative Savings (CS) (10)	Net Savings $(CS - S \times s) =$ $(CS - 12 \times 5)$
1	188	176	188 – 176 = 12	12	11	3		36	36	–24
5	180	131	180 – 131 = 49		4	2	—	24	60	0
2	146	88	146 – 88 = 58		2	1	—	12	72	12*
4*	25	25	25 – 25 = 0		8	3	–96	-	–24	–84
3*	33	19	33 – 19 = 14		7	1		12	–12	–72

Net savings = 12 Point of addition of slack is job 2

TABLE 7.18
Savings Table for Initial Optimal Sequence (Iteration 5)

Jobs Reverse Sequence (1)	D_i (2)	C_i (3)	$Y_i = D_i - C_i$ (4)	Min $(+Y_i)$ S (5)	L_i (6)	E_i (7)	Cost If $Y_i <= 0$ $S \times L_i$ (8)	Savings If $Y_i > 0$ $S \times E_i$ (9)	Cumulative Savings (CS) (10)	Net Savings $(CS - S \times s)$ (11)
1	188	200	188 – 200 = –12		11	3	–154		–154	–224
5	180	155	180 – 155 = 25		4	2	–	28	–126	–196
2	146	112	146 – 112 = 34		2	1	–	14	–112	–182
4*	25	25	25 – 25 = 0		8	3	–112	–	–224	–294
3*	33	19	33 – 19 = 14	14	7	1		14	–210	–280

Net savings = -- Point of addition of slack is NA

TABLE 7.19
Savings Table for Sequence 1-4-2-5-3 (Iteration)

Jobs Reverse Sequence (1)	D_i (2)	C_i (3)	$Y_i = D_i - C_i$ (4)	Min $(+Y_i)$ S (5)	L_i (6)	E_i (7)	Cost If $Y_i <= 0$ $S \times L_i$ (8)	Savings If $Y_i > 0$ $S \times E_i$ (9)	Cumulative Savings (CS) (10)	Net Savings (CS − $S \times s$) = (11)
3	33	137	33 − 137 = −104		7	1	−357	—	−357	−612
5	180	129	180 − 129 = 51	51	4	2	—	102	−255	−510
2	146	86	146 − 86 = 60		2	1	—	51	−204	−459
4	25	51	25 − 51 = −26		8	3	−408	—	−612	−867
1	188	45	188 − 45 = 143		11	3	—	153	−459	−714

Net savings = – Point of addition of slack is NA

Iteration 6: Best sequence 3-4-2-5-1: $K = 1$ Sequence to examine 4-3-2-5-1

Jobs 2 and 2 are exchanged giving the new sequence to examine as 4-3-2-5-1. The initial cost of this sequence is 502 and the final cost after three internal iterations is 375. The details are shown Tables 7.20 through 7.23. The exchange is accepted and sequence 3-4-1-5-2 now becomes the best sequence which is examined further.

Summary of all calculations are tabulated in Table 7.23. The best sequence remains, even after the application of forward phase, the initial sequence 3-4-2-5-1with the penalty of 318. The starting and completion times of each job are indicated in the sequence diagram.

$$0 + 11 = 11(3/8)19(4/6)25 + 14 + 14 + 12 = 77(2/35)112(5/43)155(1/45)200.$$

7.15 JOBS ARRIVING AT DIFFERENT TIMES

So far, we have assumed that all jobs are available for scheduling at time zero. If this is not the case, slight modifications to both backward and forward phases are required. The backward phase is now applied to only the jobs that are available up to the completion time of the previous job in the sequence. Initially, only the jobs available at time zero are considered for the backward phase. If all the jobs arrive at time greater than zero, then the process starts at time of the first job's arrival (there might be multiple jobs arriving at the same time). The steps are given in the following.

Backward Phase

Step 1: Set $C = 0$ and $K = 1$.
Step 2: Among all the unscheduled jobs, choose the minimum arrival time of the jobs. Let it be ES.
Step 3: Set S = maximum value of either ES or C.
Step 4: Choose all the jobs that have their arrival time values less than or equal to S. Put them in set J.
Step 5: Compute T, which is the sum of processing times of all the jobs in set J. Set $H = T + S$. Set $C = H$.
Step 6: Schedule all the jobs in set J in the backward direction based on the penalty. Penalty for job i, TD_i, is: if $(H - D_i) > 0$, $TD_i = (H - D_i) \times L_i$; otherwise, $TD_i = (D_i - H) \times E_i$.

The criterion for choosing a job is the one which has the least penalty. In case of a tie, the job with the maximum value of processing time is chosen. If there is still a tie, choose the job with the maximum value of arrival time; if there is still a tie, break it arbitrarily. If there are N jobs in set J, we assign jobs from set J starting at position $N + K - 1$, ending in position K. Once a job has been assigned, the value of H is reduced by its processing time, and penalties for the remaining jobs in set J are recalculated. This process is repeated until all the jobs in set J are assigned.
Step 7: Set $K = K + N$.
Step 8: If K is less than or equal to the total number of jobs, then go to step 2; otherwise, stop and go to forward phase.

TABLE 7.20
Savings Table for Sequence 4-3-2-5-1 (Iteration 6)

Jobs Reverse Sequence (1)	D_i (2)	C_i (3)	$Y_i = D_i - C_i$ (4)	Min $(+Y_i)$ S (5)	L_i (6)	E_i (7)	Cost If $Y_i <= 0$ $S \times L_i$ (8)	Savings If $Y_i > 0$ $S \times E_i$ (9)	Cumulative Savings (CS) (10)	Net Savings (CS − $S \times s$) = (11)
1	188	137	188 − 137 = 51		11	3	-	57	57	−38
5	180	92	180 − 92 = 88		4	2	-	38	95	0
2	146	49	146 − 49 = 97		2	1	-	19	114	19
	33	14	33 − 14 = 19		7	1	-	19	133	38
4	25	6	25 − 6 = 19	19	8	3	-	57	190	95*

Net savings = 95 Point of addition of slack before 4

TABLE 7.21
Savings Table for Sequence 4-3-2-5-1 (Iteration 2)

Jobs Reverse Sequence (1)	D_i (2)	C_i (3)	$Y_i = D_i - C_i$ (4)	Min $(+Y_i)$ S (5)	L_i (6)	E_i (7)	Cost If $Y_i <= 0$ $S \times L_i$ (8)	Savings If $Y_i > 0$ $S \times E_i$ (9)	Cumulative Savings (CS) (10)	Net Savings (CS − $S \times s$) = (11)
1	188	156	188 − 156 = 32	32	11	3	—	96	96	−64
5	180	111	180 − 111 = 69		4	2	—	64	160	0
2	146	68	146 − 68 = 78		2	1	—	32	192	32*
3*	33	33	33 − 33 = 0		7	1	-231	—	−39	−199
4*	25	25	25 − 25 = 0		8	3	-152	—	−191	−351

Net savings = 32 Point of addition of slack before 2

TABLE 7.22
Savings Table for Sequence 4-3-2-5-1 (Iteration 3)

Jobs Reverse Sequence (1)	D_i (2)	C_i (3)	$Y_i = D_i - C_i$ (4)	Min $(+Y_i)$ S (5)	L_i (6)	E_i (7)	Cost If $Y_i <= 0$ $S \times L_i$ (8)	Savings If $Y_i > 0$ $S \times E_i$ (9)	Cumulative Savings (CS) (10)	Net Savings $(CS - S \times s) =$ (11)
1	188	188	188 – 188 = 0		11	3	-407	—	-407	–592
5	180	143	180 – 143 = 37	37	4	2	—	74	-333	–518
2	146	100	146 – 100 = 46		2	1	—	37	-296	–481
3*	33	33	33 – 65 = –32		7	1	-259	—	-555	–740
4*	25	25	25 – 57 = –32		8	3	-296	—	-851	–1036

Net savings = – Point of addition of slack before NA

TABLE 7.23
Iterations Showing the Optimal Sequence Search

Iteration and K	Best Sequence	Sequence	Initial Penalty	Savings/Amount/Point of Addition			Final Penalty
				Iteration 1	Iteration 2	Iteration	
	3-4-2-5-1	3-4-2-5-1	484	55/11/3	14/14/2	26/26/2	389
1. $K = 4$		1-4-2-5-3	1527	—	—	—	1527
2. $K = 3$		5-4-2-3-1	1094	—	—	—	1094
3.		3-1-2-5-4	1482	—	—	—	1482
4. $K = 2$		2-4-3-5-1	680	—	—	—	680
		3-5-2-4-1	1036	—	—	—	1036
5.		3-4-1-5-2	610	45/9/3	4/2/3	—	561
6. $K = 1$		4-3-2-5-1	502	95/19/4	32/32/2	—	375
1. $K = 4$	4-3-2-5-1	1-3-2-5-4	1621	—	—	—	1621
2. $K = 3$		5-3-2-4-1	1149	—	—	—	1149
3.		4-1-2-5-3	1358	—	—	—	1149
4. $K = 2$		2-3-4-5-1	702	—	—	—	702
5.		4-5-2-3-1	947	—	—	—	947
6.		4-3-1-5-2	628	45/9/4	20/10/4	—	563
7. $K = 1$		4-2-3-5-1	603	—	—	—	603
8.		4-3-5-2-1	529	95/19/4	32/32/5	—	402
9.		4-3-2-1-5	541	95/19/4	24/24/2	—	422

Forward Phase

Now the forward phase is applied. The forward phase remains the same, except that the penalty is calculated differently. A_i is the arrival time for job i. CT is the total completion time for the jobs in the sequence so far examined by the forward phase. TD_i is the penalty for job i. CT is initially set to 0. We start from the first job in the sequence (i.e., from the left). If at any point in time a job does not arrive at the time the previous job was completed, the next job can only start when it arrives; therefore, the completion time of the job arriving late is given by the sum of the time at which it arrives and its processing time. This results in a delay in the processing of subsequent jobs.

The following conditions illustrate the penalty calculation.

If $A_i > \text{CT}$, then $\text{CT} = A_i + P_i$; otherwise, $\text{CT} = \text{CT} + P_i$.
If $\text{CT} - D_i > 0$, then $TD_i = (\text{CT} - D_i) \times L_i$; otherwise, $TD_i = (D_i - \text{CT}) \times E_i$.

Consider the following example shown in Table 7.24.
where P_i, A_i, D_i, L_i, and E_i represent the processing time, arrival time, due date, and late and early penalties of the jobs, respectively.

Backward Phase Application

$K = 1$ and $C = 0$
$\text{ES} = \text{minimum } (2, 5, 28, 10, 10) = 2$
$S = \text{maximum } (\text{ES} = 2, C = 0) = 2.$
$J = \{1\}$
$T = 5, H = T + S = 5 + 2 = 7, C = H = 7$
$N = 1$ since there is only one job in set J.
Assign job 1 at position $N + K - 1 (1 + 1 - 1 = 1)$.
$K = K + N = 1 + 1 = 2.$
The partial sequence is 1 - - - -.
$\text{ES} = \text{minimum } (5, 28, 10, 10) = 5$
$S = \text{maximum } (5, 7) = 7$
$J = \{2\}$
$T = 7, H = 14, C = 14$
$N = 1$
Assign job 2 at position $1 + 2 - 1 = 2$.
$K = K + N = 1 + 2 = 3$
The partial sequence is 1-2 - - -.
$\text{ES} = \text{minimum } (28, 10, 10) = 10$
$S = \text{maximum } (10, 14) = 14$
$J = \{4, 5\}$
$T = 5, H = 5 + 14 = 19, C = 19$
Penalty for job $2 = (H - D_i) = (19 - 15) \times 1 = 4$
Penalty for job $2 = (19 - 17) \times 2 = 4$

TABLE 7.24
Data for Single-Machine Penalty Minimization for Jobs Arriving at Different Times

Jobs	P_i	A_i	D_i	L_i	E_i
1	5	2	9	2	1
2	7	5	15	2	1
3	8	28	42	1	0
4	3	10	15	1	0
5	2	10	17	2	1

Since there is a tie. we look for the job with a maximum value of processing time. Since the processing time for job 4 is higher, we choose job 4. Assign job 4 at position $2 + 3 - 1 = 4$. The partial sequence now becomes 1-2-4-

$H = 19-$ Processing time for job $4 = 19 - 3 = 16$.

Since there is only one other job left (job 5) in set J, we assign it to position K, which is 3. The partial sequence is 1-2-5-4-.

$$K = 3 + 2 = 5$$

ES = 28
$S =$ maximum $(28, 19) = 28$
$J = \{3\}$
$T = 8, H = 8 + 28 = 36, C = 36.$

Note that this job can be assigned at time 28 and not 19, which was the completion time of the previous job.

$$K = K + N = 5 + 1 = 6$$

Since K exceeds the total number of jobs 2, we stop the backward phase. The optimal sequence obtained from the backward phase is 1-2-5-4-3. Now, we proceed to the forward phase.

Forward Phase Application
The total penalty for the sequence 1-2-5-4-3 is
Initially, CT = 0.

Job 2: CT = 7, $TD_1 = 2$
Job 2: CT = 14, $TD_2 = 1$
Job 2: CT = 16, $TD_3 = 1$
Job 2: CT = 19. $TD_4 = 4$
Job 2: CT = 36, $TD_5 = 0$

The total penalty for the sequence 1-2-5-4-3 is the sum of all TDi, which is 8.

Cycle 1. Iteration 1. Best sequence 1-2-5-4-3; $K =$ number of jobs $-1 = 4$. Exchange 1–3. Sequence to examine 3-2-5-4-1

Schedule: 28 – (3/8) – 36 – (2/7) – 43 – (5/2) – 45 – (4/3) – 48 – (1/5) – 53
Due dates: (3) – 42 – (2) – 15 – (5) – 17 – (4) – 15 – (1) – 9
Penalties: $0 + 56 + 56 + 33 + 88 = 233$.
Total penalty $= 233 > 8$, no exchange.

Cycle 1. Iteration 2. Best sequence 1-2-5-4-3; $K = 3$. Exchange 1–4. Sequence to examine 4-2-5-1-3.
Total penalty $= 56 > 8$, no exchange.

Cycle 1. Iteration 3. Best sequence 1-2-5-4-3; $K = 3$. Exchange 2–3. Sequence to examine 1-3-5-4-2.
Total penalty = 136 > 8, no exchange.
Cycle 1. Iteration 4. Best sequence 1-2-5-4-3; $K = 2$. Exchange 1–5. Sequence to examine 5-2-1-4-3.
Total penalty = 55 > 8, no exchange.
Cycle 1. Iteration 5. Best sequence 1-2-5-4-3; $K = 2$. Exchange 2–4. Sequence to examine 1-4-5-2-3.
Total penalty = 18 > 8, no exchange.
Cycle 1. Iteration 6. Best sequence 1-2-5-4-3; $K = 2$. Exchange 5–3. Sequence to examine 1-2-3-4-5.
Total penalty = 75 > 8, no exchange.
Cycle 1. Iteration 7. Best sequence 1-2-5-4-3; $K = 1$. Exchange 1–2. Sequence to examine 2-1-5-4-3.
Total penalty = 30 > 8, no exchange.
Cycle 1. Iteration 8. Best sequence 1-2-5-4-3; $K = 1$. Exchange 2–5. Sequence to examine 1-5-2-4-3.
Total penalty = 22 > 8, no exchange.
Cycle 1. Iteration 9. Best sequence 1-2-5-4-3; $K = 1$. Exchange 5–4. Sequence to examine 1-2-4-5-3.
Total penalty = 9 > 8, no exchange.
Cycle 1. Iteration 10. Best sequence 1-2-5-4-3; $K = 1$. Exchange 4–3. Sequence to examine 1-2-4-3-4.
Total penalty = 28 > 8, no exchange.

Applying forward phase in this case does not improve the solution. The optimal sequence remains 1-2-5-4-3, and the optimal total penalty is 8.

Alternatively, we can understand the procedure by applying it to a little expanded example. Start with the earliest arrival, and schedule that job first. If there are multiple jobs arriving at the earliest time, choose the one with shortest processing time, and schedule it first.

Jobs	P_i	A_i	D_i	L_i	E_i
1	5	2	9	2	1
2	7	5	15	2	1
3	8	15	42	1	0
4	3	10	15	1	0
5	2	10	17	2	1
6	4	9	20	1	1

Earliest arrival job is 1 at time 2. Scheduling it gives following:

$$0(1/5)5$$

The job is completed at time 5. By that time, only one new job has arrived, job 2. Schedule job 2. The schedule is as follows:

$$0(1/5)5(2/7)12$$

By time 12, jobs 2, 5, and 6 have arrived. The total time to process all three jobs is $3+2+4=9$. Apply backward phase to choose the sequence of these three jobs, but only select first job in the sequence and repeat the process.

Completion time for the last of these is $12 + 9 = 21$. The associated penalties are: job 2 $(21-15)\times 1 = 6$; for job 2 $(21-17)\times 2 = 8$; for job 2 it is $(21-20)\times 1 = 1$. The least cost is 1, and hence schedule job 2 in last position. The completion time for the other two jobs is $21 - 4 = 17$.

The completion of job 2 at time 17, costs $(17-15)\times 1 = 2$, and for job 2 the cost is $(17-17)\times 1 = 0$. Schedule job 2 in the second position from end, and therefore job 2 is scheduled next. The sequence is

$$0(1/5)5(2/7)12(4/3)15$$

By time 15, jobs 2, 5, and 6 have arrived. Since these are all jobs have arrived, applying backward phase with these three jobs results in completion of the sequence.

$$0(1/5)5(2/7)12(4/3)15(5/2)17(6/4)21(3/8)29$$

7.16 SUMMARY

The chapter illustrates many different variations of the single-machine problem. These variations are primarily due to different objectives. As the objective changes, so does the solution procedure, and ultimately the final sequence. It is, therefore, important that we understand the goal defined by management before embarking on developing a schedule.

It is possible that we may have dual criteria to evaluate. In such a case, the schedule may be a compromise schedule, based on the importance we attach to each criterion.

In some instances (and they may not be as rare as we think), it may be beneficial to keep a facility idle for some time rather than to complete the jobs before the due dates and build an inventory. The just-in-time principle is such a case. We have illustrated a procedure that adapts to this goal.

We have also seen how backward and forward phases can be modified to accommodate sequencing of jobs that arrive at different times. The procedure is simple and very effective.

7.17 PROBLEMS

7.1 A facility manufactures refrigeration compressors based on orders received from customers. The product line consists of three models (A, B, and C), each with a different BTU rating. Models A and B are manufactured in jobs of 1000 units each, and model C in jobs of 500 units each. The facility operates 20 hr/day. Models A and B can be produced at a rate of 500/day, whereas model C can be produced at a rate of only 400/day. No model or customer has priority over any other. Partial orders can be shipped to a customer, but all of the same models ordered must be shipped at the same time. The following is a list of orders from customers.

Customer	Model	Units Ordered	Due Date
1	A	2000	5
2	A	2000	8
	C	2500	16
3	B	2000	23
	C	2000	30
4	A	2000	12
	B	2000	26

i. Calculate the number of jobs for each model to be produced, along with the total processing time.
ii. Determine a schedule of jobs based on the minimization of maximum delay.
iii. Determine the maximum delay.

7.2 A defense contractor for the U.S. government manufactures circuit card subassemblies for use in weapons systems. The contractor has received the following orders from its customers:

Quantity	100	200	150	125	250	125	250	75
Due date	20	20	20	20	30	30	30	30
Late penalty	1	0	1	2	3	1	2	1

Each order is processed as a job. Each unit requires 0.05 days of processing. The company assigns a penalty of 1 for any early orders to account for inventory carrying costs. Develop a schedule for each of the common due dates, and calculate the associated penalty.

7.3 A company manufactures industrial lighting fixtures such as recessed lighting fixtures, "exit" signs, and "entrance" signs. The company receives the following order from one of its established customers:

Product Description	Quantity
XP: Exit signs—letters in red, plastic housing	1200
XM: Exit signs—letters in red, metal housing	1200
EP: Entrance signs—letters in green, plastic housing	1200
EM: Entrance signs—letters in green, metal housing	1200

Each of the products is scheduled in jobs of 1200 units. The known processing times are 3 days for XP, 5 days for XM, 6 days for EP, and 8 days for EM. The following table shows the early and late penalties that will be incurred. The customer wants delivery of all four products at the same time. Develop the schedule, and determine the shipment date and the associated penalties incurred.

Job	P_i	D_i	E_i	L_i
1	3	Common	3	2
2	5	Common	0	1
3	6	Common	4	2
4	8	Common	1	3

7.4 A chemical manufacturer that supplies a petroleum refinery has received a purchase order for four different chemicals from the refinery's purchasing department. The chemicals will be used in one of the refinery's continuous processes. Because of the refinery's safety stock, some lateness is acceptable. However, the later the shipment, the more crucial it becomes. Therefore, the following late penalties will be incurred if deliveries are late:

Order	Processing Time	Due Date	Late Penalty
1	12	20	$2x^2$
2	16	18	$3x^2$
3	8	25	x^2
4	18	30	$2x^2$

Develop the schedule for the chemicals, given the nonlinear penalty functions to be assessed if the chemicals are shipped late.

7.5 A company manufactures riding lawnmowers at one of its facilities. Model XYZ lawnmower is manufactured with the nameplate of customer 1, and model ABC is manufactured with the nameplate of customer 2. With the exception of the nameplate, the two mowers are identical. Four different size engines are available on both models: 8, 10, 12, and 16 hp. The 8-hp and 10-hp mowers have a 36" cutting deck, whereas the 12-hp and 16-hp have a 42" cutting deck. Regardless of the nameplate, the total manufacturing time (hours) for each is as follows: 0.08, 0.09, 0.11, and 0.12, for 8, 10, 12, and 16 hp, respectively. Work orders are generated from the materials department to produce batches of mowers at the following quantities: 500, 900, 560, and 455 for 8, 10, 12, and 16 hp, respectively. The following orders have been received from customers 1 and 2 with associated expected delivery dates:

Customer	Horsepower	Quantity	Expected Delivery Date
1	8	500	10
1	10	400	30
1	16	300	20
2	10	500	15
2	12	560	12
2	16	155	22

The company knows that it can actually deliver the mowers 3 days before or after the expected delivery date without any question from the customers. Outside this range, however, an early penalty of 2 and a late penalty of 3 will be incurred. Develop the production schedule for the materials department for the batches of lawnmowers (plant operation is for 8 hr per day).

7.6 Automobile tires are produced by a company and sold to automobile manufacturers. Because the automobile manufacturers are very dependent on the tire manufacturer, the relationship between the two is critical. Realizing this, the tire manufacturer's objective is to have as few late deliveries as possible to its customers. Generate the next production schedule based on the following orders received with their expected delivery dates specified by the customers and the processing times known:

Order	**1**	**2**	**3**	**4**	**5**	**6**
Processing time	18	12	6	20	13	10
Due date	25	80	40	65	15	50

7.7 Customized computer systems are built by a company for industrial customers with business application requirements. The company is in a period of growth, with an increasing number of orders being requested. Although it strives to meet all customers' requirements, it is limited by its resources and at times must reject orders. When developing its schedule, consideration is given to long-standing customers and also to larger orders. The following orders have been received. The company has listed what it believes to be an acceptable range for completion of the orders based on requested delivery dates. Also, the company has assigned early and late penalties to the jobs, indicating the relative importance of each job. The estimated processing times are listed as well.

Job	**Processing Time**	**Early Due Date**	**Late Due Date**	**Early Penalty**	**Late Penalty**
1	2	4	8	0	1
2	6	12	16	1	2
3	3	10	12	1	0
4	7	12	20	0	0
5	4	20	25	2	1
6	1	3	5	2	2

Determine which orders, if any, must be declined.

7.8 A newly built company is concerned with developing good relationships with its customers. Because it has minimum capital, it is also concerned with maximizing its cash flow. Consequently, to develop its production schedules, it desires to both minimize its cash penalties and minimize the number of jobs that it ships late. Suggest a schedule based on the following data while simultaneously considering both these objectives:

Job	Processing Time	Due Date	Early Penalty	Late Penalty
1	5	10	1	2
2	15	20	1	1
3	10	40	0	1
4	5	15	0	2

7.9 A company that competes in the industrial welding supplies market manufactures welding machines, which it sells to many different distributors and retailers. Because of the high levels of competition and low profit margins in this industry, the company desires to keep the average delay of product delivery to its customers to a minimum. Suggest a schedule based on the following orders received:

Order	1	2	3	4	5
Manufacturing time	6	9	5	12	15
Delivery date	15	35	10	50	25

The company considers each customer equally important and assesses a late penalty of 2 to itself if any order is delivered late.

7.10 An accounting firm has accepted requests from eight customers to prepare their tax returns during the upcoming week. The firm has estimated the time required for each return. Its objective is to minimize the variation in the amount of time the tax return stays in the office for processing. Given the estimated processing time for each of the following tax returns, develop the best order of processing the returns to meet the firm's objective.

Tax return	1	2	3	4	5	6	7	8
Estimated processing time	8	4	7	10	3	6	9	8

REFERENCES AND SUGGESTED READINGS

Ahmed, I. and W.W. Fisher. 1992. "Due Date Assignment, Job Order Release and Sequencing" *Decision Sciences*, 23: 633–644.

Baker, K.R. 1974. *Introduction to Sequencing and Scheduling*, New York: Wiley.

Cambadi, B. 1991. "One Machine Scheduling to Minimize Expected Mean Tardiness: Part I" *Computers and Operations Research*, 22–36.

Conway, R.W., Maxwell, W.L., and L.W. Miller. 1967. *Theory of Scheduling*, Reading, MA: Addison-Wesley.

Eilon, S. and I.E. Chowdhury. 1977. "Minimizing Waiting Time Variance in the Single Machine problem" *Management Science*, 23(6): 567–575.

Elmaghraby, S.E. 1968. "The One-Machine Scheduling Problem with Delay Costs" *Journal of Industrial Engineering*, 19: 105–108.

Emmons, H. 1969. "One Machine Sequencing to Minimize Certain Functions of Job Tardiness" *Operations Research*, 17: 701–715.

Fathi, Y. and H.W.L. Nuttle. 1990. "Heuristic for Common Due Date Weighted Tardiness Problem" *IIE Transactions*, 22: 215–225.

Hamada, T. and K. D. Glazebrook. 1993. "A Bayesian Sequential Single Machine Scheduling Problem to Minimize the Expected Weighted Sum of Flow Times of Jobs with Exponential Processing Times" *Operations Research*, 924.

Kanet, John J. December 1981. "Minimizing Variation of Flow Time in Single Machine Systems" *Management Science*, 27: 1453–1459.

Lawler, E.L. 1973. "Optimal Sequencing of a Single Machine Subject to Precedence Constraints" *Management Science*, 19: 544–546.

Lee, C.Y., Danusaputro, S.L., and C.S. Lin. 1991. "Minimizing Weighted Number of Tardy Jobs and Weighted Earliness-Tardiness Penalties about a Common Due Date" *Computers and Operations Research*, 18(4): 379–389.

Merten, A.G. and M.E. Muller. 1972. "Variance Minimization in Single Machine Sequencing Problems" *Management Science*, 18(9): 518–528.

Mittenthal, J., M. Ragavachari, and A.I. Rana. 1993. "A Hybrid Simulated Annealing Approach for Single Machine Scheduling Problems with Non-Regular Penalty Functions" *Computers and Operational Research*, 103.

Moore, J.M. 1968. "An 'N' Job, One Machine Sequencing Algorithm for Minimizing the Number of Late Jobs" *Management Science*, 15: 102–109.

Ow, P.S. and T.E. Morton. 1989. "The Single Machine Early/Tardy Problem" *Management Science*, 35: 177–187.

Panwalkar, S.S., Sluith, M.L., and A. Seidmann. 1982. "Common Due Date Assignment to Minimize Total Penalty for the One Machine Scheduling problem" *Operations Research*, 30: 391–399.

Schrage, L. 1975. "Minimizing the Time-In System Variance for a Finite Jobset" *Management Science*, 21(5): 540–543.

Sundararaghavan, P.S. and M.U. Ahmed. 1984. "Minimizing the Sum of Absolute Lateness in Single-Machine and Multimachine Scheduling" *Naval Research Logistics Quarterly*, 31: 325–333.

Taha, H.A. 1987. *Operations Research*, New York: Macmillan.

8 Flowshop Problems

In Chapters 6 and 7, we have discussed what is generally referred to as a single machine facility problem where the entire production unit or work center may be considered as the facility. Here, we shall extend the analysis to the next level of detail and complexity by evaluating a flowshop problem (Figure 8.1). The characteristic of a flowshop is: we have m machines or work centers in the facility, and all jobs are processed on these machines in the same sequence. However, the processing time for each job on each machine may vary. All jobs are assumed to be available at time zero. It is further assumed that there is sufficient physical buffer space between two successive machines. This allows each machine to release the processed jobs to the succeeding machine without being concerned about the busy or idle status of the machine. A usual objective is to develop a schedule that minimizes the makespan. Furthermore, in this case the schedule that minimizes the makespan also minimizes major alternative objectives of minimizing the sum of job waiting times and the sum of machine idle times. We use the notation p_{ij} to denote the processing time of job i on machine j. Many industrial and non-industrial applications follow the flowshop arrangement when the operations are to be performed sequentially.

8.1 TWO-MACHINE PROBLEM

The simplest form of a flowshop arrangement is when there are only two machines or work centers and each job must be processed successively on these two machines; the

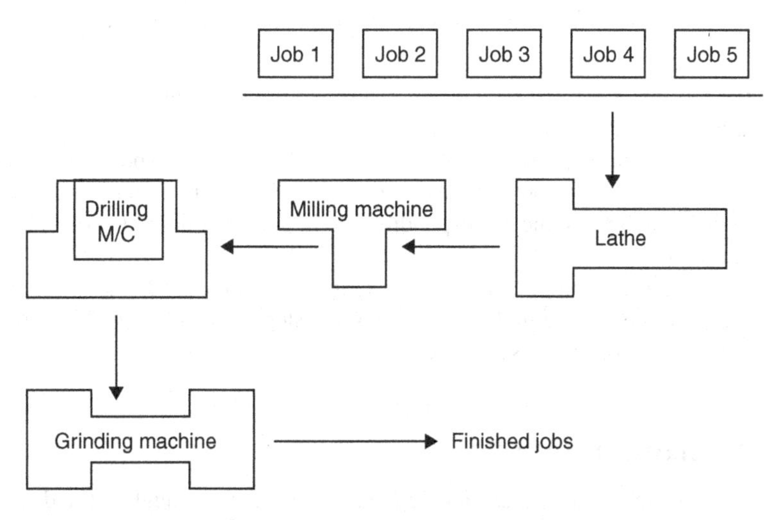

FIGURE 8.1 Scheduling in a flowshop.

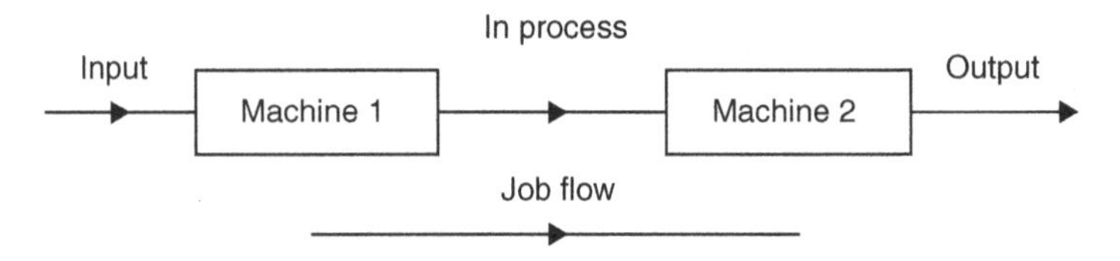

FIGURE 8.2 The two-machine flowshop model.

TABLE 8.1
Data for Example 8.1

Job Number	Processing Time on Milling Machine $P_{i1} = A_i$	Processing Time on NC Grinder $(P_{i2} = B_i)$
1	15	3
2	8	20
3	18	5
4	25	8
5	17	20
6	22	30

first operation on the first machine and the second operation on the second machine (Figure 8.2). Johnson (1954) developed a scheduling procedure that gives a minimum makespan for the jobs on which this procedure is applied. This method is well known in the scheduling literature as Johnson's rule. For making the procedure general, let us define $A_i = P_{i1}$ and $B_i = P_{i2}$. The steps of the procedure are as follows:

1. For the jobs yet to be sequenced, determine the minimum times of all A_i and B_i.
2. If the minimum is associated with A_i then place the corresponding job in the earliest possible position in the sequence. If the minimum is associated with B_i, then place the corresponding job in the latest possible position in the sequence.
3. Mark the job as being sequenced and scratch the associated A_i and B_i values.
4. If all jobs are placed in the sequence, go to step 5, otherwise go to step 1.
5. We have the optimum sequence.

8.1.1 Illustrative Example 8.1

Consider a machine shop operation where each piece is first rough cut and shaped on the milling machine and then ground to the required tolerance and polished on the NC grinder. We have six jobs that need processing, and we wish to determine the optimum sequence to minimize the makespan. The data are given in Table 8.1.

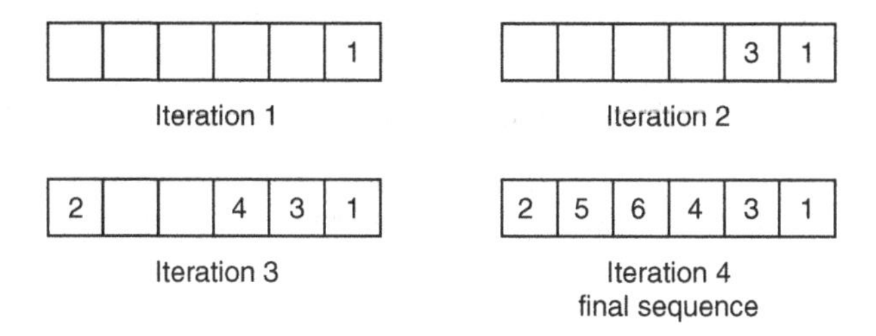

FIGURE 8.3 Solution to two-machine example.

Iteration 1: The minimum processing time is 3 on the second machine for job 1. Hence, job 1 is placed in the last position, and its processing times are scratched. The sequence developed so far is: –, –, –, –, –, 1.

Iteration 2: Processing time 5 with machine 2 associated with job 3 is the next-smallest processing value. Job 3 is placed in the fifth position, the latest possible position in the partial sequence developed so far. The new partial sequence is: –, –, –, –, 3, 1. Job 3 and the associated processing times are scratched.

Iteration 3: Minimum time now is 8 associated with job 2 (machine 1) and job 4 (machine 2). Job 2 is placed in the first position and job 4 in position 4. The partial sequence is: 2, –, –, 4, 3, 1. Both jobs are scratched.

Iteration 4: Job 5 with 17 units on machine 1 is the lowest value. Place job 5 in position 2 and the remaining job 6 is placed in position 3. The final sequence is: 2, 5, 6, 4, 3, 1.

Figure 8.3 shows the iterative developments. The detailed time schedule that shows the start and completion times of each job on each machine M1 and M2 is shown below:

M1:	0	(2/8)	8	(5/17)	25	(6/22)	47	(4/25)	72	(3/18)	90	(1/15)	105
M2:	0/8	(2/20)	28/25	(5/20)	48/47	(6/30)	78/72	(4/8)	86/90	(3/5)	95/105	(1/3)	108

On machine 1, the start and completion times are determined by adding processing times in job sequence, and are displayed using the convention from Chapters 6 and 7. Loading on machine 2 requires special attention. A job when completed in machine 1, goes to machine 2. It is processed immediately if machine 2 is available. However, it may have to wait in the buffer space if machine 2 is still processing the previous job. Thus, the starting time for a job that is in position j in the sequence on the second machine is the maximum of two values; completion time on job $j-1$ in that machine and the release time of job j from the previous machine. This is easily visualized on the second machine by writing completion time of the previous job ($C_{j-1,2}$) first followed by the release time of job j from machine one ($C_{j,1}$). For example, in machine 2 processing of job 2 can start at time 8 after it has been processed in machine 1, greater of initial time of 0 and 8. Job 5 can start on machine 2, on greater of two times, 28 when the previous job, Job 2 completes its requirements on machine 2 (time at which machine 2 is available for processing new job) and 25 when job 2 is released

TABLE 8.2
Data for Three Machine Problem: Example 8.2

Job Number	Milling Time	Grinder Processing Time	Polishing Processing Time
1	15	2	1
2	8	6	14
3	18	2	3
4	25	5	5
5	17	10	10
6	22	10	20

from machine 1 and is available for processing on machine 2. The start of job 1, for example, is delayed (and machine 2 is idle) because it is released from machine 1 at time 105 even though machine 2 is available at time 95. The makespan for this sequence is 108 time units.

8.2 THREE-MACHINE PROBLEM

Unfortunately, Johnson's rule results cannot be extended to n machine/facilities flowshop. However, in a special case, it can be extended to a three-machine flowshop, provided the second machine is not a "bottleneck" machine. For a machine not to be a bottleneck, it should not delay any job, or, in other words, as the job is released from the first machine, it should immediately be processed in the second machine. If such is the case, then develop the following two factors for each job i:

$$A_i = P_{i1} + P_{i2} \quad \text{and} \quad B_i = P_{i2} + P_{i3}$$

Then, apply the procedure described in Section 8.1, except treat factor A_i as machine 1 time and factor B_i as machine 2 time.

8.2.1 Illustrative Example 8.2

Consider Example 8.1, except now suppose the grinding and polishing operations are separated in machine 2 grinder and machine 3 polisher. The processing times now are as given in Table 8.2.

The associated A_i and B_i factors are calculated in Table 8.3.

Applying the procedure from Section 8.1 on A_i and B_i factors results in the sequence of 2-6-5-4-3-1, as illustrated in Figure 8.4.

The schedule for the sequence on three machines is as follows:

Machine 1:	0	(2/8)	8	(6/22)	30	(5/17)	47	(4/25)	72	(3/18)	90	(1/15)	105
Machine 2:	0/8	(2/6)	14/30	(6/10)	40/47	(5/10)	57/72	(4/5)	77/90	(3/2)	92/105	(1/2)	107
Machine 3:	0/14	(2/14)	28/40	(6/20)	60/57	(5/10)	70/77	(4/5)	82/92	(3/3)	95/107	(1/1)	108

TABLE 8.3
A_i and B_i Factors for the Data in Table 8.2

Job	$A_i = P_{i1} + P_{i2}$	$B_i = P_{i2} + P_{i3}$
1	17	3
2	14	20
3	20	5
4	30	10
5	27	20
6	32	30

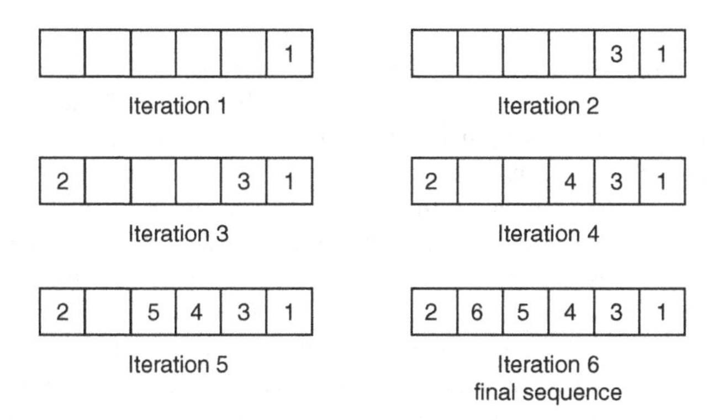

FIGURE 8.4 Solution to three-machine example.

Note the times on machine 3. The first number is when machine 3 completed processing on job $j-1$, and the next number is when job j is available to work on machine 3. For example, 60/57 in front of job 5 on machine 3 indicates that the previous job 6 is completed on time 60 and job 5 is released by machine 2 on time 57. Thus, job 5 had to wait for the machine for 3 time units. According to our convention, if the first number is greater than the second number, then the job is waiting on the machine, and if the second number is larger than the first number, then the machine is waiting on the job (machine is idle). We can easily determine the following quantities:

$$\text{Machine } k \text{ idle time} = \sum (C_{i-1,k} - C_{i,k-1}) \quad \text{for } C_{i-1,k} - C_{i,k-1} < 0$$

Total waiting time of jobs on machine k

$$= \sum (C_{i-1,k} - C_{i,k-1}) \quad \text{for } C_{i-1,k} - C_{i,k-1} \geq 0$$

For example, machine 2 idle time is $= (30-14) + (47-40) + (72-57) + (90-77) + (105-92) = 64$, and total waiting time of jobs on machine 2 is zero. Machine 2 is not a bottleneck machine, and the sequence developed by applying the modified Johnson's rule is optimum.

TABLE 8.4
Start and Completion Times Calculations of Each Job on Each Machine

Job Sequence	Machine 1		Machine 2		Machine 3	
	Process	Complete	Process	Complete	Process	Complete
2	8	8	6	14	14	28
6	22	30	10	40	20	60
5	17	47	10	57	10	70
4	25	72	5	77	5	82
3	18	90	2	92	3	95
1	15	105	2	107	1	108

Perhaps it might be easier to understand the starting and completion times for each job on each machine if we display them in a table such as Table 8.4.

The table is constructed by first listing the sequence and each job's processing times on each machine. For machine 1, since the machine is never idle, the completion time of any job is the cumulative times up to and including that job's processing times. For all other machines, we must check the availability of machine and job. For example, job 6 is available for processing on machine 2 at time 30, and the machine is available for the next job (after job 2) at time 14. The greater of these two numbers is when job 6 can start on machine 2. In short, compare the completion times for the job in the previous machine (same row, previous column) with availability time for the present machine (same column, previous row), and choose the larger of the two numbers as the starting time for the job. We shall leave it up to the reader to compare the numbers in the table with the numbers shown on each machine sequence display and also to develop a similar table for Problem 8.1.

8.3 SETUP/PROCESSING AND REMOVAL TIMES SEPARATED: ANOTHER EXTENSION OF JOHNSON'S ALGORITHM

Now, consider a two-machine flowshop problem and assume that the time taken to process a job can be divided into parts. Some parts of the processing time are dependent on the job, while others, though required, can be performed independent of the job itself. For example, consider a machine shop. Operations associated with each job on each machine may be summarized as follows:

1. *Setup time that is independent of the unit to be processed.* This may consist of activities such as obtaining the blueprints, procuring the necessary tools, fetching the required jigs and fixtures, and setting them on the machine.
2. *Setup time that is unit dependent.* This may include the time required to set the unit in jigs and fixtures and to adjust the tools as necessary.
3. *Processing time.*

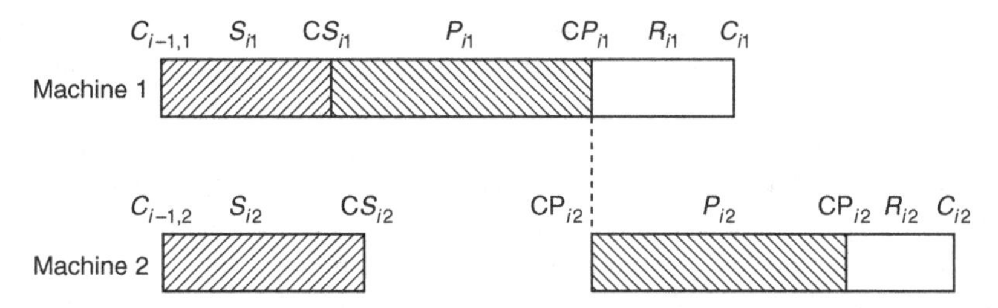

FIGURE 8.5 Illustration of two-machine problem with setup, processing, and removal times separated.

4. *Removal time that is unit dependent.* This may include activities such as disengaging the tools from the unit, and releasing the unit from jigs and fixtures.
5. *Removal time independent of the unit.* The operation includes activities such as dismantling the jigs, the fixtures and/or tools, inspecting/sharpening of tools, returning some tools to the tool room, and cleaning the machine and adjacent area.

Since activities 2, 3, and 4 are unit dependent, their times could be combined and designated as the processing time. However, times for activities 1 and 5 are independent of the unit processed. Can Johnson's rule be applied here? The answer is yes, but with some modifications.

Let us define a few more variables as follows:

S_{ik} = Setup time independent of the unit, that is, activity 1, for job i on machine k.
P_{ik} = Processing time of job i on machine k. This includes times for activities 2, 3, and 4.
R_{ik} = Removal time, that is, activity 5, for job i on machine k.

Calculate factors A_i and B_i such that: $A_i = (S_{i1} - S_{i2} + P_{i1})$ and $B_i = (P_{i2} + R_{i2} - R_{i1})$, and then apply the procedure from Section 8.1. Note that this is a generalized result, which is applicable even if independent setups or removals on one or both machines are absent (corresponding values are zeros). Figure 8.5 illustrates the separation.

8.4 TWO-MACHINE FLOWSHOP WITH TRAVEL TIME BETWEEN MACHINES

Suppose there is one transporter in the system (such as an automatic guided vehicle or an AGV), which connects machines 1 and 2. It picks a job from the first machine and delivers it to the second machine and returns to the first machine. The travel time from the first machine to the second machine is T_f and the travel time from the second machine, back to the first machine is T_b. The time to load and unload the

AGV is included in the travel time. How will the scheduling change? The problem was first introduced by Panwalker (1991). We shall study here the generalization of the problem such that unit-independent setup and removal times may also be present.

Here, if the AGV is not available to load and transport the unit when it has been processed on machine 1, the machine becomes blocked until the unit is removed by the AGV. We cannot start the unit-independent removal on machine 1 until the job is moved by the AGV, or both unit-independent removal on the first machine and travel forward by the AGV, starts at the same time. When the AGV reaches the second machine, it delivers the job to the second machine if the setup has been completed, and the machine is ready to receive the job. If not, it places the job in a waiting line (queue) and immediately starts its return to machine one. In other words, once the AGV leaves the first machine, it always returns in time $T_f + T_b$ to take the next job.

To develop the optimum sequence requires formulation of two factors:

$$A_i = \max([T_f + T_b], [S_{i1} - S_{i2} + P_{i1}])$$

$$B_i = P_{i2} + R_{i2} - R_{i1}$$

Step 1: Apply algorithm described in Section 8.1 (Johnson's algorithm) using the factors calculated above. Denote such sequence as M1 and calculate its makespan.

Step 2: For each job k, where the condition $(T_f + T_b) > (S_{k_1} - S_{k2} + P_{k1})$ is valid, place that job k first in the sequence, scratch the corresponding A_k and B_k and apply Johnson's algorithm on the remaining factors to develop the remaining sequence. Denote these sequences (one for every job that is forced first). We shall denote these sequences as M2, M3, ... Calculate the makespan for each sequence.

Step 3: From among the sequences M1, M2, M3, ..., select the sequence with minimum makespan.

The reader must note that we have introduced step 2 and step 3 here. The reason is simple. It is assumed that the AGV is initially available for transporting at machine 1 and, therefore, the first job does not wait for AGV. If $(S_{k1} - S_{k2} + P_{k1})$ is less than $T_f + T_b$, it is possible that job k completes its operations on machine 1 before AGV can make its round trip. The waiting time of the job can thus be reduced if it is processed first. Johnson's rule then applies on the remaining jobs to determine the optimum sequence of the remaining jobs. Step 3 selects the best sequence.

8.4.1 Relationships for Makespan Calculations

For job i:

ST_{i1} = Setup time on machine 1
ST_{i2} = Setup time on machine 2
RT_{i1} = Removal time on machine 1
RT_{i2} = Removal time on machine 2

PT_{i1} = Process time on machine 1
PT_{i2} = Process time on machine 2

For machine 1:
Start setup on machine 1 = $C_{i-1,1}$ (for $i = 1$, $C_{01} = 0$).
Complete setup on machine 1 (independent of the unit), $CS_{i1} = C_{i-1,1} + S_{i1}$.
Start processing on machine 1, $SP_{i1=CS_{i1}}$.
Complete processing on machine 1, $CP_{i1} = SP_{i1} + P_{i1}$.
Start removal on machine 1, $SR_{i1} = \max(CP_{i1}, CTB_{i-1})$.
CTB is complete travel backward time of AGV (for job 1, $CTB_0 = 0$).
Job completed on machine 1, $C_{i1} = CP_{i1} + R_{i1}$.

For AGV:
Start travel forward, $STF_i = SR_{i1}$
Complete travel forward, $CTF_i = STF_i + T_f$
Start travel forward, $STB_i = CTF_{i1}$
Complete travel forward, $CTB_i = STB_i + T_b$

For machine 2:
Start setup on machine 2, = $C_{i-1,2}$ (for $i = 1$, $C_{02} = 0$)
Complete setup on machine 2 (independent of the unit), $CS_{i2} = C_{i-1,2} + S_{i2}$
Start processing on machine 2, $SP_{2i} = \max(CS_{i2}, CTF_i)$
Complete processing on machine 2, $CP_{i2} = SP_{i2} + P_{i2}$
Job completed on machine 2, $C_{i2} = CP_{i2} + R_{i2}$
Start removal on machine 2, $SR_{i2} = CP_{i2}$
Complete removal and job on machine 2, $C_{i2} = SR_{i2} + R_{i2}$

8.4.2 Illustrative Example 8.3

Consider the data shown in Table 8.5 for five jobs to be processed on two machines. Each job has unit-independent setups and removals along with unit-dependent processing time. The transporter requires 15 units to travel forward from machine 1 to 2, while it takes only 12 units to travel back from machine 2 to 1. The difference occurs because on the forward travel the transporter is loaded with a unit and the time for forward travel includes the loading and unloading times of the unit.

TABLE 8.5
Data for Example 8.3

JOB #	ST_{i1}	PT_{i1}	RT_{i1}	ST_{i2}	PT_{i2}	RT_{i2}
1	10	27	17	17	46	15
2	22	52	8	5	45	19
3	8	26	9	15	35	11
4	23	70	9	5	12	9
5	3	69	11	22	48	1

TABLE 8.6
A_i and B_i Values for the Data in Table 8.5

Job Number	$A_i = \text{Max}(T_f + T_b, \text{ST}_{i1} - \text{ST}_{i2} + \text{PT}_{i1})$	$B_i = \text{PT}_{i2} + \text{RT}_{i2} - \text{RT}_{i1}$
1	$\text{Max}(27, 10 - 17 + 27 = 20) = 27$	$46 + 15 - 17 = 44$
2	$\text{Max}(27, 22 - 5 + 52 = 69) = 69$	$45 + 19 - 9 = 55$
3	$\text{Max}(27, 8 - 15 + 26 = 19) = 27$	$35 + 11 - 9 = 37$
4	$\text{Max}(27, 23 - 5 + 70 = 88) = 88$	$12 + 9 - 9 = 12$
5	$\text{Max}(27, 3 - 22 + 69 = 50) = 50$	$48 + 1 - 11 = 38$

Thus, we have: $T_f = 15$; $T_b = 12$, and therefore the total transport time, $T_f + T_b = 27$.

The first step is to develop A_i and B_i factors. Table 8.6 displays these calculations.

Applying Johnson's algorithm from Section 8.1 on the two factors, we get a sequence 1-3-2-5-4. Since job 3 also satisfies $T_f + T_b > (\text{ST}_{i1} - \text{ST}_{i2} + \text{PT}_{i1})$ we place job 3 first and develop the remaining sequence using the method from Section 8.1 and get 3-1-2-5-4 as the second sequence. Calculation of makespans shows both sequences having the makespans of 391. Details of the makespan calculations for one job sequence are shown in the following.

8.4.3 Makespan Calculations

The final step is to check the makespan for each sequence. The Table 8.7 displays a sample calculation for sequence 1-3-2-5-4. It is convenient to perform the calculations in three parts: machine l, transporter, and machine 2. Note that a job can start forward transportation (SFT_i) as soon as it completes processing on machine 1 (CP_{i1}). Machine 2 can start processing job if has received the job from transporter after the completion of transport forward (CTF_i) and has completed the unit-independent setup (CS_{i2}). Since the transporter does not wait on machine 2, CTF_i is also the time it starts back towards machine 1 (STB_i). A unit cannot be removed from machine 1 unless the transporter is available, that is, the transporter has completed its back travel (CTB_i). Start removal on machine 1 (SR_{i1}) can only start after the unit is picked up by the transporter from machine 1. The setup for next job cannot start (SS_{i1} and SSi2) until the previous job's removal has been completed (CR_{i1} and CR_{i2}).

8.5 *n* JOBS/*m*-MACHINES PROBLEM

So far, we have seen various flowshop results with two-machine problems and a specialized extension to a three-machine problem. In this section, we shall study two heuristic scheduling techniques that expand the analysis to *m* machines. The four methods are:

1. Minimize machine idle time method
2. Palmer's method

TABLE 8.7
Makespan calculations for Sequence 1-3-2-5-4.

	Machine 1					
Job	Ss_{i1}	CS_{i1}	SP_{i1}	CP_{i1}	SR_{i1}	CR_{i1}
1	0	10	10	37	37	54
3	54	62	62	88	88	97
2	97	119	119	171	171	179
5	179	182	182	251	251	262
4	262	285	285	355	355	364

	Transporter			
Job	STF_i	CTF_i	STB_i	CTB_i
1	37	52	52	64
3	88	103	113	125
2	171	186	186	198
5	251	266	266	278
4	355	370	370	382

	Machine 2					
Job	Ss_{i2}	CS_{i2}	SP_{i2}	CP_{i2}	SR_{i2}	CR_{i2}
1	0	17	52	98	98	113
3	113	128	128	163	163	174
2	174	179	186	231	231	250
5	250	272	272	320	320	321
4	321	326	370	382	382	**391**

The makespan = 391

3. Nawaz's method
4. Campbell, Dudek, and Smith (CDS) procedure

Since all four methods are heuristic, they do not guarantee optimum solution in every case, but do give good solutions that are very close to the optimum.

8.5.1 Minimize Machine Idle Time Method

Our basic principle is to schedule the jobs in such a way so as to minimize the idle times of the machines. To this extent, the information generated in Table 8.4 is most useful. In developing Table 8.4, we have listed machines on the columns and jobs in the rows. A job i can only start on the machine k if two conditions are satisfied. First, job i is available, that is, it has been processed in the previous machine (machine $k - 1$), and second, if machine k is available, (that is, the machine has completed work on the previous job). The time when job i is available is shown by the row i entry for completion time of the job in the previous machine (column $k - 1$). The time

TABLE 8.8
Processing Time of Each Jobs on Each Machine and Their Sum

Job/Machine	1	2	3	4	$\sum P_{ij}$
1	25	45	52	40	162
2	7	41	22	66	136
3	41	55	33	21	150
4	74	12	24	48	158
5	7	15	72	52	146
6	12	14	22	32	80

when the machine k is available is given by the completion time for the previous job (job $i - 1$), the completion time entry for job $i - 1$ in column k.

Thus, the idle time of machine k is given by completion time of job i on the previous machine (row i column $k - 1$) minus completion time of job $i - 1$ in machine k (row $i - 1$, column k).

8.5.1.1 Procedure

1. Find the sum of processing times for each job. Arrange the jobs in the ascending order of their sum. Schedule the job with minimum sum in the first position, and calculate its completion time on each machine.
2. Select the three jobs, if available (the number chosen somewhat arbitrarily to minimize calculations), with next minimum sums of processing times, and list them in the iteration table. These are denoted as *testing jobs*. The iteration table lists the last job (LJ) in the schedule developed so far and its completion times. Calculate the completion times of each one of the testing jobs as if each job is the next one to be scheduled.
3. Calculate the total machine idle time for each of the test jobs. Remember there is idle time on the machine only if job completion time (row i, column $k - 1$) > machine free time (completion time for LJ in column k). The value of the machine idle time is the difference in these two values.
4. The next job to be put in sequence is the one with minimum total idle time. If there is a tie, each schedule with the tie jobs may be evaluated.
5. If all jobs are scheduled, stop. If not, calculate the completion times for the job added to the schedule in step 4 and go back to step 2.

8.5.1.2 Illustrative Example

Consider the flowshop problem with four machines and six jobs. The processing times are given in Table 8.8.

The ascending job order based on the sum of processing times is 6-2-5-3-4-1.

TABLE 8.9
Search for Sequence Position 2

Job/Machine	1		2		3		4		Total Machine Idle Time $(C_{i,k-1} > C_{r,k})$
Conf. 6	12	12	14	26	22	48	32	80	
Test jobs									
2	7	19	41	67	22	89	66	155	(67 − 48) + (89 − 80) = 28
5	7	19	15	41	72	120	52	172	(120 − 80) = 40
3	41	53	55	108	33	141	21	162	(53 − 26) + (108 − 48) +(141 − 80) = 148

TABLE 8.10
Search for Sequence Position 3

Job/Machine	1		2		3		4		Total Machine Idle Time $(C_{i,k-1} > C_{r,k})$
Conf. 2		19		67		89		155	
Test jobs									
5	7	26	15	82	72	161	52	213	(161 − 155) = 6
3	41	60	55	122	33	155	21	176	(122 − 89) = 33
4	74	93	12	105	24	129	48	203	(93 − 67) + (105 − 89) = 42

The first job sequenced is job 6. The next step is to develop Table 8.9 to determine which job should follow job 6. The table lists job 6, as the confirmed job, and the next three jobs from the ascending order as the test jobs.

Some explanation of Table 8.9 is necessary. For each job, there are two entries for each machine. The first entry is the processing time and the second entry is its completion time on that machine. Each test job is compared with the last confirmed job r (job 6) for determining its completion times. For example, job 2 completion time on machine 3 is calculated as $(\max(67, 48) + 22) = 89$. Similarly, completion time for job 5 on machine 3 is $(\max(41, 48) + 72) = 120$, and for job 3, the completion time on machine 3 is $(\max(108, 48) + 33) = 141$. The idle time of each machine as a job is sequenced is calculated and added to get the total idle time (note that the idle time only occurs if the row entry for job i in column $k - 1$, $C_{i,k-1}$ is greater than the entry in row r and column $kC_{r,k}$ for the confirmed machine).

Since job 2 has the minimum machine idle time, it is scheduled next. The sequence confirmed so far is 6, 2, –, –, –, –. For the search for the next position job, job 2 becomes the confirmed job and Table 8.10 is constructed.

The minimum total idle time is with job number 5, and hence it is sequenced in position 3. The sequence thus far is 6, 2, 5, –, –, –. The next iteration is shown in Table 8.11.

TABLE 8.11
Search for Sequence Position 4

Job/Machine	1		2		3		4		Total Machine Idle Time ($C_{i,k-1} > C_{r,k}$)
Conf. 5		26		82		161		213	
Test jobs									
3	41	67	55	137	33	194	21	234	0
4	74	100	12		24		48		>0, no need to calculate
1	25	51	45	127	52	213	40	253	0

TABLE 8.12
Search for Sequence Position 5

Job/Machine	1		2		3		4		Total Machine Idle Time ($C_{i,k-1} > C_{r,k}$)
Conf. 1		51		127		213		253	
Test jobs									
3	41	92	55	182	33	246	21	274	0
4	74	125	12	139	24	237	48	301	0

TABLE 8.13
Throughput Time by Placing Job 3 in Position 5 and Job 4 in Position 6

Job/Machine	1		2		3		4	
3	41	92	55	182	33	246	21	274
4	74	166	12	194	24	270	48	322

Here, both jobs 3 and 1 have zero machine idle time. Both should be tried in the fourth position. We have not continued calculations for job 4 since the first machine itself has the idle time of $(100 - 82) = 18$, which is greater than the zero machine idle time we have obtained for the previous job. Let us continue with job 1 in the fourth position. Table 8.12 searches for sequence position 4. Table 8.13 shows the next iteration.

Again, either jobs 3 or 4 may be in position 5, and the remaining in position 6. The throughput times by placing each job in position 5 are determined in Tables 8.13 and 8.14.

Both sequences 6-2-5-1-3-4 and 6-2-5-1-4-3 have makespans of 322. Incidentally, we would have gotten the sequence of 6-2-5-3-4-1 after Table 8.11, had we chosen job 3 for position 4 instead of job 1. This sequence also has makespan of 322.

TABLE 8.14
Throughput Time by Placing Job 4 in Position 5 and Job 3 in Position 6

Job/Machine	1		2		3		4	
4	74	125	12	139	24	237	48	301
3	41	166	55	221	33	270	21	**322**

8.5.2 Palmer Procedure

Palmer (1965) suggested developing the slope index for each job given by following expression:

$$S_i = (m-1)P_{jm} + (m-3)P_{jm-1} + (m-5)P_{jm-2} + \cdots$$
$$- (m-5)P_{j3} - (m-3)P_{j2} - (m-1)P_{j1}$$

and then constructing the sequence based on the descending order of the magnitude of S_j. The idea is to progress by scheduling jobs needing short times first and then continuing in terms of the magnitude of times, with longest job being scheduled last.

For the data in Table 8.7, since we have four machines, $m = 4$, and S_j is given by:

$$S_j = 3P_{j4} + P_{j3} - P_{j2} - 3P_{j1}$$

Substituting the process time values, we get: $S_1 = 52$, $S_2 = 158$, $S_3 = -82$, $S_4 = -66$, $S_5 = 192$, and $S_6 = 68$. The job schedule developed by arranging S_j in descending order is 5-2-6-1-4-3. The makespan for this sequence is 353.

8.5.3 Nawaz Heuristic

Nawaz (1983) describes a heuristic that is easy to construct and gives good results in most cases. It does, however, construct a number of schedules that need evaluations that can be time consuming. The steps are as follows:

1. Calculate the sum of processing times for each job. Arrange the jobs in the descending order of their sum. Call this the "job list" and the job order as $a_1, a_2, a_3 \ldots a_m$.
2. Select the first two jobs from the job list. Determine the best (minimum makespan) of two sequences, first by placing job a1 in the first place and a_2 in the second place and then reversing the order. Do not change the relative positions of the two jobs with respect to each other in the remaining steps of the algorithm.

TABLE 8.15
Makespan Calculations for Sequence 6-2-5-1-3-4

Sequence/ Machine	1	2	3	4
6	12/12	14/26	22/48	32/80
2	7/19	41/67	22/89	66/155
5	7/26	15/82	72/161	52/213
1	25/51	45/127	52/213	40/253
3	41/92	55/182	33/246	21/274
4	74/166	12/194	24/270	48/**322**

3. Pick the job that is in the next position in the job list and find the best sequence by placing it in all possible positions in the partial sequence developed so far. Make sure not to change the relative positions of the jobs that are already assigned in sequence.
4. Repeat step 3 till all jobs are placed in the sequence.

For the data in Table 8.7, the list of the jobs arranged in the descending order of sum of processing time is 1, 4, 3, 5, 2, 6.

Select the first two jobs from the list, jobs 1 and 4. Two partial sequences can be formed. They are 1-4 and 4-1. The completion time for sequence 1-4 is 210 and for sequence 4-1 is 236. The best between these is 1-4, and therefore the relative positions of these two jobs will remain as job 1 ahead of job 4.

Next, job 3 is to be added to the sequence being developed. Three possible partial sequences that can be formed now are: 3-1-4-; 1-3-4; and 1-4-3 (note that the best order of jobs 1 and 4 found earlier remains and job 3 has been added from left to right). The makespan for each partial sequence is 281, 231, and 249, respectively. The best sequence is 1-3-4.

Based on the job list, job 5 is the next one to be added to the partial sequence. The sequences to examine are: 5-1-3-4; 1-5-3-4; 1-3-5-4; and 1-3-4-5. The partial makespan times are: 255, 315, 330, and 306, respectively. The partial sequence 5-1-3-4 with makespan of 255 is selected at this stage.

The next job to add is job 2. Sequences to examine are: 2-5-1-3-4; 5-2-1-3-4; 5-1-2-3-4; 5-1-3-2-4; and 5-1-3-4-2 with 304, 320, 321, 321, and 321 as their respective make spans. The partial sequence 2-5-1-3-4 with the makespan value of 304 is selected.

The LJ to add is job 6. The sequences are: 6-2-5-1-3-4; 2-6-5-1-3-4; 2-5-6-1-3-4; 2-5-1-6-3-4; 2-5-1-3-6-4; and 2-5-1-3-4-6 with the makespans of 322, 332, 336, 336, 336, and 336. The sequence 6-2-5-1-3-4 is the best sequence with the make span of 322. The makespan calculations for sequence 6-2-5-1-3-4 are shown in Table 8.15 with a/b convention where, for each machine, "a" value is the processing time and "b" value is the job completion time.

TABLE 8.16
Application of CDS Heuristics

	$l = 1$		$l = 2$		$l = 3$	
Job	$A_{i,1}$	$B_{i,1}$	$A_{i,2}$	$B_{i,2}$	$A_{i,3}$	$B_{i,3}$
1	25	40	70	92	122	137
2	7	66	48	88	70	129
3	41	21	96	54	129	109
4	74	48	86	72	110	84
5	7	52	22	124	94	139
6	12	32	26	54	48	68
Sequence	2-5-6-1-4-3		5-6-2-1-4-3		6-2-5-1-3-4	
Span	335		353		322	

8.5.4 Campbell, Dudek, and Smith (CDS) Procedure

This is a simple heuristic that in general gives a good result (Campbell et al., 1970). The method generates $m-1$ sequences, one for each value of l, where $i = 1, 2, \ldots m-1$, from which the best is chosen. The sequence number l is constructed by solving a two-machine problem using Johnson's rule where the two pseudo factors A_{il} and B_{il} are generated using the following expressions:

$$A_{i1} = \sum_{j=1}^{1} P_{i \cdot j} \quad \text{and} \quad B_{i1} = \sum_{j=1}^{1} P_{i,m-j+1}$$

Table 8.16 displays the application of CDS heuristic to the example problem. For job 1, for example, $A_{11} = 25$ and $B_{11} = 40$ for $l = 1$, $A_{12} = 24+45 = 92$ and $B_{12} = 40+52 = 92$ for $l = 2$, and $A_{13} = 25+45+52 = 122$ and $B_{13} = 40+52+45 = 137$ for $l = 3$. Applying Johnson's rule to each data set thus generates three sequences. The best among these is the sequence 6-2-5-1-3-4 with a makespan of 322.

8.6 *n*-JOB/*m*-MACHINE PROBLEM: JOBS ARRIVING AT DIFFERENT TIMES

The CDS method is extended to apply in a flowshop where jobs are not available at time 0 but have known arrival times. The objective is to minimize the makespan.

The steps of the procedure are as follows:

Phase 1

1. Schedule the job that arrives first. If there is a tie, choose the one with shortest processing time. Start developing the main sequence.
2. Determine the completion times of the selected job on machine 1.

3. Select the jobs that have arrived before previous completion time and arrange them using the CDS method.
4. Form temporary sequence(s) by attaching the sequence(s) from step 3 to the main sequence. Determine the minimum makespan for these jobs.
5. Select the sequence with minimum makespan.
6. If all jobs have arrived, the sequence from step 5 is the main sequence; go to phase 2. Otherwise, schedule the job that is first in the sequence as the next scheduled job in the main sequence and go back to step 2.

Phase 2

Apply forward phase to the sequence developed in phase 1.

Though most often the above procedure leads to minimum makespan, it is also suggested that we start with each job as the first job in the sequence (step 1) and then apply the remaining steps of the procedure. Then select the best sequence.

8.6.1 Example

Consider a flowshop problem with three machines and five jobs with different arrival times. Processing and arrival times for the jobs are tabulated in the following:

		Machine		
Job	Arrival Time	1	2	3
1	22	17	3	2
2	10	9	5	4
3	15	23	8	5
4	5	15	2	1
5	18	21	6	3

Phase I

Job 4 has the earliest arrival and is therefore selected as a starting point for the main sequence. Job 4 arrives to the shop at five time units, and its processing time on machine 1 is 15 time units. C indicates the time at which job 4 will be finished on machine 1 at 20 time units.

Jobs that arrive before time 20 are jobs 2, 3, and 5, with arrival times 10, 15, and 18, respectively. Apply the CDS method with these three jobs on all three machines and get temporary sequences. The calculations are shown in the following:

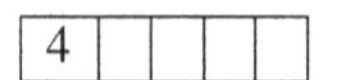

Main sequence

$C = 5 + 15 = 20$

CDS calculations for $C = 20$

Jobs	1		2	
2	9	4	14	9
3	23	5	31	13
5	21	3	27	9

3	2	5

3	5	2

Temporary sequences are:

4-3-2-5
4-3-5-2

Calculate makespan with these temporary sequences.

4-3-2-5

		1		2		3	
Job	Arrival Time	P_i	C_i	P_i	C_i	P_i	C_i
4	5	15	20	2	22	1	23
3	15	23	43	8	51	5	56
2	10	9	52	5	57	4	61
5	18	21	73	6	79	3	**82**

4-3-5-2

		1		2		3	
Job	Arrival Time	P_i	C_i	P_i	C_i	P_i	C_i
4	5	15	20	2	22	1	23
3	15	23	43	8	51	5	56
5	18	21	64	6	70	3	73
2	10	9	73	5	78	4	**82**

As both temporary sequences 4-3-2-5 and 4-3-5-2 have the same makespan, select one at random, say 4-3-2-5, and fix the second job, that is, job 3 of this temporary sequence with job 4 of the main sequence. So, the main sequence as 4-3.

Iteration 2

Completion time of job 3 on machine 1 is 43. There is only one job arriving during this time, and that is job 1. Apply the CDS method to unscheduled jobs 2, 5 and newly arrived job 1.

4	3			

Main sequence

$C = 20 + 23 = 43.$

CDS calculations for $C = 43$

Jobs	1		2	
2	9	4	14	9
5	21	3	27	9
1	17	2	20	5

2	5	1

2	5	1

Temporary sequence:

4-3-2-5-1

		1		2		3	
Job	**Arrival Time**	P_i	C_i	P_i	C_i	P_i	C_i
4	5	15	20	2	22	1	23
3	15	23	43	8	51	5	56
2	10	9	52	5	57	4	61
5	18	21	73	6	79	3	82
1	22	17	90	3	93	2	**95**

Here, there is only one temporary sequence 4-3-2-5-1 with the makespan of 95. As all jobs have been covered, this sequence will become a permanent sequence.

Permanent sequence

4-3-2-5-1 Makespan = 95.

Phase II

Forward Phase

Cycle 1. Permanent sequence 4-3-2-5-1 Makespan = 95

$K = 4$	1-3-2-5-4	Makespan = 110	Do not exchange
$K = 3$	5-3-2-4-1	Makespan = 108	Do not exchange
	4-1-2-5-3	Makespan = 105	Do not exchange
$K = 2$	2-3-4-5-1	Makespan = 100	Do not exchange
	4-5-2-3-1	Makespan = 95	Do not exchange
	4-3-1-5-2	Makespan = 99	Do not exchange
$K = 1$	3-4-2-5-1	Makespan = 105	Do not exchange
	4-2-3-5-1	Makespan = 95	Do not exchange
	4-3-5-2-1	Makespan = 95	Do not exchange
	4-3-2-1-5	Makespan = 99	Do not exchange

The final sequence with job 4 as a starting point of the solution is

4-3-2-5-1 Makespan = 95

Same way, starting first position of main sequence with different jobs leads to the following:

The final sequence with job 1 as a starting point of the solution is

4-3-2-5-1 Makespan = 95

The final sequence with job 2 as a starting point of the solution is

4-3-2-5-1 Makespan = 98

The final sequence with job 3 as a starting point of the solution is

2-3-5-1-4 Makespan = 98

The final sequence with job 5 as a starting point of the solution is

2-3-5-1-4 Makespan = 98

8.6.2 Result

The optimum sequence is 4-3-5-2-1 with makespan of 95.

8.7 SUMMARY

The chapter presents a few interesting examples in flowshop scheduling. We started by illustrating Johnson's rule for two-machine scheduling, which was first published in 1954. This rule provides the optimum solution when all jobs are processed on two machines in sequence. We have shown an extension of Johnson's rule where a job may require unit-dependent and unit-independent times. Travel time between the two machines has also been included in another extension of the results. A rule that also gives optimum schedule, under certain restrictions, for a three-machine flowshop problem is illustrated. Most generalized results are, of course, where m machines are involved. Numerous simple and effective heuristics were presented to address such problems.

8.8 PROBLEMS

8.1 A machine shop receives cast metal parts from a customer. The machine shop must turn one of the part's surfaces on a CNC lathe and then mill another surface on a CNC milling machine. All parts require both operations in the same sequence. The following orders have been received from the customer, and the milling shop has estimated the processing times of each. Determine the sequence that minimizes the makespan.

Order	1	2	3	4	5	6
Processing (lathe)	10	13	6	20	8	17
Processing (milling)	2	13	7	21	18	9

8.2 A production cell consists of a CNC milling machine, a finishing machine, and a deburring machine. These machines are programmed to process different parts. The following orders have been given to the supervisor, who has listed the known processing times for each part. Develop the best schedule for the supervisor that will minimize the makespan of five jobs.

Work Order	Mill	Finish	Debur
1	17	3	2
2	9	5	4
3	23	8	5
4	15	2	1
5	21	6	3

8.3 Using the following data, determine the optimum sequence to minimize the makespan.

Order	1	2	3	4	5	6	7	8
Processing (machine A)	30	5	23	24	9	11	18	28
Processing (machine B)	10	22	35	12	18	28	15	33

8.4 A cutoff saw and a vertical band saw are both required machines (M1 and M2) for two operations on various parts. Both machines have fixtures on which parts must be aligned. All parts delivered to the machines have drawings attached that instruct the operator as to what fixtures are needed and at what speeds the machines must be operated. Develop the sequence for the orders as given below, for installing and removing fixtures, and machine settings and adjustments.

Order	Settings Processing		Fixture Install		Fixture Removal		Adjustment	
	M1	M2	M1	M2	M1	M2	M1	M2
1	10	13	2	1	1	2	1	1
2	20	18	4	3	3	2	2	1
3	12	20	5	4	4	3	1	2
4	16	9	1	2	2	1	1	2
5	8	15	6	5	5	4	2	1

8.5 Using the data from Problem 4.3, consider AGVs that have been installed to transport parts between machines. The AGV requires eight units of time to travel from the first machine to the second, and seven units of time from the second to the first. The AGV does not have to wait for machine B to deposit the parts and return to machine A. Find the sequence with the minimum makespan.

8.6 Using the following data on the three-machine processing requirements of seven jobs, develop the best sequence to minimize the makespan.

Job Number	Machine 1 Processing	Machine 2 Processing	Machine 3 Processing
1	12	6	9
2	15	7	18
3	7	4	10
4	23	12	26
5	14	7	11
6	19	10	22
7	6	3	3

8.7 Consider the following data for six jobs processed on two machines with unit-independent setups and removals:

Job	Machine *X* Setup	Machine *X* Processing	Machine *X* Removal	Machine *Y* Setup	Machine *Y* Processing	Machine *Y* Removal
1	6	12	5	2	1	3
2	3	6	4	8	4	7
3	2	4	1	4	2	5
4	4	8	5	1	1	2
5	1	2	2	5	3	4
6	9	18	8	9	5	8

Develop the optimum schedule to minimize the makespan.

8.8 For truck drivers, the average is eight time units to deliver jobs from machine K to L and seven time units returning from machine L to K. Given the following data, develop the sequence that minimizes makespan.

Job	Machine K Setup	Machine K Processing	Machine K Removal	Machine L Setup	Machine L Processing	Machine L Removal
1	10	15	11	5	11	10
2	8	13	9	6	9	8
3	9	9	7	10	16	11
4	7	18	5	9	18	6
5	4	11	3	12	5	7

REFERENCES AND SUGGESTED READINGS

Achugbue, J.O. and F.Y. Chin. November 1982. "Complexity and Solutions of Some Three-Stage Flowshop Scheduling Problems" *Mathematics and Operations Research*, 7(4): 532–544.

Campbell, H.G., Dudek, R.A., and M.L. Smith. 1970. "A Heuristic Algorithm for the 'N' Job 'M' Machine Sequencing Problem" *Management Science*, 16: B630–B637.

Johnson, S.M. 1954. "Optimal Two- and Three-Stage Production Schedules with Setup Times Included" *Naval Research Logistics Quarterly*, 1(1): 61–68.

Nawaz, M. 1983. "A Heuristic Algorithm for the 'M'-Machine, 'N'-Job Flow-Shop Sequencing Problem" *Management Science*, 11(1): 91–95.

Palmer, D.S. 1965. "Sequencing Jobs through a Multi-Stage Process in the Minimum Total Time- A Quick Method of Obtaining a Near Optimum" *Operations Research Quarterly*, 16: 101–107.

Panwalker, S.S. 1991. "Scheduling of a Two-Machine Flowshop with Travel Time between Machines" *Journal of Operations Research Society*, 42(7): 609–613.

Pinedo, M. January-February 1982. "Minimizing the Expected Makespan in Stochastic Flow Shops" *Operations Research*, 30(1): 148–62.

Sule, D.R. 1982. "Sequencing 'N' Jobs on Two Machines with Setup, Processing and Removal Times Separated" *Naval Research Logistics Quarterly*, 29(3): 517–519.

Szwarc, W. December 1981. "Note on the Flow Shop Problem without Interruptions in Job Processing" *Naval Research Logistics Quarterly*, 28(4): 665–669.

9 Parallel Processing and Batch Sequencing

In this chapter, we introduce three different concepts. In parallel processing, jobs are processed by one of several identical machines, allowing considerable reduction in makespan. How to divide jobs among processors is the challenging issue that is addressed here. Two issues in batch sequencing are discussed in this chapter. First is the "baking problem" where a batch of jobs is subjected to identical processing. How to form job batches to minimize delays is an interesting problem. Finally, we shall study a problem where setup times are dependent on job types, and it is desired to form a job sequence to minimize the setup times while meeting due dates.

9.1 PARALLEL PROCESSING

In the previous chapters, we have assumed that we only have one facility or machine on which we can process all jobs. What if we have multiple facilities all being able to provide the same services? How will the job scheduling change? How will the makespan change? We might visualize these as being facilities/machines in parallel (they do not have to be physically parallel) because they are identical to each other and thereby performing the same services. This design is called parallel identical processors scheduling (Figure 9.1). We shall study here a few representative examples in parallel scheduling.

9.1.1 Jobs with Equal Weight and No Due Dates

Suppose we have N jobs and they can be processed on any one of M parallel processors. The objective is to develop a job schedule on each processor so as to minimize the makespan for all jobs. The procedure given in the following, in most cases, achieves this objective effectively.

9.1.1.1 Procedure

Step 1: Arrange the jobs in the descending (nonincreasing) order of the processing times. Designate this as the list.

Step 2: The lower bound of the minimum achievable makespan is given by the sum of processing times divided by the number of available parallel processors. A machine (processor) is designated as available if the sum of the processing times of the jobs assigned to the machine is less than the calculated lower bound. Initially, all the machines are available.

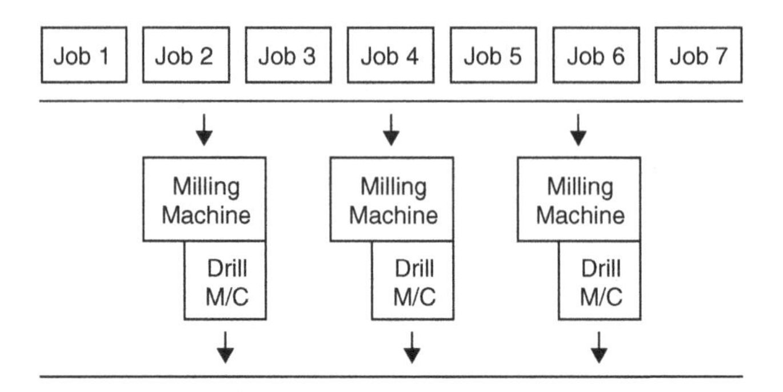

FIGURE 9.1 Parallel Processing Setup.

Step 3: Start allocating the jobs to one machine (processor) beginning with the job having the highest processing time. In case of a tie, break it randomly. Continue allocating the jobs to the machine under consideration, in decreasing order of processing times (P_i) till one of the following happens.

1. Sum of processing times of the jobs assigned to the machine under consideration becomes equal to the lower bound. If this happens, then start assigning jobs to the next available machine.
2. Sum of processing times of the jobs allocated to the machine becomes greater than the lower bound. If this happens, then the job that has caused the sum to be greater than the lower bound and subsequent jobs are allocated in the following sweeping manner:
   - Sweep across the available machines (in order 1, 2, 3, …). If the sum of processing times on the next machine is less than the lower bound and allocation of the job here will not increase the cumulative processing time on the machine beyond the lower bound, then assign the job there. If not, continue the check with the next available machine. If, on all available machines, the assignment of the present job will increase the sum beyond the lower bound, then assign the job to the machine where such increase would be minimum.

Once all the jobs are assigned, the minimum makespan is the maximum of the sum of processing times on each machine.

9.1.1.2 Illustrative Example

Suppose we have nine jobs with the processing times given in the Table 9.1 and three processors. We wish to develop the detailed schedule for each processor to minimize the makespan. We present here a slightly modified means of presentation and calculations that achieves the steps described in the procedure. The same representation will then be continued to be used in the rest of this section when we are working with other objectives.

TABLE 9.1
Processing Times for Jobs

Job	1	2	3	4	5	6	7	8	9
Processing time	10	12	5	8	7	3	5	15	12

TABLE 9.2
Jobs Arranged in Descending Order of Processing Times

Job	8	2	9	1	4	5	3	7	6
Processing time	15	12	12	10	8	7	5	5	3

TABLE 9.3
Job Assignments on Each Machine

Machine 1			Machine 2			Machine 3		
Job	Processing Time	RCT = 26	Job	Processing Time	RCT = 26	Job	Processing Time	RCT = 26
8	15	11	2	12	14	4	8	18
1	10	1	9	12	2	5	7	11
			6	3	–1	3	5	6
						7	5	1

Table 9.2 displays the application of the first step: arranging jobs in the descending order of the processing times.

The sum of the processing times is 77, and therefore with three processors the minimum makespan is $77/3 = 26$ (the numbers are rounded up). This number is noted as the remaining cumulative time (RCT) allowed in associated column for each machine in Table 9.3. Each time a job is assigned to a machine, it is noted in the table along with its processing time. The available cumulative time is then decreased by this value. This makes application of step 3 easier to implement.

Select the first job from the list, and assign it to the first available machine. The available cumulative time is reduced by 15, the processing time of job 8. The next job in the list is job 2 with the processing time of 12. It cannot be assigned to machine 1, since the available time (the present RCT value of machine 1) is only 11. It is then allocated to the next available machine, machine 2, and the RCT for machine 2 is modified. The third job from the list, job 9, has a processing time of 12 and cannot be assigned to machine 1. We check machine 2 next. The job can be specified on machine 2, and hence the allocation is made. The RCT for machine 2 is now 2. This sweeping procedure (starting from first machine and continuing in the machine order of 2, 3 ... till the assignment is made) is continued for the remaining jobs from the list. It might be interesting to note that when the last job in the list, job 6, is to be assigned, the RCTs on the machines 1, 2, and 3 are 1, 2, and 1, respectively. The job has a processing time of 3. So, we need a cumulative time on at least one machine that is greater than what is initially allocated to each machine, namely 26. The least increase in time requirement is achieved by assigning the job to machine 2, giving RCT $= -1$. The makespan for all jobs is therefore $26 + 1 = 27$.

If there are only two processors, the job distribution on each is

Machine 1: jobs 8, 2, and 9 for the total processing time of $15 + 12 + 12 = 39$.
Machine 2: jobs 1, 4, 5, 3, 7, and 6 for the total processing time of $10 + 8 + 7 + 5 + 5 + 3 = 38$, giving the makespan of 39.

9.1.2 Jobs with Priorities Ranked by Weights

Now, suppose the jobs have priorities as specified by their weights. The jobs with the higher weights should have smaller makespans than the jobs with the smaller weights. We can form the job groups based on the weights; all jobs with the same weight are grouped together. There is no specific due date for each job, but we want to minimize the makespan for each priority set.

The procedure is similar to the one developed in the previous section, except for minor variations. Develop the job sets (groups) for each weight. Rank these sets in the descending order of the weights. Calculate the minimum makespan for each weight. Apply the procedure from the previous section to assign jobs to each processor, based on the established weight priorities and continuing on the previous assignments (if any). The following example illustrates the procedure.

Suppose each job in the data Table 9.1 has a weight indicating its priority (higher the weight, higher the priority), as shown in Table 9.4. Further, assume that we now have only two processors.

There are two priority classes, and the jobs are arranged in the descending order of processing times in each weight class. The weight classes are also arranged in the descending order of their magnitude, as shown in Table 9.5.

The minimum makespan for each weight level is: weight 2 $= (12 + 10 + 7 + 5 + 5)/2 = 20$, and for weight 1 $= (15 + 12 + 8 + 3)/2 = 19$.

The Table 9.6 shows the assignments. Jobs with weight 2 are assigned first. The makespan for this weight class is 20. Jobs are assigned following sweep procedure.

TABLE 9.4
Jobs with Priorities

Job	1	2	3	4	5	6	7	8	9
Processing time	10	12	5	8	7	3	5	15	12
Weight	2	2	2	1	2	1	2	1	1

TABLE 9.5
Sets and Jobs in Descending Order

Job	2	1	5	3	7	8	9	4	6
Processing time	12	10	7	5	5	15	12	8	3
Weight	2	2	2	2	2	1	1	1	1

TABLE 9.6
Final Assignments

Machine 1				Machine 2			
Weight W2	Job	Processing time	RCT for W2 = 20	Weight W2	Job	Processing time	RCT for W2 = 20
	2	12	8		1	10	10
	5	7	1		3	5	5
					7	5	0
Weight W1			RCT for W1 $= 1 + 19 = 20$	Weight W1			RCT for W1 $= 0 + 19 = 19$
	8	15	5		9	12	7
	6	3	2		4	8	-1

After the job assignments on two machines, there is one unit of RCT left in machine 1. It is added to the RCT for the next weight jobs on that machine, and the sweep procedure is applied to the jobs with weight 1. The maximum makespan for jobs with weight 2 is 20 (on machine 2), while that for weight 1 is also 20 (machine 1: $19 + 1 = 20$).

9.1.3 Jobs with Due Dates

Suppose now that jobs have due dates, and the weights represent the lateness penalties as defined in Chapter 6. The objective is to schedule jobs to minimize the total penalty.

The procedure developed before is no longer applicable since the jobs not only have weights but also due dates and the objective has also changed from minimization of makespan to minimization of the penalty. But still, we can develop on the previous procedure as well as use the concepts from the backward procedure of Chapter 6.

The details of the procedure are as follows:

Step 1: Calculate RCT for each machine. Initially, assume that we can develop a schedule on each machine so that makespan is equal to the initial value of the RCT. With this assumption, the present value of RCT is also the time at which the last assigned job completes its processing (i.e., TT = RCT).

Step 2: Divide the jobs into each weight class. Within each class, arrange the jobs in the descending order of the due dates.
Step 3: For each processor, determine the current value of RCT. This is also the completion time of the next job to be scheduled on that processor.
Step 4: Select as many not-yet-assigned jobs (if available) from each job group as there are processors, starting with the unassigned job with the largest due date from each weight group. Calculate the penalty of assigning each selected job on each processor.
Step 5: Select the feasible combination(s) of jobs that result in minimum sum of the penalty. For each combination, perform steps 6 and 7.
Step 6: Assign each of the jobs selected in the combination in step 5 to the appropriate processor, and modify the RCT value for each processor by subtracting the processing time of the assigned job. If all jobs are assigned, go to step 7; if not, go to step 4.
Step 7: If the final RCT for any processor(s) is negative, it is due to the last job assigned to the processor in the backward phase, and therefore it is also the first job in the sequence. Evaluate by placing the job in the first place of each sequencer, and calculate the total penalty for each combination by applying the single-machine sequencing forward phase from Chapter 6. Select the combination with the minimum cost as being the best solution.

The procedure can best be illustrated with two tables developed simultaneously. The penalty table calculates the penalties on each machine, which are then transferred into an evaluation and selection table, in which we determine the jobs to be selected and assigned to the processors and determine the new values of RCTs, which are then entered back into the penalty table.

9.1.3.1 Illustrative Example

Consider nine jobs for which the data is given in Table 9.7. We have two parallel processors, and we want to schedule the jobs to minimize the tardiness penalty.

To apply the procedure, the sum of processing times is 77, and therefore, the initial value of RCT is $77/2 = 39$. We divide the jobs based on weight class, and within each class arrange the jobs in the descending order of the due dates. Table 9.8 shows the results.

TABLE 9.7
Jobs with Due Dates

Job	1	2	3	4	5	6	7	8	9
Process time	10	12	5	8	7	3	5	15	12
Weight	2	2	2	1	2	1	2	1	1
Due date	10	15	27	42	50	35	20	16	20

TABLE 9.8
Jobs Arranged in Weight Class and in Descending Order of Due Dates

	Weight 2					Weight 1			
Job	5	3	7	2	1	4	6	9	8
Process time	7	5	5	12	10	8	3	12	15
Due date	50	27	20	15	10	42	35	20	16

The procedure starts with iteration 1 in the penalty Table 9.9. Since we have two processors, the first two jobs from each weight class in Table 9.8 are selected. They are jobs 5, 3 from weight 2 class and jobs 4, 6 from weight 1 class. Initially the RCT on each machine is 39. The penalties for each job on each machine are calculated. For example, consider job 3. Its due date is 27. On machine 1, if the job is completed on the present value of RCT that is, at time 39, it would be 12 units late. The corresponding penalty is $12 \times 2 = 24$ units, both values are noted in the table. Since the RCT for the second machine is the same, the associated numbers are the same on the second machine.

The penalty values are inserted into the evaluation and selection Table 9.10. Each job's penalty value for assigning the job to that machine now is noted below each machine in the evaluation phase. We require that each job be assigned to one machine and only to one machine. A combination of jobs that gives the minimum cost is selected. If there is more than one such combination, then every combination of jobs that gives the minimum cost must be selected and evaluated. In this illustration, a total of six options (combinations) is generated through the iterations that need checking. For example, in iteration 1, the two possible combinations are: job 5 on machine 1 with job 4 on machine 2 and job 5 on machine 2 with job 4 on machine 1. Both combinations have the same penalty value of zero. The first combination is shown as option 1, and the second is shown as option 4. We shall continue the description of the procedure on option 1. Job 5 is assigned to machine 1, and job 4 is assigned to machine 2 and recorded in the selection section of the Table 9.10 (option 1). RCTs of each machine are modified by reducing it by the processing time of the job assigned to the machine. These values are then entered in the penalty table and are used to calculate the penalties for the selected jobs in the next iteration.

In iteration 2, cost combination of 0 and 8 gives the minimum total cost for that iteration, and therefore the associated job 6 is assigned to machine 1 and 3 is assigned to machine 2. The RCT for each machine is modified by subtracting processing time of the assigned job from the previous value of the RCT, and the resulting values are entered in the penalty table for the next iteration.

Iteration 5 is of interest. RCT values from iteration 4 are: 5 for machine 1, and 6 for machine 2. Job 1 has a processing time of 10, which may not be assigned to any machine without making RCTs negative. Therefore, according to step 7, job 1 should be assigned to each machine and checked for cost as shown in the following

TABLE 9.9
Penalty Table

Iteration	Weight 2 Job Group						Weight 1 Job Group					
	Job/ Processing Time	Due Date	Machine 1 RCT = 39		Machine 2 RCT = 39		Job/ Processing Time	Due Date	Machine 1 RCT = 39		Machine 2 RCT = 39	
Option 1												
			Late	Pen	Late	Pen			Late	Pen	Late	pen
1	5/7	50	0	0	0	0	4/8	42	0	0	0	0
	3/5	27	12	24	12	24	6/3	35	4	4	4	4
			RCT = 32		RCT = 31				RCT = 32		RCT = 31	
2	3/5	27	5	10	4	8	6/3	35	0	0	0	0
	7/5	20	12	24	11	22	9/12	20	12	12	11	11
			RCT = 29		RCT = 26				RCT = 29		RCT = 26	
3	7/5	20	9	18	6	12	9/12	20	9	9	6	6
	2/12	15	14	28	11	22	8/15	16	13	13	10	10
			RCT = 17		RCT = 11				RCT = 17		RCT = 11	
4	7/5	20	0	0	0	0						
	2/12	15	2	4	4	8						
			RCT = 5		RCT = 6							
5	1/10		PRCT = −5		PRCT = −4							
Option 2												
3			RCT = 14		RCT = 14							
4	7/5	20	0	0	0	0						
	2/12	15	0	0	0	0						
			RCT = 9		RCT = 2							
5	1/10	10	0	0	0	0						
			RCT = −1		RCT = −8							

Option 3												
4			RCT = 2		RCT = 9							
5	1/10	10	0	0	0	0						
			RCT = −8		RCT = −1							
Option 4												
			Late	pen	Late	Pen			Late	Pen	Late	pen
1	5/7	50	0	0	0	0	4/8	42	0	0	0	0
	3/5	27	12	24	12	24	6/3	35	4	4	4	4
			RCT = 31		RCT = 32				RCT = 31		RCT = 32	
2	3/5	27	4	8	5	10	6/3	35	0	0	0	0
	7/5	20	11	22	12	24	9/12	20	11	11	12	12
			RCT = 26		RCT = 29				RCT = 26		RCT = 29	
3	7/5	20	6	12	9	18	9/12	20	6	6	9	9
	2/12	15	11	22	14	28	8/15	16	10	10	13	13
			RCT = 14		RCT = 14				RCT = 14		RCT = 14	
4	7/5	20	0	0	0	0						
	2/12	15	0	0	0	0						
			RCT = 9		RCT = 2							
5	1/10	10	0	0	0	0						
			PRCT = −1		PRCT = −8							
Option 5												
4			RCT = 2		RCT = 9							
5	1/10	10	0	0	0	0						
			PRCT = −8		PRCT = −1							
Option 6												
3			RCT = 11		RCT = 17							
4	7/5	20	0	0	0	0						
	2/12	15	0	0	0	0						
			RCT = −1		RCT = 12							
5	1/10	10	0	0	2	4						
			RCT = −11		RCT = 2							

TABLE 9.10
Evaluation and Selection Table

Iteration	Evaluation			Selection				Penalty
				Machine 1		Machine 2		
	Job	Mach 1	Mach 2	Job	RCT	Job	RCT	
Option 1								
1	5	**0**	0	5	$39-7=32$	4	$39-8=31$	$0+0=0$
	3	24	24					
	4	0	**0**					
	6	4	4					
2	3	10	**8**	6	$32-3=29$	3	$31-5=26$	$0+8=8$
	7	24	22					
	6	**0**	0					
	9	12	11					
3	7	18	12	9	$29-12=17$	8	$26-15=11$	$9+10=19$
	2	28	22					
	9	**9**	6					
	8	13	**10**					
4	7	0	**0**	2	$17-12=5$	7	$11-5=6$	$0+4=4$
	2	**4**	8					
5	1							
Option 2								
3	7	18	12	8	$29-15=14$	9	$26-12=14$	$13+6=19$
	2	28	22					
	9	9	**6**					
	8	**13**	10					
4	7	**0**	0	7	$14-5=9$	2	$14-12=2$	0
	2	0	**0**					
5	1							
Option 3								
4	7	0	**0**	2	$14-12=2$	7	$14-5=9$	0
	2	**0**	0					
5	1							
Option 4								
1	5	0	**0**	4	$39-8=31$	5	$39-7=32$	$0+0=0$
	3	24	24					
	4	**0**	0					
	6	4	4					
2	3	8	10	3	$31-5=26$	6	$32-3=29$	$8+0=8$
	7	22	24					
	6	0	0					
	9	11	12					
3	7	12	18	9	$26-12=14$	8	$29-15=14$	$6+13=19$

TABLE 9.10
Continued

	2	22	28					
	9	**6**	9					
	8	10	**13**					
4	7	**0**	0	7	$14-5=9$	2	$14-12=2$	$0+0=0$
	2	0	**0**					
5	1							
Option 5								
4	7	0	**0**	2	$14-12=2$	7	$14-5=9$	$0+0=0$
	2	**0**	0					
5	1							
Option 6								
3	7	12	18	8	$26-15=11$	9	$29-12=17$	$9+10=19$
	2	22	28					
	9	6	**9**					
	8	**10**	13					
4	7	0	**0**	2	$11-12=-1$	7	$17-5=12$	$0+0=0$
	2	**0**	0					
5	1							

Note: Bold entries in columns in each option provide the minimum sum. The associated jobs are selected and loaded on respective machine

combination calculation (note that in Table 9.9, jobs are developed applying backward phase, and therefore they are listed in the reverse order of actual schedule).

Combination I

Assign job 1 to machine 1. The sequence to check on the first machine is 1-2-9-6-5. Applying the procedure from Section 6.4 results in no change in the sequence. The cost calculations for the sequence are shown in the following.

Machine 1

Schedule	0	(1/10)	10	(2/12)	22	(9/12)	34	(6/3)	37	(5/7)	42
Due dates		(1)	10/T	(2)	15/L	(9)	20/L	(6)	35/L	(5)	50/E
Penalty				$7\times 2+14\times 1+2\times 1=30$							

On machine 2, the sequence to check is 7-8-3-4. The application of the procedure in Section 6.4 results in 8-7-3-4 as being the optimum sequence. The cost calculations are as follows:

Schedule	0	(8/15)	15	(7/5)	20	(3/5)	25	(4/8)	33
Due dates		(8)	16/E	(7)	20/ T	(3)	27 /E	(4)	50/E

Machine 2

Schedule	0	(7/5)	5	(8/15)	20	(3/5)	25	(4/8)	33
Due dates		(7)	20/E	(8) 16/L	(3)	27/E	(4)	42/E	
Penalty		0							

Total penalty for combination 1 is $30 + 0 = 30$

Combination 2

Assign job 1 to machine 2. The job sequence on machine 1 now is 2-9-6-5. Application of the Section 6.4 procedure does not change the sequence. The cost calculations are as follows:

Machine 1

Schedule:	0	(2/12)	12	(9/12)	24	(6/3)	27	(5/7)	34
Due dates:		(2)	15/E	(9)	20/L	(6)	35/E	(5)	50/E
Penalty:					$4 \times 1 = 4$				

The initial sequence on machine 2 is 1-7-8-3-4. It changes to 1-7-3-8-4 after the application of the Section 6.4 procedure. The corresponding cost calculations are

Machine 2

Schedule	0	(1/10)	10	(7/5)	15	(3/5)	20	(8/15)	35	(4/8)	43
Due dates		(1)	10/T	(7)	20/E	(8)	27/E	(3)	16/L	(4)	42/L
Penalty					$19 \times 1 + 1 \times 1 = 20$						

The total penalty for combination 2 is $4 + 20 = 24$. Hence, we choose combination 2 as our assignments for the first possible sequence.

Calculations for all six combinations developed as the alternate job combinations give the same minimum cost. They are also shown in Tables 9.9 and 9.10. The tables are constructed such that each option only shows the calculations that have changed from the previous option. The resulting sequence penalties are shown in Table 9.11. Combination 1 for sequence 2 and combination 2 for sequence 5, both yield a minimum penalty of 22.

9.2 SINGLE OPERATION JOB-RELATED EARLINESS/TARDINESS PENALTIES WITH MACHINE ACTIVATION COST

There are n jobs available at time zero, each with early and late penalties to be delivered with a common due date specified by us. There are number of identical parallel machines that can be activated for an activation cost of D dollars/machine. The objectives are to determine the optimum number of machines to activate and the common due date for the jobs on each machine (which may be different from machine to machine), and to minimize the total of activation cost and late and early penalties.

TABLE 9.11
Sequence Penalties

Sequence Number	Combination Number	Machine Number	Initial Sequence (Before Forward Pass)	Final Penalty (After Forward Pass)
1	1	1	1-2-9-6-5	30
		2	7-8-3-4	
	2	1	2-9-6-5	24
		2	1-7-8-3-4	
2	1	1	1-7-8-6-5	22
		2	2-9-3-4	
	2	1	7-8-6-5	38
		2	1-2-9-3-4	
3	1	1	1-2-8-6-5	38
		2	7-9-3-4	
	2	1	2-8-6-5	23
		2	1-7-9-3-4	
4	1	1	1-7-9-3-4	23
		2	2-8-6-5	
	2	1	7-9-3-4	38
		2	1-2-8-6-5	
5	1	1	1-2-9-3-4	38
		2	7-8-6-5	
	2	1	2-9-3-4	22
		2	1-7-8-6-5	
6	1	1	1-2-8-3-4	48
		2	7-9-6-5	
	2	1	2-8-3-4	23
		2	1-7-9-6-5	

The procedure is simple. Let P_i, E_i, and L_i be the processing time and early and late penalties per unit of time for job i, and then follow the steps given in the following:

1. Take the ratios of P_i/E_i and P_i/L_i for each job.
2. Compare all ratio values. If the highest ratio is P_i/E_i, place the corresponding job in the earliest available position in the sequence; if highest value is associated with P_i/L_i, place the associated job in the latest available position in the sequence. If same job has highest ratios for both ratios simultaneously, develop two sequences, one by placing job earliest and the second by placing the job latest in the sequence.

The number of sequences depends on the number of parallel machines. If there are K parallel machines, then there are at least K sequences developed. In assigning a job to a sequence, follow the rule stated earlier; in addition, assign the next job to the machine sequence that has least job time assigned so far.

3. Determine the common due date on each machine. We can apply the algorithm described in Chapter 7, or we can make a good approximation by evaluating penalty at the end of completion time for each job, assuming that time to be the common due date for all jobs and then selecting the time with minimum total penalty as the common due date.

If we are allowed to have one common due date per machine, each sequence is evaluated independently. If only one common due date is allowed for all jobs, completion times for each job on each machine should be used as the common due date.

To determine the optimum number of parallel processors, repeat the preceding process till incremental savings are achievable, starting with one processor and increasing by one in every iteration.

9.2.1 Example

A four-job problem with data is given in Table 9.12. The activation cost for a parallel machine is $15. We wish to determine the optimum number of parallel machines and common due date.

The associated P_i/E_i and P_i/L_i for each job are as follows (Table 9.13):

9.2.2 Single Machine

Start with one machine. The largest value among the ratios is 17, associated P_3/L_3 or with job $3P_i/L_i$. Place job 3 in the last position. Scratch job 3 from further considerations.

TABLE 9.12
Data for the Problem

Job i	P_i Processing Time	E_i Early Penalty	L_i Late Penalty
1	24	3	8
2	30	2	5
3	17	8	1
4	9	5	3

TABLE 9.13
P_i/E_i and P_i/L_i Ratios

Job i	P_i Processing time	E_i Early Penalty	L_i Late Penalty	P_i/E_i	P_i/L_i
1	24	3	8	8	3
2	30	2	5	15	6
3	17	8	1	2.12	**17**
4	9	5	3	1.8	3

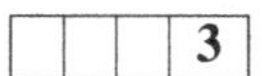

Of the remaining, job 2 has the maximum value for both ratios, namely 15 and 6. Two partial sequences are developed, assigning job 2 the earliest and latest possible positions.

After job 1 is assigned, the next highest ratio is associated with job 1, again for both P_i/E_i and P_i/L_i factors. Develop two partial schedules for each schedule available so far.

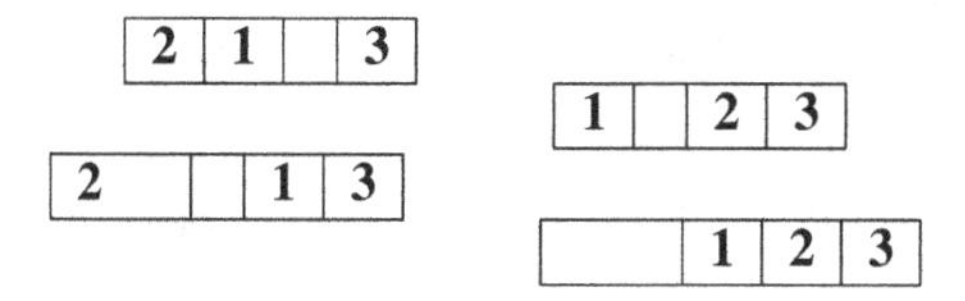

Because job 4 is the last job to be assigned, place it in the available position in each sequence.

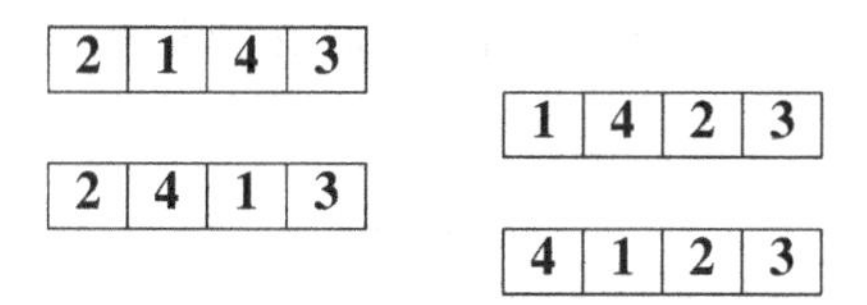

The initial schedules are

2-1-4-3
2-4-1-3
1-4-2-3
4-1-2-3

To determine the common due and penalty dates, let us examine one sequence, 2-1-4-3.

0(2/30)30(1/24)54(4/9)63(3/17)80

By considering completion times of 80, 63, 54, and 30 as the common due dates, determine the penalties (Table 9.14).

The minimum cost for 2-1-4-3 is 101, with due date of 54. Although we are not doing it in this illustration, it should be noted that the cost curve has only one minimum value. Therefore, it is not necessary to evaluate any due dates once the minimum is obtained. For other sequences, the common due date and penalties are as follows in Table 9.15.

TABLE 9.14
Common Due Date Calculations—Single Machine

Due Date	Penalty
80	$(80 - 30) \times 2 + (80 - 54) \times 3 + (80 - 63) \times 5 + (80 - 80) = 263$
63	$(63 - 30) \times 2 + (63 - 54) \times 3 + (63 - 63) + (80 - 63) = 110$
54	$(54 - 30) \times 2 + (54 - 54) + (63 - 54) - 3 + (80 - 54) - 1 = 101$
30	$(30 - 30) + (54 - 30) \times 8 + (63 - 30) \times 3 + (80 - 30) \times 1 = 341$

TABLE 9.15
Common Due Date Costs

Schedule	Common Due Date	Minimum Cost
2-1-4-3	54	101
2-4-1-3	63	203
1-4-2-3	33	269
4-1-2-3	33	317

9.2.3 Two Machines

Assign first job selected, job 3 to machine 1. Each time a job is placed in the sequence, the processing time is noted. Scratch job 3 from further considerations.

Machine 1

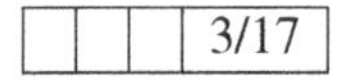

The next job selected is job 2. Since both ratios are highest in the respective columns, job 2 should be placed both at the start and end available positions in the sequence. It is placed in machine 2 since it has the least processing time assigned so far.

Machine 2

Sequence 1

2/30			

Sequence 2

			2/30

Next job to schedule is Job 1. Again, it has largest ratios in both columns (after job 3 and 2 are scratched). So, job 1 should be at the beginning and end of schedules. Machine 1 has the least processing time, and hence job 1 is assigned there.

Machine 1

Sequence 1

1/24			3/17

Sequence 2

		1/24	3/17

Job 4 is the last job to be assigned. Assign it to Machine 2 since it has the least processing time.

Machine 2

Sequence 1			
2/30	4/9		

Sequence 2			
		4/9	**2/30**

The sequences to examine are

Machine 1 1–3 0(1/24) 24 (3/17) 41
Machine 2 2–4 0(2/30) 30 (4/9) 39
4–2 0 (4/9) 9(2/30) 39

The penalties are

For One Common Due Date

Due Date	Penalty	Total Penalty
41	$(41-41)+(41-24)\times 8+(41-39)3+(41-9)\times 5$	302
	$(41-41)+(41-24)\times 8+(41-39)5+(41-9)\times 3$	242
39	$(41-39)\times 1+(39-24)\times 3+(39-39)+(39-30)\times 5$	163
	$(41-39)\times 1+(39-24)\times 3+(39-39)+(39-9)\times 3$	212
30	$(41-30)\times 1+(30-24)\times 3+(39-30)\times 3+(30-30)$	**80**
	$(41-30)\times 1+(30-24)\times 3+(39-30)\times 5+(30-9)\times 5$	203
24	$(41-24)\times 1+(24-24)+(39-24)\times 3+(30-24)\times 5$	92
	$(41-24)\times 1+(24-24)+(39-24)\times 5+(24-9)\times 5$	167
9	$(41-9)\times 1+(24-9)\times 8+(39-9)\times 3+(30-9)\times 5$	347
	$(41-9)\times 1+(24-9)\times 8+(39-9)\times 5+(9-9)$	302

The minimum cost is 80, associated with the due date of 30 and associated sequences:

Machine 1 1–3 0 (1/24) 24 (3/17) 41
Machine 2 2–4 0 (2/30) 30 (4/9) 39

The cost decrease from single machine to two machines is $101-80=\$21$. Since the activation cost of new machine is \$15, activate the second machine.

If we were allowed to have two different common due dates, one for each machine, the minimum cost for each machine would be

Machine	Common Due date	Penalty
1	41	51
2	39	18

For the total penalty of $51+18=69$, there is much less than one common due date.

Three Machines

For three machines, the resulting sequences are as follows:

Machine 1 4–3 0 (4/9) 9 (3/17) 26

Machine 2	2	0 (2/30) 30
Machine 3	1	0 (1/24) 24

With a single common due date, the minimum cost due date is 17 with cost of $114, while the common due dates for jobs on each machine results in following:

Machine	Due Date	Penalty
1	9	17
2	30	0
3	24	0

with the total penalty of 17. With an activation cost of $15, it is obvious that we should go for the third machine if three common due dates are possible. If only one common due date is allowed, a two-machine solution is optimum.

9.3 NONIDENTICAL PARALLEL PROCESSORS

Now, consider a case where a job may be performed on any of the m available processors. Because of the differences between the processors, however, the time requirement for the job may vary from processor to processor. For example, in a machine shop, out of two machines that can perform the same task, one might be the "old reliable" while the other is fairly new and therefore comes equipped with the latest technological improvements. The variations in design and technology might contribute towards the speed and accuracy, and ultimately the differences in processing times, as the work is performed on one of these two machines. We will illustrate here a procedure to distribute jobs among such nonidentical parallel processors to minimize the makespan. In addition, we allow for jobs to arrive at the processing shop at different times. The procedure is as follows:

9.3.1 Procedure

1. Let $T_1, T_2 \ldots, T_m$ be the times to which processors $1, 2 \ldots, m$ are engaged or loaded.
2. Determine the maximum time to which any of the processors are loaded. $T\max = \max(T_1, T_2, \ldots, Tm)$. Determine all unscheduled jobs available till $T\max$.
3. Develop a table (illustrated in the example) showing jobs from step 2, their arrival times, processing time (P) for each job on each processor, and the available time for each processor when it is available for loading the next job.
4. Determine the completion time (C) for each job listed if it is loaded next on that processor. It is equal to: maximum of (processor available time, job arrival time) + process time.
5. Determine the minimum completion time. Mark the associated processor as the efficient processor in this iteration. If there is more than one job

tied for minimum completion time on the same processor, go to step 6; otherwise, go to step 7.

6. Calculate penalty for each job. Penalty is defined as: (next efficient completion time for the job time for the job - completion time on the efficient processor). Schedule the job that has maximum penalty on the efficient processor. Go to step 8.
7. Schedule the job on efficient processor.
8. If all bobs are scheduled, stop. Determine the maximum load time from the processors that gives the minimum value of makespan. If there are some jobs yet to be scheduled, go to step 2.

Example

Consider a 12-job problem with data as shown in Table 9.16.

Initially, $T\max = 0$. All three processors and six jobs are available at time 0. Table 9.17 displays the arrival times, processing time (P), and completion time (C) of these jobs on each processor, if they are loaded at this time.

In the initial table, all processors are loaded simultaneously in this example, but they can be loaded one at a time following the procedure described before. Load the least completion time job on each processor if possible. Since only job 1 is a candidate on P1, assign it to P1. Scratch job 1 from further considerations. On P2 and P3, job 5 is the earliest completer. However, it is more efficient on P2. So load job 5 on P2, and scratch it from further considerations. There are two candidate jobs, 3 and 4, with minimum completion time of 4 on P3. Choose one using the maximum penalty rule.

TABLE 9.16
Data for the Problem (Reproduced)

Job i	P_i Processing Time	E_i Early Penalty	L_i Late Penalty	P_i/E_i	P_i/L_i
1	24	3	8	8	3
2	30	2	5	15	6
3	17	8	1	2.12	**17**
4	9	5	3	1.8	3

TABLE 9.17
Data for with 3 Nonidentical Parallel Processors and 12 Jobs

Job number	1	2	3	4	5	6	7	8	9	10	11	12
Arrival time	0	0	0	0	0	0	5	5	8	10	12	13
Processor 1 P1	2	6	5	10	3	8	2	8	4	3	5	1
Processor 2 P2	4	5	6	6	1	6	3	7	1	5	4	3
Processor 3 P3	3	8	4	4	2	5	7	6	6	2	8	2

TABLE 9.18
Calculations for T max = 0

	Processor Available	Job Number	1		2		3		4		5		6	
		Arrival time	0		0		0		0		0		0	
			P	C	P	C	P	C	P	C	P	C	P	C
Processor 1 P1	0			2*	6	6	5	5	10	10	3	3	8	8
Processor 2 P2	0		4	4	5	5	6	6	6	6	1	1*	6	6
Processor 3 P3	0		3	3	8	8	4	4	4	4**	2	2*	5	5

*Minimum completion times on each processor.

TABLE 9.19
Calculations for T max = 4

	Processor Available	Job Number	2		3		6	
		Arrival Time	0		0		0	
			P	C	P	C	P	C
Processor 1 P1	2		6	8	5	7	8	10
Processor 2 P2	1		5	6	7	8	6	7
Processor 3 P3	4		8	12	4	8	5	9

The penalty for job 3 is $(5 - 4) = 1$ and for job 4 is $(6 - 4) = 2$. Job 4 has the higher penalty, so it is chosen to load on P3. The loading on the processors are

P1 0 (1/2) 2
P2 0 (5/1) 1
P3 0 (4/4) 4

The maximum time a processor is engaged is max $(2, 1, 4) = 4$. Unscheduled jobs available by time 4 are 2, 3, and 6 (Table 9.18).

The earliest completion time is that of job 2 on P2. Load job 2

P1 0 (1/2) 2
P2 0 (5/1) 1 (2/5) 6
P3 0 (4/4) 4

By time 6, the jobs available are 3, 6, 7, and 8 (Table 9.19).

The least completion time is 7; therefore, load job 3 on P1.

P1 0 (1/2) 2 (3/5) 7
P2 0 (5/1) 1 (2/5) 6
P3 0 (4/4) 4

By time 7, the jobs available are 6, 7, and 8 (Table 9.20).

TABLE 9.20
Calculations for T max = 6

	Processor Available	Job Number	3		6		7		8	
		Arrival Time	0		0		5		5	
			P	C	P	C	P	C	P	C
Processor 1 P1	2		5	7*	8	10	3	8	8	13
Processor 2 P2	6		7	13	6	12	3	9	7	13
Processor 3 P3	4		4	8	5	9	7	12	6	11

TABLE 9.21
Calculations for T max = 7

	Processor Available	Job Number	6		7		8	
		Arrival Time	0		5		5	
			P	C	P	C	P	C
Processor 1 P1	7		8	15	2	9	8	15
Processor 2 P2	6		6	12	3	9	7	13
Processor 3 P3	4		5	9*	7	12	6	11

We could choose job 6 or 7 to load at this time. Since they are to be loaded on different processors, there is no need to calculate penalty values unless we decide to load 7, since it can be loaded on P1 or P2 with same completion time. Let us select job 6, since it involves less calculations and load it on P3. The results are

P1 0 (1/2) 2 (3/5) 7
P2 0 (5/1) 1 (2/5) 6
P3 0 (4/4) 4 (6/5) 9

By time 9, we have jobs 7, 8, and 9, yet to be scheduled (Table 9.21).

Jobs 7 and 9 are tied with $C = 9$. P2 can load both jobs 7 and 9 while in P1 only job 7 can be loaded. It is therefore, efficient at this stage load job 7 in P1. By the way job 7 is also most efficient in P1. (Table 9.22 and 9.23).

P1 0 (1/2) 2 (3/5) 7 (7/2) 9
P2 0 (5/1) 1 (2/5) 6
P3 0 (4/4) 4 (6/5) 9

Load job 9 on P2

P1 0 (1/2) 2 (3/5) 7 (7/2) 9
P2 0 (5/1) 1 (2/5) 6 - 8(9/1) 9
P3 0 (4/4) 4 (6/5) 9

TABLE 9.22
Calculations for T max = 9

	Processor Available	Job Number	7		8		9	
		Arrival Time	5		5		8	
			P	C	P	C	P	C
Processor 1 P1	7		2	9*	8	15	4	12
Processor 2 P2	6		3	9	7	13	1	9
Processor 3 P3	9		7	16	6	15	6	16

TABLE 9.23
Calculations for T max = 9; II

	Processor Available	Job Number	8		9	
		Arrival Time	5		8	
			P	C	P	C
Processor 1 P1	9		8	17	4	13
Processor 2 P2	6		7	13	1	9*
Processor 3 P3	9		6	15	6	15

TABLE 9.24
Calculations for T max = 9; III

	Processor Available	Job Number	8	
		Arrival Time	5	
			P	C
Processor 1 P1	9		8	17
Processor 2 P2	9		7	16
Processor 3 P3	9		6	15*

Load job 8 on P3.

P1 0 (1/2) 2 (3/5) 7 (7/2) 9
P2 0 (5/1) 1 (2/5) 6 - 8(9/1) 9
P3 0 (4/4) 4 (6/5) 9 (8/6) 15

By time 15, jobs 11 and 12 have arrived (Table 9.24).
Load job 12 on P1. The resulting loading table is shown in Table 9.25.

P1 0 (1/2) 2 (3/5) 7 (7/2) 9 - 13 (12/1) 14
P2 0 (5/1) 1 (2/5) 6 - 8 (9/1) 9

TABLE 9.25
Calculations for *T* max = 15

	Processor Available	Job Number	11		12	
		Arrival Time	12		13	
			P	C	P	C
Processor 1 P1	9		5	17	1	14*
Processor 2 P2	9		4	16	3	16
Processor 3 P3	15		8	22	2	17

	Processor Available	Job Number	11	
		Arrival Time	12	
			P	C
Processor 1 P1	14		5	19
Processor 2 P2	9		4	16*
Processor 3 P3	15		8	22

P3 0 (4/4) 4 (6/5) 9 (8/6) 15

Load job 11 on P2.

P1 0 (1/2) 2 (3/5) 7 (7/2) 9 - 13 (12/1) 14
P2 0 (5/1) 1 (2/5) 6 - 8 (9/1) 9 - 12 (11/4) 16
P3 0 (4/4) 4 (6/5) 9 (8/6) 15

All jobs are loaded. The makespan is 16.

9.4 PARALLEL MACHINES IN A FLOWSHOP

It is difficult to build a generalized procedure that gives optimum solution to a problem of minimization of makespan in a multistage flowshop when there are number of identical parallel machines at a bottleneck stage. Gupta and Tunc (1997) have developed a heuristic for a two-stage flowshop, but multistage is difficult.

The heuristic presented here combines and extends thoughts from CDS and Gupta's algorithm. Consider first a two-stage problem and parallel machines are in stage 1. The procedure is simple.

1. Apply Johnson's rule to obtain the initial sequence.
2. Move the job with the smallest processing time at stage one to the first position in the sequence.
3. Use the resulting sequence to schedule the jobs.

TABLE 9.26
Data for Parallel Flowshop

Jobs	Stages	
	1	2
1	4	1
2	11	7
3	14	6
4	10	11

TABLE 9.27
Makespan Calculations

Sequence	Machines					
	1/1		1/2		2	
	P	C	P	C	P	C
1	4	4			1	5
4			10	10	11	21
2			11	21	7	28
3	14	18			6	34

Example

Consider a flowshop with four jobs with two identical parallel machines at stage one. The processing times are as follows (Table 9.26):

With Johnson's rule, the sequence is 4-2-3-1. But job 1 has the smallest processing time in stage one. Move it in front, giving the final sequence of 1-4-2-3. Makespan calculations are illustrated in Table 9.27, which shows the processing times, P, and completion times, C, for each job on each machine. The makespan is 34

When multiple parallel machines are in stage 2, the procedure is same as before, except in step 2 move the job with the smallest processing time in stage 2 to the front of the sequence. Gupta's method in this case is very effective, giving the correct answer each time in every problem tested.

When the number of stages are more than two, the problem becomes more challenging. One of the solutions from following four alternatives gave on average within 5% of true optimum, obtained by exhaustive enumerations for maximum of eight jobs, about 85% of the time.

Policy A: Apply the CDS method on total processing times at each stage for each job to generate the initial sequence. Move the job with the least time on stage 1 to the front of each sequence to evaluate the makespan.

Policy B: Same as A, except move job with the second smallest time in stage 1 to the front of the sequence.

TABLE 9.28
Five stage Flowshop Problem

Jobs	Stages				
	1	2	3	4	5
1	5[a]	51	1	4	11
2	10[b]	6	6	9	20
3	12	39	2	19	6
4	18	21	16	16	12
Total	45	117[c]	25	48	49

[a] Job with minimum processing time in stage one.
[b] Job with second smallest time in stage one.
[c] Bottleneck stage.

TABLE 9.29
A_{ij} and B_{ij} Factor Calculations and Policy A and B Sequences

Job	A_{i1}	B_{i1}	A_{i2}	B_{i2}	A_{i3}	B_{i3}	A_{i4}	B_{i4}
1	5	11	56	15	57	16	61	67
2	10	20	16	29	22	35	31	41
3	12	6	51	25	53	27	72	66
4	18	12	39	28	55	44	71	65
Sequence	1-2-4-3		2-4-3-1		2-4-3-1		2-1-4-3	
Policy A Sequence	1-2-4-3		1-2-4-3		1-2-4-3		1-2-4-3	
Policy B Sequence	2-1-4-3		2-4-3-1		2-4-3-1		2-1-4-3	

Policy C: Same as A, except calculate the modified processing time for each job in a stage where multiple processors are present, by dividing the process time by the number of processors.

Policy D: Same as B with same modification as C.

Example:

Suppose there are two processors planned at stage two (Table 9.28). We wish to determine the best schedule and makespan. For policies A and B, the A_{ik} and B_{ik} factors and the corresponding sequences are given in the Table 9.29.

For policies C and D, the A_{ik} and B_{ik} factors and the corresponding sequences are given in Table 9.30.

Because we have two processors in stage 2, in calculating A_{ik} and B_{ik} factors, the times in stage 2 for jobs are taken as follows: job 1 $51/2 = 25.5$, job 2 $6/2 = 3$, for job 3 $39/2 = 19.5$, and for job 4 $21/2 = 10.5$.

There are three sequences to examine. They are 1-2-4-3, 2-1-4-3, and 2-4-3-1. For example, the makespan for sequence 1-2-4-3 is calculated as follows. P is the processing time and C is the completion time (Table 9.31).

TABLE 9.30
A_{ij} and B_{ij} Factor Calculations and Policy C and D Sequences

Job	A_{i1}	B_{i1}	A_{i2}	B_{i2}	A_{i3}	B_{i3}	A_{i4}	B_{i4}
1	5	11	30.5	15	31.5	16	35.5	41.5
2	10	20	13	29	19	35	28	38
3	12	6	31.5	25	335	27	52.5	46.5
4	18	12	28.5	28	44.5	44	60.5	54.5
Sequence	1-2-4-3		2-4-3-1		2-4-3-1		2-1-4-3	
Policy C Sequence	1-2-4-3		1-2-4-3		1-2-4-3		1-2-4-3	
Policy D Sequence	2-1-4-3		2-4-3-1		2-4-3-1		2-1-4-3	

TABLE 9.31
Makespan Calculations for Sequence 1-2-4-3

Job Sequence	Machines											
	1		2/1		2/2		3		4		5	
	P	C	P	C	P	C	P	C	P	C	P	C
1	5	5	51	56			1	57	4	61	11	72
2	10	15			6	21	6	63	9	72	20	92
4	18	33			21	54	16	79	16	95	12	107
3	12	45			39	93	2	95	19	114	6	120

Similarly, the makespan for 2-1-4-3 is 124; and for 2-4-3-1, it is 126. The minimum is for 1-2-4-3.

9.5 BATCH SCHEDULING FOR A LIMITED-CAPACITY, FIXED-PERIOD PROCESS PROBLEM

We shall now address another problem type that is common in industry. We call this the "baking problem" because of its similarity to the baking operation. The problem is described in detail in the following.

Consider a problem of scheduling N jobs for baking, in an oven with limited capacity, that is, no more than "C" jobs at a time. The term "job capacity" is used in the same sense as the "unit load" in material handling. If a job is large in terms of units or in terms of size of the units, the real job can be broken down into number of jobs, each having the identical characteristics as the real (or parent) job.

Once placed in the oven, a job must be baked continuously for B time periods (days). A new job *cannot* be added to the batch while the oven is in operation, even though the number of jobs in the batch may not have filled the oven to its capacity. Associated with each job is its known arrival date, that is, the date when the job is available for baking. Also known is the promised delivery date for the job. If the job is delayed, then there is a tardiness penalty, proportional to the number of days the job is delayed.

The problem is to find the optimal schedule for baking that would minimize the tardiness penalty.

The problem can thus be stated as: minimize the total tardiness subject to the constraints.

1. The number of jobs scheduled for baking in one batch cannot exceed the capacity of the oven.
2. Once a baking cycle starts, no new jobs can be added to the batch or any existing jobs removed from the batch until the batch is completely baked (i.e., after B days).
3. Each job must be processed exactly once.
4. A job cannot be scheduled for baking before the day of its arrival.

The problem is quite common for processes that require a fixed time interval to complete their operation. The baking operation, for example, is essential in making bread and other simple products; in manufacturing china, pottery, and bricks; in production of pharmaceutical products; and in processing microchips for the computer industry. Laminating and heat treatment operations also require a fixed time process. Determining when and for which jobs a cupola should be charged in a foundry operation poses a similar problem of scheduling.

9.5.1 Integer Programming Model (Optional, May Go directly to 9.5.2)

To illustrate a mathematical approach, we shall first formulate the problem as an integer programming problem using the following notations:

N: Total number of jobs to be scheduled.
T: Total number of time periods for which the schedule is to be developed or the planning horizon.
D_i: The due delivery time for job i starting from present (in days).
J_i: The arrival time (in days) of job i.
C: Capacity of the oven.
B: Number of time periods (days) each batch must be baked.
X_i: The state of job i in the time period t,
= 1 if baking for job i starts at time t.
0 otherwise
(it implies that $X_{it} = 0$ for $t < J_i$)
A_{it}: Delay in delivery for job i, if it begins processing at time t (days).

1. Objective Function is to minimize the tardiness penalty, which is given by

$$\text{Min} \sum_{t=1}^{T} \sum_{i=1}^{N} A_{it} \times X_{it}$$

The explanation of the expression is as follows. Since the penalty for delay for each job is the same, the total penalty is proportional to the total delay of all jobs. Tardiness of each job can be determined by checking when the job is scheduled for baking.

X_{it} indicates when the job is scheduled for operation. It has a value of 1 if the job i is scheduled for baking in time t. The associated A_{it} would be the delay caused in delivery of job i when it starts in period t. This value can be calculated as

$$A_{it} = t + B - D_i$$

where D_i is the due delivery time (day) for job i. If A_{it} is negative, then there is no delay, and its value is taken as zero.

2. Processing requirements

   Each job must be scheduled exactly once during a planning period. This can be represented as

$$\sum_{t=J_i}^{T} X_{it} = 1 \quad \text{for} i = 1, 2, \ldots, N$$

   where J_i is the time period or the day when job i arrives for processing.

3. Capacity constraints

   The oven cannot hold more than C jobs (its capacity) at any time and only the jobs that are available at time t can be placed in the oven. This can be represented as

$$\sum_{i=1}^{N} X_{it} + C \times y_t <= C \quad \text{for} t = 1, 2, \ldots, T$$

   y_t is an artificial integer variable taking a value of zero or one that is used in the next expression for continuity of operation. Remember $X_{it} = 0$ for the job i if $t \quad < \quad J_i$, and thus it need not be included in the above equation.

4. Continuous processing restriction

   Each batch must be baked for B days continuously. This can be formulated as

$$\sum_{t=g}^{t+B} y_t - B \times Z_{1,g} - (B - 1) \times Z_{2,g} = 0 \quad \text{for} g = 1, 2, \ldots, T$$

$$Z_{1,g} + Z_{2,g} = 1$$

   $Z_{1,g}$ and $Z_{2,g}$ are artificial variables taking a value of zero or one that ensures a minimum gap of B days between two batches scheduled for baking.

A sample problem is illustrated in Appendix A.

Computationally, even a small problem of scheduling 9 jobs over 12 time periods has 124 0/1 variables and 41 constraints. The computational time depends on the software and computer used, but is substantial. The number of variables and constraints increases rapidly with an increase in

1. Time period under consideration
2. Number of jobs to be scheduled
3. Baking capacity of the oven

An alternate method must be developed to work with any realistic size problem that one observes in the industry. A simple heuristic procedure discussed in the next section provides a pragmatic and suitable alternative for the problems of this type.

9.5.2 Heuristic Approach

The following are the steps for the heuristic procedure that have been found to be extremely efficient in solving even a large size problem.

Step 1: Rank the jobs in ascending order of their arrival dates, and within each arrival date in ascending order of their promised delivery dates.

Step 2: Schedule a job that has not yet been scheduled in a batch based on the following three conditions:

1. The job has arrived on or before the starting time of the batch (making the job eligible for scheduling).
2. The job has the earliest promised delivery date of all the jobs that are eligible for scheduling in the present batch.
3. The batch is not full.

Step 3: We have the best solution if one of the following two conditions is valid.

1. If the total tardiness in the schedule is zero.
2. If there is some tardiness in the schedule, then each batch, except perhaps the last one, is filled to the capacity and each has been scheduled to start as early as possible.

If the best schedule is attainable at this stage, the procedure is terminated. If not, proceed to step 4.

Step 4: If none of the conditions from step 3 are satisfied, then it may be possible to improve the solution by delaying the start of a batch that is not completely filled. Let us define the schedule from step 3 as the master schedule (or basic schedule) at this point. Check delaying possibility, one batch at a time, starting from the first batch that is not full. Delaying the start of an unfilled batch may not necessarily fill the batch completely, but would fill it with more jobs than what were scheduled earlier. Repeat the following two steps until no further improvement is possible.

1. If there is a reduction in the total tardiness, use the new schedule as the master schedule and recheck the first unfilled batch in the new schedule for delay and so on.

TABLE 9.32
Scheduling of Jobs in the Oven

			Schedule								
			I			II			III		
Job Number	Arrival Date	Promise Delivery Date	St	Fin	Del	St	Fin	Del	St	Fin	Del
1	1	3	1	3	—	2	4	1	2	4	1
2	1	4	1	3	—	2	4	—	2	4	—
3	1	10	1	3	—	2	4	—	2	4	—
4	2	4	3	5	1	2	4	—	2	4	—
5	4	6	5	7	1	4	6	—	5	7	1
6	4	10	5	7	—	4	6	—	5	7	—
7	4	15	7	9	—	4	6	—	7	9	—
8	5	7	5	7	—	6	8	1	5	7	—
9	5	12	5	7	—	6	8	—	5	7	—
10	5	15	7	9	—	6	8	—	7	9	—
11	5	15	7	9	—	8	10	—	7	9	—
12	5	19	9	11	—	8	10	—	9	11	—
13	6	8	7	9	1	6	8	—	7	9	1
Total tardiness					3			2			3

2. If there is no reduction in the total delay, using the original master schedule as the base, go to the next unfilled batch and examine the possibility of delaying it, and so on.

Step 5: From all the schedules developed so far, select the best schedule as one with minimum total tardiness.

9.5.3 Illustrative Example

Consider a problem where arrival dates and promised delivery dates for the jobs planned for baking are as given in the first two columns of Table 9.32. In accordance with step 1 of the procedure, the jobs are arranged in ascending order of arrival date and within each arrival date in ascending order of the promised delivery dates. In this data, two jobs have the same arrival and promised delivery dates, namely 10 and 11. They represent a job with a large production volume, which requires two job loads in the oven. The capacity of the oven is for four jobs per load, and the baking time is two days per batch.

Schedule I is developed by allowing the earliest possible start times for each batch. For example, batch 1 starts at time 1 (with only three jobs available) and finishes at time 3. At time 3, only one additional job is available; therefore, it is the only job scheduled in the second batch. Similarly, other batches are developed. Observe that the jobs are selected to form a batch following step 2 of the procedure. The total delay in the schedule is three units, and at least one batch, for example, batch 1, is not completely full. Therefore, there may be some room for improvement in this schedule.

Schedule II is developed by delaying the start of batch 1 from the previous schedule (master schedule at this point) by 1 day. Now, it can accommodate four jobs. However, because of the arrival times of the remaining jobs, the next batch can only accommodate three jobs. The total tardiness in the schedule is 2 days. Since the tardiness in schedule II is less than schedule I, schedule II becomes the master schedule. Again, the total tardiness is not zero, and there exists a batch, which is not the last batch, and is filled less than the capacity of oven. Hence, some improvement in the solution may be possible.

In schedule III, the second batch of the master schedule (schedule II) is delayed by 1 day. The resulting schedule has the tardiness of 3 days. Since tardiness has not decreased, schedule II remains the master schedule.

The third batch of the master schedule, that is, the one with start time of 6, has the oven filled to capacity (jobs 8, 9, 10, and 13). The fourth batch is the last batch, and it starts as early as possible, even though it is not full. Therefore, no further improvement in the present master schedule is possible. Schedule II is the optimum. This solution is the same as the one obtained by analytical means.

9.6 BATCH SCHEDULING FOR LIMITED-CAPACITY PROCESSORS IN SEQUENCE WITH VARYING JOB REQUIREMENTS

It is possible to extend further the procedure developed in Section 9.5 by allowing a number of ovens in sequence, where each job goes from one oven to the next, requiring different times in each oven. The additional considerations are as follows:

1. All jobs that run in one batch must have same processing time.
2. All jobs have equal importance.
3. A job cannot be processed in the succeeding oven until it is released by the previous oven. For example, a job cannot be processed in oven 2 until it is available from oven 1, and so on.

The procedure is a minor modification of the heuristic presented in Section 9.3. For each variation in the previous oven, all variations in the succeeding oven must be checked. A policy that gives the optimum result is then selected as the best policy. This procedure is best explained in Section 9.6.1.

9.6.1 Illustrative Example

Ten jobs will arrive at a heat treatment facility the next week. Each job must be processed in two heat-treating ovens in sequence. The expected arrival time of each job, the processing times in each oven, and the promised delivery date are known. Table 9.33 shows the data, already ranked in ascending order of the job arrival dates,

TABLE 9.33
Jobs Arranged in the Ascending Order of Arrival Dates and within It by Due Dates

Job	Arrival Time	Oven 1 Process Time	Oven 2 Process Time	Due Date
1	1	1	1	3
2	1	1	2	4
3	1	2	2	5
4	2	1	1	6
5	2	2	2	6
7	4	2	2	7
6	4	1	2	8
9	5	1	1	7
8	5	2	1	8
10	5	1	1	8

TABLE 9.34
Initial Assignments in Oven 1 and Job Arrival Times in Oven 2

Batch Number	Jobs in the Batch	Assignment Time	Process Time	Arrival Time for Oven 2
1	1, 2	1	1	2
2	3, 5	2	2	4
3	4, 6	4	1	5
4	7, 8	5	2	7
5	9, 10	7	1	8

and within it in ascending order of the due dates. Each oven has a capacity for four jobs at a time.

The initial assignment in the first oven is made using the algorithm in Section 9.3. The results are shown in Table 9.34. Note that each batch contains only those jobs that require the same processing times. This is made sure by defining the duration of a batch as being equal to the duration of the first unassigned job in the list, and then selecting the unassigned jobs that have the same durational requirement for the oven. The table also shows the batch (job) arrival times in oven 2, which is the batch assigned time in oven 1 plus the processing time for that batch in oven 1.

Using the job arrival times from Table 9.33 and the other information from the problem definition, the initial assignments in oven 2 are as shown in Table 9.35. Again, the rules for assignments are followed. For example, jobs 1 and 2 are available on day 2, but they have different processing times on the second oven; job 1 is ranked higher in the initial listing, so it is the one assigned on day 2. It completes its processing on day 3, the time when it is due, thus causing no delay.

TABLE 9.35
Initial Assignments in Oven 2

Batch Number	Jobs in the Batch	Assignment Time	Process Time	Due Date	Delay
1	1	2	1	3	0
2	2	3	2	4	1
3	3, 5, 6	5	2	5, 6, 8	2, 1, 0
4	4, 8	7	1	6, 7	2, 0
5	7	8	2	6	3
6	9, 10	10	1	8	4, 3

TABLE 9.36
First Improvement in Oven 2

Batch Number	Jobs in the Batch	Assignment Time	Process Time	Completion time	Due Dates	Delay
1	1	2	1	3	3	0
2	2, 3, 5	4	2	6	4, 5, 6	2, 1, 0
3	4	6	1	7	6	1
4	6, 7	7	2	9	7, 8	2, 1
5	8, 9, 10	9	1	10	8, 7 ,8	3, 2, 2

On day 3, only job 2 is available, so it is scheduled next. It takes 2 days to get processed, thus making the oven available on day 5.

On day 5, jobs 3, 4, 5, and 6 are available for scheduling. The duration of the batch is defined by job 3's duration, namely 2 days. Job 4 has different processing time, and therefore, batch 3 consists of jobs 3, 5, and 6. The process is continued. The total delay in this schedule is 16 days.

9.6.2 Improvement Routine

Beginning with the current oven 1 assignment, each possible variation in oven 2 will be checked. Then, each possible oven 2 variation will be checked for each of the remaining oven 1 variations.

1. Oven 1 initial assignment
   a. Oven 2: Delay job 2. It results in an improvement with the total delay of 14 as shown in Table 9.36. Since the solution has improved, it becomes the basic solution for further improvements.
   b. Batch 1 in Table 9.27 is delayed. The delay in the schedule is 15. No improvement.
   c. Batch 2 is delayed. It yields no improvement.
   d. Batch 3 in Table 9.27 is delayed. The delay in schedule is 17. No improvement.

TABLE 9.37
Assignments in Oven 2

Batch Number	Jobs in the Batch	Assignment Time	Process Time	Completion Time	Due Dates	Delay
1	3	3	2	5	5	0
2	1, 4	5	1	6	3, 6	30
3	2, 5, 7	6	2	8	4, 6, 7	4, 2, 1
4	6	8	2	10	10, 8	0, 2
5	8, 9, 10	10	1	11	8, 7, 8	3, 4, 3

TABLE 9.38
Summary of Remaining Calculations for the Example Problem

Oven 1 Action	Oven 2 Action/Batches Delayed in Sequence One at a Time	Schedule Delay	Oven 2 Improvements
Batch 1 is delayed	1	18	New basic (not shown)
	2 , 3	19, 26	—
Batch 2 is delayed	New basic; Table 9.21	18	
	1 , 2, 3, 4	22, 24, 21, 24	—
	5	14	New basic; Table 9.20
	1, 2, 3, 4, 5	23,22,21,24,18	
Batch 3 is delayed	New basic; Table 9.22	18	
	1, 2, 3, 4, 5	19, 21, 21, 21, 21	
Batch 4 is delayed	New basic; Table 9.23	13	
	1	15	
	2	12	New basic; Table 9.24
	1,2,	14,12	

e. Batch 4 in Table 9.27 is delayed. The delay in schedule is 16. No improvement.

All checks on the basic solution of the second oven are performed. Go to the first oven.

2. Batch 1 in oven 1 is delayed.

The new initial schedule of oven 2 is given by Table 9.37. The delay is 22.

We now try to improve the assignments of Table 9.37 and continue to apply the procedure from there on. The entire results are displayed in Table 9.38. Where necessary, the basic solutions for oven 2 are shown in the supporting Tables 9.39 through 9.43. Each batch in the current basic table is delayed (if possible) in sequence. If the delay decreases, the corresponding arrangement in oven 2 becomes the new basic solution, and the batch delaying routine is applied to the new basic solution.

TABLE 9.39
Assignments in Oven 2

Batch Number	Jobs in the Batch	Assignment Time	Process Time	Completion Time	Due Dates	Delay
1	1	2	1	3	3	0
2	2	3	2	5	4	1
3	3, 5	5	2	7	56	21
4	48	7	1	8	6, 8	20
5	6,7	8	2	10	8, 7	2, 3
6	9, 10	10	1	11	7, 8	4, 3

TABLE 9.40
Assignments in Oven 2

Batch Number	Jobs in the Batch	Assignment Time	Process Time	Completion Time	Due Dates	Delay
1	1	2	1	3	3	0
2	2	3	2	5	5	1
3	3, 5	5	2	7	5,6	2, 1
4	4, 9	7	1	8	6, 7	2, 1
5	6, 7	8	2	10	8,7	2, 3
6	8, 10	10	1	11	8, 8	3, 3

TABLE 9.41
Assignments in Oven 2

Batch Number	Jobs in the Batch	Assignment Time	Process Time	Completion Time	Due Dates	Delay
1	1	2	1	3	3	0
2	2	3	2	5	4	1
3	3, 5, 6	5	2	7	5, 6, 8	2, 1
4	4, 9, 10	7	1	8	6, 7, 8	2, 1
5	7	8	2	10	7	3
6	8	10	1	11	8	3

9.7 BATCH SEQUENCING

Now, consider a problem where different part types are manufactured on the same machine. We define the two parts as belonging to the same part type if the machine does not require a change in setup when these two parts are produced one after the other in a sequence. The problem is common in numerically controlled manufacturing where,

TABLE 9.42
Final assignments in Oven 2

Batch Number	Jobs in the Batch	Assignment Time	Process Time	Completion Time	Due Dates	Delay
1	1	2	1	3	5	0
2	1, 4	5	1	6	3, 6	3, 0
3	2, 5, 7	6	2	7	4, 6, 7	4, 2, 1
4	6	8	2	10	10, 8	0, 2
5	8, 9, 10	10	1	11	8, 7, 8	3, 4, 3

TABLE 9.43
Final Assignments in Oven 2

Batch Number	Jobs in the Batch	Assignment Time	Process Time	Arrival at Oven 2
1	1, 2	1	1	5
2	3, 5	2	2	4
3	4, 6	4	1	5
4	9, 10	5	1	6
5	7, 8	6	2	8

because of the universal nature of the setup, the same setup may be used to produce orders from a number of customers. The same is true in the electronic industry and in printed circuit board manufacturing. Cellular manufacturing also presents similar conditions. In cellular manufacturing, one or more part families are produced in a single cell (which may consist of one or more machines) with one setup. Change of part family may require changeover from present setup to another setup, requiring considerable time.

The problem can be stated as follows: There are N jobs to be produced on a single facility. Each job belongs to a part type, and we know the processing time and due date for each job. Each part type requires a setup, the time for which is known. We want to develop a schedule that meets the due dates and also minimizes the makespan.

Batch sequencing is a decision problem and not an optimization problem. There may or may not exist a sequence that meets the due dates of all jobs. We shall discuss the decision making aspect further in the illustrated procedure.

9.7.1 Procedure and Analysis

The procedure for analysis is best explained by means of an example. Suppose that we have six jobs for which the data is as given in Table 9.44. Note that we have three job types, and each type has its own setup time.

TABLE 9.44
Data for Batch Sequencing Problem

Job	1	2	3	4	5	6
Process time	2	3	5	5	7	6
Due dates	9	14	16	26	32	42
Job type	1	2	1	3	1	2
Setup time for the job type	3	1	3	2	3	1

The steps of the procedure areas follows:

Step 1: Group the jobs by job types and within each job type arrange the jobs in the ascending order of the due dates.
Step 2: Construct a row that gives the minimum start time (MST) for each job. MST is defined as: due date - (process time + setup time).
Step 3: Start the job sequence with a job that has the minimum value of MST. Calculate the sequence time, that is, the time associated with the sequence so far.
Step 4: Continue the sequence by assigning the next sequential job from the same group until either all jobs are assigned or the sequence time exceeds the MST of an unassigned job in some other group. Let us denote this job as job K. Go to step 5.
Step 5: If all the jobs are assigned, then stop. If not, remove the last job that had caused the sequence time to exceed the MST of job K and assign job K in the sequence instead. Calculate the new value of sequence time, and go back to step 4.

9.7.2 Application

Because we have setup times associated with each job type, we need to modify the sequence notation that we have been using so far. We shall display (job number/job type/setup time/processing time) whenever a new job type is scheduled. If a job with the same job type as the immediately preceding job is scheduled next, we shall continue to use the old notation (job number/processing time). The setup time is also included in our sequence time calculations wherever appropriate.

The application of steps 1 and 2 results in Table 9.45. The minimum MST is associated with job 1, and we start building our sequence by choosing job 1 first. It completes the setup and processing at time 5. Next, job 3 is placed since it is the next job in group type 1. It completes processing at time 10 since it does not need the additional setup. Job 2 of type 2 has an MST of 10, and hence it is scheduled next. With setup and processing, the job is completed at time 14. Continuing with the next job (job 6) from the group type 2 leads to a completion time of 20, which exceeds the MST value for job 4. Hence, we go back and do not schedule job 6 yet; instead, we schedule job 4. Since that is the only job in group 3, we go back to the jobs from

TABLE 9.45
Application of Step 1 and 2

	Type 1		Type 2		Type 3	
Job	1	3	5	2	6	4
Process time	2	5	7	3	6	5
Due date	9	16	32	14	42	26
Setup time	3	3	3	1	1	2
MST	4	8	22	10	35	19

TABLE 9.46
Application of Steps 1 and 2

	Type 1			Type 2			Type 3	
Job	1	3	5	2	6	7	4	8
Process time	2	5	7	3	6	4	5	10
Due date	9	16	32	14	42	43	26	36
Setup time	3	3	3	1	1	1	2	2
MST	4	8	22	10	35	38	19	24

groups 1 and 2 and select a job with the minimum MST, which is job 5. The sequence is completed by assigning job 6 in the end. The makespan is 38 time units, and no job is delayed.

0 (1/1|3/2) 5 (3/5) 10 (2/2|1/3) 14 (6/6) 20 exceeds
14 (4/3|2/5) 21 (5/1|3/7) 31 (6/2|1/6) 38

Such nonconflicting assignments are not always possible. For example, suppose we have two additional jobs, one of type 2 and one of type 3, as given in Table 9.46.

The sequence development up to time 14 (i.e., jobs 1, 3, and 2) is identical to the previous example. Assignment of job 6 exceeds the MST of job 4, and job 4 is therefore placed in the next position. Since time 21 is less than the MST values for all unassigned jobs, we continue in the same group and assign job 8. The sequence time is now greater than the MST for job 5. So,we back up and assign job 5 instead of job 8. The completion for job 5 is 31, which exceeds the MST of job 8. In other words, it is not possible to have both jobs 5 and 8 complete on time. We must make a decision as to which job is more important to us and complete that job on time, while realizing that the other job is going to be late.

0 (1/1|3/2) 5 (3/5) 10 (2/2|1/3) 14 (6/6) 20 exceeds
14 (4/3|2/5) 21 (8/10) 31 exceeds
21 (5/1|3/7) 31 exceeds

Suppose that we choose job 8 as the next job to schedule. It completes its processing at time 31, and there is no unassigned job with an MST of less than 31 except job 5. So, we choose job 5 next, which results in the following.

14 (4/3|2/5) 21 (8/10) 31 (5/1|3/7) 41

However, both jobs 6 and 7 are now late. To avoid three jobs being late (i.e., jobs 5, 6, and 7). Let us check what would happen if we scheduled job 6 after job 8.

14 (4/3|2/5) 21 (8/10) 31 (6/2|1/6) 38 (7/4) 42

It is clear that we can schedule both job 6 and 7 and complete them on time. Job 5 can be placed in the end to give the completion time as shown in the following.

14 (4/3|2/5) 21 (8/10) 31 (6/2|1/6) 38 (7/4) 42 (5/1|3/7) 52

Thus, in all these calculations, we have decisions to make as we schedule the jobs. What is preferable, delaying one job for a long time or three jobs for a short time is a situation-dependent question. Factors such as who the customers are and what the profit margin of each job is may provide the answer.

9.8 SUMMARY

This chapter included sequencing jobs on parallel machines and batch sequencing. We have seen three different objectives in dividing and sequencing jobs for processing on identical (parallel) machines. They are (1) all jobs have equal importance and we wish to minimize the makespan, (2) jobs have priorities indicated by their weights and we wish to minimize the makespan for each job group formed based on the priorities, (3) jobs have due dates and penalty values and we wish to minimize the total of early and late penalties, and (4) where nonidentical processors provide an opportunity for alternate routing and scheduling.

Batch sequencing is also further divided into three distinct problems. They are: (1) Forming job batches when the processing facility, can handle only a limited number of jobs at one time. The jobs may arrive at various times and may have different due dates. (2) Jobs still arrive at different times and have different due dates, but now there are number of limited-capacity facilities in sequence. Each facility may require different processing times, and all jobs must be processed through all the facilities. The objective in both cases is to minimize the tardiness penalty.

The third variation of batch sequencing we have discussed is that of forming a sequence to minimize setup times. Jobs may belong to different classes, and each job requires a setup unless a job from the same class follows the job just completed. The objective is to find a sequence that best meets the due dates.

All the procedures depend on the successive development of tables.

9.9 PROBLEMS

9.1 A manufacturer needs to place orders for materials to meet its production requirements. For materials A, B, C, D, E, F, and G, the company has suppliers 1, 2, and 3 that can each supply the materials. Each of the suppliers has the same lead time for each of the materials. From the following data on the materials and lead times, develop the ordering policy that will yield the best minimum makespan for the delivery of materials.

Job	A	B	C	D	E	F	G
Process time	9	17	14	7	11	13	21

9.2 Develop a departmental schedule for the production manager of a machine shop that will allow all of the following jobs to be processed in the minimum amount of time in the punch press department. The department has three identical punch presses. The data are as follows:

Job	1	2	3	4	5	6	7	8	9	10
Process time	3	9	13	7	5	19	6	15	11	9

9.3 Develop the minimum makespan schedule for the following data, assuming three machines are to be used in parallel.

Job	1	2	3	4	5	6	7	8	9	10
Process time	10	23	14	6	21	15	8	9	17	14

9.4 The following orders have arrived in the scheduling department of a manufacturer and have been assigned priorities as listed.

Job	1	2	3	4	5	6	7	8
Process time	19	9	10	13	20	17	12	14
Priority	2	1	1	2	2	1	2	1

All jobs are to be processed on one of two identical machines (1 and 2). Develop the schedule that minimizes the makespan.

9.5 Given the following data on jobs to be processed on parallel machines with known processing times, tardy penalties, and due dates, develop the best schedule.

Job	1	2	3	4	5	6	7	8
Process time	20	16	13	15	11	8	12	18
Penalty	2	2	1	1	2	1	2	1
Due date	55	44	32	39	24	28	18	20

9.6 In a particular workshop, 10 jobs run through three nonidentical parallel machines. Their processing times are as follows:

Job	1	2	3	4	5	6	7	8	9	10
Machine 1	10	5	10	9	12	5	10	8	11	13
Machine 2	7	4	5	7	9	4	7	7	9	10
Machine 3	18	12	12	15	19	10	14	9	20	15

Also, run this problem on the computer (program namenonidpp.exe) and check your results.

9.7 The following nine jobs are to be produced at either of the two available manufacturing facilities. Knowing that each of the facilities requires the same processing time for each job, develop the best schedule using the due dates and tardiness penalties given. (Hint: Only one minimum-cost solution is required.)

Job	1	2	3	4	5	6	7	8	9
Process time	2	9	5	4	8	6	10	3	7
Penalty	1	1	2	2	2	1	1	2	2
Due date	40	18	33	26	21	38	15	29	36

9.8 In Problem 5.7, change the processing time as follows:

Job	1	2	3	4	5	6	7	8	9
Process time	12	15	19	14	18	16	10	13	17

9.9 The following ten jobs are to be processed through a baking oven. The data include the arrival date and promised delivery date for each. The oven has a capacity of three jobs with a baking time of 3 days per batch. Develop the best schedule.

Job	1	2	3	4	5	6	7	8	9	10
Arrival	1	1	2	2	3	4	4	4	5	5
Delivery	3	4	4	5	6	6	7	8	8	10

9.10 The following seven jobs consist of orders for three different part types. Each part type has a unique setup time associated with it. The corresponding processing times and due dates are given as well. Develop the best schedule that will meet the due dates and minimize the makespan.

Job	1	2	3	4	5	6	7
Process time	2	3	6	4	5	1	2
Due dates	40	22	14	33	19	27	35
Job type	1	2	3	1	2	2	3
Setup time	2	1	3	2	1	1	3

9.11 Eight jobs need to be scheduled to be processed through two heat treatment ovens. The arrival times, processing times, and due dates for each are listed. Each oven has a capacity for three jobs at a time. Develop the best schedule.

	Arrival Time	Oven 1 Processing Time	Oven 2 Processing Time	Due Date
Job 1	1	1	2	3
Job 2	1	2	1	4
Job 3	2	2	1	6
Job 4	2	3	2	6
Job 5	2	1	2	5
Job 6	3	2	1	8
Job 7	4	1	2	8
Job 8	4	2	2	7

REFERENCES AND SUGGESTED READINGS

Ashour, S., T.E. Moore, and K.V. Chiu. 1974. "An Implicit Enumeration Algorithm for the Nonpreemptive Shop Scheduling Problem" *AIIE Transaction*, 6: 62–72.

Barens, J.W. and L.K. Vanston. 1981. "Scheduling Jobs with Linear Delay Penalties and Sequence Dependent Setup Costs" *Operations Research*, 29(1): 146–160.

Bruno, J., and O. Downey. 1978. "Complexity of Task Sequencing with Deadlines, Setup Times and Changeover Costs" *SIAM Journal of Computers*, 7(4): 393–404.

Cheng, T.C.E. 1990. "A Note on a Partial Search Algorithm for the Single-Machine Optimal Common Due Date Assignment and Sequencing Problem" *Computers and Operations Research*, 17(3): 321–324.

Cho, Y. and S. Sahani. 1981. Preemptive Scheduling of Independent Jobs with Release and Due Dates in Operation Flow and Job Shops" *Operation Research*, 29(3): 511–522.

Crabill, T.B. and W.L. Maxwell. 1969. Single Machine Sequencing with Random Processing Times and Random Due Dates" *Naval Research Logistics Quarterly*, 6(4): 549–552.

Emmons, H. 1969. "One Machine Sequencing to Minimize Certain Functions of Job Tardiness" *Operations Research*, 17(4): 701–715.

Erschler, J., F. Roubellat, and J.P. Vernhes. 1980. "Characterizing the Set of Feasible Sequences for 'N' Jobs to Be Carried Out on a Single Machine" *European Journal of Operational Research*, 4: 189–194.

Erschler, J., G. Fontan, C. Merce, and F. Roubellat. 1983. "A New Dominance Concept in Scheduling 'N' Jobs on a Single Machine with Ready Times and Due Dates" *Operations Research*, 31(1): 114–127.

Grabowski, J. 1980. "One Two-Machine Scheduling with Release and Due Dates to Minimize Maximum Lateness" *Opsearch*, 17(4): 133–154.

Graham, R.L., E.L. Lawler, J.K. Lenstra, and A.H.G. Rinnooy Kan. 1979. "Optimization and Approximation in Deterministic Sequencing and Scheduling: A Survey" *Annals of Discrete Mathematics*, 5: 287–326.

Graves, S.C. 1981. "A Review of Production Scheduling" *Operation Research*, 29(4).

Gupta and Tunc. 1997. "Scheduling a two-stage hybrid flow shop with parallel machines at first stage" *Annals of Operations Research*, Vol. 69, Springer, Netherlands.

Laxminarayanan, L., L. Lakashmanan, R.L. Papineau, and R. Rochette. 1978. "Optimal Single Machine Scheduling with Earliness and Tardiness Penalties" *Operation Research*, 26(6): 1079–1082.

McMahon, G. and M. Florian. 1975. "One Machine Scheduling with Ready Times and Due Dates to Minimize Maximum Lateness" *Operation Research*, 23(3): 475–482.

Monma, C.L. and C.N. Potts. 1989. "On the Complexity of Scheduling with Batch Setup Times" *Operations Research*, 37(5): 798–804.

Nakamura, N., T. Yoshida, and K. Hitomi. 1978. "Group Production Scheduling for Minimum Total Tardiness" *Part I, AIIE Transection*, 10(2): 152–162.

Posner, M.E. 1988. "The Deadline Constrained Weighted Completion Time Problem: Analysis of a Heuristic" *Operations Research*, 36(5): 742–746.

Ramasesh, R. 1990. Dynamic Job Shop Scheduling: A Survey of Simulation Research" *Omega International Journal of Management Science*, 18(1): 43–57.

Sahni, S. 1979. "Preemptive Scheduling with Due Dates" *Operation Research*, 27(5): 925–934.

Schrage, L. 1970. "Solving Resource-Constrained Network Problem by Implicit Enumeration-non-Preemptive case" *Operation Research*, 18: 263–278.

Sen, T. and R.K. Gupta. 1984. "A State of Art Survey of Static Scheduling Research Involving Due Dates" *The International Journal of Science*, 12(1): 67–76.

Sen, T.T. and L.M. Austin. 1980. An Efficient Algorithm for Minimizing Total Tardiness in the n-job, One Machine Sequencing Problem, Working Paper, College of Business Administration, Texas Tech. University, Luhbock, TX.

Shwimer, J. 1972. "On the n-Job, One-Machine, Sequence-Dependent Problem with Tardiness Penalties: A Branch-and-Bound Approach" *Management Science*, 18: B301–B313.

Sidney, J.B. 1977. "Optimal Single Machine Scheduling with Earliness and Tardiness Penalties" *Operation Research*, 25(7): 62–69.

Srinivasan, V. 1971. "A Hybrid Algorithm for the One-Machine Sequencing Problem to Minimize total Tardiness" *Naval Research Logistics Quarterly*, 18: 317–327.

Sule, D.R. and S. Saxena. 1992. "Batch Scheduling for Limited Capacity Fixed Period Process Problem" *Production Planning and Control*, 3(1): 47–52.

Tang, C.S. 1990. "Scheduling Batches on Parallel Machines with Major and Minor Setups" *European Journal of Operational Research*, 46(3): 171–175.

Unal, A.T. and A.S. Kiran. September 1992. "Batch Sequencing" *IIE Transactions*, 24(4): 73–83.

Wittock, R.J. 1986. "Scheduling Parallel Machines with Setups" Research Report, IBM Thomas J Watson Research Center, Yorktown Heights, NY.

10 Network-Based Scheduling

So far, we have analyzed scheduling methods in which jobs are independent of each other. Advance sequencing of one job is not a prerequisite for follow-up sequencing of another. There are sets of problems, however, where the order of job sequencing is important. For example, a product may require processing in three different machines with a specific sequence of operations, cutting–milling–drilling. Here, the operation on the cutting machine must be finished before the work on the milling machine can start, and the job on the milling machine must be completed before processing on drilling machine can start. These predecessor–successor relationships can often be made clear by a network representation of the problem, one similar to the critical path method (CPM) diagram. Furthermore, the analysis similar to the one performed in CPM is often used to as a guide in making sequencing decisions. We will briefly review the CPM before illustrating its application in scheduling.

10.1 CRITICAL PATH METHOD

There are two basic elements that comprise a network — activity and event. An activity is represented by an arrow and an event by a node. An activity is the time (and resource)-consuming portion of the operation, while a node indicates start or completion of the activity. The logical connection, indicating the precedence relationships, of nodes and activities forms a network. The arrowhead of the activity describes the precedence relationship. For example, in the network shown in Figure 10.1, activity "a" must be completed before the activity "e" can start, while both activities "c" and "d" must be completed before the activity "g" can start. Some activities can be performed simultaneously, for example, activities "a", "b," and "c." The project is completed when activities leading to the end nodes are completed; in our example, activities "e," "f," and "g." All the activities "a, b, c..." consume resources, generally time, shown within parenthesis next to the associated activity. The duration for the activity I-j is also noted as d_{i-j}. Occasionally, an activity is introduced in the network just to show the precedence relationship. Such activity does not consume any resources and is shown in the diagram by a dotted line. For example, nodes 2 and 3 are connected with a dotted line (activity h) with the arrowhead on node 3. This indicates that for activities from node 3 to start (i.e., activities "f" and "d"), both "b" and "a" must be completed. Activity h consumes no time. There is also a popular convention where the activities are identified by their start and completion nodes. For example, activity "a" can also be designated as activity 1-2, indicating that it starts in node 1 and completes in node 2.

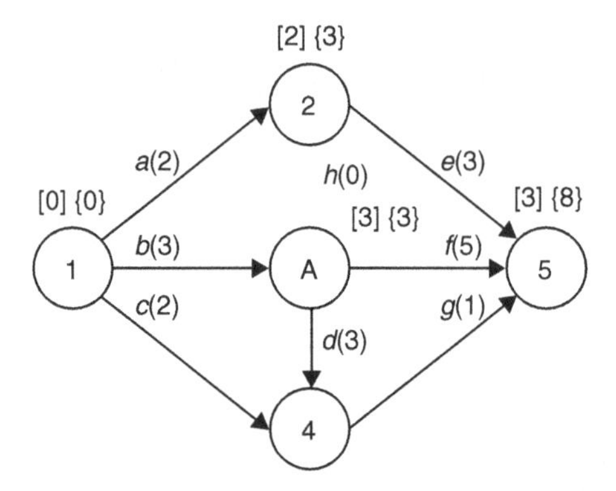

FIGURE 10.1 Critical path diagram.

There are some basic time calculations that can be performed on the nodes and the activities in the network. These calculations lead us to identify the "critical activities," that is, the activities that must start and finish on time for the project to remain on time. These time elements are

1. Earliest start time from node I, ET_i
2. Earliest expected projected completion time (C_E)
3. Latest allowable completion time for a node I, LT_i
4. Allowable delay in reaching a node I or slack in the node I, S_i
5. Allowable delay in the starting or completion of an activity I-j, that is, slack in the activity S_{i-j}

Earliest start time (ET_i): This is the earliest possible time when the processing of activity(ies) emerging from node I can commence. The activity can only start if all activities that are immediate predecessors to the activity under consideration (i.e., all activities leading into node I) have been completed. Thus, ET_i for node I is the latest of the completion times of all the immediate predecessor activities.

Earliest expected project completion time (C_E): This is the maximum required time in the network to complete all jobs.

The earliest start times are obtained by making a forward pass through the network, starting from the start node and proceeding towards the final node. Each node is evaluated, and the earliest an activity emerging from the node can start is when all the activities that are merging in that node are completed. Thus, the earliest start time of the activity emerging from a node is the maximum of completion times of the activities incident to the node.

Latest start time from node I, (LT_i): This is the latest time that processing of an activity emerging from node I can commence such that the expected project completion time, "C_E," is not delayed. This time is calculated by making a backward pass in the network, starting from the last node and proceeding backward to the start node. The earliest expected completion time C_E is used as the reference point as to when the project is expected to finish. For each node, the latest time is calculated as the maximum of time that node can be reached (the event can occur), so that immediately following activities can be completed by their latest times if they consume their estimated times.

Slack (S): This is the permissible delay in the completion of a node (arrival at a node) such that the earliest expected completion time, "C_E," is not delayed. It is the difference between LT_i and ET_i for the node. Thus, $S_i = LT_i - ET_i$. The nodes with zero slack are identified as critical nodes. If these nodes are delayed from their earliest completion time, it will delay the entire project.

Slack in the activity I-j (Si-j): Slack in activity I-j is defined as (LT of node j - ET of node I - the duration of the activity I-j). This is the maximum value that the activity can be delayed without delaying the entire project. An activity with the zero slack is a critical activity.

In our example, the ET_i and LT_i for node I are shown in the figure in [ ] and { }, respectively. The values are also shown in Table 10.1. In forward pass, we can start from node 1 at time 0. It takes 2 time units to complete activity 1-2 and therefore the earliest we can start from node 2 is 2. The earliest we can start from node 3 is the latest of completion times for activities emerging into node 3, that is, 1-3 and 2-3. Since 2-3 does not consume any time, the earliest time for node 3 is, Max $(ET_1 + d_{1-3}, ET_2 + d_{2-3}) = \text{Max}(0 + 3, 2 + 0) = 3$. Similarly, the earliest completion times for node 4 is Max $(0 + 2, 3 + 3) = 6$, and for node 5 is Max $(2 + 3, 3 + 5, 6 + 1) = 8$.

The earliest completion time for end node, that is, node 5, is also the project completion time C_E.

The latest completion times are calculated by following a backward pass. The latest we wish to complete the project is also the latest completion time for the end node. Here, $LT_5 = C_E = 8$.

To calculate the next LT_i, we should proceed backwards to nodes 4, 3, and 2. Both nodes 2 and 3 each have two emerging activities, and therefore, we cannot calculate their latest times till the latest times of the nodes where the activities terminate are known. In his case, for node 2, we must have latest times for nodes 3 and 5 resolved, and for node 3, the latest times for nodes 5 and 4 must be known. The only node that

TABLE 10.1
Node Calculations

Node	Earliest Start Time, ET_i	Latest Completion Time, LT_i	Slack in Node, S_i
1	0	0	0
2	2	3	1
3	3	3	0
4	6	7	1
5	8	8	0

TABLE 10.2
Activity Calculations

Activity	Duration	Early Time on the Tail Node I	Latest Time on the Head Node j	Slack in the Activity
1-2	2	0	3	1
1-3	3	0	3	0
1-4	2	0	7	5
2-3	0	2	3	1
2-5	3	2	8	3
3-4	3	3	7	1
3-5	5	3	8	0
4-5	1	6	8	1

can be resolved next is node 4, since the latest time for the end node of the activity emerging from node 4 is known. The LT_3 is obtained by subtracting the appropriate activity durations from LT_5, that is, $LT_4 = LT_5 - d_{4-5} = 8 - 1 = 7$. Since the latest times for the end nodes for the activities emerging from node 3 are now known, we can determine $LT_3 = \text{Min}\ (LT_5 - d_{3-5}, LT_4 - d_{3-4}) = \text{Min}\ (8 - 5, 7 - 3) = 3$. Other values are shown in Table 10.1 and also in Figure 10.1.

The slack in the node can easily be calculated by subtracting the earliest time from the latest time at each node. The nodes with zero slack are also nodes on the critical path, indicating activities between when these nodes must start and complete on time for the project to remain on time. However, if a node has more than one activity emerging or terminating in it, not all these activities are necessarily on critical path. This is where the calculation for slack in the activity becomes important. Only the activities with zero slack are on critical path. For example, nodes 1, 3, and 5 are on critical path.

For node 3, the question may be, which activity, 1-3, 2-3 or both are on critical path? Slack in activity 1-3 is $LT_3 - ET_1 - d_{1-3} = 3 - 0 - 3 = 0$ and $S_{2-3} = LT_3 - ET_2 - d_{2-3} = 3 - 2 - 0 = 1$. Only activity 1-3 has zero slack, indicating that is the activity on critical path. The rest of activity slack calculation results are shown in Table 10.2. The critical path consist of activities 1-3 and 3-5.

10.2 SCHEDULING A NETWORK OF JOBS ON A SPECIFIED NUMBER OF PARALLEL PROCESSORS

We have seen in Chapter 9, the challenging nature of the problem when there are *n* jobs that may be processed on *m* identical (parallel) machines. The objective was to develop a schedule on each machine that would minimize the makespan for completing all jobs. The problem could be further extended to include predecessor–successor relationships between some jobs. This means the start of processing of some jobs may depend on completion of processing of some other jobs. Furthermore, the number of parallel machines available may be limited in number.

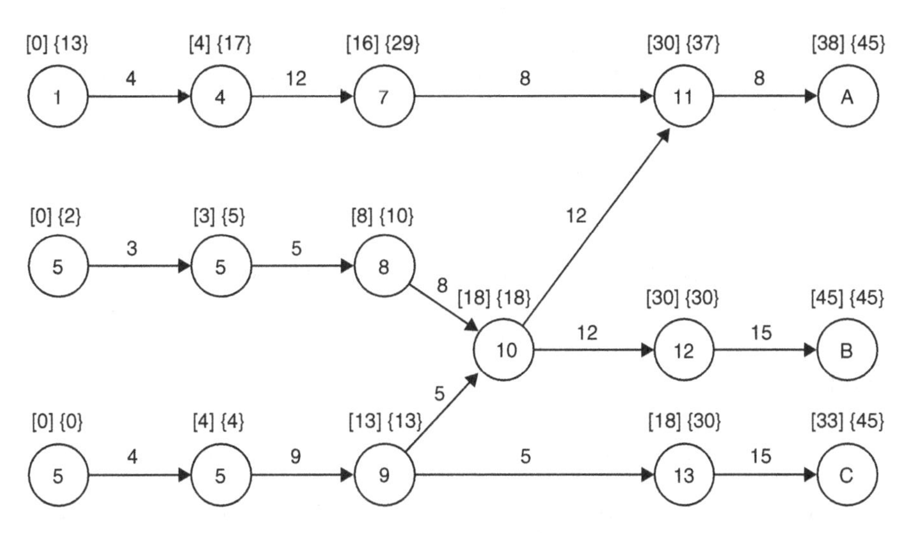

FIGURE 10.2 Job precedence relationships.

Consider the network shown in Figure 10.2. In displaying the sequence-dependent scheduling problem by a network, the CPM definition of network is slightly modified. Now, a node represents a job, and the duration of the activity emerging from that node indicates the processing time for the job. For example, job 1, shown by node 1, takes 4 time units to processes. The figure shows the precedence relationships of all jobs. For example, job 1 must be completed before job 4 can start. Some jobs have no other jobs as their predecessor (e.g., jobs 1, 2, and 3). These jobs are called root jobs. Other jobs havea number of predecessors. For example, job 4 has 1 (job 1), job 7 has 2 (jobs 1 and 4), and job 10 has 6 (jobs 2, 5, 8, 3, 6, and 9). There is no single end point to the network. The network is completed when all jobs have been processed. In our example, there are three branches terminating in dummy nodes A, B, and C, respectively. The path taking the longest time to reach these nodes will define the makespan for the system.

There are three root jobs, and each has a chain of dependent jobs. If we have three parallel machines available, then we can assign the jobs in each single branch to one machine and thus easily resolve the scheduling problem. If there are more than three machines, because of the precedence relationship within the jobs, we will not be able to use more than three machines. It is only when the number of machines are less than the number of root jobs that we get a problem the solution to which is not very obvious. The procedure to solve such problems is illustrated next.

10.2.1 Solution Procedure

Let us define following terms:

m_a = the number of machines available for processing at time T_{cur}.
n_a = the number of jobs available for processing at time T_{cur}.

T_i = the time till machine I is occupied at the end of the present iteration.
T_{cur} = The time at which at least one machine is available for loading the next job.

The candidate job is defined as a job which has all predecessor jobs assigned.

We have also seen that the project completion time, C_E would be equal to maximum value of completion times of all end jobs, that is, maximum of (ET_i of the end job + the processing time of the corresponding job). In our network terms, it would be Max $(ET_{11} + 8, ET_{12} + 15, ET_{13} + 15)$.

The steps of the procedure are as follows:

Step 1

1. Mark current time, $T_{cur} = 0$, since all machines are available for loading at time 0.
2. Calculate ET_i, LT_i for each job (node), along with C_E for the network. Also calculate slack for the candidate jobs. The slack is defined as: $LT_i - T_{cur}$.
3. Identify the critical jobs that is, nodes with slack = 0.

Step 2

1. Identify candidate jobs and candidate machines that is, jobs and machines available for processing at T_{cur}.
2. Calculate the slack *available* for each candidate job. This is the difference between LT_i for job I and T_{cur}, that is, $S_i = LT_i - T_{cur}$. This is because T_{cur} is also the time at which the candidate machine is available for processing and the candidate jobs must wait for loading until this time, consuming any real slack they may have had.
3. Check for critical job(s) if any, among the candidate jobs. These are the job(s) with zero or negative available slack. Negative slack indicates that the job is already delayed and will cause a delay in the completion of the entire network.

Step 3

1. If the number of critical jobs, $n_{critical} > m_a$ the available machines, assign critical jobs to the machines based on increasing order of negative slack (i.e., −5 first, −4 next, and so on). If the slacks are equal, then assign a critical job using the shortest processing rule. The critical job with the shortest processing time is assigned to the first available machine, the critical job with the next value of the shortest processing time is assigned to the second machine, and so on. Go to step 6.
2. If the number of critical jobs $n_{critical} > 0$ and less than or equal to m_a, the number of available machines, and $n_a >= m_a$, give priority to critical jobs, assign them first using the SPT rule described in step a. For the remaining machine assignments, go to step 4.

3. If, in the jobs available, $n_{\text{critical}} = 0$ and $n_a > m_a$, go to step 4.
4. If $n_a = m_a$, assign one available job to one available machine. Go to step 5.

Step 4

1. Assign an available job "j" on the machine, and determine its effect on the slack values of the remaining available jobs. This is obtained by subtracting the processing time of job j from current slack times of all other available jobs. The resultant value of slack, for each of the available jobs, shows the possible effect of placing job j on the present machine, if no other machine is available for other job placement. This quantity measures the delay for the job. If the value is positive there is no delay, on the other hand the negative value indicates the amount of delay. If the value is zero, the job status has changed from noncritical to critical.
2. Perform step "a" for all available jobs.
3. Select for placement, the job that results in the least delay in the schedule. If there is a tie, select a job that will have the least remaining slack value if other jobs are placed on machine first.

Step 5: Delete the assigned job and the associated machine from appropriate available lists. If all the available machines are now assigned, go to step 6. If not go to step 3.

Step 6: If all jobs are assigned, stop. If not, make T_{cur} = minimum of the completion times of the jobs being processed at the end of the cycle. If due to the assignment made, the C_E calculated in step 1 is extended, that is, the network can no longer be completed at the earliest possible time, then go to step 7, otherwise go to step 2 (going to step 2 starts the next cycle).

Step 7: Recalculate LT for each unassigned job (node) using the present value of C_E. Go to step 2.

10.2.1.1 Illustrated Example

Consider the problem of scheduling 13 jobs with precedence relationships shown in Figure 10.2. The processing times for the jobs are shown in Figure 10.2 as well as noted in Table 10.3. Suppose we have two parallel processors to schedule these jobs.

Applying first step, we calculate ET and LT for each job. The values are displayed in Table 10.3. The present T_{cur} is equal to zero.

The earliest completion time for the network $C_{E,}$ is 45 time units. Only jobs 1, 2, and 3 are available jobs since they have no other predecessor jobs; hence $n_a = 3$. The slack values for jobs 1, 2, and 3 are 13, 2, and 0, respectively, since T_{cur} is zero. Job 3 is critical; it has zero slack, giving $n_{critical}$= 1. Both machines 1 and 2 are available for sequencing, that is, $m_a = 2$.

We start scheduling arbitrarily on the first machine. Using step 3.2, the first machine is assigned the critical job 3. The machine is occupied for (T_{cur}+ processing time of job 3), that is, $(0 + 4) = 4$. Next, we proceed to machine 2, with jobs 1 and 2. Both jobs are not critical. Based on step 4.1, if 1 is assigned to the machine

TABLE 10.3
Early and Late Start Times for Each Job

Job	Duration	ET	LT
1	4	0	13
2	3	0	2
3	4	0	0
4	12	4	17
5	5	3	5
6	9	4	4
7	8	16	29
8	8	8	10
9	5	13	13
10	12	18	18
11	8	30	37
12	15	30	30
13	15	18	30

the slack in job 2 will reduce to, the present $S_2 - d_1 = 2 - 4 = -2$, indicating the project will be delayed by 2 time units. If job 2 is scheduled first, the slack of job 1 will reduce to the present $S_1 - d_2 = 13 - 3 = 10$. The project still remains on time, and hence schedule job 2 on machine 2. Job 1, since it is not scheduled, becomes the candidate job. We go to step 6. Since all jobs (there are 13 jobs in total) are not assigned, we modify the T_{cur} = minimum of times that each machine is occupied to, that is, min $(0 + 4, 0 + 3) = 3$.

Cycle 2

At $T_{cur} = 3$, there is one available machine (machine 2) and two available jobs (jobs 1 and 5). According to step 2, the new values for the slacks are $S_1 = 13 - 3 = 10$, since the processing of job 1 instead of starting at time zero could not start till time 3 (note that job 1 is the candidate job), it may consume three units of slack from available 13 units. S_5 remains 2 since $LT_5 = 5$ and $T_{cur} = 3$ (job 5 can start as early as possible). None of the jobs are critical, and we are in condition in step 3.3, which directs us to go to step 4.

1. The assignment is made after checking for probable resulting delay due to placing each candidate job first. Placing job 1 first, delays job 5 for the moment, by four time units and since $S_5 = 2$, the network could get late by two time units $(2 - 4 = -2)$. Putting job 5 first, delays job 1 for the moment by five time units. Since $S_1 = 10$, the probable resulting lateness would be zero (the slack is, $10 - 5 = 5$, which is positive). Hence job 5 is scheduled on machine 2. C_E remains unchanged.

$T_{cur} = 4$, which is the completion time on machine 1 since job 3 get completed first among the jobs being processed at end of cycle 2.

We will now note only the important points in each cycle that goes in decision making.

Cycle 3

1. Candidate job is 1, and job 6 is the job that becomes available now. Machine 1 is the candidate *machine*.
2. $S_1 = 9$ instead of the earlier ten time units since $T_{cur} = 4$ (or $13 - 4 = 9$). $S_6 = 0$.
3. Thus, among these jobs, job 6 is critical.
4. $n_{cril} = 1$ and $m_a = 1$. Hence, give priority to the critical job. Job 6 is scheduled on machine 1. C_E remains unchanged. Machine 1 is occupied till $(4 + 9) = 13$.

Next, $T_{cur} = 8$, the completion time of job 5 since it has the earliest completion time among the jobs being processed.

Cycle 4

1. Candidate jobs are jobs 1 and 8. Machine 2 is the candidate machine.
2. S_8 is equal to two *time* units, while S_1 has been further reduced to five time units.
3. None of the candidate jobs are critical.
4. $n_{critical} = 0$ and $n_a > m_a$. Hence, the *probable* total resulting lateness due to putting each of the candidate jobs first, on the candidate machine, is calculated before making the assignment. The lateness resulting from placing job 1 first is two time units, while the same due to placing job 8 first is three time units. Thus, job 1 is scheduled on machine 2. C_E remains unchanged. Machine 2 is occupied till (T_{cur}+ processing time of job 1) $= (8 + 4) = 12$. In cycle 3, machine 2 is occupied till time 13.

The next $T_{cur} = \min(12, 13) = 12$, the completion time of job 1.

Cycle 5

1. Since job 4 is dependent on job 1, and job 1 was scheduled, job 4 becomes a candidate job. Job 8 is also a candidate job since it was available in the previous cycle but was not scheduled. Since both jobs 4 and 8 are candidate jobs, their present slack times can be calculated as $LT_i - T_{cur}$. Machine 2 is the candidate machine.
2. $S_4 = (17 - 12) = 5$ time units and $S_8 = (10 - 12) = -2$. The negative slack means that, job 8 used up more slack than was available to it. In other words, job 8 became critical at time 10, when it is available, slack is reduced to zero and then it is delayed further by two time units when the job is scheduled on time 12. This will extend the C_E by at least two time units.

TABLE 10.4
Modified Early, Late, and Slack Times for Unscheduled Jobs at T_{cur} = 12

Job	Duration	T_L
4	12	19
7	8	31
9	5	15
10	12	20
11	8	39
12	15	32
13	15	32

3. $n_{critical} = 1$ and $m_a = 1$, that is, $n_{critical} = m_a$. Schedule job 8 on machine 2. The new C_E is now $45 + 2 = 47$ time units. Since the C_E has changed, step 7 must be executed.
4. $T_{cur} = 13$, the completion time of job 6.
5. Recalculate the slacks for all unscheduled jobs using the $C_E = 47$. The results are shown in Table 10.4.

Notice that the status of job 9, which was earlier critical, is now noncritical. This change brings out the importance of recalculating slacks when C_E changes.

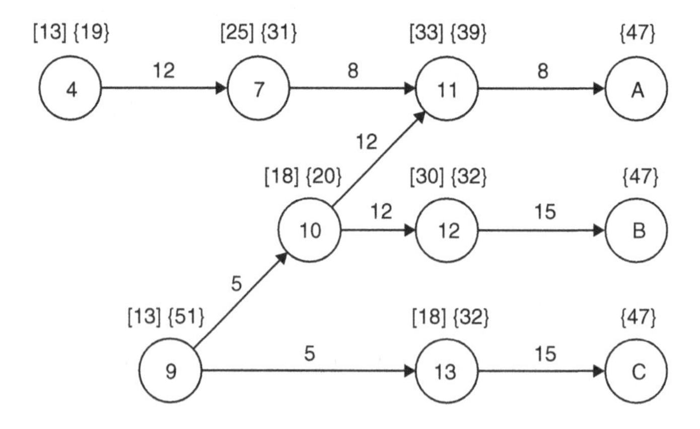

Cycle 6

1. $T_{cur} = 13$.
2. Candidate jobs are jobs 4 and 9, while the available machine is machine 1.
3. $S_4 = 6$ and $S_9 = 2$.

4. None of the candidate jobs are critical.
5. $n_{\text{critical}} = 0$ and $n_a > m_a$, hence the probable resulting lateness due to scheduling each job first must be calculated. The slack in job 9 when job 4 is scheduled first is $2 - 12 = -10$, that is, a delay of 10 units. The slack in job 4 when job 9 is scheduled first is $6 - 5 = 1$. Hence, job 9 is scheduled on machine 1C_E remains unchanged at 47 time units.

Cycle 7

1. $T_{cur} = 18$.
2. Candidate jobs are jobs 4, 10, and 13. The candidate machine is machine 1.
3. Slack available for job 4 = 1, for job 10 = 2, and that for job 13 = 14.
4. None of the candidate jobs are critical.
5. $n_{\text{critical}} = 0$ and $n_a > m_a$, hence the probable resulting lateness due to scheduling each candidate job first, must be calculated. The slacks available to jobs 10 and 13, when each is scheduled immediately after job 4, are -10 and 2, respectively. Slacks available to jobs 4 and 13, when each is scheduled immediately after job 10, are -11 and 2, respectively. Slacks available to jobs 4 and 10, when each is scheduled immediately after job 13, are -14 and -13, respectively. Choosing the job that results in the least probable delay, we schedule job 4 on machine 1. C_E remains unchanged at 47 time units.

Cycle 8

1. $T_{cur} = 20$.
2. Candidate jobs are jobs 10 and 13, while the candidate machine is machine 2.
3. Slack available for job 10 is 0 and that for job 13 is 12.
4. Job 10 is the critical job.
5. $n_{\text{critical}} = 1$ and $n_a = m_a$, that is, 1. Thus, we schedule job 10 on machine 2.

Cycle 9

1. $T_{cur} = 30$.
2. Candidate jobs are jobs 7 and 13. The candidate machine is machine 1.
3. Slack available for job 7 is 1 and that for job 13 is 2.
4. None of the candidate jobs are critical.
5. $n_{\text{critical}} = 0$ and $n_a > m_a$. Thus, the probable resulting lateness is calculated by putting each job first. The slack available to job 13 by putting job 7 first is -6 time units, while that for job 7 by putting job 13 first is -14. Job 7 is thus scheduled on machine 1.

Cycle 10

1. $T_{cur} = 32$.
2. The candidate jobs are jobs 12 and 13. The candidate machine is machine 2.
3. The available slack for job 12 is 0, and that for job 13 is also 0.
4. Both jobs are critical, and thus $n_{critical} = 2$.
5. $n_{critical} = 2$, and $n_a > m_a$. The scheduling decision will thus be based on the resulting probable delay, that is, resultant slacks by putting each of the candidate jobs first. In this case, either job results in an available slack of -15 for the candidate jobs. Making an arbitrary choice, job 12 is scheduled on machine 2.

Cycle 11

1. $T_{cur} = 38$.
2. Candidate jobs are jobs 11 and 13. The candidate machine is machine 1.
3. Available slack for job 11 is 1 and that for job 13 is -6, indicating that the final completion

Continuing the procedure till all the jobs are scheduled, we have the following schedules on machines 1 and 2.

Machine 1: Job 3; Job 6; Job 9; Job 4; Job 7; Job 13.
Machine 2: Job 2; Job 5; Job 1; Job 8; Job 10; Job 12; Job 11.

The resulting makespan is 55 time units which is the least possible with two parallel processors.

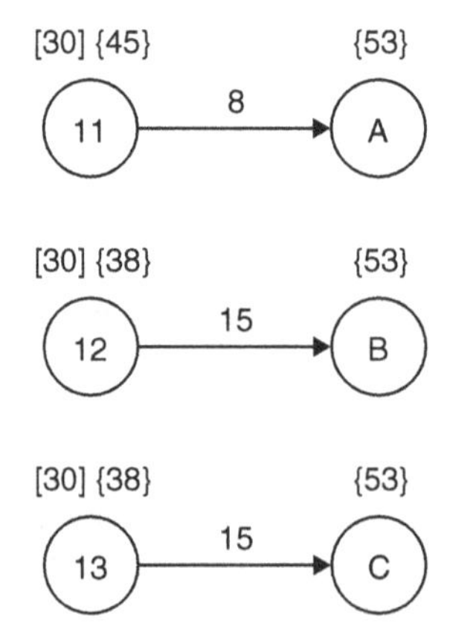

10.3 SCHEDULING n JOBS ON m PARALLEL MACHINES WHEN EACH JOB CAN BE SCHEDULED ON p MACHINES, "p" BEING A SUBSET OF m, THAT IS, $p \leq m$

Though not directly related to network-based scheduling, one of the variations of the parallel machine problems discussed in the literature is that of scheduling n jobs on m parallel machines when some (or all) of the jobs can only be processed on a small subset of machines "p." This may be because the external requirements such as tools, fixtures, or material handling equipment required for the job are only available on certain machines.

Thus, each job can be processed on p of the m available machines, such that $p \leq m$. Job i can be processed on p_i machines, job j can be processed on p_j machines and so on.

Any one of the following relations between p_i and p_j could exist:

1. $p_i = p_j$ that is, jobs i and j share the same machines.
2. p_i is a subset of p_j.
3. p_j is a subset of p_i.
4. p_i and p_j are distinct.

10.3.1 Solution Procedure

Step 1: Develop a job–machine matrix showing which job can be processed on which machines.

Step 2: Calculate the total number of jobs that a machine can process and the total number of machines on which a job can be processed. This gives the job and machine flexibility indexes. The more jobs that a machine can process, the greater is its flexibility. Similarly, the more the number of machines on which a job can be processed, the greater is its flexibility. The calculation is shown in Table 10.5.

Step 3: Develop a table as shown in Table 10.6. The table shows the individual job assignment based on the objective of an even distribution of cumulative jobs among all machines (the procedure can be easily modified for objective of even distribution of total processing time on each machine).

Select the current least flexible job and determine the current least flexible machine on which this job can be assigned. If there is a tie for the least flexible job, break it based on the minimum value of current least flexible machine index. Schedule the job selected based on the objective of even distribution of the job load on the machines, checking in the ascending order of the magnitude of the flexibility of machine index for the machines where the job can be assigned.

Delete the assigned job from further selection. Enter it in the bottom half of Table 10.5 as an assigned job. Modify the flexibility index by subtracting the

TABLE 10.5
Job-Machine Use Matrix

Job/Machine	M1	M2	M3	M4	Job Flexibility
1	1	1	1		3
2	1				1
3		1			1
4			1	1	2
5	1	1			2
6	1	1	1	1	4
7		1	1		2
Total machine flexibility	4	5	4	2	
Jobs assigned/ rem machine flexibility					
2	3	5	4	2	
3	3	4	4	2	
4	3	4	3	1	
5	2	3	3	1	
7	2	2	2	1	
1	1	1	1	1	
6	0	0	0	0	

TABLE 10.6
Job Assignments

		Assign Job to Machine				Cumulative Job Assignment on Machine			
Iteration	**Job**	**1**	**2**	**3**	**4**	**1**	**2**	**3**	**4**
1	2	x				1	0	0	0
2	3		x			1	1	0	0
3	4				x	1	1	0	1
4	5	x				2	1	0	1
5	7		x			2	2	0	1
6	1			x		2	2	1	1
7	6				x	2	2	2	1

job-machine entry (in this case, 1) for all machines on which this job could have been processed. The resultant row gives the current value of flexibility index for the next iteration.

Repeat step 3 of the process till all the jobs are scheduled.

10.3.2 Illustrated Example

Schedule seven jobs on four parallel processors with the following job–machine relationships. The notation indicates that, for example, to process job 1 either machine 1, 2, or 3 can be used.

Job 1: M1, M2, and M3
Job 2: M1
Job 3: M2
Job 4: M3 and M4
Job 5: M2 and M1
Job 6: M1, M2, M3, and M4
Job 7: M2 and M3

The objective is to distribute the job load to four machines as evenly as possible.

The job-machine matrix is shown in Table 10.5. The presence of one in cell C_{ij} indicates that job i can be processed on machine j.

The least flexible jobs are jobs 2 and 3, each with the job flexibility index of 1. The flexibility indexes of the machines on which these jobs are performed, M1 and M2, are 4 and 5, respectively. Since the flexibility index on M1 is smaller, we choose the associated job, job 2, for the first assignment. It is assigned to machine 1, in Table 10.6. It is noted as the assigned job in Table 10.5, and the remaining machine flexibility index is modified by subtracting 1 from the flexibility indexes of the machines on which job 2 could have been assigned, in this case only machine 1. In iteration 2, job 3 is selected based on job flexibility index and assigned to machine 2. It is noted in Tables 10.5 and 10.6. In Table 10.5 job assigned/rem machine flexibility index is further modified. Next, a job from among jobs 4, 5, and 7, all with a job flexibility index of 2, must be selected. The least current machine flexibility indexes for each one of these jobs are as follows (iteration 3, Table 10.5):

For job 4 = Min (4, 2) = 2.
For job 5 = Min (3, 4) = 3.
For job 7 = Min (4, 4) = 4.

The least value is associated with job 4, and hence, we choose job 4 as the next job in the sequence. Job 4 could be assigned to machine 3 or 4, based on cumulative assignment so far. It is scheduled on machine 4 since M4 have least flexibility index (iteration 2, Table 10.6). The machine flexibility indexes for both machines 3 and 4 are modified since job 4 could have been placed on either of the machines (iteration 4). Next, between jobs 5 and 7, the minimum machine index for both is 3. We break the tie randomly and choose job 5 for the assignment. The job could be loaded on M1 or M2 and, since both machines have the same cumulative frequency so far, we choose to load on machine 1 based on least machine flexibility index. Again, the machine flexibility index for both M1 and M2 are modified. Job 7 is similarly loaded and machine indexes modified. Next, job 1 with the flexibility index of 3 is chosen. It could be loaded on machines 1, 2, or 3. Since cumulative loading on these machines is 2, 2, and 1, respectively, job is loaded on M3 to even out the loading. Similarly the last job, job 6, could be loaded on all machines but M4 is chosen to even out the loading among the machines. The final assignments are shown in Table 10.6.

10.4 ASSEMBLY LINE BALANCING

One of the industrial scheduling problems where the precedence relationships shown by network play an important role is in assembly line balancing. An assembly line is a method of production in which the parts are assembled and made into the final product as the unit progresses from station to station. In order to develop a balanced assembly line, it is necessary to distribute the total work content of the product among all the workstations, so that stations can complete their assigned task in approximately the same time. If the line is perfectly balanced, each station would require identical time to complete all its assigned tasks. However, such perfect balance is rarely possible, and the longest station time becomes the cycle time for the assembly line.

The first step in developing an assembly line is to break the total job content into small task or work elements and determine the precedence relationship among the task. Such a relationship, shown in a diagram or in a table, indicates which work elements must be performed before others can start and which work elements can be worked on simultaneously. Once the precedence relationships are established, the next step is to distribute the total work content among the stations in a manner that will give a balance load to each station.

There are a number of methods available for developing a balanced assembly line. We present here two procedures: (1) largest candidate rule (LCR), and (2) ranked positional weighted (RPW) method.

Both procedures determine the minimum number of stations necessary to obtain the desired cycle time. The steps of the procedures are as follows:

10.4.1 Largest Candidate Rule

Step 1: List the tasks in descending order of magnitude of task times. Also list the corresponding immediate predecessor task(s) for each task.

Step 2: Start with the first station, and number the remaining stations consecutively as we apply step 3.

Step 3: Beginning at the top of the task list, assign the first feasible task to the station under consideration. Once the task is assigned, all reference to it is removed from the predecessor task list. A task is feasible only if it does not have any predecessor or if all the predecessors have been deleted. It may be assigned only if it does not exceed the cycle time for the station. This condition can be checked by comparing the cumulative time for all jobs so far assigned to that station, including the task under consideration, with the cycle time. If the cumulative time is greater than the cycle time, the task under consideration cannot be assigned to the station. Go to the next feasible task. If no task is feasible, proceed to step 5.

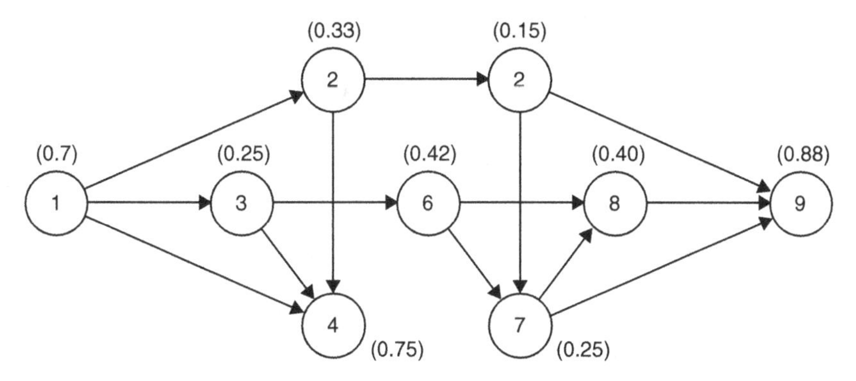

FIGURE 10.3 Precedence relationships with task times

TABLE 10.7
Task Precedence Relations and Immediate Predecessors

Task	Task time	Immediate Predecessor(s)
1	0.70	-
2	0.33	1
3	0.25	1
4	0.75	1, 2, 3
5	0.15	2
6	0.42	3
7	0.52	4, 5, 6
8	0.40	6, 7
9	0.88	5, 7, 8

Step 4: Delete the task that is assigned from the first column of the task list. If the list is now empty, go to step 6; otherwise, return to step 3.

Step 5: Create a new station by increasing the station count by one. Return to step 3.

Step 6: All jobs are assigned, and the present station number reflects the number of stations required. The procedure also shows the job assignments for each station. The largest cumulative time for the individual station is the cycle time.

We will illustrate the procedure by applying it to a job that can be broken into nine tasks. The precedence relationships and task times are displayed in Figure 10.3 and in Table 10.7. Assume we desire a cycle time of 0.95 min or the production rate of $500/0.95 = 526$ units, for a 500-min working shift.

The application of the first step leads to Table 10.8, cycle 1. In the table, the tasks are ranked in the descending order of their time requirements. Also listed are their precedence tasks. Initially, the first station cumulative time is zero. A task with

TABLE 10.8
Task Ranked in Descending Order of Task Times for LCR

Task	Task Time	Remaining Immediate Predecessor(s) per Cycle									
		1	2	3	4	5	6	7	8	9	10
9	0.88	5, 7, 8	5, 7, 8	5, 7, 8	5, 7, 8	5, 7, 8	5, 7, 8	7, 8	8	- *	- -
4	0.75	1, 2, 3	2, 3	2	2	- *	- -	- -	- -	- -	- -
1	0.7	- *	- -	- -	- -	- -	- -	- -	- -	- -	- -
7	0.52	4, 5, 6	4, 5, 6	4, 5, 6	4, 5	4, 5	5	- *	- -	- -	- -
6	0.42	3	3	- *	- -	- -	- -	- -	- -	- -	- -
8	0.4	6, 7	6, 7	6, 7	7	7	7	7	- *	- -	- -
2	0.33	1	-	-	- *	- -	- -	- -	- -	- -	- -
3	0.25	1	- *	- -	- -	- -	- -	- -	- -	- -	- -
5	0.15	2	2	2	2	-	- *	- -	- -	- -	- -

- : No prerequisites, assignment can be made.
- - : Job has been assigned.

TABLE 10.9
The Task Assignments

Station	Task Assigned	Task Time	Cumulative Task Time/Station
1	1	0.70	0.70
	3	0.25	0.95
2	6	0.42	0.42
	2	0.33	0.75
	5	0.15	0.90
3	4	0.75	0.75
4	7	0.52	0.52
	8	0.40	0.92
5	9	0.88	0.88

no existing prerequisites or with all the prerequisite being assigned to station(s), is marked by the "-" symbol.

The procedure begins (cycle 1) by screening from top of the task list in Table 10.8. Tasks 9 and 4 both have immediate precedence tasks that are yet to be sequenced. Task 1 can be assigned to the first station since it does not have any precedence task. The task time is 0.70, which is less than the cumulative time (cycle time = 0.95) allowed for the station. The task is assigned to the first station in Table 10.9, which is being developed simultaneously and also indicated by an * in Table 10.8. All references to task 1 in the precedence list are scratched since task 1 is now assigned. For convenience, the resulting remaining precedence relations are shown as cycle 2 relationships in Table 10.8. A task that is assigned and is no longer in contention is shown by - -.

To decide which task to assign next, we go back to the top of the Table 10.8 and examine the task list along with the cycle 2 precedence relationships. Tasks 9, 4, 7, 6, or 8 cannot be assigned yet, since they still have the predecessor tasks. Task 2 can now be assigned, but addition of task time to the present station 1 cumulative time, results in the increase of cumulative time to $0.70 + 0.33 = 1.03$, exceeding the permissible cycle time of 0.95. Hence, task 2 cannot be assigned to station 1. The next task on the list with no predecessor(s) is task 3. It can be assigned to station 1 without exceeding the cycle time, and hence the assignment is made in Table 10.9, and the precedence list is modified in Table 10.8.

Going back to the top of the list with cycle 3 predecessors list, we see that job 6 can be assigned to the station 1, but this will exceed the cycle time. In fact, no job can be assigned to station 1, so we must start with a new station, station 2 with the cumulative cycle time of zero. The procedure is continued till all the tasks are assigned to stations as shown in Table 10.9.

The actual cycle time for the assembly is the largest cumulative time of any station. In this case, station 1 has the largest cumulative time of 0.95 and hence 0.95 is the actual cycle time. Not all stations are completely utilized. For example, station 2 has the work contain of only 0.75 unit within the cycle time of 0.95. An efficiency measure e indicates how well balanced and utilized an assembly line is. It is defined as $(1 - p) \times 100$ where p is

$$p = \frac{\text{(Total worker time required/unit)} - \text{(Time necessary per unit of job)}}{\text{Total worker time required/unit}}$$

For our example, we have five stations, each manned by one worker, hence the total worker time required per unit is 5×0.95, since one unit is produced every cycle. The actual time necessary is the sum of all task times. Here, it is equal to 4.4. Therefore, the efficiency is

$$e = \left[1 - \frac{5 \times 0.95 - 4.4}{5 \times 0.95}\right] \times 100 = 92.63\%$$

10.4.2 Ranked Positional Weight Method

In the previous method, the tasks were ranked in the descending order of their magnitude. In this method, they are ranked according to how important they are in completing all the tasks that depend on them. The importance is measured based on the RPW of each element, which is the sum of the times for all elements that directly follow it in the precedence diagram, plus the time for the particular task itself. For example, RPW of element 6 is the sum of times for elements 7, 8, and 9 plus the time for the element 6; that is $0.52 + 0.40 + 0.88 + 0.42 = 2.22$.

TABLE 10.10
Task Ranked in Descending Order of RPW

Task	RPW	Task Time	Remaining Immediate Predecessor(s)									
			1	2	3	4	5	6	7	8	9	10
1	4.4	0.7	- *	- -	- -	- -	- -	- -	- -	- -	- -	- -
2	3.45	0.33	1	-	- *	- -	- -	- -	- -	- -	- -	- -
3	3.22	0.25	1	- *	- -	- -	- -	- -	- -	- -	- -	- -
4	2.55	0.75	1, 2, 3	2, 3	2	-	-	- *	- -	- -	- -	- -
6	2.22	0.42	3	3	-	- *	- -	- -	- -	- -	- -	- -
5	1.95	0.15	2	2	2	-	- *	- -	- -	- -	- -	- -
7	1.80	0.52	4, 5, 6	4, 5, 6	4, 5, 6	4, 5, 6	4, 5	4	- *	- -	- -	- -
8	1.28	0.40	6, 7	6, 7	6, 7	6, 7	7	7	7	- *	- -	- -
9	0.88	0.88	5, 7, 8	5, 7, 8	5, 7, 8	5, 7, 8	5, 7, 8	7, 8	7, 8	8	- *	- -

- : No prerequisites, assignment can be made.
- - : Job has been assigned.

TABLE 10.11
Task Assignments Using RPW

Station	Task Assigned	Task Time	Cumulative Task Time/Station
1	1	0.70	0.70
	3	0.25	0.95
2	2	0.33	0.33
	6	0.42	0.75
	5	0.15	0.90
3	4	0.75	0.75
4	7	0.52	0.52
	8	0.40	0.92
5	9	0.88	0.88

Once the elements are listed in the descending order of RPW along with their predecessor elements in the first column as shown in Table 10.10, from there on we apply the same steps 2–6 as in the previous procedure. The results of the application of RPW method are given in Table 10.11. The cycle time here is also 0.95, and therefore the efficiency of this procedure is

$$e = \left[1 - \frac{0.95 \times 5 - 4.4}{95 \times 5}\right] \times 100 = 92.63\%$$

We might note that although in this example both procedures result in the same cycle times, it is not necessarily true in all examples.

10.4.3 Cycle Time Less Than Task Time

Now, suppose the required cycle time is less than task time of one or more tasks, the previous assignments are then not possible. One way to develop an assembly line is to try to reexamine the tasks that exceed the cycle time and see if they can be further divided so that each subtask requires less than cycle time. Another way is to .develop parallel stations in the assembly line, where each station performs the same task. Thus, for example, if we have two parallel stations each taking t time units, we have two units produced in that time or we can have $t/2$ as the cycle time.

Suppose in the previous data the cycle time is 0.6 min. Tasks 9, 4, 1 cannot be in 0.6 time units and must have at least two parallel stations for each task. Apply the same logic as longest candidate rule or RPW methods with one modification. Assign sufficient parallel work centers at each stage, if needed, to accommodate the first assignable task in the list, the one without any precedence task remaining. For example, the first task in the list that can be assigned in following table using LCR is task 1. It requires 0.7 min. The way to archive that time is to have two parallel stations, each with 0.6 min. Similar analysis leads to assignments shown in the tables.

		Remaining Immediate Predecessor(s) per Cycle									
Task	**Task Time**	**1**	**2**	**3**	**4**	**5**	**6**	**7**	**8**	**9**	**10**
9	0.88	5, 7, 8	5, 7, 8	5, 7, 8	7, 8	7,8	7, 8	7, 8	8	*	- -
4	0.75	1, 2, 3	2, 3	3	3	-*	- -	-	- -	- -	- -
1	0.7	-*	- -	- -	- -	- -	- -	- -	- -	- -	- -
7	0.52	4, 5, 6	4, 5, 6	4, 5, 6	4, 6	4, 6	6	*	- -*	- -	- -
6	0.42	3	3	3	3	- -	- *	- -	- -	- -	- -
8	0.4	6, 7	6, 7	6, 7	6, 7	6, 7	6, 7	7	*	- -	- -
2	0.33	1	-*	-	-	- -	- -	- -	- -	- -	- -
3	0.25	1	-	-	*	- -	- -	- -	- -	- -	- -
5	0.15	2	2	*		-	- *	- -	- -	- -	- -
Task Assignments		1	2	5	3	4	6	7	8	9	

- : No prerequisites, assignment can be made.
- - : Job has been assigned.

Station	Number of Parallel Centers	Time Available At the Center	Task Assigned	Task Time	Cumulative Task Time/Station
1	2	1.2	1	0.70	0.70
			2	0.33	1.03
			5	0.15	1.18
2	1	0.6	3	0.25	0.25
3	2	1.2	4	0.75	0.75
			6	0.42	1.17
4	1	0.6	7	0.52	0.52
5	1	0.6	8	0.40	0.40
6	2	1.2	9	0.88	0.88

10.5 MIXED-MODEL ASSEMBLY LINE BALANCING

A mixed model assembly (MMA) line is an assembly line on which number of similar products and/ or number of different models of an entity, are manufactured on the same line during the course of a day. This is in contrast to a single product assembly line where the line is responsible for only one product. In MMA, like operations for each model are assigned to the same worker. Thus, he/she may work on different models, but the job contained in the task(s) requires the similar job skills. MMA may be useful in situations where the demands for different models are fairly constant over a long run; however, there may be slight daily variations in production needs. The daily irregularity can best be handled by having workers work on a single line that produces various models. Workers are familiar with the work content of various products and can adjust their times between models to compensate for the demand variations. A unique assembly line devoted to each model may not be able to handle such production changes without pain.

We should be aware that due to design differentials, the work elements (or tasks) necessary to build one model may or may not be essential to build another model. There may even be some variations in the similar tasks, requiring varying task times in different models. Generally though, there are many common elements between the models (requiring the same skills and knowledge base) to make the development of a MMA line feasible.

The problem of designing an MMA line can be approached in two stages: (1) balancing and (2) scheduling. In balancing, we determine which products or models should be grouped together, and in scheduling, we design the assembly line for that group. Balancing is a decision process based on our knowledge and judgement of the problem. It may also be influenced by the number of assembly lines we expect to develop. Ultimately, it is possible to compare the performances of different balanced groups and then make the appropriate selection. The procedure is explained while simultaneously solving an example.

10.5.1 PROCEDURE AND AN ILLUSTRATIVE EXAMPLE

Consider a diesel engine factory that manufactures air- and water-cooled versions of diesel engines. The air-cooled engines known as the "VA" series, are available in 4-cylinder, 6-cylinder, and 6-cylinder turbo-charged models, namely VA4, VA6, and VA6 TC. Similarly, the water-cooled engines, known as the "VW" series, are available in 2- and 3-cylinder models, namely VW2 and VW3. The production required per shift of these diesel engines is VA4—2 units, VA6—3 units, VA6 TC—1 unit, VW2—2 units, and VW3—2 units. The technical requirements to build the various models are in the precedence relationship tables (displayed later). The various tasks involved in assembling one unit of a model and the associated times in minutes, are as shown in Table 10.12.

The total time necessary to produce the required production of each model is 2 (271) + 3(246) + 1(437) + 2(464) + 2(526) = 3697 min per shift. Assuming the total available time per shift to be 500 min per operator, theoretically the minimum

TABLE 10.12
Tasks and Associated Times in Minutes

Task	VA4	VA6	VA6 TC	VW2	VW3
1	88	65	68	58	60
2	0	0	6	6	6
3	0	0	125	100	158
4	7	7	18	21	21
5	14	14	25	25	35
6	0	0	30	34	30
7	13	13	13	11	11
8	48	50	48	88	87
9	67	63	70	87	84
10	34	34	34	34	34
Total	271 min	246 min	437 min	464 min	526 min

number of operators needed to satisfy the production is $3697/500 = 7.39$, that is, 8. Theoretical efficiency is $3697/(8 \times 500) = 92.42\%$.

The first step is that of grouping the products into cohesive groups so that an assembly line for each group can be developed. Some guidelines for grouping may be

1. Models requiring exactly the same tasks should be grouped together
2. The groups should be well balanced as regards the work content.

"How many groups should one form?" will depend on how many mixed-mode assembly lines we wish to develop and how similar the products are. Each group will be processed in one assembly line. We could try grouping different models together and selecting the number that gives the most efficiency and is the most practical. For the problem under consideration, the models are grouped based on the similarity of tasks involved. Considering the limited number of models and the total workload of only 2417 min, dividing the jobs in further smaller groups was thought unnecessary. Accordingly, engine models VA4 and VA6 formed Group 1, while models VA6 TC, VW2, and VW3 formed Group 2. (As mentioned earlier, the effects of forming groups, in ways other than the one chosen here are discussed in the sections which follow.). Thus, we have two groups:

Group 1—models VA4 and VA6
Group 2—models VA6 TC, VW2, and VW3

Step 2: For each group, list the tasks associated total time requirements for the planned production per shift, that is, the number of units produced per shift × time required per unit and the cumulative total of time for the task.

TABLE 10.13
Task, Total Time, and Cumulative Time List for Group 2

Task	VW2	VW3	VA6 TC	Total
1	116	120	68	304
2	12	12	6	30
3	200	316	125	641
4	42	42	18	102
5	50	70	25	145
6	68	60	30	158
7	22	22	13	57
8	176	174	48	398
9	174	168	70	412
10	68	68	34	170
Total	928	1052	437	2417

TABLE 10.14
Task, Total Time, and Cumulative Time List for Group 1

Task	VA4	VA6	Total
1	176	195	371
2	0	0	0
3	0	0	0
4	14	21	35
5	28	42	70
6	0	0	0
7	26	39	65
8	96	150	246
9	134	189	323
10	68	102	170
Total	542	738	1280

For our example, Tables 10.13 and 10.14 show the results for groups 1 and 2, respectively.

Step 3: Form workstations using either LCR or RPW method assuming available cycle time to be equal to shift time. This is because, in general, no two models are expected to have the same workload (in minutes) as regards assembly per unit. This feature of the MMA Line does not allow us to calculate the cycle time per unit by dividing the available time by the required production as is done in the models for the single-model assembly line.

TABLE 10.15
Group 1: Models VA4 and VA6 Precedence Relationships

Task	Immediate Predecessor	Time (Minutes)
1	-	371
9	8, 5	323
8	7	246
10	9	170
5	4	70
7	1	65
4	1	35

TABLE 10.16
Workstation Formation for Group 1

Workstation	Tasks Assigned	Number of Operators Needed	Loading Efficiency (%)
1	1, 7, 4	1	94.2
2	8, 5	1	63.2
3	9, 10	1	98.6
Overall		3	85.33

The method of forming workstations for an assembly line may be slightly modified. Its application in the current context is more of a decision than a rigid rule. Thus, however, the number of tasks that may be attached to a workstation has a limit; this limit is entirely a decision of the end-user. This is because the more tasks attached to a workstation, more is the training required for the operators of that workstation. If we assume that in pump manufacturing one operator can effectively handle up to four different tasks, then this fact can be used intelligently to improve the loading efficiency of the various workstations. Whatever the number of tasks attached to a workstation, the total available time per operator remains fixed, which is 500 min for the current example, but we can increase the station time by placing more operators per station. Thus, if a task or a combination of tasks requires time that exceeds the time available for one worker, we can increase the available time by hiring multiple workers on that station. For example, if the required task time is 553, we must assign two workers to the station, giving us 1000 min of time on that station. It may improve the efficiency if we can add a few more tasks to the station to utilize the excess capacity.

The precedence relationships of the tasks involved for the models in group 1 are shown in Table 10.15. The resulting workstation formations using LCR are shown in Table 10.16. Similarly, group 2 results are shown in Tables 10.17 and 10.18.

TABLE 10.17
Group: Models VW2, VW3, and VA6
TC Precedence Relationships

Task	Immediate Predecessor	Time (Minutes)
3	2	641
9	8, 6	412
8	7	398
1		304
10	9	170
6	5	158
5	3, 4	145
4	2	102
7	3	57
2	1	30

TABLE 10.18
Workstation Formation for Group 2

Workstation	Tasks Attached	Number of Operators	Loading Efficiency (%)
1	1, 2, 3	2	97.5
2	4, 5, 6, 7	1	92.4
3	8, 9, 10	2	98.0
Overall		5	95.96

TABLE 10.19
Summary

Group	Models	Number of Stations	Number of Operators	Loading h (%)
1	VA4 and VA6	3	3	85.33
2	VA6 TC, VB2, and VB3	3	5	95.96
Overall	—	—	8	90.64

Table 10.19 displays the summary of two groups requirements. It is noted that even though the number of operators is equal to the theoretical minimum, that is, 8, the overall loading efficiency as seen in the preceding table is 90.64%, which is the simple average of the two lines. The actual efficiency is the weighted value, three operators are working on average of 85% efficiency, while five others are working with 96% efficiency (with some round-off errors).

TABLE 10.20
Requirements and Efficiencies of the Independent Assembly Lines

Group	Models	Number of Stations	Number of Operators	Loading h (%)
1	VA4	2	2	54.2
2	VA6	2	2	73.8
3	VA6 TC	3	3	29.13
4	VW2	3	3	61.0
5	VW3	3	3	70.13
Overall	—	—	13	57.65

TABLE 10.21
A Single Assembly Line for All Models

Number of Stations	Number of Operators	Loading h (%)
5	9	76.47

10.5.2 Effects of Other Ways of Balancing (Grouping)

10.5.2.1 An Assembly Line for Each Model

With the same production rate for each model, it might be of interest to see what effect it would have if we develop an assembly line for each individual model. Using the cycle time of 500 min per station and assuming no operator can be trained for more than four tasks, we get the final requirements for each model as shown in Table 10.20.

The average efficiency is quite low. The efficiency could be improved by assigning all ten tasks for model VA6 TC, to a single operator at a single workstation. The efficiency for the model would then be 87.4% but one can imagine the variety of training the particular operator would have to undergo. Also, we could improve the efficiency by assigning five instead of four tasks per operator for model VW2. The efficiency for the model would then be 92.8%, and the number of operators required would be two. Even if one was to implement both the preceding modifications, the number of operators required would be ten, which is greater than eight we had obtained in Table 10.19 assignments.

10.5.2.2 A Single Assembly Line for All Models

Now, suppose we develop a single assembly line to produce all models simultaneously, with no more than four tasks per operator as the restriction, we will need nine operators on five stations as shown in Table 10.21. Again, the efficiency is not as high as the mixed-mode assembly.

10.5.3 Advantages of the Suggested Approach

The approach that has been suggested here to design MMA lines offers the following advantages over the conventional methods:

1. The approach is very simple and easy to understand.
2. Operators handle fewer models and fewer tasks, which reduces the training requirements.
3. Fewer operators handling fewer models and tasks, and improves tractability and accountability since responsibility is more specific.
4. Responsibility being more specific, quality is expected to improve.
5. The production system becomes more flexible, and thus absenteeism and breakdowns can be managed effectively.
6. The number of stations required per model is minimized, and thus managing the production becomes easy.
7. Since models from the various groups are scheduled simultaneously, the waiting time or idle inventory is reduced.
8. Optimal results as regards number of operators and loading efficiency are usually achieved.
9. Overall, the production is more organized and easy to control.
10. The suggested approach gives better results as the size of the problem approaches real-life proportions.

10.5.3.1 Problem Areas

1. The suggested approach may not always result in the minimum possible manpower required to satisfy production, although the difference in such instances is usually to the extent of one, two, or three operators.
2. More decisions are required to be made by the management, although this need not be considered a problem.

10.6 MIXED-MODEL ASSEMBLY—METHOD TO MINIMIZE STATIONS

The mixed-model assembly (MMA) line balancing technique also presents a two-step procedure for solving the MMA line balancing problem with parallel workstations and zoning constraints. The goal is to minimize the number of workstations for a given cycle time. This method also allows the user to define a limit on the maximum number of replicas of a workstation and the limiting condition under which the workstation can be replicated to increase the available total time for performance in that workstation.

This method has been developed based on the following parameters:

1. The planning horizon has a fixed length, L.
2. A set of similar models, $i = 1, 2, 3, N$ can be simultaneously assembled.
3. The forecast demand, over the planning horizon, for model i is D_{i}.

4. The line cycle time, C, is constant for all models considered and can be computed as follows,

$$C = \frac{L}{\sum_{i=1}^{N} D_i}$$

5. The overall proportion of the number of units of model, i, being assembled, Q_i, is then,

$$Q_i = \left[\frac{D_i}{\sum_{i=1}^{N} D_i}\right] \times 100$$

6. Each model has its own set of precedence relationships, but there is a subset of tasks common to all the models. The resulting precedence diagrams for all the models can be combined into one with "K" tasks.
7. The time required to perform task k on model i, T_{ki}, may vary among models, where $k = 1, 2, 3, K$ and $i = 1, 2, 3, N$. Zero processing time implies that a particular task is not required to be performed for that model.
8. A task can be assigned to only one workstation and, consequently, the tasks that are common to several models need to be performed on the same workstation.
9. A set of assignment constraints that forbid the assignment of different tasks to the same workstation is defined. This set of constraints is called zoning constraint.
10. The average weighted task time, *AWTT*, for all models is computed from the given values of T_{ki}, can be computed as follows,

$$AWTT = \sum_{i=1}^{N}\sum_{k=1}^{K}(Q_i \times T_{ki})$$

11. The limiting value of the maximum number of replications of workstations possible is user defined. The user may define the minimum task time that triggers the replication of a workstation. The default value of this minimum task time is 100% of the cycle time, which indicates that only workstations performing the tasks whose duration is larger than the cycle time, for at least one of the models, can be replicated.

10.6.1 PROCEDURE

Step 1: From the available values of L, i, D_i, and T_{ki}, compute C using (4). Also compute *AWTT* by using (11). Tabulate the list of tasks from the precedence diagram, in the ascending order of tasks, their immediate precedence relationship task(s), and average weighted task times $(A.W.T.T)$.

Step 2: Beginning with the first task tabulated in step 1, assign task(s) to workstations, ensuring that the *AWTT* for task(s) is less than or equal to the line cycle time, C. While assigning task(s), also consider zoning constraints. Continue assigning tasks until no more tasks can be assigned to that workstation without violating the precedence relationship. Increment the workstation index, "WS #," by 1, each time an assignment is complete. If a task(s) has an *AWTT* greater than the line cycle time, replicate the workstation. The maximum number of replications of a workstation possible is limited to the user-defined limiting value. This step terminates when the final task from the precedence diagram has been feasibly assigned. The sum of all the workstations (including replicas) is the total number of workstations required for completing all the tasks.

10.6.2 Numerical Illustration

Data from published literature is used to illustrate a numerical example. The data is shown in Table 10.22. The example has two models and 25 tasks. In addition, the following parameters are also considered:

1. The models 1 and 2 are simultaneously assembled in a line and over a planning horizon, L, of 480 time units. The demand of model 1, D_1, is 20 units, and the demand of model 2, D_2, is 28 units. Thus, Q_1, which represents the proportion of number of units of model 1 being assembled, is 42% $(20/(20 + 28) \times 100)$, and Q_2, which represents the proportion of number of units of model 2 being assembled, is 58% $(28/(20+28) \times 100)$.
2. Tasks 9 and 10 cannot be processed on the same workstation.
3. Only workstations performing tasks with a processing time greater than the line cycle time can be replicated.
4. The maximum number of replicas of a workstation is limited to two replicas.

10.6.2.1 Solution

Step 1: Computing line cycle time and AWTT

Using eq (4) line cycle time, C, is computed as,

$$C = \frac{480}{(20 + 28)} = 10$$

which remains constant.

From the task times for each model enumerated in Table 10.22 and calculated values of Q_1 and Q_2, the average weighted task times (*AWTT*) are calculated using (11) and enumerated in Table 10.23. Sample calculations for *AWTT* for six tasks are also shown.

Task 1: $0\ (0.42) + 2(0.58) = 1.2$
Task 2: $7.7\ (0.42) + 7.7(0.58) = 7.7$

TABLE 10.22
Data from VS: Processing Times for Model 1 and 2, and Immediate Precedence

Task (k)	1	2	3	4	5	6	7	8	9	10	11	12	13
T_{k1}	0	7.7	7.3	15.0	8.8	6.2	3.6	0	6.6	2.5	5.5	7.1	5.9
T_{k2}	2.0	7.7	7.3	15.0	8.8	0	0	2.0	6.6	2.5	5.5	7.1	5.9
Immediate precedence	—	—	1	3	3	3	3	4, 5	5	2, 6	5, 6	8, 9	11

Task (k)	14	15	16	17	18	19	20	21	22	23	24	25
T_{k1}	1.3	5.5	1.9	3.7	9.4	1.3	0	2.0	4.7	9.6	4.1	12.5
T_{k2}	0	5.5	2.0	0	9.4	1.3	9.0	2.0	4.7	8.2	3.7	0
Immediate precedence	9	12, 13	10, 13	16	16	14, 18	7, 18	17	21	15, 19, 21	20, 22, 23	24

TABLE 10.23
Task(s), *AWTT*, and Immediate Precedence

Task (k)	1	2	3	4	5	6	7	8	9	10	11	12	13
AWTT	1.2	7.7	7.3	15	8.8	2.6	1.5	1.2	6.6	2.5	5.5	7.1	5.9
Immediate precedence	—	—	1	3	3	3	3	4, 5	5	2, 6	5, 6	8, 9	11

Task (k)	14	15	16	17	18	19	20	21	22	23	24	25
AWTT	0.5	5.5	2.0	1.6	9.4	1.3	5.2	2.0	4.7	8.8	3.9	5.3
Immediate precedence	9	12, 13	10, 13	16	16	14, 18	7, 18	17	21	15, 19, 21	20, 22, 23	24

TABLE 10.24
Task(s), *AWTT*, Workstation Index and Number of Workstations

WS #	1	2	3	4	5	6	7	8	9	10	11	
Task(s)	1,2	3	4, 6, 7	5	9	8, 10, 11	12, 13, 14, 15	16	17, 18, 19, 20, 21	22, 23,24	25	
AWTT	8.9	7.3	19.1	8.8	6.6	9.2	19.0	2.0	19.5	17.4	5.3	
Number of Workstations	1	1	2	1	1	1	2	1	2	2	1	15

Task 3: 7.3 (0.42) + 7.3(0.58) = 7.3
Task 4: 15 (0.42) + 15(0.58) = 15
Task 5: 8.8 (0.42) + 8.8(0.58) = 8.8
Task 6: 6.2 (0.42) + 0(0.58) = 2.6

Step 2: Assignment of tasks

Beginning with the first task in Table 10.23, assign task(s) to workstations such that *AWTT* is less than or equal to $C = 10$. Tasks 1 and 2 can be combined and assigned to WS # 1 since the sum of *AWTT* associated with them is less than the line cycle time, $C = 10$. Similarly, assign the remaining task(s) listed in Table 10.23 maintaining precedence relationships. Task 3 is assigned to WS # 2. Tasks 4, 6, and 7 assigned to WS # 3 have a combined *AWTT* almost twice that of C. Therefore, WS # 3 is replicated confirming to the user-defined limit on the maximum number of replications allowed, being 2. Task 5 is assigned to WS # 4. Due to the zoning constraint forbidding Tasks 9 and 10 from being assigned to the same workstation, Task 9 is assigned to WS # 5. Tasks 8, 10, and 11 are combined and assigned to WS # 6. Tasks 12, 13, 14, and 15 are assigned to WS # 7, which is replicated. Task 16 is assigned to WS # 8. Tasks 17, 18, 19, 20, and 21 are assigned to WS # 9, which is replicated. Tasks 22, 23, and 24 are assigned to WS # 10, which is replicated. Task 25 is assigned to WS # 11.

The sum of all the WS # including replicas is the total number of workstations required, in this case, 15 (Table 10.24).

Line cycle time, $C = 10$.

10.7 NETWORK SCHEDULING WITH RESOURCE CONSTRAINT

So far, we have been assuming that the completion of a project on time is the major consideration in scheduling. We determined the activities on critical path and suggested that each of these operations must start and complete on time for the project to finish in minimum possible time. This is achievable when the resources are not limited. However, when the resources are limited, the starting of an activity may depend not only on the precedence relationships (i.e., completion of precedence activities), but also on availability of the resources.

Besides time, the resources required for an operation may include manpower, material, capital, and/or working space. Resources such as manpower and working space may be renewable. For example, after the completion of an activity, men working on that activity may be released and may be used by the operations that follow. Recourse such as money may not be renewable, that is, once spent, it is not available. The additional money may be obtained from different sources (such as bank loans and product sales) at different time periods to complete the project. There are two additional variations in the way the activities may be performed. In non-preemptive mode, the activities once started cannot be interrupted till they are completed, while in the preemptive mode, the activities may be stopped and restarted at any stage of completion.

TABLE 10.25
Heuristic Rules for Resource Scheduling

Rule	Description
MINSLK	Select the activity with minimum slack.
ACT	Select the activity based on maximum value of the critical path (CP). The CP for each activity is calculated by imagining the tail node of the activity as the starting node of the network.
MINLFT	Select the activity with minimum latest completion time(finish time) on the tail.
RAN	Randomly select an activity.
SA	Select the shortest duration activity first.
LA	Select the activity with longest duration.
GRD	Select the activity with largest ratio of recourse per unit time.
SRD	Select the activity with smallest ratio of recourse per unit time.

We shall discuss here most commonly faced non-preemptive, renewable single-resource constraint problem: such as that of manpower limitation. There is no single heuristic rule that guarantees an optimum solution for each problem. Authors have suggested different rules that work good for one set of problems than other. Table 10.25 list a number of such rules. Within each rule, another rule may be used for breaking ties. For example, after applying the MINLFT rule, if there is still a tie for selection of activities, then the SA rule may be used to break the tie.

In the discussion, a node from where an activity emerges is designated as the tail of the activity, and the node where the activity merges is called the nose of the activity. The slack in the activity is defined as: latest time on the nose - earliest time on the tail - activity duration.

10.7.1 Illustrative Example

We shall illustrate couple of scheduling rules by applying them to the activities with the precedence relationship, as shown in Figure 10.4. The first three rows of the Table 10.26 show the data including the man power required for each activity. The next four rows in the table illustrate the results of application of activity time (ACT) rule. The ACT value for each activity is calculated by assuming that the network only exist from the beginning node of that activity. For example, ACT for activity 2–4 is the critical path value for the network with node 2 as the starting node. ACT for a node can also be calculated as LT for the last node - LT for the present node. For example ACT for all activities starting from node 2 is $LT_6 - LT_2 = 14 - 5 = 9$. Let us assume that we have maximum of five workers available during the project.

Table 10.25 shows the scheduling calculations based on ACT. At time 0, activities 1–2, 1–3, and 1–5 all have an ACT value of 14 and have early time equal to the present start time. These activities also satisfy the precedence relationship and, therefore, are marked as eligible activities. The following rules are applied in sequence to eligible activities for placing priorities. The activities are selected based on the priorities until the resources are consumed or no activity can be assigned.

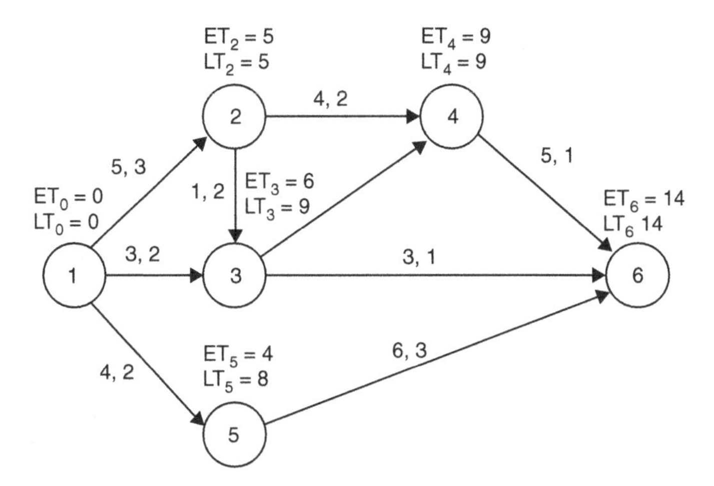

FIGURE 10.4 Precedence diagram for resource-scheduling example

TABLE 10.26
Data and Assignment Based on ACT

Activity	1-2	1-3	1-5	2-3	2-4	3-4	3-6	4-6	5-6
Duration	5	3	4	1	4	0	3	5	6
Manpower required	3	2	2	2	2	0	1	1	3
ACT value	14	14	14	9	9	5	5	5	6
T early	0	0	0	5	5	6	6	9	4
T start	0	0	3	5	6	6	6	10	9
T complete	5	3	7	6	10	6	9	15	15

1. An eligible activity having higher ACT value is selected first.
2. If there is a tie in step 1, an eligible activity requiring no recourse has a higher priority.
3. After step 2 assignment, an activity requiring maximum manpower has the next higher priority.
4. If there is a tie in rule 3, the tie is broken randomly.

At time 0, since all eligible activities have the same ACT value and rule 2 does not apply, the priorities are assigned based on rule 3. Assignment of manpower is made based on the priorities until all the available manpower is utilized.

Activity completion time indicates when additional activities may become eligible for scheduling and when some of the presently assigned resources are becoming free for the next allocation. Choose the next imminent event time, that is, time 3 in our case, to reexamine the network. It is easier to identify the activities that may become eligible, as we progress through the scheduling process, if along with working on Table 10.27, we also simultaneously develop the last three rows of Table 10.26 and visualize the precedence relationships in Figure 10.4. In Table 10.23, T early is earliest

TABLE 10.27
Scheduling Based on ACT

	Release			Eligible Activities						Assignments				
Start Time	Activity	Men	Manpower Available	Early Times	Precedence	ACT value	Dur.	Manpower	Priority	Activity	Dur.	Manpower	Rem. Manpower	Activity Complete Time
0			5	1-2	Y	14	5	3	1	1-2	5	3	2	5
				1-3	Y	14	3	2	2	1-3	3	2	0	3
				1-5	Y	14	4	2	3					
3	1-3	2	2	1-5	Y	14	4	2	1	1-5	4	2	0	7
5	1-2	3	3	2-3	Y	9	1	2	1	2-3	1	2	1	6
				2-4	Y	9	4	2	2					
6	2-3	2	3	5-6	Y	6	6	3	2	2-4	4	2	1	10
				2-4	Y	9	4	2	1	3-4	0	0	1	6
				3-4	Y	5	0	0	3	3-6	3	1	0	9
				3-6	Y	5	3	1	4					
6	3-4	0	0	No check										
7	1-5	2	2	5-6	Y	6	6	3	1					
9	3-6	1	3	5-6	Y	6	6	3	1	5-6	6	3	0	15
10	2-4	2	2	4-6	Y	5	5	1	1	2-4	5	1	1	15

TABLE 10.28
Activity Slack Calculation

Activity	T Late	T Early	Duration	Slack
1-2	5	0	5	0
1-3	9	0	3	6
1-5	8	0	4	4
2-4	9	5	4	0
2-3	9	5	1	3
3-4	9	6	0	3
3-6	14	6	3	5
4-6	14	9	5	0
5-6	14	4	6	4

time an activity may start, the value obtained during critical path analysis. T start and T complete times start and completion times of an activity as we make the assignments in Table 10.25.

The network completed in 15 time units.

It might be interesting to note that at time 5, though the activity 5–6 satisfies the eligibility based on early start time, the precedence activity, activity 1–5 has not been completed and, therefore, 5–6 is not eligible. With three men released from just completed activity 1–2, only one assignment, that to activity 2–3, is possible.

At time 6, we make the first assignment to priority 1 activity, 2–4. Next, 5–6 is eligible and has a priority of 2, but it cannot be started because the remaining manpower is less than what is needed for the activity. We continue the assignments by going to the next priority activity 3–4, which requires no resources and is completed immediately. It is marked as such, and the next evaluation is again done at time 6. At that point, since no manpower is available, no check on eligible activities is required.

At time 7, only two men are released by the activity 1–5, while three men are required by only eligible activity 5–6. Therefore, no assignment is made. The schedule is completed in 15 time periods.

Tables 10.28, 10.29, and 10.30 show the application of minimum slack rule (MINSLK). It now takes 16 time periods to complete the project. Thus, we see that the application of different scheduling rules may lead to different schedule completion times.

The network is completed in 14 time units.

10.7.2 Multiple Resources

If the activities are controlled by the availability of multiple resources, the method for scheduling essentially follow the same procedure as the one that was illustrated earlier. In this case, however, it is necessary to check the availability of all resources before an activity can commence, and when the activity is completed, all the resources assigned to the activity must be accounted for. Any one or a combination of the heuristics rules displayed in Table 10.23 may be used in selecting an activity for scheduling, provided

TABLE 10.29
Scheduling Based on MINSLK

	Release			Eligible Activities						Assignments				
Start Time	Activity	Men	Manpower Avail.	Early Times	Precedence	Dur.	Manpower	Slack	Priority	Activity	Dur.	Manpower	Remaining Manpower	Activity Complete Time
0			5	1-2	Y	5	3	0	1	1-2	5	3	2	5
				1-3	Y	3	2	6	3	1-5	4	2	0	4
				1-5	Y	4	2	4	2					
4	1-5	2	2	1-3	Y	3	2	6	2	1-3	3	2	0	7
				5-6	Y	6	3	4	1					
5	1-2	3	3	2-3	Y	1	2	3	2	2-4	4	2	1	9
				2-4	Y	4	2	0	1					
				5-6	Y	6	3	4	3					
7	1-3	2	3	2-3	Y	1	2	3	1	2-3	1	2	1	8
				3-4	N	6	3	4	2					
				3-6	N									
				5-6	Y									
8	2-3	2	3	3-4	Y	0	0	3	1	3-4	0	0	3	8
				3-6	Y	3	1	5	3	5-6	6	3	0	14
				5-6	Y	6	3	4	2					
8	3-4	0	0	On check										
9	2-4	2	2	3-6	Y	3	1	5	2	3-6	3	1	0	12
				4-6	Y	5	1	0	1	4-6	5	1	1	14

TABLE 10.30
Assignments Based on MINSLK

Activity	1-2	1-3	1-5	2-3	2-4	3-4	3-6	4-6	5-6
Duration	5	3	4	1	4	0	3	5	6
Manpower	3	2	2	2	2	0	1	1	3
Slack	0	6	4	3	0	0	2	0	2
T Early	0	0	0	5	5	6	6	9	4
T Start	0	4	0	7	5	8	9	9	8
T Completion	5	7	4	8	9	8	12	14	14

the activity is eligible and all the resources required by the activity are accessible. The rules for prioritizing the recourse allocation may also be developed based on which resources we think are critical.

10.8 SUMMARY

Jobs that have precedence relationships add an additional dimension to scheduling. Drawing a network makes the visualization and manipulation of such constraints easier. The CPM illustrates how we can determine the available slack times in jobs and identify the jobs that are critical. This knowledge is used in developing a scheduling rule for parallel processors when jobs have precedence constraints. The objective is to minimize the makespan. We have also seen a procedure for scheduling on parallel machines when the jobs can only be performed on a limited subset of the available machines. This may happen because of limited availability of external resources such as jigs and fixtures.

We have also observed how to develop an assembly line for a single product. If there are multiple products to be made on the same assembly line, we have suggested a procedure for grouping the products and for developing the assembly line. The procedure is called MMA line balancing, which may have a number of advantages over the single-product assembly line.

10.9 PROBLEMS

10.1 The ten jobs shown in this network have precedence relationships and processing times as indicated. Assuming that two parallel processors can be used to process these jobs, develop the corresponding schedule.

10.2 Develop the schedule for processing the jobs depicted in this network with the associated precedence relationships and processing times, assuming that two parallel processors are to be used.

10.3 Schedule eight jobs on four parallel processors with the following job-machine relationships.

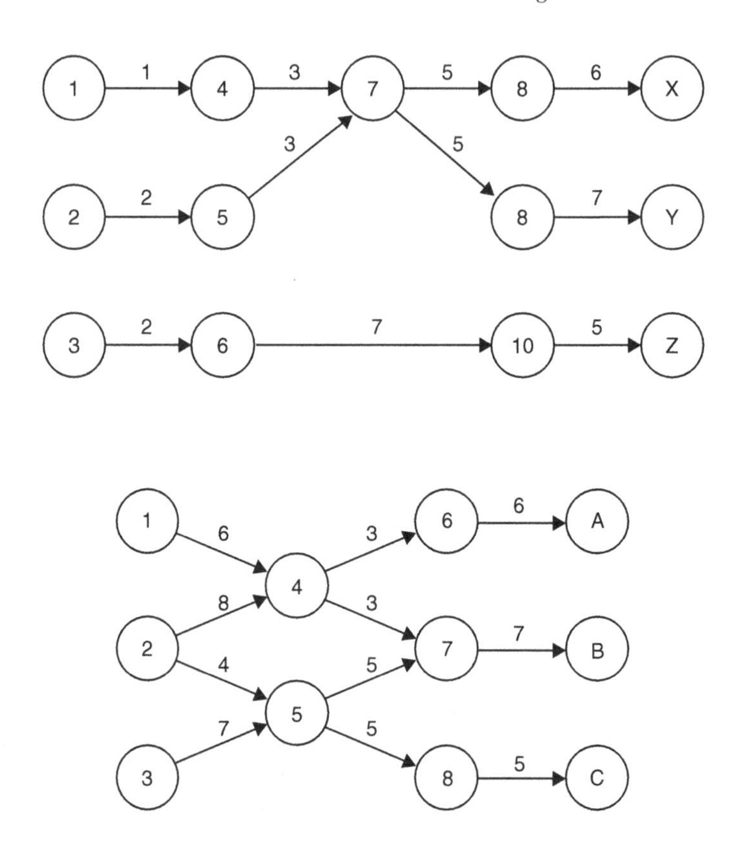

Job	1	2	3	4	5	6	7	8
Machine	M3	M2, M4	M1	M1, M2, M3	M1, M3, M4	M3, M4	M1, M3	M1, M2, M3, M4

10.4 Develop the following seven jobs on three parallel processors with the job-machine relationships as indicated.

Job	1	2	3	4	5	6	7
Machine	M1, M3	M1, M2, M3	M2	M1, M2	M2, M3	M1	M1, M2

10.5 Develop a balanced assembly line for a job that can be broken into ten tasks with precedence relationships and task times as displayed in the following network. Assume a desired cycle time of 0.95. Use the LCR.

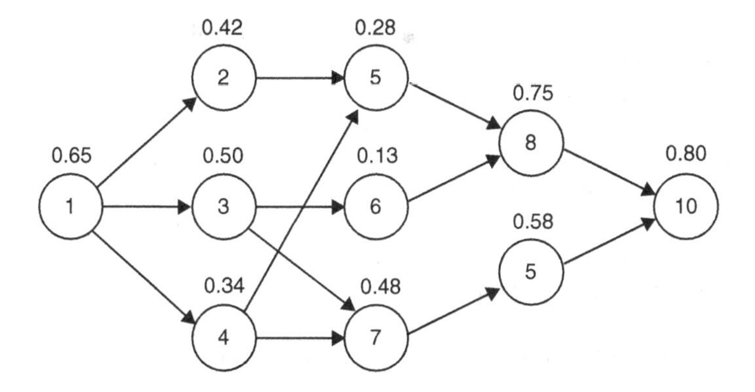

10.6 Develop the schedule for the network in Problem 6.5 using the RPW method.

10.7 Balance the schedule production for a MMA line given the following data: items A, B, C, and D are to be produced at quantities 2, 1, 3, and 3, respectively, per shift; each requires the following times to complete each of eight tasks necessary to produce the item; grouping of the items is based on task similarity; sift time per operator is 500 min; an operator can handle up to four tasks.

Tasks	A	B	C	D
1	70	65	0	0
2	50	55	90	80
3	60	40	30	30
4	0	0	50	45
5	20	20	0	0
6	10	10	0	0
7	0	0	65	65
8	0	0	25	35
Total	210	190	260	255

The precedence relationships for the two groups are as follows:

Group 1

Tasks	Immediate Predecessor
1	—
2	1
3	1
5	3
6	2, 3

Group 2

Tasks	Immediate Predecessor
2	—
3	2
4	2
7	3
8	4,3

10.8 Compare the efficiencies achieved in Problem 6.7 with the efficiency achieved by balancing the same problem using
   a. An assembly line for each model
   b. A single assembly line for all models

10.9 For the single-resource problem, solve the network when the number of resources available is seven:
   a. Determine the early and late start times for each node.
   b. Determine the critical path and its duration.
   c. Which of the two methods (ACT or MINSLK) will you use to schedule the network? Justify your answer.
   d. Repeat part c with the number of resources as five.

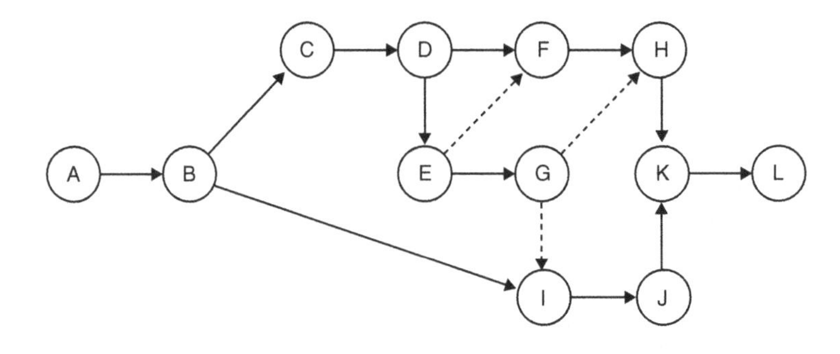

Activity	1–2	1–3	1–4	2–5	2–3	3–5	3–6	4–7	5–8	6–8	6–7	7–8
Duration	6	8	4	4	4	12	14	6	8	16	2	12
Resources	2	3	3	4	2	3	1	4	2	1	1	3

10.10 Schedule the activity network in the following figure. The duration and number of resources required for each activity are given on the node links. *D* is the duration and *R* is the number of resources required. The total number of resources available is five.

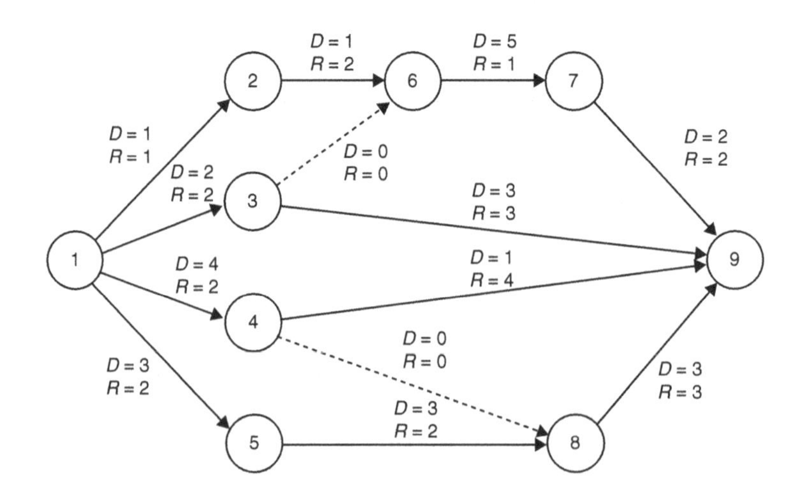

10.11 A stretch of land alongside a major highway needs water pipelines installed. Solve the resource scheduling problem for the following network, and data by

a. The ACT method
b. The MINSLK method, and compare the two results

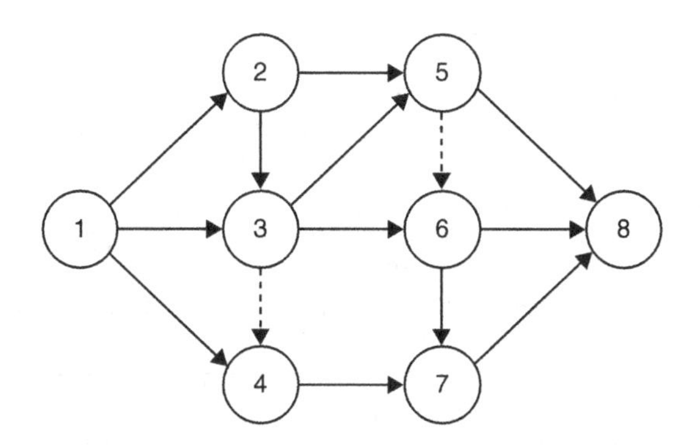

Activity	Activity Description	Duration	Resources
A-B	Start job	1	1
B-I	Relocate electric	6	4
B-C	Excavate	5	3
C-D	Install pipe section #1	10	4
D-F	Backfill section #1	5	3
D-E	Install pipe section #2	12	5
F-H	Backfill section #2	8	2
E-G	Test Pipe Sections	3	4
H-K	Grade and sod	7	3
I-J	Install Manhole	3	3
J-K	Test relocated electric	2	5
K-L	Final inspection	1	2

Resource available: 5.

10.12 For the data in Problem 6.11, consider a two-resource scenario and solve the problem:

Activity	A-B	B-I	B-C	C-D	D-F	D-E	F-H	E-G	H-K	I-J	J-K	K-L
Resource type	1	2	1	2	1	2	1	2	1	1	2	2

Resource available: Type 1 = 5, and Type 2 = 4.

REFERENCES AND SUGGESTED READINGS

Atabaksh, H. 1991. "A Survey of Constraint Based Scheduling Systems Using an Artificial Intelligence Approach" *Artificial Intelligence in Engineering*, 6(2): 58–73.

Baker, K.R. 1974. *Introduction to Sequencing and Scheduling*, New York: John Wiley & Sons.

Bard, J.F., A. Shtub, and S.B. Joshi. 1994. "Sequencing Mixed-Model Assembly Lines to Level Parts Usage and Minimize Line Length" *International Journal of Production Research*, 32(10): 2431–2454.

Blazewicz, J., W. Cellary, R. Slovinski, and J. Weglarz. 1986. "Scheduling under Resource Constraints- Deterministic Models" *Annals of Operations Research*, 7.

Burns, L., and C.F. Daganzo. 1987. "Assembly Line Job Sequencing Principles" *International Journal of Production Research*, 25, 71–99.

Dar-El, E.M. and A. Nadivi. 1981. "A Mixed-Model Sequencing Application" *International Journal of Production Research*, 19(1): 69–84.

Du, J. and J.Y.-T. Leung. 1989. "Scheduling Tree-Structured Tasks on Two Processors to Minimize Schedule Length" *SIAM Journal of Discrete Mathematics*, 2: 176–196.

Du, J., J.Y.-T. Leung, and G.H. Young. 1991. "Scheduling Chain-Structured Tasks to Minimize Makespan and Mean Flow Time" *Information and Computation*, 92: 219–236.

Frostig, E. 1988. "A Stochasitc Scheduling Problem with Intree Precedence Constraints" *Operations Research*, 36: 937–943.

Hu, T.C. 1961. "Parallel Sequencing and Assembly Line Problems" *Operations Research*, 9: 841–848.

Kampke, T. 1989. "Optimal Scheduling of Jobs with Exponential Service Times on Identical Parallel Processors" *Operations Research*, 37: 126–133.

Lenstra, J.K. and A.H.G. Rinnooy Kan. 1978. "Computational Complexity of Scheduling under Precedence Constraints" *Operations Research*, 26: 22–35.

McCormick, S.T., M.L. Pinedo, S. Shenker, and B. Wolf. 1990. "Transient Behaviour in a Flexible Assembly System" *International Journal of Flexible Manufacturing Systems*, 3: 27–44.

McCormick, S.T., M.L. Pinedo, S. Sheker, and B. Wolf. 1989. "Sequencing in an Assembly Line with Blocking to Minimize Cycle Time" *Operations Research*, 37: 925–936.

Pinedo, M. 1995. *Scheduling-Theory, Algorithms, and Systems*, Englewood Cliffs, NJ: Prentice Hall.

Pinedo, M. and G. Weiss. 1984. "Scheduling Jobs with exponentially Distributed Processing Times and Intree Precedence Constraints on Two Parallel Machines" *Operations Research*, 33: 1381–1388.

Sule, D.R. 1992. *Manufacturing Facilities- Location, Planning and Design*, 2d ed. Boston: PWS Publishing.

11 Job Shop Scheduling

Most scheduling problems associated with n jobs on m machines can be divided into four broad categories. Job operational flow (processing order) determines the type as being one of the following.

1. Flow shop
2. Job shop
3. Dependent shop
4. Open or general shop

We have seen flowshop in Chapter 9 and dependent shop in Chapter 10. In flowshop, the processing order of every job is identical. The arrangement of machines is up to the planner, and the machines can be placed so that the flow of the jobs through machines is unidirectional. In job-shop processing, the order of different jobs may be different, but is fixed and independent of each other. If the processing order of any one or more jobs depends on other jobs, the shop is called a dependent shop. A general or open shop is one in which there are no precedence constraints, that is, jobs can be processed in any order. As with the single-machine problem, variations in each type exist as setup times, due dates, and arrival dates are included. We shall study in this chapter a few basic models of job shop, while open shop is discussed in Chapter 12.

11.1 JOB SHOP

The Job-shop problem is considered important because it reflects the actual operation for several industries. In a job shop on any given day, we may have several jobs requiring scheduling, each with a different processing sequence and different processing times on the machines. Jobs may or may not have the promised delivery dates, and the solution procedures differ as the objective of scheduling changes.

For an n job m machine scheduling problem, there are $(n_1)!(n_2)!\ldots(n_m)!$ theoretically possible sequences, where n_k is the number of operations to be performed on machine k; however, not all of them are feasible. The best sequence then must satisfy the following conditions:

1. One that is technologically feasible, that is, the one that satisfies the machine precedence constraints.
2. Optimize the effectiveness measure.

Evaluation of all possible combinations is an impossible task even for a moderate-size problem and, therefore, many heuristic rules are developed by various researchers to determine the priority by which a job should be processed. These rules can also be classified as either static or dynamic, based on when the decisions referred to are made. In the *static* (sometimes referred to as global) rule, job priorities are determined in advance of processing jobs. For example, rules based on due date or arrival time, such as first come first serve (FCFS), or based on the penalty weight of the jobs, are all static rules. The priorities remain the same throughout job processing. In the *dynamic* (some times referred to as local) rule, on the other hand, the job priorities may change from machine to machine, based on the status of a predefined condition that is examined before a job is assigned to the machine. Typically, the job priorities are calculated for each machine using a rule such as COVERT, which we have seen in Chapter 6. Dynamic rules, in general, perform better than the static rules. We shall study some additional rules in this chapter.

Because of the challenging nature of the problem, many researchers have analyzed job shop and flow shop scheduling. The interest has mainly been in developing scheduling rules that relate to either in improving shop flow performance or adhering to the compliance with due dates. Shop flow time is the time a job spends in the shop. It also indirectly reflects on other shop-related measures such as work in progress, inventory cost, and how swiftly we can handle the customer's order. Due-date-related measures reflect on the tardiness of the jobs, penalty cost and number of jobs tardy. As with the single-machine problem, researchers have not found a common single rule that optimizes all goals at the same time since some goals can be conflicting. For example, shortest processing time (SPT), which gives a preference to a job with minimum immediate processing requirement, is a simple yet dominant rule used in reducing shop flow time. The same rule, however, can produce jobs that are very late if the due dates for the jobs are very tight.

Many dispatching rules exist in both static and dynamic environments. Table 11.1 lists some of them.

TABLE 11.1
Job Shop Dispatching Rules

Code	Choose the Job Based On
EDD	Earliest due date
FCFS	First come first serve
WSPT	Highest ratio of weight divided by processing time
LWKR	Least work remaining
WLWKR	Highest value of weight divided by least work remaining
WTWK	Largest value of weight divided by total work remaining
MST	With minimum slack time, i.e., due date - present time - process time
OPNDD	With earliest operational due date, i.e., due date - present time
Critical ratio	Highest value of total remaining time divided by total time until the due date
COVERT	Explained in Chapter 6
SCR	With smallest critical ratio, remaining allowance divided by work remaining
A/OPN	With smallest remaining allowance per remaining operations
S/OPN	With smallest slack per remaining operation

Bowmen [6], Manne [16] and Balas [3], and Conway et al. [10] regarded scheduling as a conventional programming problem and suggested linear programming formulations with integer constraints on its solution. The computational effort required to solve the problem with these approaches makes them impractical. Also, the scope of integer programming problems is very limited. Naser [17] developed the problem of minimizing the mean flow time as a mixed integer programming model. Then he developed an efficient algorithm based on this formulation to solve the problem by decomposing it into sub-problems that are easier to solve. He also compared the performance of a second algorithm based on SPT with the performance of a mixed-integer programming model. All these models are outside the scope of this book.

One of the pioneering studies of the effectiveness of dispatching rules in a job shop situation was published by Conway and Maxwell [8]. They noted that, in a single-server environment, the smallest processing time (SPT) rule was optimum for a special type of problem (see Chapter 6). They also found that in a multiple-machine environment also, the SPT retained the advantage of throughput maximization (under fairly evenly loaded machines). Elvers [12] studied the performance of ten dispatching rules over five variations of the SPT due date assignment methods. He showed that when the due date is set at six times the total processing time or less, the SPT rule performed the best. Due-date-based rules do not perform well. Similarly, the FIFO rule performs worse than good-processing-time rules.

Schultz [20] reported results relating to CEXSPT, a new heuristic relating to the SPT rule in the context of the dynamic job shop. We shall explain this heuristic later in this chapter.

11.2 JOB SHOP SCHEDULING TO MINIMIZE MAKESPAN (SPT)

The problem can be stated as follows: There are *n* jobs, each of which is to be processed one at a time on *m* or less machines. Each job follows a predefined machining order and has a specified processing time; however, the machine order is random from job to job. The jobs do not have due dates, and the objective is to minimize makespan.

The techniques illustrated here have the following assumptions:

1. Assumptions concerning jobs
   a. All *n* jobs are simultaneously available at the beginning of the planning period.
   b. A single job cannot be processed simultaneously by more than one machine.
   c. The processing time for each job is known and is deterministic.
   d. Set-up and transportation time is independent of the sequence and is included in the process time of the jobs.
   e. Jobs are processed as soon as possible or as planned.
   f. All jobs are of equal importance

TABLE 11.2
Data for a Job Shop Example

Job Number	Machine Sequence	Processing Times
1	1, 3	1, 2
2	2, 3	3, 1
3	3, 1, 2	2, 1, 1
4	1, 2, 3	3, 2, 1
5	3, 1	2, 1

2. Assumptions concerning machines
   a. All m machines are available at the beginning of the planning period and are ready to work on any of the n jobs requiring that machine for its first operation.
   b. At most, one job can be processed on a specific machine at any given time.
   c. There is only one machine of each type in the shop.
3. Other
   a. In-process inventory is allowed

11.2.1 Shortest Processing Rule

Since the shortest processing rule preforms well under a semi-evenly loaded shop, we shall illustrate it first by applying it to the data in Table 11.2. The table gives, for each job, the machines used in its production sequence and the processing time required on each machine. For example, job 1 requires machines 1 and 3, in that order, and 1 unit of processing time on machine 1 and 2 units of processing time on machine 3.

It is convenient to construct a table such as Table 11.3, displaying for each machine the jobs needing the machine arranged in SPT order. We refer to this arrangement as the list for each machine or simply list. In developing the day-to-day schedule (assuming unit time is a day) in Table 11.4, we shall refer to Table 11.3 frequently. The table displays for each machine a job and its processing time. For example, 3/1 on machine 1 indicates job 3 requiring 1 day to process on that machine.

At time 1, all machines are available. From Table 11.3, we observe that job 1 is the first job on the list for machine 1. From Table 11.2, we see that it is also the first machine required by job 1, and hence, we load job 1 on machine 1 on the first day. On machine 2, according to the list, job 3 should be loaded, but job 3 requires machine 3 first. Similarly, for the next job from the list, job 4 requires machine 1 first and hence, cannot be loaded. The next job (job 2) can be loaded on machine 2, and hence, it is scheduled next. It takes 3 days, and therefore, the machine is busy with job 2 for days 1, 2, and 3. Similar analysis for machine 3 results in loading of job 3 for two time periods.

At time 2, machine 1 is free, and the next job on the list for the machine is job 3. But, at that time, job 3 is being processed on machine 3 and is not available; therefore, job 4 is loaded for the next 3 time units.

TABLE 11.3
Jobs/Processing Times Arranged in SPT

Machine 1	Machine 2	Machine 3
1/1	3/1	2/1
3/1	4/2	4/1
5/1	2/3	—
4/3		3/2
		5/2

TABLE 11.4
Day-to-Day Schedule: Jobs on Each Machine

Time	Machine 1	Machine 2	Machine 3
1	1	2	3
2	4	2	3
3	4	2	1
4	4	—	1
5	3	4	2
6	—	4	5
7	—	3	5
8	5	—	4

The procedure is now clear. We load (using the list) the first job that can be loaded which is available and has completed all its precedence processing. If no such job is available (e.g., machine 2, time 4) then the machine is idle for that period. We continue day by day planning till all the jobs receive their required processing. We shall leave it to the reader to verify the schedule. The makespan is 8 days.

The associated Gantt chart is illustrated in Figure 11.1.

11.3 NETWORK APPROACH TO JOB SHOP SCHEDULING

The precedence relationships in a job shop with no recirculation (i.e., each job processed on a machine at most once only) can be represented by a project network. Here, a node represents a job on a machine and the branch emerging from the node represents the time required for the job on that machine and the sequence in which the nodes are connected for a job represents the precedence relationship of the operations.

For example, the data in Table 11.2 is represented by the network shown in Figure 11.2. Each node represents a specific operation for a job on the machine indicated. For example, node 1/1 represents operation for job 1 on machine 1. The

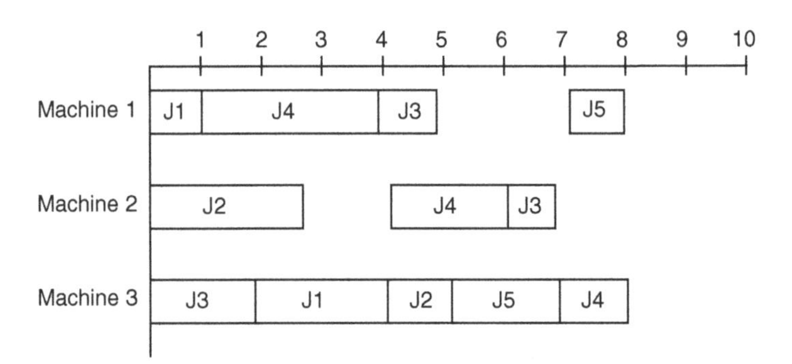

FIGURE 11.1 Gantt chart for a daily schedule using SPT.

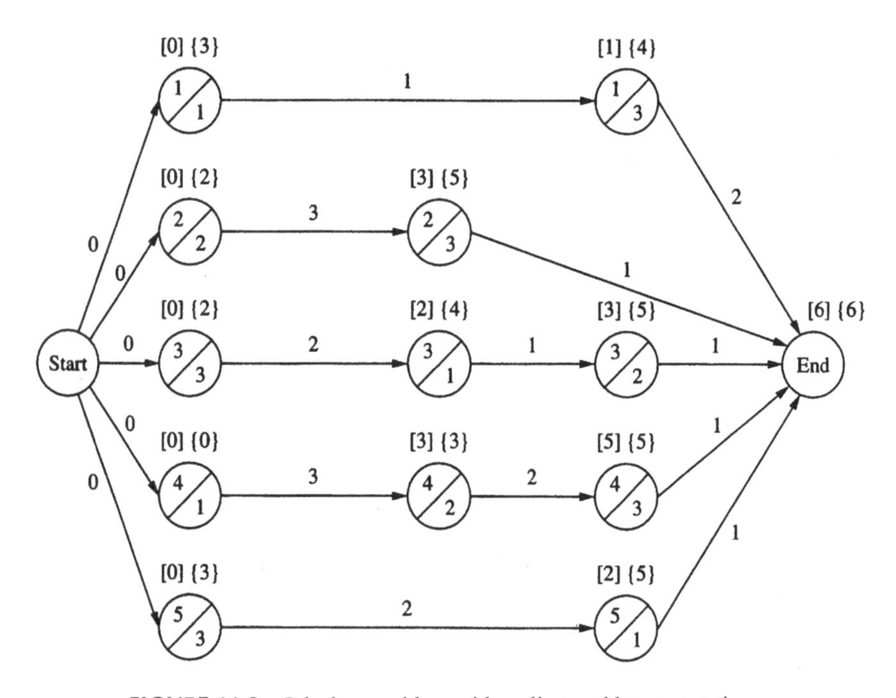

FIGURE 11.2 Job shop problem with earliest and latest start times.

processing time is 1 as shown on the arrow coming out of node 1/1. Similarly, node 1/3 represents operation for job 1 on machine 3. Its processing time is 2 units. That is the end of all processing for job 1, and hence, the arrow ends in the end node, which is a dummy node. A similar dummy node, a start node, is introduced at the beginning of the network with branches leading to each job's first operation with zero time consumption on each branch. With all jobs similarly drawn, the network is complete.

Operations belonging to different jobs that are to be processed on the same machine may be connected to one another by a dotted line. In this example, operations

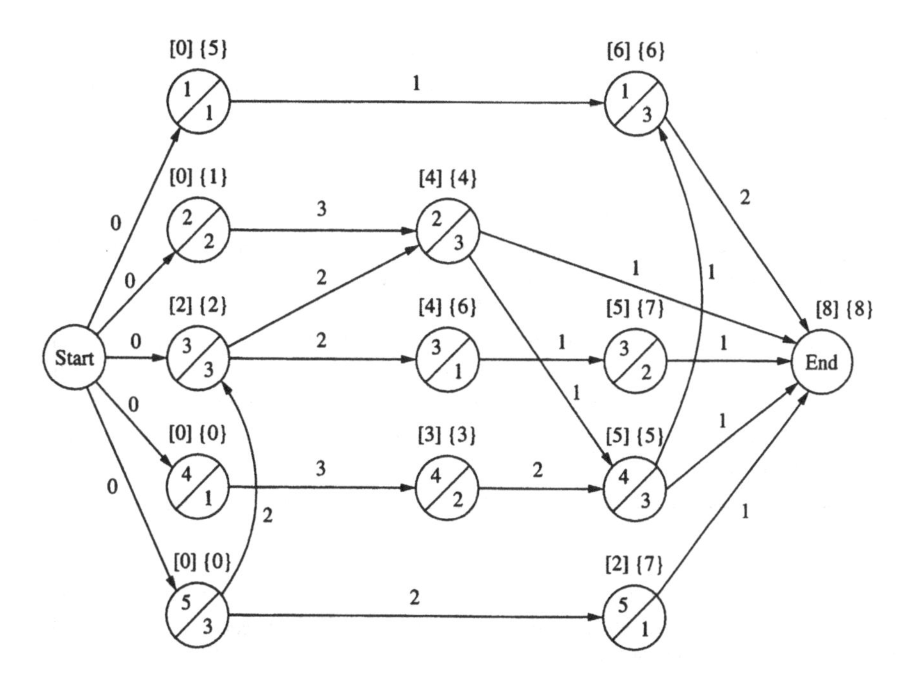

FIGURE 11.3 Machine 3 solved; Iteration 1.

1/1, 3/1, 4/1, and 5/1 may all be connected by dotted lines (the lines are not shown on the figure so as not to clutter the figure). Similarly, operations 2/2, 3/2, and 4/2 may be connected. Note that each operation is connected to every other operation on the same machine with a dotted line. This is to imply that any two operations could precede one another on this machine. After the job sequence has been established on the machine, a single solid line would emerge with an arrow pointing to either one of the two nodes showing the appropriate precedence relationship. For example, in Figure 11.3, the solid line shows the connection between node 5/3, 3/3, 2/3, 4/3, and 1/3 with the arrowhead showing the precedence on machine 3, job 5 is done first then job 3 then 2, and so on. All solid arcs (lines with an arrow on one end) emanating from a node have the processing time displayed by a number on the arc. For example, again in Figure 11.3, each branch emerging from node 5/3 is given the time that is equal to the processing time of job 5 on machine 3.

A feasible schedule is the one in which all nodes indicating the operation on a machine are connected by solid arcs, indicating the job precedence relationship on that machine. The makespan C_{max} of a feasible schedule is the longest path (i.e., the critical path) from the START node to the END node. The objective is to develop a feasible schedule with a minimum value for critical path.

11.3.1 The Modified Shifting Bottleneck Heuristic

One of the successful heuristic for minimizing the makespan in a job shop problem is the shifting bottleneck heuristic (Adams et al. 1988). In this procedure, each unscheduled machine is considered as the separate single machine, and the machine that gives the maximum delay in a single job is identified as a "bottleneck machine" and scheduled first.

We present a procedure here that is a minor modification of the original shifting bottleneck procedure. In the new procedure, modified shifting bottleneck heuristic (MODSB), each unscheduled machine is considered as a separate single-machine problem, and the machine that has the largest maximum tardiness, is designated as the bottleneck machine and sequenced first. The advantage of MODSB over shifting bottleneck procedure is, in MODSB, the total delay for jobs arriving at different times can be determined by using the heuristic from Section 7.15, while the original shifting bottleneck machine procedure requires solution by branch and bound method to ascertain the job with maximum delay and hence the bottleneck machine. The MODSB deviates very little from the final answer obtained by the original shifting bottleneck procedure.

The procedure for the MODSB heuristic can be described as follows:

Step 1: Let "M" denote the set of all the machines. Let M_r denote the set of all the machines in "M" for which the sequence of the jobs is already known. Initially, $M_r = 0$. The earliest and latest start times are calculated for all the nodes in the network.

Step 2: All unscheduled machines in set $M–M_r$ are solved separately as a single-machine problem. All the operations that require services of that machine are being treated as the jobs in the sense used in single-machine problems. The jobs (operations) arrive at the machine at different times based on the sequence of previous operations. The arrival time for a job and its due date are taken directly from the network. The arrival time "r_i" for an operation is the "earliest start time" for the corresponding node (operation) determined in step 1. The due date "d_i" for the operation is the (minimum of) latest start time(s) of the immediately succeeding node(s). Use the procedure described in Section 7.15 to develop the optimum single-machine sequence for each machine. Calculate the tardiness for each machine. This is equivalent to calculating the total penalty, assuming a late penalty of 1 and an early penalty of 0 for each job.

Step 3: Select the machine that serves as a bottleneck (i.e., the one with the largest tardiness T_{max}, which is also the maximum penalty value). Let us say that this machine is "k." The network is now modified by adding solid arcs between nodes (operations) that are using machine k based on the job sequence that was obtained in step 2 for machine "k." Machine "k" would be the new machine that would be added to set M_r, in step 6.

Step 4: Recompute the earliest and latest start times for the modified network from step 3 (with the solid arcs added).

Step 5: Before adding machine k to set M_r, a check must be made of all the machines that presently are in M_r. The check consists of determining if the directed arcs between operations that were made earlier for each machine are still optimum. The optimality may have changed due to the sequence of the operations being specified on machine k.

TABLE 11.5
Job Shop Example Data

Jobs	Machine Sequence	Processing Times
1	1, 3	1, 2
2	2, 3	3, 1
3	3, 1, 2	2, 1, 1
4	1, 2, 3	3, 2, 1
5	3, 1	2, 1

All the machines from set M_r are resequenced one by one to check if there are any improvements in the makespan (in the first iteration, M_r is zero). This is done by temporarily removing a machine, say "J," from the set M_r. The network is modified by removing the solid arcs connecting operations on machine "J" and solving for T_{max} (maximum tardiness) for this machine, with the modified arrival times and due dates based on the temporary modification. The new job sequence is determined for machine J, and solid lines are constructed between the nodes associated with machine J based on the new sequence. The makespan for the entire network is calculated based on this new directional lines added for machine J. If the new makespan is less than or equal to the previous makespan, then the new lines are maintained; if not, the original operational sequence on machine J is retained. The checking of each machine from the set M_r completes this step.

Step 6: Add machine k to M_r. If M_r is equal to M, then STOP the procedure, otherwise go to step 2.

11.3.1.1 Illustrative Example

Consider the same problem that was illustrated with the SPT rule. For convenience, the data is reproduced in Table 11.5.

Since we are only interested in determining the bottleneck machine(s), the late and early penalties for all the jobs (l and e), are 1 and 0, respectively.

The data is represented by Figure 11.2, and the CPM procedure is applied to the network. The numbers within square brackets "[ ]" represent the earliest start time for that node or operation. The numbers within flower braces "{ }" represent the latest start time for that node or operation. The earliest or latest start time for the END node denotes the makespan C_{max}.

For example, for machine 1, (see Figure 11.2) the processing times for jobs 1, 3, 4, and 5 are 1, 1, 3, and 1, respectively, and their arrival times are 0, 2, 0, and 2, respectively. The due date for operation 1/1 is 4, the latest time for node 1/3, since there is only one branch leading from 1/1 that goes to the next eminent node 1/3. Similarly, the due dates for jobs 3, 4, and 5 are 5, 3, and 6, respectively. Initially, $M = \{1, 2, 3\}$.

Iteration 1

In the first cycle, the set M_r is empty; therefore, we have to determine the minimum tardiness for all the machines in set M. Since we have assigned the tardiness penalty

TABLE 11.6
Data for Machine 1

Jobs	1	3	4	5
p_i	1	1	3	1
r_i	0	2	0	2
d_i	4	5	3	6
l_i	1	1	1	1
e_i	0	0	0	0

TABLE 11.7
Data for Machine 2

Jobs	2	3	4
p_i	3	1	2
r_i	0	3	3
d_i	5	6	5
l_i	1	1	1
e_i	0	0	0

TABLE 11.8
Data for Machine 3

Jobs	1	2	3	4	5
p_i	2	1	2	1	2
r_i	1	3	0	5	0
d_i	6	6	4	6	5
l_i	1	1	1	1	1
e_i	0	0	0	0	0

of 1/unit of tardiness, for all jobs, minimization of tardiness penalty is equivalent to the minimization of the tardiness. We will apply the procedure in Section 7.15 to determine the penalty value for each machine. The data for each machine is obtained from the CPM application on Figure 11.2. For each machine, the optimum sequence and total tardiness are shown next (Tables 11.6, 11.7, and 11.8).
Optimal sequence for machine 1: 4-1-3-5; penalty = 0.
Optimal sequence for machine 2: 2-4-3; penalty = 0.
Optimal sequence for machine 3 is 5-3-2-4-1; penalty = 2.

Choose the machine that has the maximum penalty as the bottleneck machine. In this case, machine 3 has the maximum penalty, so it is the bottleneck machine. Using

TABLE 11.9
Data for Machine 1

Jobs	1	3	4	5
p_i	1	1	3	1
r_i	0	4	0	2
d_i	6	7	3	8
l_i	1	1	1	1
e_i	0	0	0	0

TABLE 11.10
Data for Machine 2

Jobs	2	3	4
p_i	3	1	2
r_i	0	5	3
d_i	4	8	5
l_i	1	1	1
e_i	0	0	0

the optimal sequence for the bottleneck machine as the guide for the precedence, we connect the operations performed on machine 3 by arcs. We also note the processing time for each operation on the corresponding arc as shown by the modified network in Figure 11.3. Add machine 3 to M_r. Now, $M_r = \{3\}$. Recompute the earliest and latest start times for the entire network using Figure 11.3. The new C_{max} is (8).

Iteration 2

Since $M - M_r = \{1, 2\}$, we develop the optimum sequences for machines 1 and 2 that would minimize the tardiness penalty. The corresponding p_i, r_i, d_i values are taken from Figure 11.3 (Tables 11.9 and 11.10).

Optimal sequence for machine 1: 4-1-3-5; penalty = 0.

Optimal sequence for machine 2: 2-4-3; penalty = 0.

Choose either machine 1 or machine 2 as the bottleneck machine, since they have the same penalty. Let us choose machine 2 at random. Using the optimum sequence for machine 2, modify the network as shown in Figure 11.4, by connecting in sequence the operations associated with machine 2. Recompute the earliest and latest start times on Figure 11.4. C_{max} remains 8.

At this stage, for the application of step 5, $M_r = \{3\}$. The optimum sequence for machine 3 in the last iteration was 5-3-2-4-1, shown by the connected arcs in Figure 11.4. The network is now modified by removing machine 3 connections (all solid lines between jobs on machine 3 are removed), as shown in Figure 11.5. Recompute the early and late start times on Figure 11.5. The data from the solution is used as the input data for machine 3 reevaluation (Table 11.11).

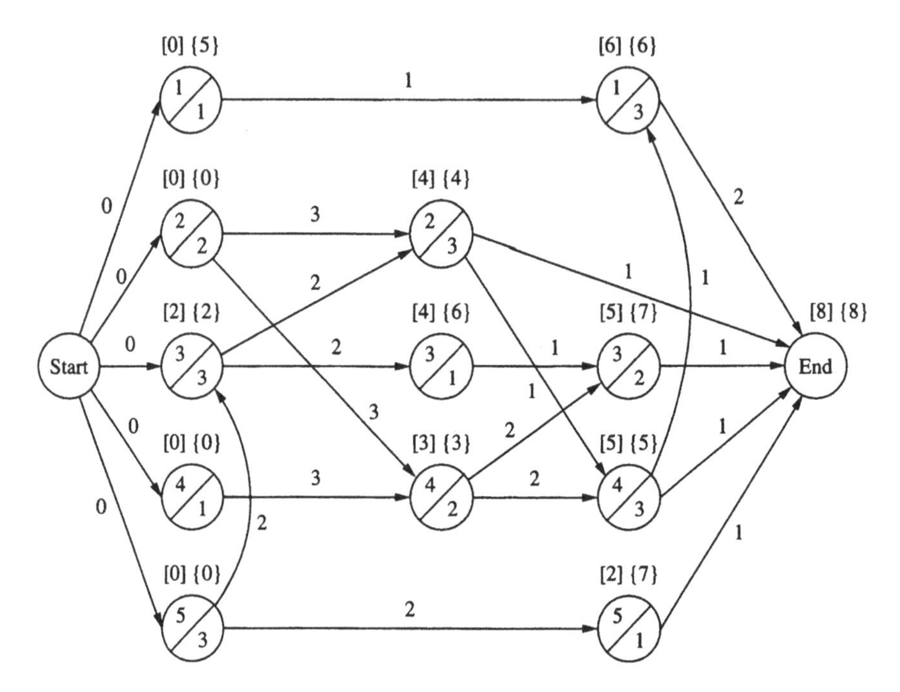

FIGURE 11.4 Machine 2 and 3 solved; Iteration 2.

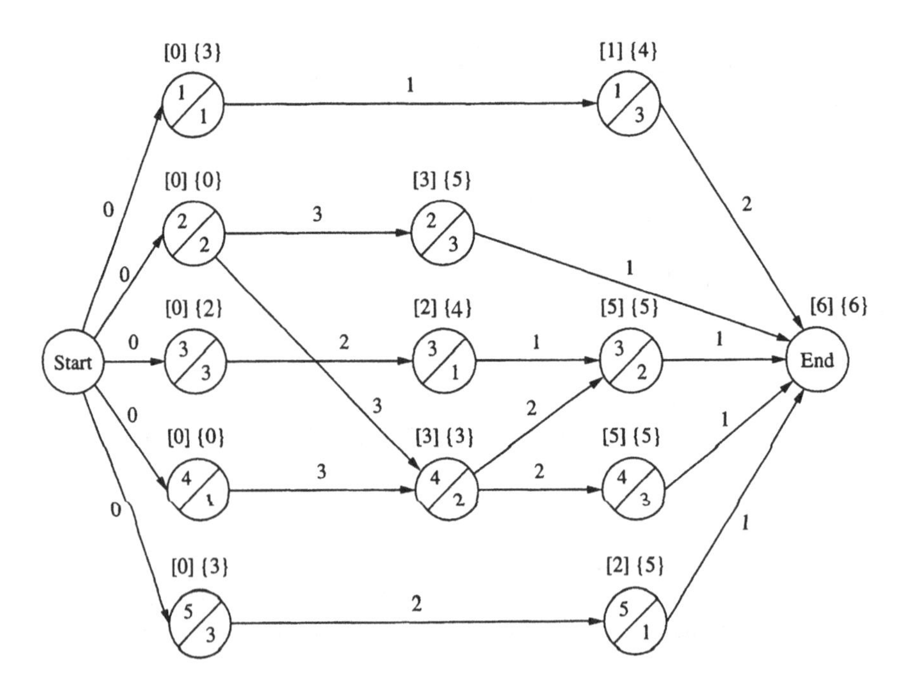

FIGURE 11.5 Machine 2 solved; Machine 3 removed; Iteration 2.

TABLE 11.11
Data for Machine 3

Jobs	1	2	3	4	5
p_i	2	1	2	1	2
r_i	1	3	0	5	0
di	6	6	4	6	5
l_i	1	1	1	1	1
e_i	0	0	0	0	0

TABLE 11.12
Data for Machine 1

Jobs	1	3	4	5
p_i	1	1	3	1
r_i	0	4	0	2
d_i	6	7	3	8
l_i	1	1	1	1
e_i	0	0	0	0

Optimal sequence: 5-3-2-4-1; penalty = 2.

Since we get the same job sequence as before, we join the operations on machine 3 again, and our network remains the same as in Figure 11.4. There is no improvement (i.e., C_{max} is still 8). Add machine 2 to M_r. Now, $M_r = \{2, 3\}$.

Iteration 3

Since machine 1 is the only machine left in M-M_r, it is solved for minimum total penalty. The data for machine 1 is taken from Figure 11.4 (Table 11.12).
The optimal sequence: 4-1-3-5; penalty = 0.

Modify the network as shown in Figure 11.6 by adding solid lines between operations for machine 1. Recompute the earliest and latest start times. C_{max} remains 8. At this stage, since $M_r = \{2, 3\}$, we resequence machines 2 and 3 for minimum total penalty and check for any improvement. This is done by removing the solid lines between operations for machine 3 first as shown Figure 11.7 and determining the minimum penalty for machine 3 from the resulting condition (Table 11.13).
Optimal sequence on machine 3 is 5-3-2-4-1; penalty = 2.

Since the job sequence does not change, the makespan remains 8. We go back to the original network in Figure 11.6. Now, remove the solid lines between operations for machine 2 as shown in Figure 11.8. Recompute the earliest and latest start times. Machine 2 is now solved for minimum penalty using the data for machine 2 from Figure 11.8 (Table 11.14).
Optimal sequence on machine 2 is 2-3-4; penalty = 0.

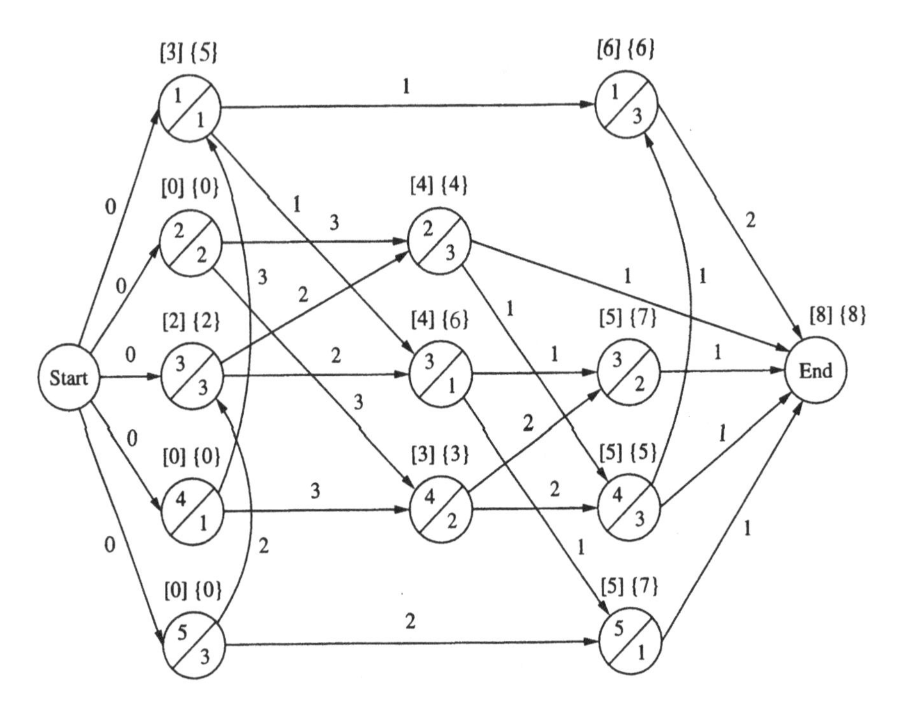

FIGURE 11.6 Machine 1, 2, and 3 solved; Iteration 3.

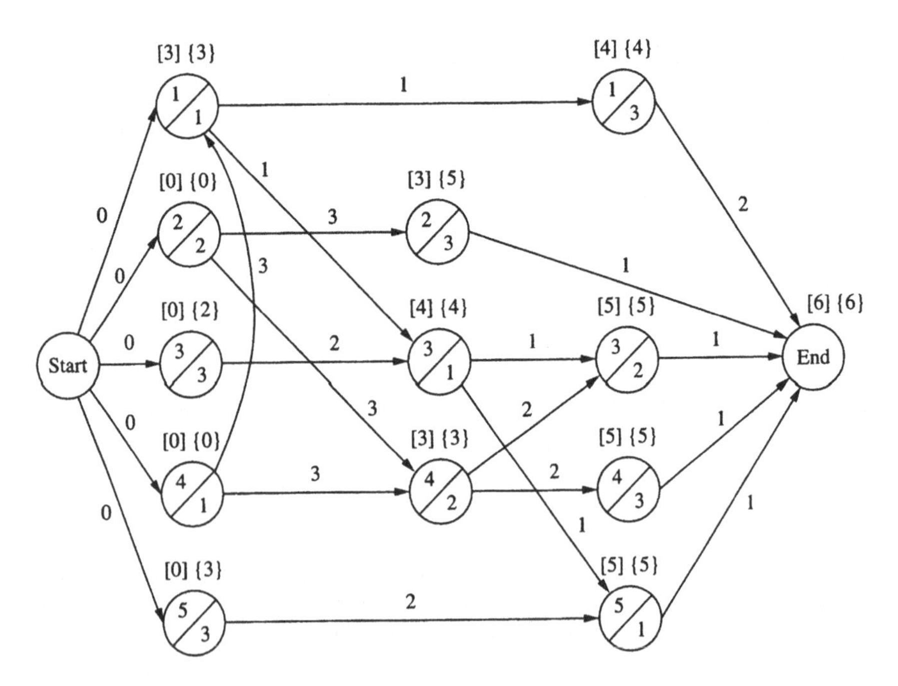

FIGURE 11.7 Machine 1 and 2 solved; Machine 3 removed; Iteration 3.

TABLE 11.13
Data for Machine 3

Jobs	1	2	3	4	5
p_i	2	1	2	1	2
r_i	4	3	0	5	0
Di	6	6	4	6	5
l_i	1	1	1	1	1
e_i	0	0	0	0	0

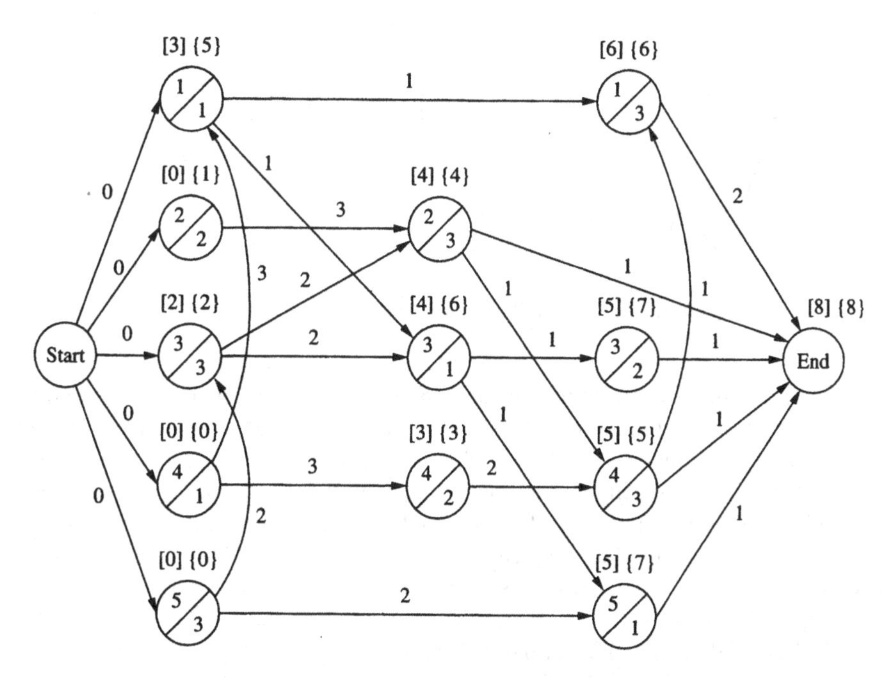

FIGURE 11.8 Machine 1 and 3 solved; Machine 2 removed; Iteration 3.

TABLE 11.14
Data for Machine 2

Jobs	2	3	4
p_i	3	1	2
r_i	0	5	3
d_i	4	8	5
l_i	1	1	1
e_i	0	0	0

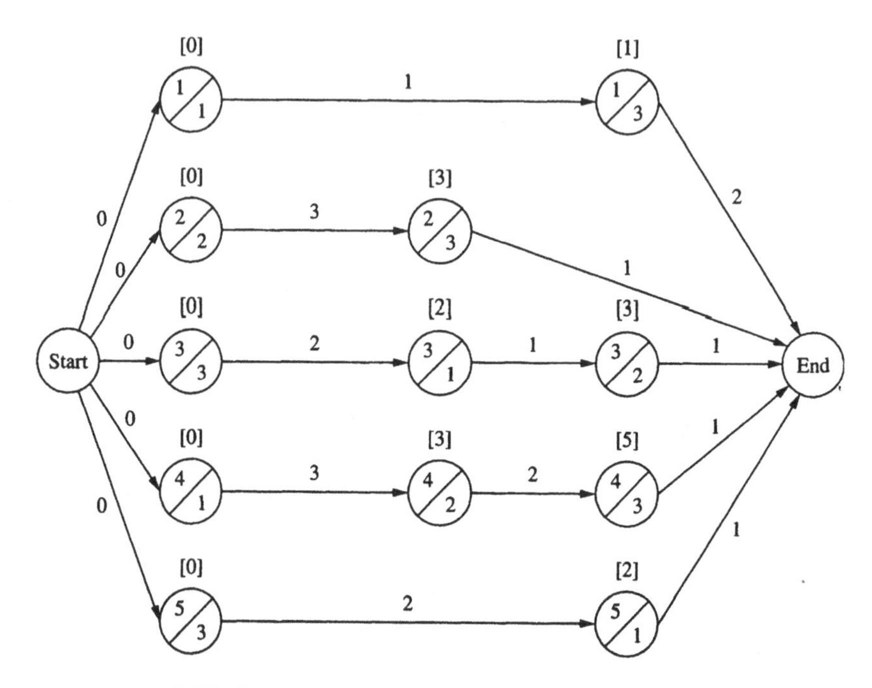

FIGURE 11.9 Initial network; Machine 1 to be resolved.

Again, the sequence does not change, and the makespan remains 8. We modify the network as it was in Figure 11.6. All machines are in set M_r so we have achieved the optimum job sequences for each machine in Figure 11.6. The optimum makespan C_{max} is 8, which is the completion time of the network in Figure 11.6. The sequence for each machine is obtained by following job—sequence-connecting arcs for that machine. In our example, the sequences are machine 1, 4-1-3-5; machine 2, 2-4-3; and machine 3, 5-3-2-4-1. The corresponding machine loadings are as follows.

Machine 1: 0 (4/3) 3 (1/1) 4 (3/1) 5 (5/1) 6
Machine 2: 0 (2/3) 3 (4/2) 5 (3/1) 6
Machine 3: 0 (5/2) 2 (3/2) 4 (2/1) 5 (4/1) 6 (1/2) 8

11.3.2 Two-Stage Job Shop Scheduling Heuristic

The next heuristic presented here is developed by Vancheeswaran and Townsend [1993]. It also has a goal to minimize the makespan of a job shop schedule. This is a two-stage heuristic; the first stage develops the initial feasible near-optimum sequence, and the second stage improves it. The procedure starts with network representation of all jobs similar to Figure 11.2, shown here as Figure 11.9. There are two phases to the procedure. In the first phase, the preliminary sequence of jobs on each machine is established by using an "urgency criteria." In the second phase, the sequences may be improved by analyzing the "critical jobs."

Phase 1

The steps of the phase 1 are as follows:

1. Calculate earliest times of all nodes in the network.
2. Machine-by-machine, calculate the urgency criterion for all the jobs. Start with machine 1.

   Urgency criterion for job i is defined as job remaining/(job done+PT_{pi}) where

$$\text{Job remaining} = (ES_{li} + PT_{li}) - ESpi$$

$$\text{Job done} = ESei + PTei$$

   and

$$ES_{li} = \text{Earliest start time for the last node of job i}$$

$$PT_{li} = \text{Processing time for the last node of job i}$$

$$ES_{pi} = \text{Earliest start time for the present node of job i}$$

$$PT_{pi} = \text{Processing time for the present node of job i}$$

$$ES_{ei} = \text{Earliest start time for the earlier node of job i}$$

$$PT_{ei} = \text{Processing time for the earlier node of job i}$$

3. Arrange jobs based on the descending order of the urgency criterion. This gives the job sequence on the machine under consideration.
4. Modify the network. This is done by connecting the nodes for the machine by arcs in the direction shown by the job sequence established in step 4.
5. Recalculate the earliest start times.
6. Choose the next machine and repeat steps 3–7 until all machines are sequenced.
7. The earliest start time of the END node would be the makespan, and the job(s) that has its path equal to the earliest start time of the END node would be the critical job (the job that forms the critical path is the critical job).

Phase 2

1. Draw up the rank matrix. This is done by ranking jobs on a machine according to their position in the sequence on that machine.
2. For the critical job, determine on which machine it has the highest rank.
3. On this machine, swap the critical job with the job that has one lower rank in the list.

TABLE 11.15
Job Shop Example Data

Jobs	Machine Sequence	Processing Times
1	1, 3	1, 2
2	2, 3	3, 1
3	3, 1, 2	2, 1, 1
4	1, 2, 3	3, 2, 1
5	3, 1	2, 1

4. Redraw the network. Recalculate the earliest start time and hence, the makespan based on the modification performed in step 4.

The following rules apply thereafter:

1. If the new makespan is lower than the earlier makespan, then this is the new schedule on the machine where jobs were swapped. Repeat steps 2 through 5.
2. If the new makespan is higher than the present makespan, revert back to the earlier schedule and repeat steps 2 through 5 with the next highest rank for the critical job (it need not be on the same machine where the earlier calculations were made).
3. If the new makespan is higher than OR equal to the present makespan, AND if the critical job does NOT change, revert back to the earlier schedule and repeat steps 2 through 5 with the next highest rank for the critical job.
4. If the new makespan is EQUAL to the present makespan, AND if the critical job CHANGES, keeping this job as the new critical job, repeat steps 2 through 5. Sometimes, this might result in an infinite loop. If that happens, the iterations are stopped using one of the following rules:
   a. After a fixed number of iterations
   b. If the critical job is preformed first on every machine

11.3.2.1 Illustrative Example

We shall continue with the same example as shown earlier in this chapter. The data has been reproduced here in Table 11.15 for convenience.

Phase 1

The data in Table 11.15 is transferred as a network. The network is shown in Figure 11.9. Processing times are shown on the links connecting the nodes. Earliest start times are calculated and shown in square brackets on top of the node.

Urgency criterion for the jobs on machine 1 are calculated as follows:

Job 1: $((1+2)-0)/(0+1) = 3.0$
Job 3: $((3+1)-2)/(0+2+1) = 0.66$

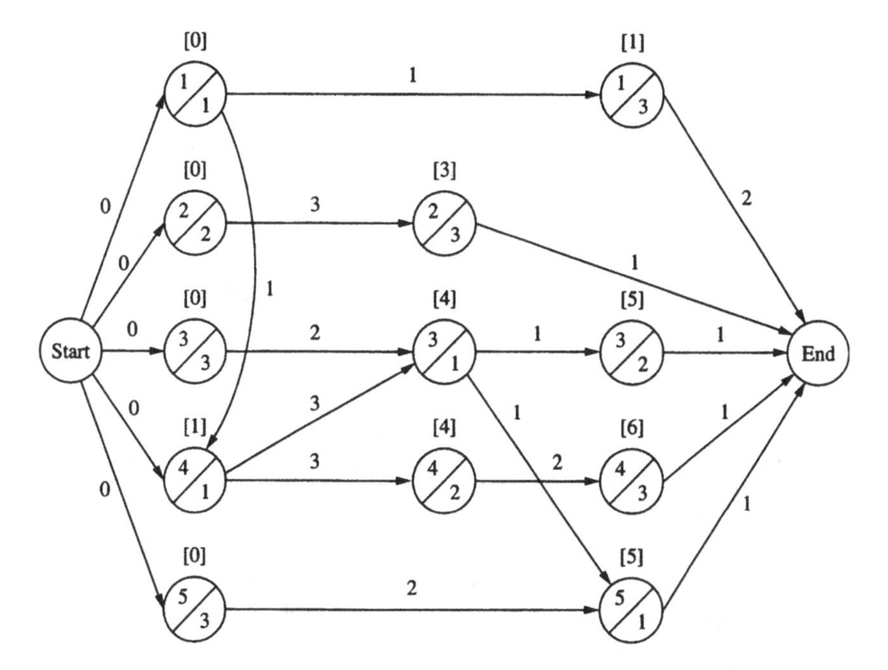

FIGURE 11.10 Phase 1, Machine 1 resolved; Machine 2 to be resolved.

Job 4: $((5+1)-0)/(0+3)=2.0$
Job 5: $((2+1)-2)/(0+2+1)=0.33$

Arranging the jobs based on the descending order of urgency criteria, we get the job sequence of 1-4-3-5.

The network is modified by connecting nodes of machine 1 by solid lines, with the arrows showing the precedence relationship corresponding to the sequence. This is shown in Figure 11.10. Recalculate the earliest start times for all the nodes as shown in [ ] in Figure 11.10. The next step is to calculate the urgency criteria for all the jobs on the next machine in the order. The urgency criteria calculations for the jobs on machine 2 are as follows:

Job 2: $((3+1)-0)/(0+3)=1.33$
Job 3: $((5+1)-5)/(4+1+1)=0.16$
Job 4: $((6+1)-4)/(1+3+2)=0.50$

Arranging the urgency criteria in the descending order, the job sequence for machine 2 is, 2-4-3. The associated modified network is shown in Figure 11.11. The network is solved, and the earliest start times for all nodes are determined. The urgency criteria calculations for the next machine, machine 3, are as follows:

Job 1: $((1+2)-1)/(0+1+2)=0.66$
Job 2: $((3+1)-3)/(0+3+1)=0.25$

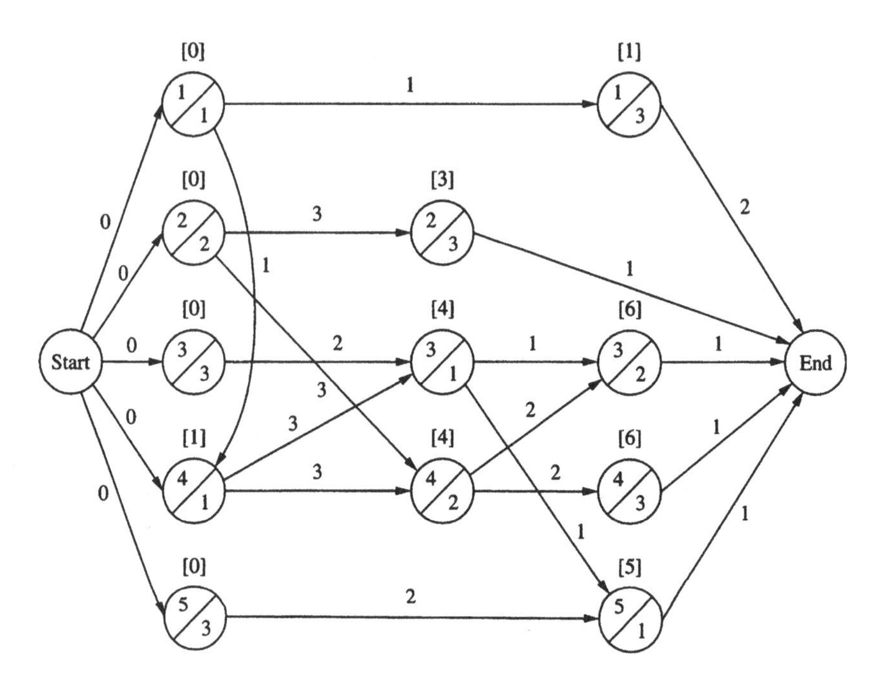

FIGURE 11.11 Phase 1, Machines 1 and 2 resolved; Machine 3 to be resolved.

Job 3: $((6+1)-0)/(0+2) = 3.50$
Job 4: $((6+1)-6)/(4+2+1) = 0.14$
Job 5: $((5+1)-0)/(0+2) = 3.0$

Arranging the urgency criteria in a descending order, the sequence on machine 3 is 3-5-1-2-4. The modified network is shown in Figure 11.12.

Since the sequences for all machines have been obtained, the earliest start time of the last node in Figure 11.12 is the makespan. In this case, the makespan is 8. The job with the longest path, that is, the critical job, is job 4. This is because the last node of path for job 4 is machine 3 with early time of 7. Add to that the processing time of job 4 on machine 3, 1 makes the value of critical path 8.

Phase 2

Note: This phase is unnecessary if the critical job is processed first on all machines. The rank for a critical job is 1.

Draw the rank matrix. This is done by ranking jobs on a machine according to their sequence. For example, the job sequence for machine 1 is 1-4-3-5. Hence, as noted in Table 11.16, the rank for job 1 is 1, the rank for job 4 is 2, the rank for job 3 is 3, and so on. Similarly, the ranks are assigned to jobs on other machines.

Pick the highest rank for the critical job. In this case, the critical job is 4, and the highest rank is 5 on machine 3. On this machine, swap the critical job with the job with the next lower ranking. In this case, job 4 is swapped with job 2 on machine 3.

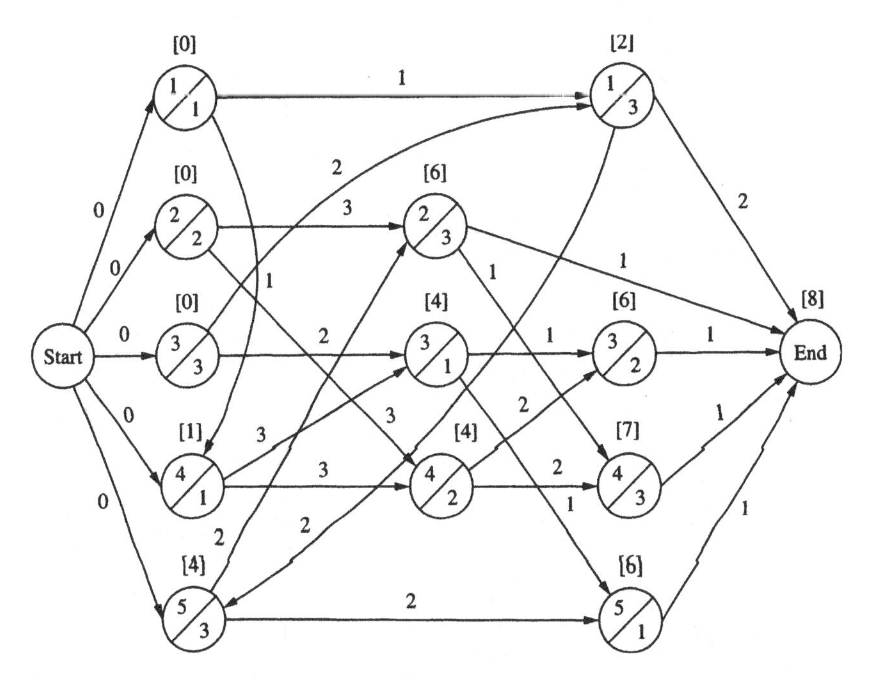

FIGURE 11.12 Phase 1, All Machines resolved; Critical job 4; Makespan=8.

TABLE 11.16
Job Ranks on Machines

	Machine 1	Machine 2	Machine 3
Job 1	1	—	3
Job 2	—	1	4
Job 3	3	3	1
Job 4	2	2	5
Job 5	4	—	2

Redraw the network (Figure 11.13). Recalculate the earliest start times and the makespan. In our example, in Figure 11.13, the makespan remains unchanged at 8, but the critical job changes from job 4 to job 2.The ranking of jobs after swap is shown in Table 11.17. The highest rank on the critical job (job 2) is 5, on machine 3. The critical job is swapped with the one with the next lower ranking. In this case, job 4 is swapped with job 2. Earlier, we had swapped jobs 2 and 4, so this swap should lead us back to the solution in Table 11.16. This is an example of a loop; therefore, the procedure is stopped with selecting sequences on each machine from either Table 11.16 or 11.17.

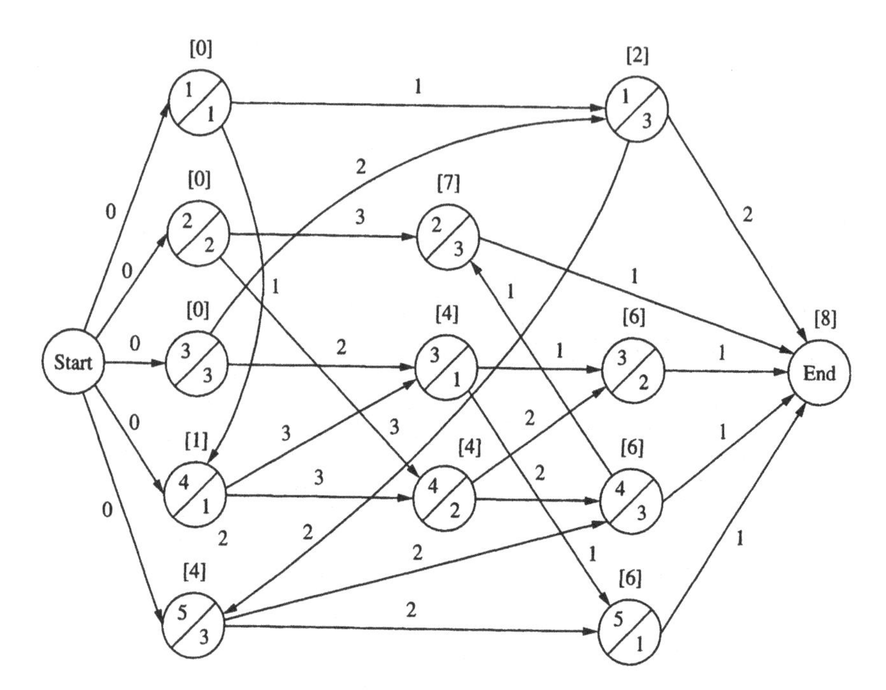

FIGURE 11.13 Phase 2, Critical job changes; Network changes; Makespan=8.

TABLE 11.17
Job Ranks with Critical Job Swapped

	Machine 1	Machine 2	Machine 3
Job 1	1	—	3
Job 2	—	1	5
Job 3	3	3	1
Job 4	2	2	4
Job 5	4	—	2

11.4 JOB SHOP SCHEDULING TO MINIMIZE TARDINESS

We now have for each job a known promised delivery date. If a job is completed after its due date, it is considered late. If the job is completed before the due date, it is considered to be on time. There is no penalty attached for early completion, while there is a penalty for late completion that is proportional to the amount of delay.

The objective is to determine, for each machine, the job sequence, so that all the jobs can be completed within the due date. If this is not possible, then the objective is to complete all the jobs with minimum total delay, that is, minimum sum of positive deviation for all the jobs from their respective due dates.

A number of approaches have been developed to address this problem. Since SPT has been found to be very effective in minimization of makespan, many approaches for tardiness minimization try to take advantage of the SPT rule in a due date environment. The approach is to develop different queues based on some criteria and then select a job from a specific queue based on SPT. Six-rule (Eilon and Cotterill, 1968, Eilon and Chowdhury, 1975), for example, develops two queues. A job whose slack time (calculated as remaining time minus remaining work) is less than a control parameter U is placed in the priority queue, while others are placed in non-priority queue. A job is selected from the priority queue based on SPT; however, if no job is available in the priority queue, then a job from the non-priority queue is selected based on SPT. A similar logic is employed by Baker and Kanet (1983) in the MOD rule synonym for modified operation due date. The MOD value for a job on the imminently needed machine is selected as the larger of two values: original operation due date (explained later in the chapter) and its earliest operation completion time (calculated as the current time plus the operation time). For each job, the MOD is calculated on the machine job needed next in each time period, and a job is processed using the SPT rule on the MOD values.

The CEXSPT (stands for conditionally expedited by SPT) approach proposed by Schultz [20] is an efficient strategy to minimize tardiness. The procedure is also not too difficult to apply, and it is illustrated in the following.

11.4.1 The CEXSPT Rule

The CEXSPT rule tries to balance between the SPT rule, which may make jobs with long processing times late, and rules that are directly related to tardiness, which often gives priority to jobs with large processing times, making a large number of jobs with small processing times late.

The rule divides jobs available for scheduling on a machine in three queues.

Queue 1 consists of jobs that are late, that is, those jobs that are already behind schedule.

Queue 2 consists of jobs that are operationally behind schedule. The job is operationally behind schedule if the present date is past the milestone date for the particular operation. The milestone date is set by dividing the available slack equally for all the operations, and then adding the previous operation time. This gives an indication of the theoretical latest starting date for the operation, such that the job has a good chance of being on time.

Queue 3 consists of jobs that are ahead of or on time.

The available job selected with the SPT rule from queue 1 is processed next, unless doing so creates at least one new late job, in which case the available job selected with the SPT rule from queue 2 is processed next, provided it does not create a new operationally late job. If it does, then an available job with the SPT rule from queue 3 is processed next. Jobs that are tied with respect to processing time, than select a job amongst tied jobs, based on operational due date.

TABLE 11.18
Data for CEXSPT Rule Application

Job	Due Date	Machine Sequence	Processing Time	Operational Due Date
1	5	1, 3	1, 2	2, 5
2	10	2, 3	3, 1	6, 10
3	5	3, 1, 2	2, 1, 1	3, 4, 5
4	8	1, 2, 3	3, 2, 1	4, 7, 8
5	15	3, 1	2, 1	8, 15

The CEXSPT rule is illustrated next by applying it to the modified data from Table 11.2. Table 11.18 shows the data along with the due dates and operational due dates for the jobs.

An operational due date is the estimated time when an operation should be completed for a job to remain on schedule. It includes the processing times of all prior operations on the job, plus the slack time that may have been allowed so far, plus the processing and slack time for the present operation. Generally, to calculate operational due date, the total slack (defined as due date - total operational time) is distributed equally (as far as possible) among the number of operations associated with the job. For example, for job 2, the total slack is equal to $(10 - (3 + 1)) = 6$. It is equally distributed between two operations, giving the operational due date of the first operation, that is, operation 2 as: process time + slack $= 3 + 3 = 6$ and of operation 3 as: previous operational due date + present operation time + slack in the operation $= 6 + 1 + 3 = 10$.

It might be of interest to examine operation (job) 3. The due date is 5, and the total processing time required is $2 + 1 + 1 = 4$. Therefore, there is only one unit of slack in the operation. Assuming we are working in an integer unit, the slack could be distributed in any one operation. We choose to assign the slack to the first task for the job, operation 3.

The schedule based on the CEXSPT rule is shown in Table 11.19. The evaluation part shows the status of different queues Q1, Q2, and Q3 for each machine. The entries in each queue represent the job in that, along with its required processing time on the machine. Q1 lists the jobs that are late, Q2 lists the jobs that are operationally late, and Q3 list the jobs that are operationally ahead or operationally on time.

The second part of the table lists the job schedule on each machine M1, M2, and M3.

At time 1, jobs 1 and 4 are available for machine 1, job 2 for machine 2, and jobs 3 and 5 for machine 3. They are all ahead of their operational due dates and are placed in their respective Q3s. All machines are available. A job from each queue is assigned to the respective machine based on the SPT rule. The procedure is continued for time 2 and 3. At time 4, machine 2 is free, but no job requiring machine is available, and therefore the machine is idle. At time 5, machine 1 is free, and jobs 3 and 5 can be processed. The operational due date for job 3 on machine 1 is 4, and therefore the job is behind schedule. It is, however, not late since the due date for the job is 5. Job 3

TABLE 11.19
Application of CEXSPT

	Evaluations											
	Machine 1			Machine 2			Machine 3			Schedule		
Time	Q1	Q2	Q3	Q1	Q2	Q3	Q1	Q2	Q3	M1	M2	M3
1			1/1, 4/3			2/3			3/2, 5/2	1	2	3
2			4/3							4	2	3
3									1/2, 5/2	4	2	1
4										4	—	1
5		3/1				4/2			2/1, 5/2	3	4	2
6									5/2	—	4	5
7				3/1						—	3	5
8			5/1						4/1	5	—	4

TABLE 11.20
Job Completion Times and Due Dates

Job	1	2	3	4	5
Completion time	4	5	7	8	8
Due date	5	10	5	8	15

is placed in Q, and that being the only job in Q, is selected next. At time 6, machine 1 is free, but there is no eligible job available. Job 5, which needs machine 1, must be processed in machine 3 first and cannot be assigned to machine 1. All the jobs are processed by time 8, giving a makespan of 8. The completion times of the jobs and their due dates are shown in Table 11.20. The last operation completion time of a job gives its completion time. The total delay in the schedule is for 2 time units. The Gantt chart is given in Figure 7.14.

11.4.2 Minimizing Penalty Using Modified Shifting Bottleneck Procedure

Consider the same example as shown in Table 11.1, except now we add the due date and late and early penalties for each job. The data are shown in Table 11.21.

In the example shown in Table 11.1, three additional parameters (due date and late and early penalties) are introduced with each job. The objective here is to minimize the total penalty, that is, if the last operation of job i finishes at time T (the sum of T_E of the last operation and its processing time), then its penalty is given by,

If $T - D_i > 0$, then penalty $= (T - D_i) \times L_i$; else penalty $= (D_i - T) * E_i$, where D_i, L_i, and E_i are the due date and late and early penalties for job I, respectively. The sum of penalties for all the jobs is the total penalty.

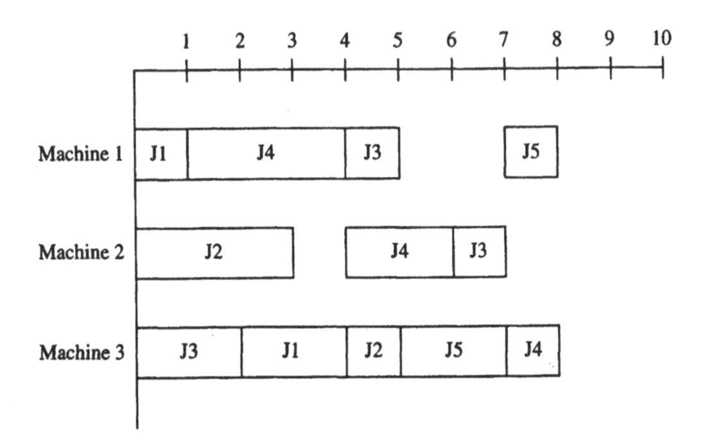

FIGURE 11.14 Illustration of a daily schedule using the CEXSPT rule.

TABLE 11.21
Data for MODSB Penalty Minimization with Due Date and Late and Early Penalties

Jobs	Machine Sequence	Processing Times	Due date	Late Penalty	Early Penalty
1	1, 3	1, 2	5	3	1
2	2, 3	3, 1	7	2	1
3	3, 1, 2	2, 1, 1	6	5	2
4	1, 2, 3	3, 2, 1	8	1	4
5	3, 1	2, 1	4	2	1

For example, in Figure 11.15 the last operation for job 1 is on machine 3. It can start at time 1 (T_E) and has a processing time of 2 units. Therefore, its completion time T is $1+2=3$. Since the due date for job 1 is 5, the penalty on this job is $(5-3)\times 1=2$.

The MODSB procedure is now modified to include different penalties for different jobs. All the steps in the procedure remain the same except for two changes. When solving individual machines for minimum penalty sequence, the early and late penalties of corresponding jobs are used. In the example shown in Table 11.21, the late and early penalties for job 1 are 3 and 1, respectively.

The second modification is associated with calculations of the due date for each operation. We shall now call this due date an operational due date d_i for operation I. These values are only determined once (at the beginning) in the entire procedure.

The operational due dates are determined in the same manner as was described for the CEXSPT rule. An operational due date is the estimated time when an operation should be completed for a job to remain on schedule. It includes the processing times of all prior operations on the job plus the slack time that may have been allowed so far plus the processing and slack times for the present operation.

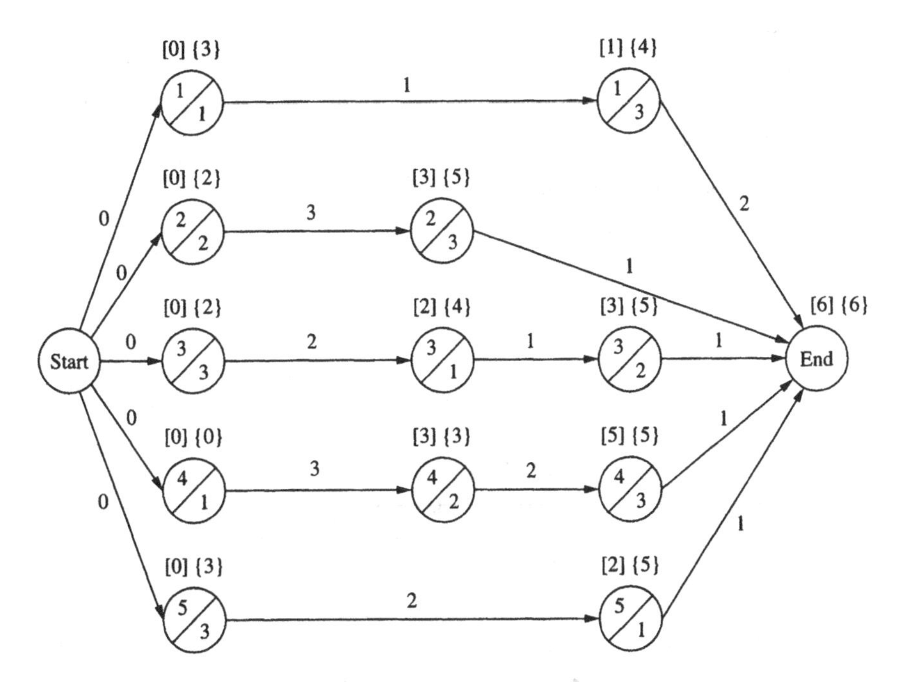

FIGURE 11.15 Job shop problem with earliest and latest start times (last operations for all the jobs shown in thick circles).

To calculate the operational due date, the total slack (defined as due date − total processing time) is distributed equally among the number of operations associated with the job. For example, for job 2, the total slack is equal to $(7 - (3 + 1)) = 3$. It is equally distributed between two operations (1.5 each), giving the operational due date of the first operation, that is, operation 2 as: process time + *slack* $= 3 + 1.5 = 4.5$ and of operation 3 as: previous operational due date + present operation time + slack in the operation $= 4.5 + 1 + 1.5 = 7$. Unlike the CEXSPT rule, this method considers both integer as well as fractional numbers for operational due dates as in this case, that is, 4.5 and 7 for operations 2 and 3, respectively. Table 11.22 shows the operational due dates for all the operations.

We now solve the above-mentioned problem based on the modified procedure.

Iteration 1

Initially, the set M_r is an empty set; therefore, we have to solve all the machines in M for the minimum total penalty. Compute the early and late start times for the initial network in Figure 11.15. The corresponding values for p_i and r_i are taken from Figure 11.15. Operational due dates d_i are taken from Table 11.22. For example, for job 1, operation 1, the operational due date is 2 (Tables 11.23, 11.24, and 11.25).

Optimal sequence on machine 1 is 1-5-3-4; penalty = 6.
Optimal sequence on machine 2 is 2-4-3; penalty = 6.8.
Optimal sequence on machine 3 is 3-5-1-2-4; penalty = 7.3.

TABLE 11.22
Data Showing Operational Due Dates for the MODSB Penalty Minimization

Jobs	Machine Sequence	Processing Times	Due date	Operational Due Date
1	1, 3	1, 2	5	2, 5
2	2, 3	3, 1	7	4.5, 7
3	3, 1, 2	2, 1, 1	6	2.7, 4.4, 6
4	1, 2, 3	3, 2, 1	8	3.7 ,6.4, 8
5	3, 1	2, 1	4	2.5, 4

TABLE 11.23
Data for Machine 1

Jobs	1	3	4	5
p_i	1	1	3	1
r_i	0	2	0	2
d_i	2	4.4	3.7	4
l_i	3	5	1	2
e_i	1	2	4	1

TABLE 11.24
Data for Machine 2

Jobs	2	3	4
p_i	3	1	2
r_i	0	3	3
d_i	4.5	6	6.4
l_i	2	5	1
e_i	1	2	4

TABLE 11.25
Data for Machine 3

Jobs	1	2	3	4	5
p_i	2	1	2	1	2
r_i	1	3	0	5	0
d_i	5	7	2.7	8	2.5
l_i	3	2	5	1	2
e_i	1	1	2	4	1

Choose the machine that has the maximum penalty. In this case, machine 3 has the maximum penalty, so we choose machine 3. Modify the network by adding solid lines between operations of machine 3, as shown in Figure 11.16. Recompute the early and late start times. C_{max} is 8. Add machine 3 to M_r. Now, $M_r = \{3\}$. The total penalty for the partial schedule (all jobs have not been processed by all machines) is 9 as shown in the following.

Job	Due Date	Completion Time	Early Penalty	Late Penalty
1	5	6		3
2	7	7		
3	6	4	4	
4	8	8		
5	4	5		2

Iteration 2

Since M - $M_r = \{1, 2\}$, we solve machine 1 and 2 for minimum total penalty. The corresponding p_i and r_i, are taken from Figure 11.16. The corresponding operational due dates d_i are taken from Table 11.22.

Optimal sequence on machine 1 is 1-4-3-5; penalty = 8.7.

Optimal sequence on machine 2 is 2-4-3; penalty = 6.8.

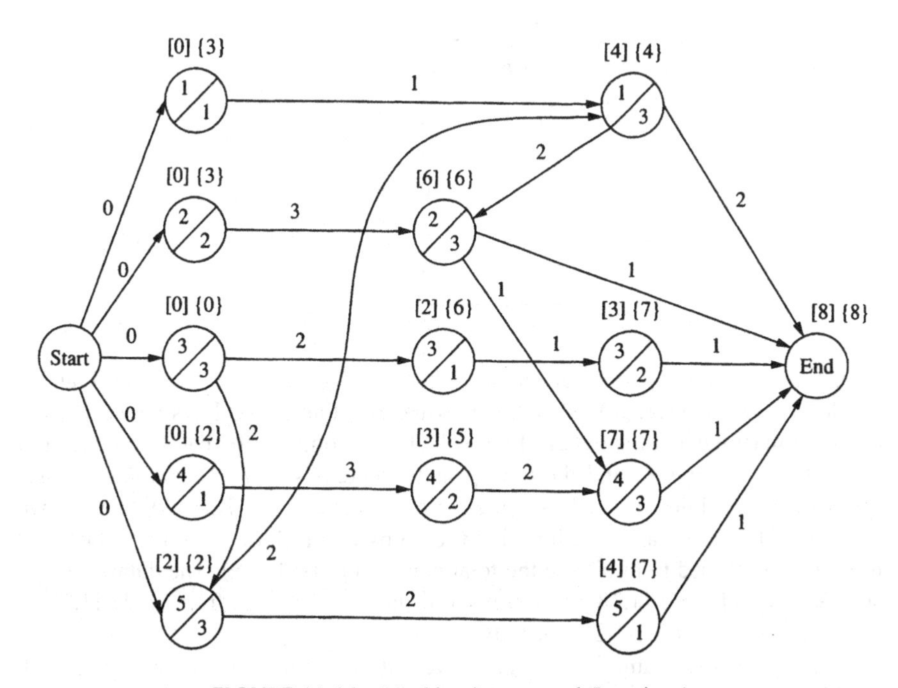

FIGURE 11.16 Machine 3 connected; Iteration 1.

TABLE 11.26
Data for Machine 1

Jobs	1	3	4	5
p_i	1	1	3	1
r_i	0	2	0	4
d_i	2	4.4	3.7	4
l_i	3	5	1	2
e_i	1	2	4	1

TABLE 11.27
Data for Machine 2

Jobs	2	3	4
p_i	3	1	2
r_i	0	3	3
d_i	4.5	6	6.4
l_i	2	5	1
e_i	1	2	4

TABLE 11.28
Data for Machine 3

Jobs	1	2	3	4	5
p_i	2	1	2	1	2
r_i	1	3	0	6	0
d_i	5	7	2.7	8	2.5
l_i	3	2	5	1	2
e_i	1	1	2	4	1

Choose machine 1 since it has a higher penalty between machines 1 and 2 (Tables 11.26 and 11.27). Modify the network by adding solid lines between operations of machine 1, as shown in Figure 11.17. Recompute the early and late start times. C_{max} remains 8, and the total penalty decreases to 7. At this stage, since $M_r = \{3\}$, operations for machine 3 are removed, the network is modified as shown in Figure 11.17 (i.e., all solid lines between jobs on machine 3 are removed), and machine 3 is solved to minimize the total penalty (Table 11.28). The data for r_i, p_i, are taken from Figure 11.18. Operational due dates d_i are taken from Table 11.22. Optimal sequence is 3-5-1-2-4; penalty = 7.3.

Since we get the same job sequence, we join the operations on machine 3 with solid lines again, and our network remains the same as in Figure 11.17. There is no

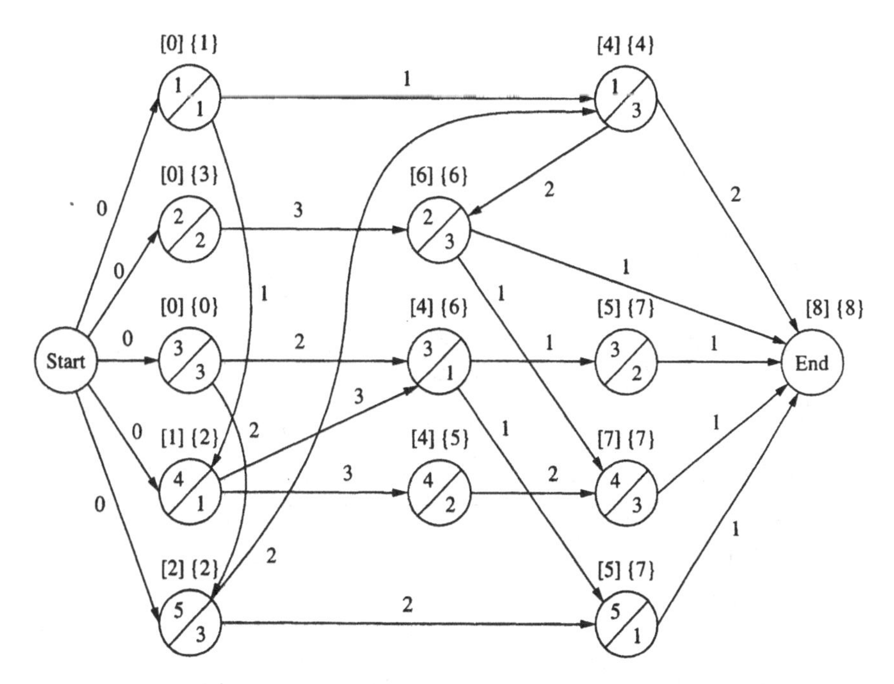

FIGURE 11.17 Machines 1 and 3 connected; Iteration 2.

TABLE 11.29
Data for Machine 2

Jobs	2	3	4
p_i	3	1	2
r_i	0	5	4
d_i	4.5	6	6.4
l_i	2	5	1
e_i	1	2	4

improvement (i.e., C_{max} is still 8, and the total penalty is 7). Add machine 1 to M_r. Now, $M_r = \{1, 3\}$.

Iteration 3

Since machine 2 is the only machine left in $M - M_r$, it is solved for minimum total penalty (Table 11.29). The data for machine 2 is taken from Figure 11.16 and Table 11.22.

Optimal sequence on machine 2 is 2-3-4; penalty = 3.7.

Modify the network by joining with solid lines the sequence of operations of machine 2, as shown in Figure 11.19. Recompute the early and late start times. C_{max} increases to 9, and the total penalty increases to 8. At this stage, since $M_r = \{1, 3\}$ we resequence machine 3 and 1 for minimum total penalty and check for

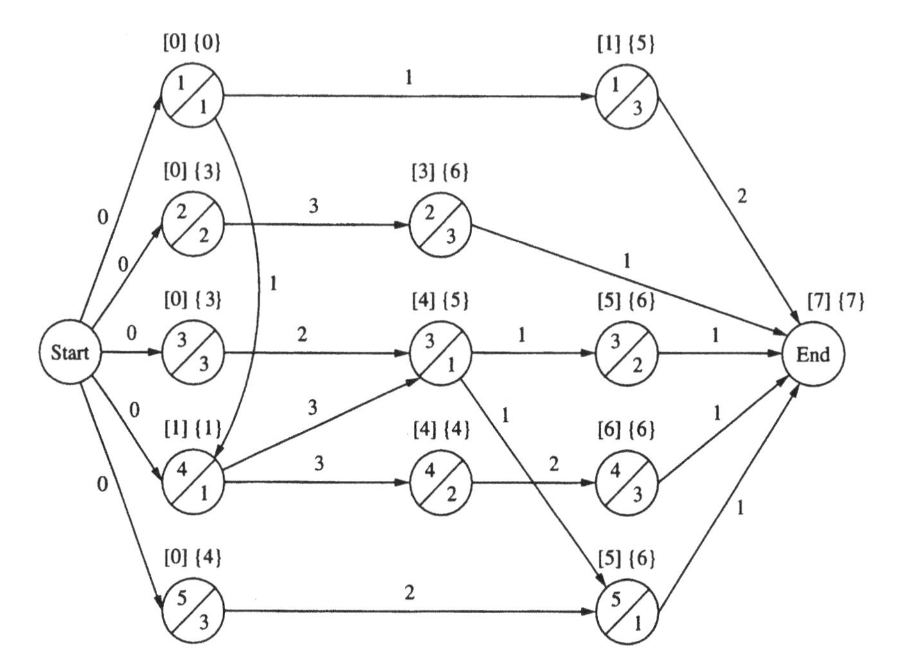

FIGURE 11.18 Machine 1 connected; Machine 3 removed; Iteration 2.

TABLE 11.30
Data for Machine 3

Jobs	1	2	3	4	5
p_i	2	1	2	1	2
r_i	1	3	0	8	0
d_i	5	7	2.7	8	2.5
l_i	3	2	5	1	2
e_i	1	1	2	4	1

any improvement. This is done by removing the solid lines between operations for machine 1 first as shown Figure 11.20. Recompute the early and late start times. Then, machine 3 is solved for minimum total penalty (Table 11.30). The data for machine 3 is taken from Figure 11.20 and Table 11.22.
Optimal sequence on machine 3 is 3-5-1-2-4; penalty = 8.3.

Since the job sequence does not change, the makespan remains 9 and the total penalty 8. We modify the network as shown in Figure 11.19. Now, remove the solid lines between operations for machine 3 as shown in Figure 11.21. Recompute the early and late start times. Then, machine 1 is solved for minimum penalty (Table 11.31). The data for machine 1 is taken from Figure 11.21 and Table 11.22.
Optimal sequence on machine 1 is 1-4-3-5; penalty = 8.7.

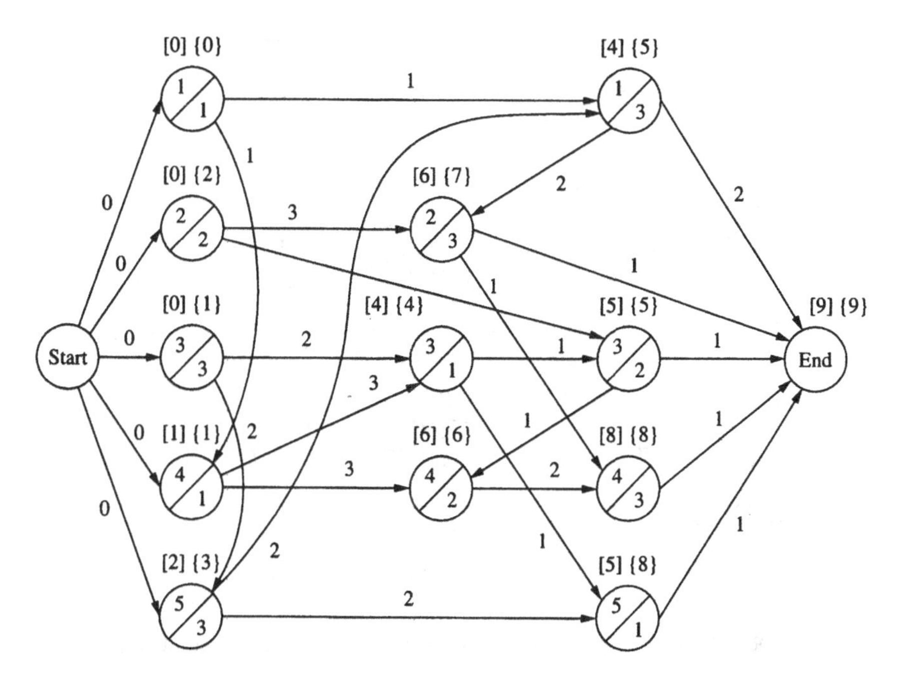

FIGURE 11.19 Machines 1, 2, and 3 connected; Iteration 3 (the last operations for all the jobs shown in thick circles).

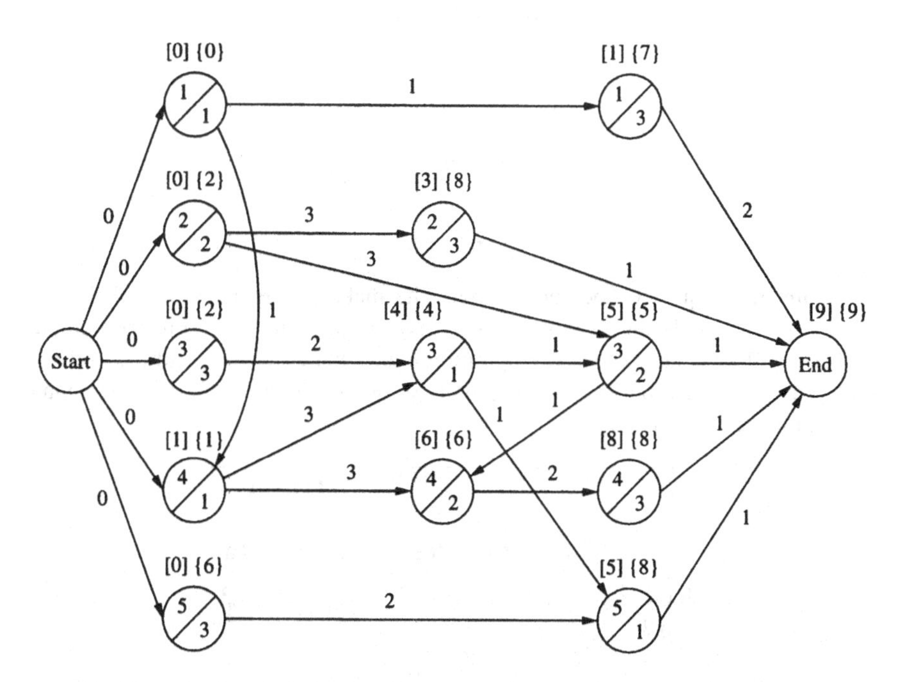

FIGURE 11.20 Machines 1 and 2 connected; Machine 3 removed; Iteration 3.

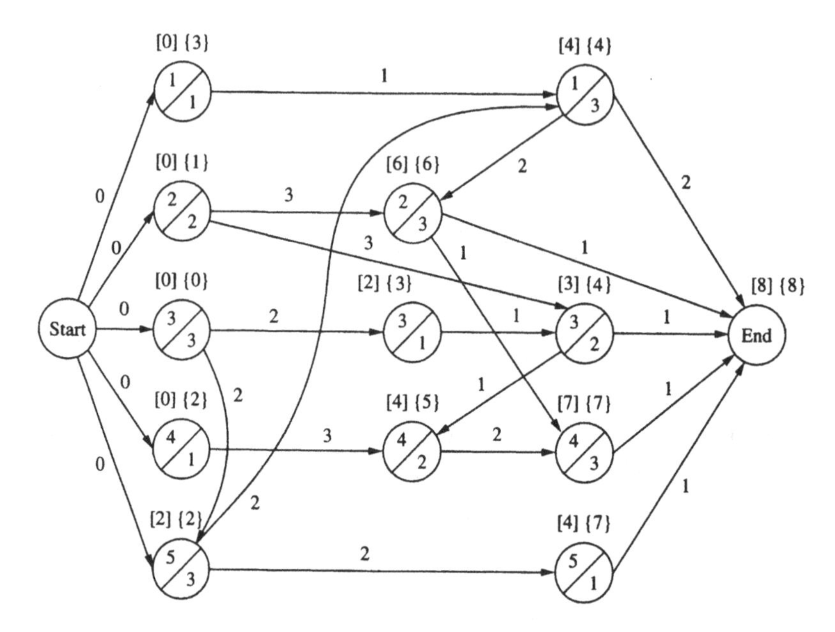

FIGURE 11.21 Machines 2 and 3 connected; Machine 1 removed; Iteration 3.

TABLE 11.31
Data for Machine 1

Jobs	1	3	4	5
p_i	1	1	3	1
r_i	0	2	0	4
d_i	2	4.4	3.7	4
l_i	3	5	1	2
e_i	1	2	4	1

Again, the sequence does not change; the makespan remains 9, and the total penalty 8. We modify the network back as shown in Figure 11.19. Finally, compute the early and late start times. The optimum makespan C_{max} is 9, and the optimum total penalty is 8. Associated Gantt chart is shown in Figure 11.22 and the penalty calculations are as follows:

			Penalty	
	Due	Completed	Early	Late
Job 1	5	6		3
Job 2	7	7	—	—
Job 3	6	6	—	—
Job 4	8	9	—	1
Job 5	4	6		4
			Total	8

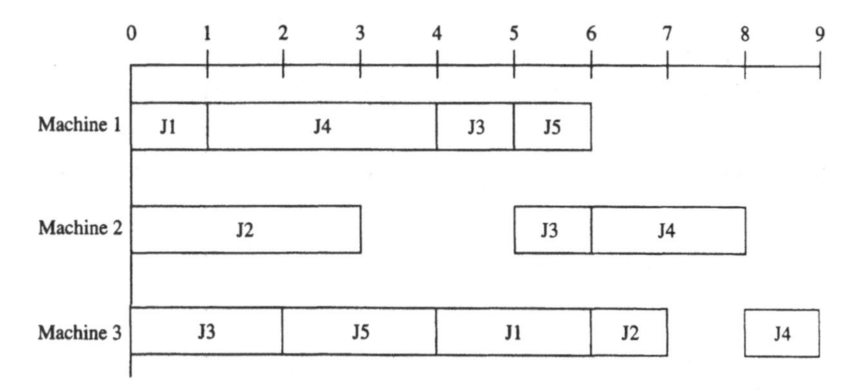

FIGURE 11.22 Gantt chart for MODSB penalty minimization example.

11.4.3 Optimizing Total Penalty and Makespan Using MODSB (Dual Criteria)

The job shop problems that have been discussed so far have focused their objective on either minimization of total tardiness (penalty) or makespan minimization. Sometimes, we may come across a situation where more than one criterion needs to be examined simultaneously. For example, in a job shop if there is a penalty associated with each job being late or early, and there is also a fixed daily overhead cost based on the makespan, then both total penalty costs as well as the overhead costs become equally important. In such a case, an optimum mix of the dual criteria is required.

The MODSB procedure described earlier, for both makespan minimization as well as total penalty minimization, is combined in this section. The objective is to achieve a good mix of both total penalty as well as makespan.

The procedure remains the same as described for MODSB for penalty minimization, except that the due date for the operations (when solving individual machines for minimum penalty sequence) or operational due date is calculated as the minimum of equal slack due date (ESD) or network-based due date, which are defined below.

The operational due date obtained when we considered penalty function alone is now denoted as ESD. The due date that is obtained based on the network when solving MODSB makespan minimization problem is now called network-based due date (NBD). Note that the NBD changes every time the network changes, whereas the ESD remains the same throughout.

Every time a machine is solved for minimum penalty sequence, the NBD is determined. The operational due dates for all the operations of the machine under consideration are given by the minimum of either NBD or ESD.

Applying this procedure for the example shown in Table 11.32 gives a makespan of 36 and a total penalty of 59. The job schedule on each machine is as follows. The numbers enclosed within parentheses indicate start time for the job.

Machine 1: 6 (5/7) 13 (6/6) 19 (3/3) 22 (4/6) 28
Machine 2: 0 (1/7) 7 13 (5/6) 19 (4/1) 20 28 (3/6) 34
Machine 3: 0 (6/4) 4 (4/3) 7 (2/6) 13 19 (5/1) 20
Machine 4: 0 (3/15) 15 (1/11) 26 (6/3) 29 (4/4) 33 (5/3) 36
Machine 5: 0 (2/3) 3 (5/3) 6 19 (6/15) 24 (3/4) 28 (1/2) 30 33 (4/2) 35

TABLE 11.32
Six-Job-Five-Machine Job Shop Data for Dual-Criteria Optimization Problem

Jobs	Machine Sequence	Processing Times	Due Date	Late Penalty	Early Penalty
1	2, 4, 5	7, 11, 2	33	11	4
2	5, 3	3, 6	11	4	1
3	4, 1, 5, 2	15, 3, 4, 6	32	1	0
4	3, 2, 1, 4, 5	3, 1, 6, 4, 2	47	2	2
5	5, 1, 2, 3, 4	3, 7, 6, 1, 3	27	1	0
6	3, 1, 5, 4	4, 6, 5, 3	28	4	3

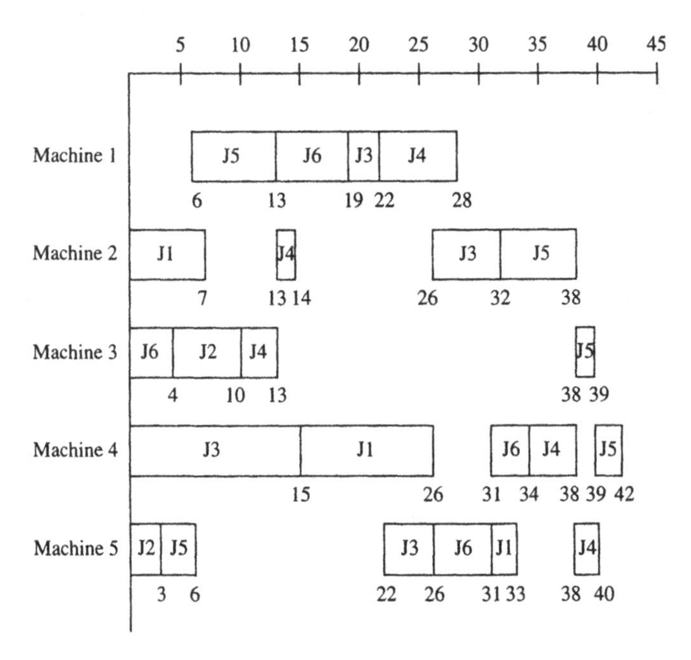

FIGURE 11.23 Illustration of a schedule obtained using MODSB (dual criteria).

Applying the procedure described in MODSB penalty minimization gives a better total penalty of 54, but the makespan increases to 42. The schedule for each machine is as follows.

Machine 1: 6 (5/7) 13 (6/6) 19 (3/3) 22 (4/6) 28
Machine 2: 0 (1/7) 7 13 (4/1) 14 26 (3/6) 32 (5/6) 38
Machine 3: 0 (6/4) 4 (2/6) 10 (4/3) 13 38 (5/1) 39
Machine 4: 0 (3/15) 15 (1/11) 26 31 (6/3) 34 (4/4) 38 39 (5/3) 42
Machine 5: 0 (2/3) 3 (5/3) 6 22 (3/4) 26 (6/5) 31 (1/2) 33 38 (4/2) 40

The Gantt chart is given in Figure 11.23.

11.5 SUMMARY

Job shop and open shop are the most generalized and complex production systems. There are a number of heuristic procedures or rules to schedule these systems, and this chapter illustrates few old and few new approaches. One of the popular rules, which is also easy to apply, is the SPT rule. The SPT rule is found to perform well in minimizing makespan for a job shop that is mostly evenly loaded.

There are network-based methods that perform good in minimizing makespan for an evenly or not so evenly loaded job shop. Two methods are introduced here. They are: shifting bottleneck heuristic and two-stage scheduling heuristic by Vancheeswaran and Townsend. Both methods are illustrated by a simple example, and computer programs are provided on the size mentioned in introduction for larger size problems.

Just like a single-machine problem, minimization of tardiness in job shop requires different logic than the one used in the minimization of makespan. Two procedures are illustrated. The first is the CEXSPT rule, and a new procedure using shifting bottleneck as the base. Unlike the CEXSPT rule, where all jobs have the same weights and only tardiness is measured, the new procedure allows different penalty values for each job and minimizes the total penalty, which includes both the early and late penalties. Another extension of the shifting bottleneck procedure is introduced that can work with the dual criteria of makespan and penalty.

11.6 PROBLEMS

11.1 Using the SPT rule, develop a schedule that minimizes makespan for the following jobs:

	Machine Sequence	Processing Times
Job 1	2, 3	1, 1
Job 2	3, 1	3, 2
Job 3	1, 2, 3	2, 3, 3
Job 4	2, 1, 3	1, 1, 3
Job 5	1, 3, 2	2, 3, 2

11.2 Using the SPT rule, develop a schedule that minimizes makespan for the following jobs:

	Machine Sequence	Processing Times
Job 1	1, 2, 3, 4	1, 3, 5, 4
Job 2	2, 4, 1	6, 2, 4
Job 3	4, 3, 2	5, 1, 2
Job 4	3, 2, 1	3, 4, 1

11.3 Using the modified shifting bottleneck heuristic, minimize the makespan for the jobs given in Problem 11.1. The late and early penalties for the jobs are 1 and 0, respectively.

11.4 Using the modified shifting bottleneck heuristic, minimize the makespan for the jobs given in Problem 11.2. The late and early penalties for the jobs are 1 and 0, respectively.

11.5 Using the two-stage job shop heuristic by Vancheeswaran and Townsend, minimise the makespan of the jobs in the following network:

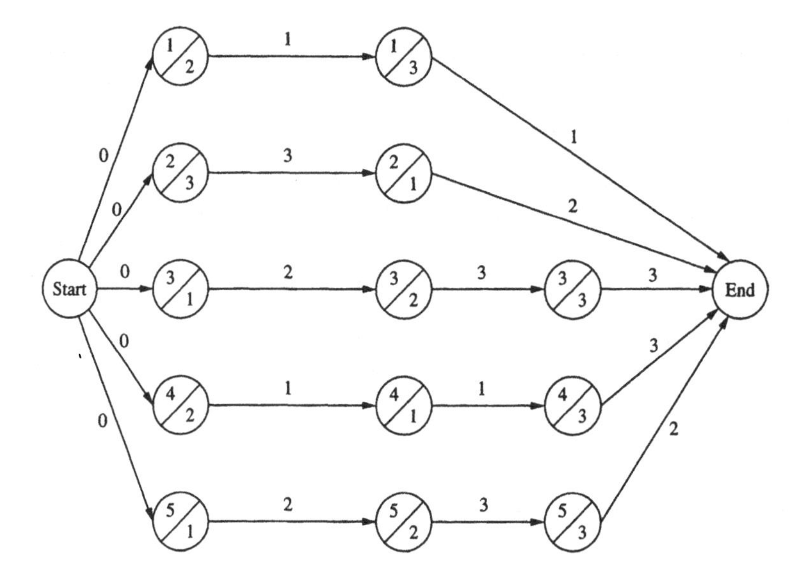

11.6 Using the two-stage job shop heuristic by Vancheeswaran and Townsend, minimize the makespan of the jobs in the following network:

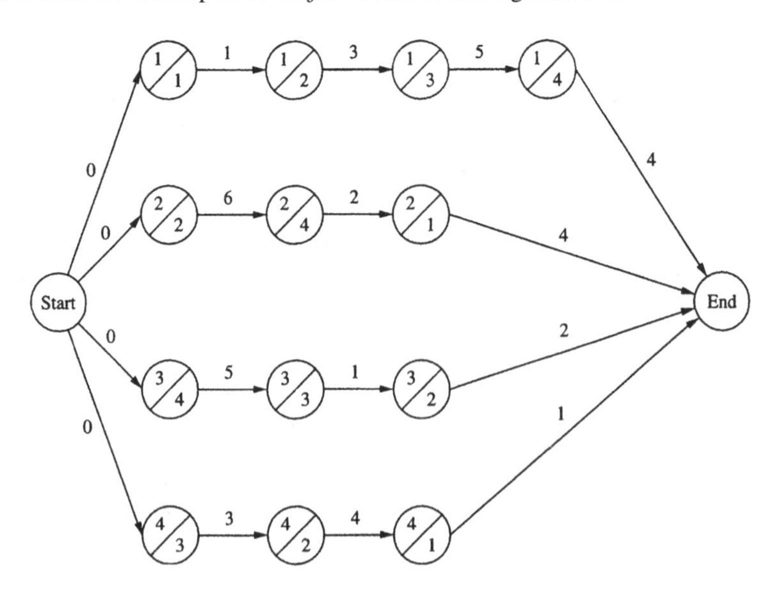

11.7 Using the CEXSPT rule, develop a schedule to minimize tardiness for the following jobs:

	Due Date	Machine Sequence	Processing Time
Job 1	7	2, 3	1, 1
Job 2	15	3, 1	3, 2
Job 3	18	1, 2, 3	2, 3, 3
Job 4	6	2, 1, 3	1, 1, 3
Job 5	23	1, 3, 2	2, 3, 2

Compare your results with those obtained by using the modified shifting bottleneck heuristic.

11.8 Develop a schedule to minimize tardiness for the jobs listed below using the CEXSPT rule.

	Due Date	Machine Sequence	Processing Time
Job 1	21	1, 2, 3, 4	1, 3, 5, 4
Job 2	29	2, 4, 1	6, 2, 4
Job 3	38	4, 3, 2	5, 1, 2
Job 4	10	3, 2, 1	3, 4, 1

Compare your results with those obtained by using the modified shifting bottleneck heuristic.

11.9 M/s Smith & Co. manufactures jobs to specifications at its factory. Its machinery includes a milling machine, a lathe, and a radial drilling machine. As of the current quarter, the company has either completed all orders or intends to do so by the end of the quarter. The marketing department has been working with current customers and has identified a total of four jobs that will have a constant demand for at least two years. In view of the promising trend, product development suggests replacing the lathe (still in good condition) with a new CNC lathe so as to reduce the makespan. With the help of the data given in the following table, make your recommendation as to whether the company should replace the lathe. Solve this case using the two-stage job shop scheduling heuristic.

	Sequence of Operations	Processing Time
Job 1	Turning, milling, drilling	5, 7, 2
Job 2	Turning, milling	6, 6
Job 3	Milling, drilling, turning	4, 8, 5
Job 4	Drilling, milling	5, 6

Processing times for the jobs on the new CNC lathe are estimated to be job 1, 4; job 2, 5; and job 3, 5.

11.10 Lock Fasteners Co. manufactures fasteners on a mass scale. Its machinery includes a multispindle drilling machine, a milling, a lathe, and an SPM (special purpose machine) for thread cutting. Because of diminishing market demands, the company has decided to manufacture only per order. This has brought in an excess capacity of 25 minutes per cycle. One of its customers has two assembly lines. Line 1 needs jobs 1, 3, and 4, and line 2 needs jobs 2, 5, and 6. Lock Fasteners can feed either one of the lines. For every minute beyond the available 25 minutes, there is a penalty of $1.25 per minute accounting for overtime and other expenses. If the company does not take up the order, it stands to lose $0.40 per minute from idle time costs. Make your recommendations as to whether Lock Fasteners should take up the order, and if they have to take up an order, which line should they feed? Solve this problem using the modified shifting bottleneck heuristic.

	Sequence of Operations	Processing Time
Job 1	Milling, drilling, threading	10, 5, 9
Job 2	Turning, drilling, threading	7, 4, 10
Job 3	Drilling, milling	7, 12
Job 4	Drilling, milling, threading	7, 8, 12
Job 5	Milling, turning, drilling	8, 5, 12
Job 6	Drilling, threading	7, 5

11.11 Visor Industries has recently started a small project that involves an extensive market survey. There are four salespeople on the team taking charge of four different subprojects. Each salesperson has a set of customers. Because of vicinity and other prioritizing criteria, the salespeople conduct business per the following schedule:

Salesperson 1 Sector 1, Sector 3, Sector 2
Salesperson 2 Sector 2, Sector 1, Sector 3
Salesperson 3 Sector 1, Sector 3
Salesperson 4 Sector 3, Sector 2, Sector 1

Although they may discuss different business, there cannot be two salespeople at any one sector. They use company vans to go from the home office to a sector and back. If a salesperson wants to go to a sector, he or she must use the van scheduled for that particular sector. The vans are scheduled as follows:

Van A To and from Sector 1
Van B To and from Sector 2
Van c To and from Sector 3

The salespeople's time is spent as shown:

	Sector #1		Sector #2		Sector #3	
	One-Way Travel Time	Time Spent on Business	One-Way Travel Time	Time Spent on Business	One-Way Travel Time	Time Spent on Business
Salesperson 1	10	30	7	20	12	10
Salesperson 2	12	20	8	20	15	25
Salesperson 3	8	30	5	15	10	20
Salesperson 4	10	22	5	12	10	30

Identiy a feasible and efficient way to schedule these salesperson on the sectors. (*Hint:* Salesperson = Jobs; Sectors = Machines; Operation time = Total travel time + Business time.)

REFERENCES AND SUGGESTED READINGS

Adams, J., Balas, E., and D. Zawack. (March 1988). "The Shifting Bottleneck Procedure for Job Shop Scheduling," *Management Science*, 34 (3), 391–401.

Anderson, E.J., and J.C. Nyirenda. (1990). "Two New Rules to Minimize Tardiness in a Job Shop," *International Journal of Production Research*, 28, 2277–2292.

Baker, K.R. (1984). "Sequencing Rules and Due-Date Assignments in a Job Shop," *Management Science*, 30, 1093–1104.

Baker, K.R., and J.J. Kanet. (1983). "Job Shop Scheduling with Modified Due-Dates," *Journal of Operations Management*, 4,11–21.

Balas, E. (November 1969). "Machine Sequencing via Disjunctive Graphs: An Implicit Enumeration Algorithm," *Operations Research*, 17, 941–957.

Bowman, F.H. (1959). "The Schedule-Sequencing Problem," *Operations Research*, 7, 621–624.

Carlier, J. (1982). "The One Machine Sequencing Problem," European Journal of *Operations Research*, 11, 42–47.

Conway, R.W. (1965). "Priority Dispatching and Job Lateness in a Job Shop," *Journal of Industrial Engineering*, 16(4), 228–237.

Conway, R.W., Maxwell, W.L., and L. Miller. (1967). *Theory of Scheduling*, Addison-Wesley, Reading, MA.

Eilon, S., and I.G. Chowdhury (1976). "Due Dates in Job Shop Scheduling," *International Journal of Production Research*, 14 (2), 223–237.

Eilon, S., and D.J. Cotterill. (1968). "A Modified SI Rule in Job Shop Scheduling," *International Journal of Production Research*, 7, 135–145.

Eilon, S., and R.M. Hodgson. (1967): "Job Shop Scheduling with Due Dates," *International Journal of Production Research*, 6, 1–13.

Elvers, D.A. (1973). "Job Shop Dispatching Using Various Due Date Setting Criteria," *Production and Inventory Management*, 14(4), 62–69.

Manne, A.S. (1960). "On the Job-Shop Scheduling Problem," *Operations Research*, 8, 219–223.

Muth, J.F., and G.L. Thompson. (1963). "Industrial Scheduling," Prentice Hall, Englewood Cliffs, NJ.

Nasr, N., and E.A. Elsayed, (1990). "Job Shop Scheduling with Alternative Machines," *International Journal of Production Research*, 28(9), 1594–1609.

Schultz, C.R. (1989). "An Expediting Heuristic for the Shortest Processing Time Dispatching Rule," *International Journal of Production Research*, 27, 31–41.

Spachis, A.S., and J.R. King. (1979). "Job Shop Scheduling Heuristics with Local Neighborhood Search," *International Journal of Production Research*, 17(6), 507–526.

Vancheeswaran, R., and M.A. Townsend. (1993). "TwoStage Heuristic Procedure for Scheduling Job Shops," *Journal of Manufacturing Systems*, 12 (4), 315–325.

12 Open-Shop Scheduling

Open shop is the processing of jobs where there is no definite sequence of operations that a job must follow. As long as all the operations needed for the job are done, the job is done. This gives much flexibility in scheduling, but also makes it difficult to develop rules that gives an optimum sequence for every problem.

12.1 MINIMIZE MAKESPAN: TWO-MACHINE PROBLEM

A simple rule that has proven to give optimum results for two-machine open-shop problems is given by Pinedo (1995). The rule is called "longest alternate processing time first" (LAPT) rule. According to LAPT rule, whenever a machine is available, apply the appropriate job selection procedure: for machine 1, select the job with the largest processing time on machine 2 and for machine 2 select the job with the largest processing time on machine 1.

For example, consider the two-machine open-shop problem for which the data is given in Table 12.1. The total times required on machine 1 and 2 are 20 and 13, respectively, with the maximum of 20. The maximum time for the jobs is associated with job 2 and is equal to 13. Therefore, the makespan cannot be less than the maximum of 20 and 13, that is, 20.

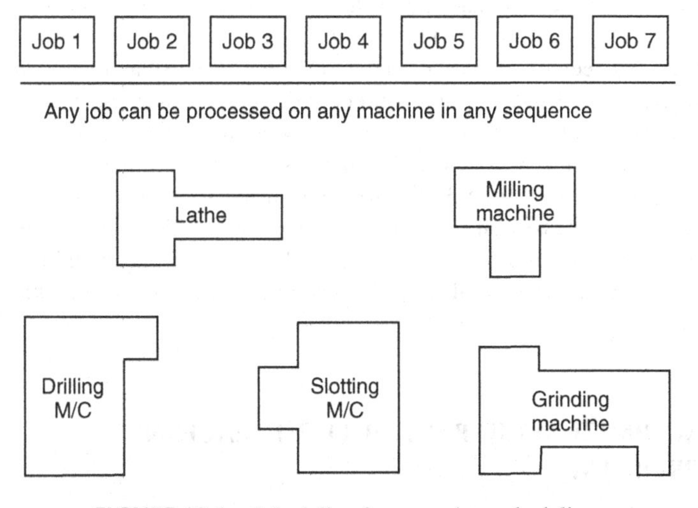

FIGURE 12.1 Scheduling for open shop scheduling.

TABLE 12.1
Data for an Open-Shop Problem

	Machines		
Jobs	**1**	**2**	**Total Time Required for Jobs**
1	2	3	5
2	5	8	13
3	6	2	8
4	7	0	7
Total time on machine	20	13	

Based on the data, the sequence of operations in machine 1 and 2 should be as follows, if the jobs are available for processing when the machines are free.

Machine 1: 2-1-3-4
Machine 2: 4-3-2-1

The jobs are placed in the sequence on each machine using the descending order of processing times in the other machine. For example, on machine 2, the descending order of processing times is 8, 3, 2, and 0. The corresponding sequence of jobs on machine 1 is 2-1-3-4.

Based on the sequence, the jobs are scheduled if they are available. Since machine 1 has the largest time requirement, that is, it is bottleneck machine, it is sequenced first. A time-job diagram (Gantt chart) helps in identifying availability or non-availability of a job when they are suppose to be scheduled. The time is displayed on x-axis, and the numbers between two bars and on the top of the line display the job number.

In the data, job 4 has the largest processing time on machine 1 but it does not require processing on machine 2. So the next job with the largest processing time on machine 1, that is, job 3, is selected as the first job on machine 2. The processing of job 2 is completed at time 3. The next job in the sequence is job 2. The job is presently being processed on machine 1 and is not available; therefore, the next job in the sequence, job 1, is selected. The entire schedule is displayed in Figure 12.2, as well as in Table 12.2 (the numbers in the parenthesis indicate the completion times of the jobs).

12.2 MINIMIZE MAKESPAN: MULTIPLE-MACHINE PROBLEM

In the case of a multiple machines (including two machines), scheduling jobs based on the longest total processing time (LTPT) rule, described next, leads to schedule that provides minimum makespan. The minimum makespan still cannot be less than

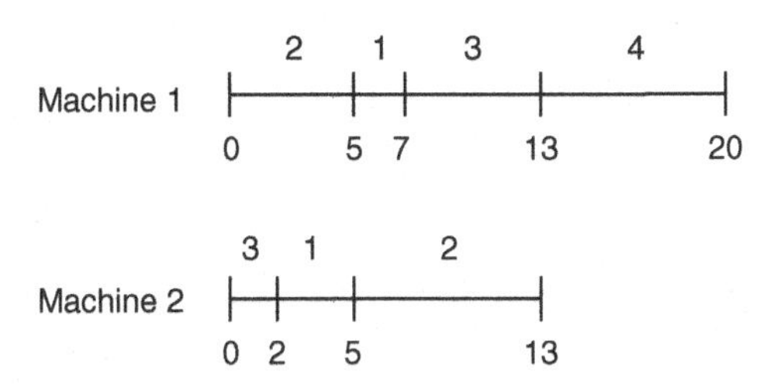

FIGURE 12.2 Gant charts for scheduling on two machines.

TABLE 12.2
Solution to the Two-Machine Problem

Machines		
Jobs	**1**	**2**
	2 (5)	3 (2)
	1 (7)	1 (5)
	3 (13)	2 (13)
	4 (20)	—

TABLE 12.3
Job-Machine Table (Iteration 1)

	Machines					Time When Job Is Available for Next Assignment (TJA-Next)
	1	**2**	**3**	**4**	**Job Total**	
Jobs 1	**7**	5	2	5	19	7
2	7	1	**8**	8	24	8
3	6	2	4	**8**	20	8
4	2	8	9	5	24	0
5	2	1	6	3	12	0
6	6	3	7	3	19	0
Time when machine is available for next assignment(TMA-next)	7	0	8	8		

the maximum of either maximum processing time on any machine or the maximum time required for a job.

The steps of LTPT procedure are as follows:

Step 1: Construct a job-machine table as shown in Table 12.3. The table shows the processing time of each job on each machine it uses. Calculate the total processing time for each job by summing the processing times on the associated machines.

TABLE 12.4
Work Table I

		First Half					Second Half			
	Machines					TJA-Next	Minimum Time When Jobs Can be Placed on Machines			
	1	2	3	4	Total (RTP)		M1	M2	M3	M4
Jobs 1	–	5	2	5	12	7	–	7	8	8
2	7	1	–	**8**	16	16	8	8	–	**8**
3	6	2	**4**	–	12	12	8	8	**8**	–
4	2	**8**	9	5	24	8	8	**0**	8	8
5	2	1	6	3	12	0	7	0	8	8
6	6	3	7	3	19	0	7	0	8	8
TMA-next	7	8	12	16						

Step 2: Select the job with the maximum total processing time. Schedule that job on the machine on which it has the largest processing time, if the machine is available. If the machine is not available, go to the next job in descending order of the total processing time. Repeat the step for all jobs. Once the jobs have been assigned, find the time when the jobs will be available for the next assignment on the remaining machines. This will be the processing time for the jobs on the machines on which they have been assigned. If a job has not been assigned, this time will be zero. Similarly, find the time when the machine is free to accommodate the next job assigned to it. This will be the processing time for the job that has been assigned on that machine. If no job has been assigned to a machine, this time will be zero. These values are recorded in the job-machine table.

Step 3: Develop a new work table by eliminating the processing times for the jobs from the machines on which they have been assigned and recording other processing times as shown in Table 12.4. Record the total of remaining time of processing (RTP) still required for each job. This new table should also include the minimum time when the jobs can be placed on the machines on which they are yet to be assigned. Suppose a particular job i is to be assigned on machine k, then the minimum time when job i can be placed on machine k, will be Max {time when job i is available, time when machine k is free}. The values for these comparisons are obtained from the last column and last row of job-machine table (Table 12.3) for the first iteration and immediate previous work table in the successive iterations. Record these entries in the second half of the work table associated with when the jobs can be placed on the machines (e.g., Table 12.4).

Step 4: Select the job with the maximum remaining total processing time. For that job, find the minimum time when it can be placed on the machines. Check whether this is also the minimum time when the machine can be assigned to a job. This is equivalent to checking each row and each column of the second half of the work table and checking to see if the minimum for both is the same element. If yes, then assign that job to the machine. Continue the check for the job with the next highest remaining processing time by repeating step 4. Go to next step when all the jobs with positive value for RCT are examined.

Step 5: Once all the possible assignments are made, find the time when the jobs will be available for assignment on the remaining necessary machines (TJA-next). This can be obtained as follows:

If a job has been assigned in the current iteration, this time will be the sum of the processing time for the job on the machine on which it has been assigned and the corresponding minimum time when the job can be placed on that machine. This requires looking at the data points in both halves of the work table as will become clear from the illustrative example.

If the job has not been assigned in the current iteration, this time will be the same as in the previous iteration. This data is obtained from the previous work table (use job-machine table for the first iteration).

The data is recorded in the central column in the work table that separates the two halves of the table.

Step 6: Find the time when the machine will be free to accommodate the next job assigned to it (TMA-next). This can be obtained as follows:

1. If a job has been assigned to the machine in the current iteration, this time will be the sum of the processing time for the job on that machine and the time when the job was placed on that machine.
2. If a job has not been assigned to the machine in the current iteration, this time will be the same as in the previous iteration.

Step 7: Repeat steps 3–7 until all the jobs have been assigned.

12.2.1 Illustrative Example

Consider an open-shop problem of scheduling six jobs on four machines. The processing time for each job on each machine is given in the job-machine Table 12.3. We wish to develop a sequence of operation on each machine that will minimize the makespan for all six jobs.

The total processing time for each job is indicated in the column job total. Similarly, the total processing time required on each machine is indicated by row Machine Total. The makespan cannot be less than 36, the maximum of these numbers. The sequencing procedure is illustrated next, as we follow the iterations.

Iteration 1: Select either job 2 or 4 since both have the maximum total processing time. In our example, let us select job 2 arbitrarily. We can assign job 2 to either machine 3 or 4 since it has the largest processing times on these machines. We again assign the job 2 arbitrarily on machine 3. The next job to sequence is job 4. It also has the largest processing time on machine 3, but since that machine is already assigned, we move to the next job, Job 3. It can be assigned to machine 4. Continuing the procedure, the only other job that can be assigned is job 1 to machine 1.

Next, we find the time when the jobs are available for the next assignment. Since jobs 1, 2, and 3 are presently assigned on machines 1, 3, and 4, respectively, their available time is the processing time for those jobs on the corresponding machines. The remaining jobs are not assigned to any machines presently, and therefore, they are available right now, at time zero. These values are recorded in the last column of Table 12.3.

Now, we find the time when the machines are free to accommodate the next job assignment. Since machines 1, 3, and 4 are loaded with jobs 1, 2, and 3, respectively, this time is the processing time for those jobs on the corresponding machines, or 7, 8, and 8 for machines 1, 3, and 4, respectively. Since, machine 2 is not loaded with any job yet, it is available at time zero. These values are recorded in the last row of Table 12.3.

Iteration 2: Develop the work table, Table 12.4. Note, the column TJA-next defines when job will be available to the next table calculations and entries are determined at the end or completion of first and second half of the present table. The column TJA-next is shown in the center because it uses information from first and second halves of the table. The entries in the first half of the table are the processing times of the jobs on the machines that are yet to be processed. These are obtained by eliminating the processing times for the jobs that have already been placed on the machines in the previous iteration. In this case, we eliminate the processing times for jobs 1, 2, and 3 from machines 1, 3, and 4, respectively. The remaining total processing time for each job is obtained and displayed in the column RTP (remaining total processing time).

Next, we construct the second half of Table 12.4. These entries display the minimum time when the jobs can be placed on machines. Suppose we want to assign job 2 on machine 1. The minimum time when job 2 can be assigned to machine 1 is Max {time when job 2 is available, time when machine 1 is free}, which is Max {8, 7}, which is 8. Both these values are obtained from the entries in the last column and last row of Table 12.3. Thus, the minimum time when job 2 can be assigned to machine 1 is 8. This is the entry in the second half of the table under M3 (machine 3) and row for job 2. Similarly, we find the minimum times when the other jobs can be placed on the machines and record the data appropriately in the table.

Select job 4 since it has the maximum total remaining processing time of 24. The minimum time when it can be placed on the machines is 0, the time on M2 in the second half of the table for job 4. Also, the minimum time when machine 2 can be assigned to any job is 0 (the minimum value in column M2). Since these two values are equal, assign job 4 to machine 2. Indicate it by bolding the associated numbers in both halves of the table. Check for the next job, that is, job 6. The minimum time when it can be placed on a machine is 0, that is, the time on M2. But machine 2 is already assigned to job 4, and therefore, we cannot assign job 6 to it. Similarly, we check for the other jobs and find that job 2 may be assigned to machine 4 and job 3 to machine 3.

Now, we must find the time when the jobs are available for the next assignment. Since jobs 2, 4, and 3 have been assigned in the current iteration, this time will be the

TABLE 12.5
Work Table II (Iteration 3)

	First Half						Second Half			
	Machines					TJA-Next	Time When Jobs Can be Placed on Machines			
	1	2	3	4	Total (RTP)		M1	M2	M3	M4
Jobs 1	—	**5**	2	5	12	13	—	**8**	12	16
2	7	1	—	—	8	16	16	16	—	—
3	6	2	—	—	8	12	12	12	—	—
4	2	—	9	5	16	8	8	—	12	16
5	2	1	6	3	12	0	7	8	12	16
6	**6**	3	7	3	19	13	**7**	8	12	16
TJM-next	13	13	12	16						

sum of the processing time for these jobs on the machines on which they have been assigned, and the corresponding minimum time when these jobs can be placed on those machines. The numbers are obtained by looking at bold entries on both halves of the table. For example, this time for job 2 is $8+8 = 16$, for job 4 is $8+0 = 8$, and for job 3 it is $4+8 = 12$. For the other jobs that are not assigned in this iteration, that is, jobs 1, 5, and 6, these times are the same as in the previous iteration. The times are noted in the *central* or *TJA-next* column of Table 12.4.

Now, find the time when each machine will be free to accommodate the next job. Since machines 2, 3, and 4 have been loaded with jobs 4, 3, and 2, respectively, this time is the sum the processing time for those jobs on these machines and the corresponding minimum time when the jobs can be placed on those machines. Thus, the time is 8 + 8 = 16 for machine 4, 4 + 8 = 12 for machine 3, and 8 + 0 = 8 for machine 2. For the other machines on which assignments are not made in the current iteration, the times remain the same as in the previous iteration. All the values are recorded in the *last row* or *TMA-next row*, of the first half of Table 12.4.

Iteration 3: Develop work Table 12.5. Again, we eliminate the processing times for the jobs assigned in the previous iteration, that is, jobs 2, 3, and 4 from machines 4, 3, and 2, respectively. The remaining total processing time for each job is obtained and displayed in the column RTP.

Next, we construct the second half of Table 12.5. These entries display the minimum time when the jobs can be placed on the machines. Suppose we want to assign job 4 on machine 3. The minimum time when job 4 can be assigned to machine 3 will be Max {time when job 4 is available, time when machine 3 is free}, which is Max {8, 12}, which is 12. Both of these values are obtained from the entries in the last column (TJA-next column) and last row (TMA-next) of Table 12.4. Thus, the minimum time when job 4 can be assigned to machine 3 is 12. This is the entry in the second half of the table under M3 (machine 3) and row for job 4. Similarly, we find the minimum time when the other jobs can be placed on the machines and record the data appropriately in Table 12.5.

Select job 6 since it has the maximum total processing time of 19. The minimum time when it can be placed on the machines is 7, the time on M1 in the second half of the table for job 6. Also, the minimum time when machine 1 can be assigned to any job is 7 (the minimum value in column M1). Since these two values are equal, assign job 6 to machine 1, indicate it by bolding of associated numbers in both halves of the table. Check for the next job, that is, job 4. The minimum time when it can be placed on a machine is 8, that is, the time on M1. But machine 1 is assigned to job 6, and hence job 4 cannot be placed on machine 1. Similarly, we check other jobs and find that job 1 can be assigned to machine 2 and job 6 can be assigned to machine 1.

Now, we must find the time when the jobs are available for the next assignment. Since jobs 1 and 6 have been assigned in the current iteration, this time will be the sum of the processing time for these jobs on the machines on which they have been assigned and the corresponding minimum time when these jobs can be placed on those machines. The numbers are obtained by looking at bold entries on both halves of the table. For example, this time for job 1 is $5 + 8 = 13$, and for job 6 it is $6 + 7 = 13$. For the other jobs that are not assigned in this iteration, these times are the same as in the previous iteration. The times are noted in TJA column of the table.

Now, find the time when the machines will be free to accommodate the next job assigned to them. Since machines 2 and 1 have been loaded with jobs 1 and 6, respectively, this time is the sum of the processing time for those jobs on these machines and the corresponding minimum time when the jobs can be placed on those machines. This time is 5 + 8 = 13 for machine 2, and 6 + 7 = 13 for machine 1. For the other machines, on which assignments are not made in the current iteration, the time remains the same as in the previous iteration. All the values are recorded in the TMA-next row of the first half of the table.

Iteration 4: Develop the work table, Table 12.6. As done in the previous iteration, we eliminate the processing times for jobs 1 and 6 from machines 2 and 1, respectively. The remaining total processing time for each job is obtained and displayed in the column RTP.

Next, we construct the second half of the Table 12.6. Again, these entries display the minimum time when the jobs can be placed on the machines. Suppose we want to assign job 2 on machine 1. The minimum time when job 2 can be assigned to machine 1 will be Max {time when job 2 is available, time when machine 1 is free}, which is Max {16, 13}, which is 16. Both of these values are obtained from the entries in the TJA-next and TMA-next of Table 12.5. Thus, the minimum time when job 2 can be assigned to machine 1 is 16. This is the entry in the second half of the table under M1 (machine 1) and row for job 2. Similarly, we find the minimum time when the other jobs can be placed on the machines and record the data appropriately in the table.

Select job 4 since it has the maximum remaining total processing time of 21. The minimum time when it can be placed on a machine is 12, the time on M3 in the second half of the table for job 4. Also, the minimum time when machine 3 can be assigned

TABLE 12.6
Work Table III (Iteration 4)

	First Half						Second Half			
	Machines					TJA-Next	Time When Jobs Can be Placed on Machines			
	1	2	3	4	Total (RTP)		M1	M2	M3	M4
Jobs 1	—	—	2	5	7	13	—	—	13	16
2	7	1	—	—	8	16	16	16	—	—
3	**6**	2	—	—	8	19	**13**	13	—	—
4	2	—	**9**	5	16	21	13	—	**12**	16
5	2	1	6	3	12	0	13	13	12	16
6	—	**3**	7	3	13	16	—	**13**	13	16
TJM-next	19	16	21	16						

to any job is 12 (the minimum value in column M3). Since these two values are equal, assign job 4 to machine 3; indicate it by bolding of associated numbers in both halves of the table. Check for the next job, that is, job 3. The minimum time when it can be placed on a machine is 13, that is, the time on M1 or M2. Also, the minimum time when machine 1 can be assigned to any job is 13 (the minimum value in column M1). Since these two values are equal, assign job 3 to machine 1. Similarly, we check for the other jobs and find that only job 6 can be assigned on machine 2.

Now, we must find the time when the jobs are available for the next assignment. Since jobs 3, 4, and 6 have been assigned in the current iteration, this time will be the sum of the processing time for these jobs on the machines on which they have been assigned and the corresponding minimum time when these jobs can be placed on those machines. The numbers are obtained by looking at bold entries on both halves of the table. For example, this time for job 3 is $6 + 13 = 19$, for job 4 it is $9 + 12 = 21$, and for job 6 it is $3 + 13 = 16$. For the other jobs that are not assigned in this iteration, these times are the same as in the previous iteration. The times are noted in the TJA-next column of the table.

Now, find the time when the machines will be free to accommodate the next job assigned to them. Since machines 1, 2, and 3 have been loaded with jobs 3, 4, and 6, respectively, this time is the sum of the processing time for those jobs on these machines and the corresponding minimum time when the jobs can be placed on those machines. This time is 6 + 13 = 19 for machine 1, 3 + 13 = 16 for machine 2, and 9 +12 = 21 for machine 3. For the other machines, on which assignments are not made in the current iteration, the time remains the same as in the previous iteration. All the values are recorded in the TMA-next row of the first half of Table 12.6.

By now, the reader is used to the construction of tables and the decision making process. From here on, we display only the final table in each iteration and the decisions made in the iteration (Tables 12.7 through 12.10).

TABLE 12.7
Work Table IV (Iteration 5)

	First Half					Second Half				
	Machines					TJA-Next	Time When Jobs Can be Placed on Machines			
	1	2	3	4	Total (RTP)		M1	M2	M3	M4
Jobs 1	—	—	2	5	7	13	—	—	21	16
2	7	**1**	—	—	8	17	19	**16**	—	—
3	—	2	—	—	2	19	—	19	—	—
4	2	—	—	5	7	21	21	—	—	21
5	2	1	6	**3**	12	19	19	16	21	**16**
6	—	—	**7**	3	10	16	—	—	21	16
TMA-next	19	17	21	19						

TABLE 12.8
Work Table V (Iteration 6)

	First Half					Second Half				
	Machines					TJA-Next	Time When Jobs Can be Placed on Machines			
	1	2	3	4	Total (RTP)		M1	M2	M3	M4
Jobs 1	—	—	2	5	7	13	—	—	21	19
2	**7**	—	—	—	7	26	**19**	—	—	—
3	—	2	—	—	2	19	—	19	—	—
4	2	—	—	5	7	21	21	—	—	21
5	2	**1**	6	—	9	20	19	**19**	21	-
6	—	—	7	**3**	10	22	—	—	21	19
TMA-next	26	20	21	22						

Iteration 5: Assign job 2 to machine 2 and job 5 to machine 4.

Iteration 6: Assign job 2 to machine 1, job 5 to machine 2 and job 6 to machine 4.

Iteration 7: Assign job 3 to machine 2, job 4 to machine 4, job 4 and job 5 to machine 3.

Iteration 8: Assign job 1 to machine 4, job 5 to machine 1, and job 6 to machine 3.

Iteration 9: Since job 1 to machine 3 and job 4 to machine 1.

Results: Total makespan = 36 and total idle time of machines = 3.

TABLE 12.9
Work Table VI (Iteration 7)

		First Half						Second Half			
		Machines					TJA-Next	Time When Jobs Can be Placed on Machines			
		1	2	3	4	Total (RTP)		M1	M2	M3	M4
Jobs	1	*v*	—	2	5	7	13	—	—	21	22
	2	—	—	—	—	—	26	—	—	—	—
	3	—	**2**	—	—	2	22	—	**20**	—	—
	4	2	—	—	**5**	7	27	26	—	—	**22**
	5	2	—	**6**	—	8	27	26	—	**21**	—
	6	—	—	7	—	7	22	—	—	22	
TMA-next		26	22	27	27						

TABLE 12.10
Work Table VII (Iteration 8)

		First Half						Second Half			
		Machines					TJA-Next	Time When Jobs Can be Placed on Machines			
		1	2	3	4	Total (RTP)		M1	M2	M3	M4
Jobs	1	—	—	2	**5**	7	32	—	—	27	**27**
	2	—	—	—	—	—	26	—	—	—	—
	3	—	—	—	—	—	22	—	—	—	—
	4	2	—	—	—	2	27	27	—	—	—
	5	**2**	—	—	—	2	29	27	—	—	—
	6	—	—	**7**	—	3	32	—	—	—	**27**
TMA-next		29	22	34	27						

Sequence on machine 1: 1 - 6 - 3 - 2 - 5 - 4 with idle time = 1 and total time of 31.
Sequence on machine 2: 4 - 1 - 6 - 2 - 5 - 3 with idle time = 2 and total time of 22.
Sequence on machine 3: 2 - 3 - 4 - 5 - 6 - 1 with idle time = 0 and total time of 36.
Sequence on machine 4: 3 - 2 - 5 - 6 - 4 - 1 with idle time = 0 and total time of 32.

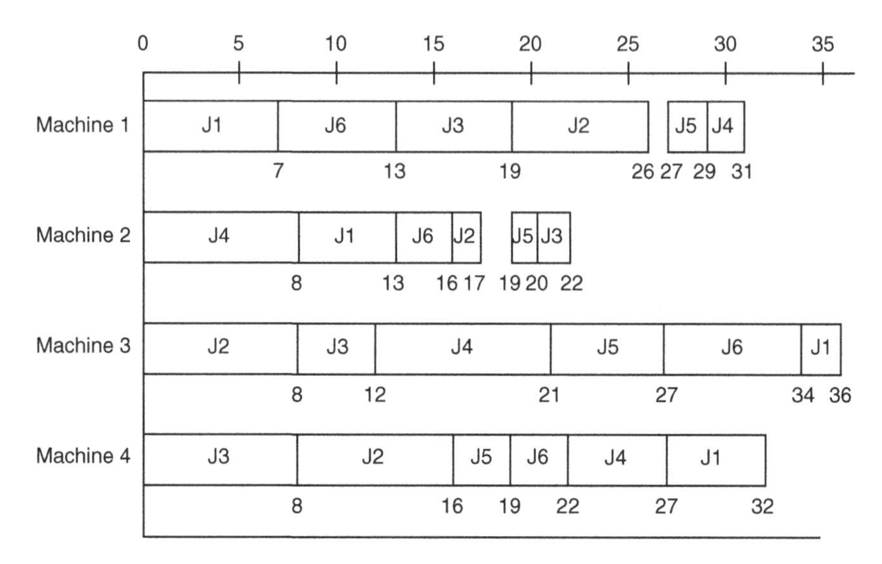

FIGURE 12.3 The two-machine flowshop model.

12.3 MINIMIZATION OF TOTAL TARDINESS–OPEN SHOP (MTT-OP)

In the previous section, we discussed a method to minimize the makespan in an open-shop scheduling. In this section, we will modify the procedure to minimize the total tardiness in the schedule, given the due date for each job.

12.3.1 Solution Procedure

The procedure is divided into two sections. The initial assignment phase is only applied once, while the subsequent assignment phase is repeated till all the jobs are scheduled. Most of the steps in the subsequent assignment phase are similar to the steps in Section 12.2.

The main objective is to complete the jobs before the deadlines (due dates) so as to minimize the total tardiness. The procedure is explained by applying it to the data shown in Table 12.11.

12.3.2 Initial Assignment

Step 1: Construct a job-machine table as shown in Table 12.12. The table shows the processing time for each job on the required machine and the due date for each job (data from Table 12.11). Calculate the total processing time for each job by adding the processing times on the associated machines. Similarly, calculate the total processing time on each machine for the jobs that require that machine. Find the slack for each job by subtracting the total processing time for that job from its due date.

Step 2: An important point to note is that we apply this step only to as many jobs as there are machines (exception being the Special Cases in Section 12.3.3). In other

TABLE 12.11
Job-Machine Table (iteration 1)

		Machines				Job Total	Due Date	Stack	TJA-Next
		1	**2**	**3**	**4**				
Jobs	1	7	6	9	7	29	40	11	0
	2	**4**	3	1	—	8	15	7	4
	3	5	7	—	**6**	18	25	7	6
	4	8	3	**2**	—	13	21	8	2
	5	—	**9**	1	3	13	18	5	9
	6	4	6	3	—	13	24	11	0
Machine total		28	34	16	16				
TMA-next		29	22	34	27				

TABLE 12.12
Work Table I (Iteration 2)

		First Half								Second Half			
		Machines				Job Total (RTP)	Due Date	Stack	TJA-Next	Minimum Time When Jobs Can be Placed on Machines			
		1	**2**	**3**	**4**					**M1**	**M2**	**M3**	**M4**
Jobs	1	7	6	9	7	29	40	11	0	4	9	2	6
	2	—	3	**1**	—	4	15	11	5	—	9	**4**	—
	3	5	7	—	—	12	25	13	6	6	9	—	—
	4	**8**	3	—	—	11	21	10	12	**4**	9	—	—
	5	—	—	1	**3**	4	18	14	12	—	—	9	**9**
	6	4	6	3	—	13	24	11	0	4	9	2	—
Machine total		24	25	14	10								
TMA-next		12	9	5	12								

words, if there are n jobs and m machines with $n > m$, then we apply step 2 to only m jobs.

Select a job using following priority rules:

1. Select the job with the minimum slack.
2. If more than one job have the same slack, select the one which has the earliest due date.
3. If more than one job have the same slack and the same due date, select randomly any one of these jobs.

12.3.3 Special Cases

1. If there is a tie when selecting the mth job using rule 1, then one must select all the tied jobs even if the total number of selected jobs exceed m. Then, among the tied jobs, select the one with the earliest due date (rule 2), assign it to a machine and then move to the next phase, that is, step 3. If the selected job cannot be assigned, go to the next phase.
2. If there is a tie when selecting the mth job using the above step, then we must select all the tied jobs even if the total number of selected jobs exceeds m. Make a random selection among the tied jobs (rule 3). If the selected job cannot be assigned to a machine, select another tied job. If none of the tied jobs can be assigned, go to the next phase.

Once the job has been selected, schedule that job on the machine on which it has the largest processing time, if the machine is available. If the machine is not available, select the machine on which the job has the next largest processing time. Repeat this procedure till that job is assigned to a machine. If at any point, no machine is available for assigning the selected job, move to the next job. Repeat this step for the selected m jobs.

Once all the m jobs have been assigned, find the time when all the jobs will be available for next assignment on the remaining machines. This procedure is similar to the one in described in Section 12.2.

12.3.4 Subsequent Assignments

Step 3: Develop a new work table as per step 3 described in Section 12.2. The only difference between this table and the one in Section 12.2 is the addition of the slack column as shown in Table 12.2. The slack for each job is calculated by subtracting the job total for each job from its due date. This value is then recorded in the slack column for that particular job.

Step 4: For the next iteration, select the next job with the minimum slack as per the rules mentioned in step 2. For that job, find the minimum time when it can be placed on the machines and assign the job to that machine. If there are more than two machines on which the job can be assigned, place the job on the machine on which it has the largest processing time. If the machine is not available for the assignment to be made, place the job on the machine on which it has the next largest processing time. If no machines are available, move to the next job by selecting the job with the next lowest slack and repeating step 4.

Step 5: Once all the possible assignments are made, find the time when the jobs will be available for next assignment by following the same procedure as in step 5 of Section 12.2.

Step 6: Find the time when the machine will be free to accommodate the next job by following the same procedure as in step 6 of Section 12.2.

Step 7: Repeat steps 3–6 till all the jobs are assigned.

12.3.5 Illustrative Example

Consider an open-shop problem in which six jobs have to be assigned on four machines. The processing time for each job on each machine and the due date for each job are given in the job-machine table, Table 12.11. We wish to develop a sequence of operation on each machine that will minimize the total tardiness. The total processing time for each job is indicated in the column job total. Similarly, the total processing time required on each machine is indicated by the column machine total. The sequencing procedure is illustrated next, as we follow the iterations.

Iteration 1: Calculate the slack for each job by subtracting the job total for each job from its due date. The value is entered in the under the slack column of Table 12.11.

As stated in step 2, we can apply step 2 to only as many jobs as there are machines. In this case, there are 4 machines and hence, we can apply step 2 to only 4 jobs. As per step 2, select the job with the lowest slack from Table 12.11. Job 5 with a slack of 5 is selected. We assign job 5 to machine 2, on which as it has the largest processing time of 9. Now, we select the job with the next lowest slack. There are two jobs, job 2 and 3, which satisfy this criterion. Between these two, we should select the one that has the earliest due date. We select job 2 with a due date of 15. Job 2 is assigned to machine 1 as it has the largest processing time of 4 on that machine. Job 3 is selected next. Job 3 has the largest processing time on machine 2. But machine 2 is occupied by job 5, and so we cannot assign job 3 to machine 4. We check whether the machine on which job 3 has the next largest processing time is free. We find that machine 4 is free for assignment and so we assign job 3 to it. Job 4 has the next lowest slack of 8. The machines on which it has the largest processing time of 8, that is, machine 1, and on which it has the next largest processing time of 3, that is, machine 2, both are occupied by jobs 1 and 2, respectively. So, we cannot assign job 4 to either of them. Job 4 has the next largest processing time of 2 on machine 3 and, since this machine is free, we assign job 4 to it. We have checked for four jobs and hence we stop here.

Next, we find the time when the jobs are available for next assignment. Jobs 2, 3, 4, and 5 being presently assigned on machines 1, 4, 3, and 2, respectively, the time when they are available for next assignments are the processing time for those jobs on the corresponding machines, or 4, 6, 2, and 9 for jobs 2, 3, 4, and 5, respectively. The remaining jobs are not assigned to any machines presently, and therefore, they are available right now, at time zero. These values are recorded in the last column of Table 12.11.

Next, we find the times when the machines are free to accommodate the next job assignment. Machines 1, 4, 3, and 2 are loaded with jobs 2, 3, 4, and 5, respectively, in this iteration and hence these times are the processing times for those jobs on the corresponding machines, or 4, 6, 2, and 9 for machines 1, 4, 3, and 2, respectively. These values are recorded in the last row of Table 12.11.

Iteration 2: Develop a work table, Table 12.12, in the same way as we did in Section 12.2. The entries in the first half of the table are the processing times of the jobs on the machines on which they are yet to be processed. The entries in the second half of the table display the minimum time when each job can be placed on each machine.

Select the job with the lowest slack. We select job 4 with a slack of 10. The minimum time when job 4 can be placed on a machine is 4, the time on machine 1. So assign job 4 to machine 1. Select the jobs with the next lowest slack. In this case, there are three jobs; jobs 1, 2, and 6 with the lowest slack of 11. Among them, we select the job with the earliest due date, in this case, job 2. The minimum time when job 2 can be placed on a machine is 3, which is the time on machine 3. We assign job 2 to machine 3. We select the next lowest slack. Again, we have two jobs with the same slack of 11. Between them, we select job 6, which has the earliest due date. The minimum time when it can be placed on the machines is 2, the time on M3 in the second half of Table 12.12 for job 6. But machine 3 is loaded with job 2, and hence we cannot assign job 6 to machine 3. We now select job 1 with a slack of 11. The minimum time when job 1 can be placed on a machine is 2, the time on machine 3. But machine 3 is occupied, so we cannot assign job 1 to it. We select the next job, job 3 with a slack of 13. The minimum time when job 3 can be placed on machines is 6, the time on machine 1. But machine 1 is occupied by job 4, and so it cannot accommodate job 3. The next job to be selected is job 5. The minimum time when it can be placed on machines is 9, the time on machines 3 and 4. Between them, we select the machine on which it has the largest processing time. Job 5 has the largest processing time of 3 on machine 4, and so we assign job 5 to machine 4.

Once the jobs are assigned to the machines, we must find the time when the jobs are available for the next assignment. Since jobs 2, 4, and 5 have been assigned in the current iteration, this time will be the sum of the processing time for these jobs on the machines on which they are assigned and the corresponding minimum time when these jobs can be placed on those machines. The numbers are obtained by looking at bold entries on both halves of the table. For example, this time for job 2 is $1+4=5$, for job 4 is $8+4=12$, and for job 5 is $3+9=12$. For the other jobs that are not assigned in this iteration, these times are the same as in the previous iteration. The times are noted in the central column TJA-next column of the Table 12.12.

Now, find the time when the machines will be free to accommodate the next job assigned to them. Since machines 3, 1, and 4 have been loaded with jobs 2, 4, and 5, respectively, this time is the sum of the processing time for those jobs on these machines and the corresponding minimum time when the jobs can be placed on those machines. This time is 1 + 4 = 5 for machine 3, 8 + 4 = 12 for machine 1, and 3 + 9 = 12 for machine 4. For the other machines, on which assignments are not made in the current iteration, the time remains the same as in the previous iteration. All the values are recorded in the last row of the first half of the Table 12.12.

Iteration 3: Following steps 3–6 and as shown in iteration 2, we assign job 2 to machine 2 and job 6 to machine 3 (Table 12.13).

Iteration 4: Following steps 3–6 and as shown in iteration 2, we assign job 1 to machine 3, job 3 to machine 2 and job 6 to machine 1 (Table 12.14).

Iteration 5: Following steps 3–6 and as shown in iteration 2, we assign job 1 to machine 4, job 3 to machine 1, job 4 to machine 2, and job 5 to machine 3 (Table 12.15).

Iteration 6: We assign job 1 to machine 1 and job 6 to machine 2 (Table 12.16).

Iteration 7: We assign job 1 to machine 2 (Table 12.17).

TABLE 12.13
Work Table II (Iteration 3)

	First Half								Second Half			
	Machines				Job Total (RTP)	Due Date	Stack	TJA-Next	Minimum Time When Jobs Can be Placed on Machines			
	1	2	3	4					M1	M2	M3	M4
Jobs 1	7	6	9	7	29	40	11	0	12	9	5	12
2	—	**3**	—	—	3	15	12	12	—	**9**	—	—
3	5	7	—	—	12	25	13	6	12	9	—	—
4	—	3	—	—	3	21	18	12	—	9	—	—
5	—	—	1	—	1	18	17	12	—	—	12	—
6	4	6	**3**	—	13	24	11	8	12	9	**5**	—
Machine total	16	25	13	7								
TMA-next	12	12	8	12								

TABLE 12.14
Work Table III (Iteration 4)

	First Half								Second Half			
	Machines				Job Total (RTP)	Due Date	Stack	TJA-Next	Minimum Time When Jobs Can be Placed on Machines			
	1	2	3	4					M1	M2	M3	M4
Jobs 1	7	6	**9**	7	29	40	11	17	12	12	**8**	12
2	—	—	—	—	—	15	—	**12**	—	—	—	—
3	5	**7**	—	—	12	25	13	19	12	**12**	—	—
4	—	3	—	—	3	21	18	12	—	12	—	—
5	—	—	1	—	1	18	17	12	—	—	12	—
6	**4**	6	—	—	10	24	14	16	**12**	12	—	—
Machine total	16	22	10	7								
TMA-next	16	19	17	12								

Results: Total makespan = 37, and total penalty = 5.

Sequence on machine 1: 2 - 4 - 6 - 3 - 1 with idle time = 3.
Sequence on machine 2: 5 - 2 - 3 - 4 - 6 - 1 with idle time = 3.
Sequence on machine 3: 4 - 2 - 6 - 1 - 5 with idle time = 2.
Sequence on machine 4: 3 - 5 - 1 with idle time = 8.

TABLE 12.15
Work Table IV (Iteration 5)

		First Half								Second Half			
		Machines				Job Total (RTP)	Due Date	Stack	TJA-Next	Minimum Time When Jobs Can be Placed on Machines			
		1	2	3	4					M1	M2	M3	M4
Jobs	1	7	6	—	**7**	22	40	18	24	17	19	—	**17**
	2	—	—	—	—	—	15	—	**12**	—	—	—	—
	3	**5**	—	—	—	5	25	13	24	**19**	—	—	—
	4	—	**3**	—	—	3	21	18	22	—	**19**	—	—
	5	—	—	**1**	—	1	18	17	18	—	—	**17**	—
	6	—	6	—	—	6	24	18	16	—	19	—	—
Machine total		12	15	1	7								
TMA-next		24	22	18	24								

TABLE 12.16
Work Table V (Iteration 6)

		First Half								Second Half			
		Machines				Job Total (RTP)	Due Date	Stack	TJA-Next	Minimum Time When Jobs Can be Placed on Machines			
		1	2	3	4					M1	M2	M3	M4
Jobs	1	7	6	—	—	13	40	27	31	**24**	24	—	—
	2	—	—	—	—	—	15	—	**12**	—	—	—	—
	3	—	—	—	—	—	25	—	**24**	—	—	—	—
	4	—	—	—	—	—	21	—	**22**	—	—	—	—
	5	—	—	—	—	—	18	—	**18**	—	—	—	—
	6	—	**6**	—	—	6	24	18	28	—	**22**	—	—
Machine total		7	12	—	—	—							
TMA-next		31	28	**18**	**24**								

Job 1 is early by 3.
Job 2 is early by 3.
Job 3 is early by 1.
Job 4 is late by 1.
Job 5 is on time.
Job 6 is late by

TABLE 12.17
Work Table VI (Iteration 7)

	First Half								Second Half			
	Machines				Job Total (RTP)	Due Date	Stack	TJA-Next	Minimum Time When Jobs Can be Placed on Machines			
	1	2	3	4					M1	M2	M3	M4
Jobs 1	—	**6**	—	—	6	40	34	37	—	**31**	—	—
2	—	—	—	—	—	15	—	**12**	—	—	—	—
3	—	—	—	—	—	25	—	**24**	—	—	—	—
4	—	—	—	—	—	21	—	**22**	—	—	—	—
5	—	—	—	—	—	18	—	**18**	—	—	—	—
6	—	—	—	—	—	24	—	**28**	—	—	—	—
Machine total	—	6	–	—								
TMA-next	**31**	37	**18**	**24**								

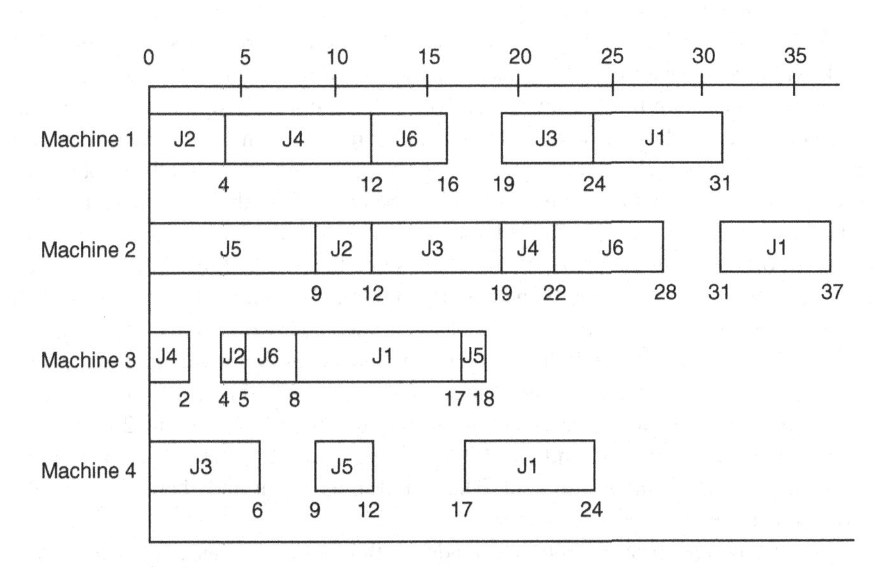

FIGURE 12.4 Gantt chart for the problem illustrated in section 12.3

12.3.6 Special Cases

12.3.6.1 Special Case 1 (Example)

Consider an open-shop problem in which four jobs are to be assigned on three machines. The processing time for each job on each machine and its due date are given in the job-machine table, Table 12.18. The job total and slack for each job are found and entered in the table. We wish to develop a sequence of operation on

TABLE 12.18
Job-Machine Table

		Machines			Job Total	Due Date	Slack	Time When Job Is Available for Next Assignment
		1	2	3				
Jobs	1	2	7	—	9	14	5	0
	2	3	—	7	10	15	5	0
	3	**4**	6	1	11	14	3	4
	4	2	**5**	—	7	10	3	5
Machine Total		11	19	8				
TMA-next		4	5	0				

each machine that will minimize the total tardiness. The sequencing procedure is illustrated.

As there are three machines, only three jobs can be selected. There is a tie between jobs 3 and 4 when selecting the jobs with the smallest slack. Break this tie by selecting job 4, which has earlier due date, and by assigning it to machine 2 on which it has the largest processing time. Then, select job 3. It cannot be assigned to machine 2, on which it has the largest processing time since that machine is not free. Assign it on machine 1, on which it has the next largest processing time. Already, 2 jobs have been selected and only one more selection can be made. But, there are two jobs (jobs 1 and 2) with the same slack. In such a case, both the tied jobs have to be selected, which is contradicting the previous statement since the total number of jobs selected becomes four, which is greater than the maximum number allowed, that is, 3. In such a special case, screening of all the tied jobs is allowed. However, only one would be checked further, since we only have one more possible assignment. Applying rule 2 of step 2 between the tied jobs, select the job with the earliest due date. In this case, select job 1. This job cannot be assigned to either machine 1 or machine 2 since both the machines are occupied. And since it does not require machine 3, job 1 remains unassigned in the initial assignment. The initial assignment ends here since 3 jobs have been checked.

The subsequent assignments are made in the same way as explained in the illustrative example (Section 12.3). The final solution to this problem is given below.

Total makespan = 18 and total penalty = 4.

Sequence on machine 1: 3 - 1 - 4 - 2 with idle time = 0.
Sequence on machine 2: 4 - 3 - 1 with idle time = 0.
Sequence on machine 3: 2 - 3 with idle time = 4.
Job 1 is late by 4.
Job 2 is early by 4.
Job 3 is early by 2.
Job 4 is early by 2.

TABLE 12.19
Job-Machine Table

		Machines			Job Total	Due Date	Slack	Time When Job Is Available for Next Assignment
		1	2	3				
Jobs	1	2	8	—	10	14	4	0
	2	3	—	**7**	10	14	4	7
	3	**4**	6	1	11	14	3	4
	4	2	**5**	—	7	10	3	5
Machine Total		11	19	8				
TMA-next		4	5	7				

12.3.6.2 Special Case 2 (Example)

Modify the previous example for this case. Change the processing time of job 1 on machine 2 is 8 instead of 7. Also change the due date of job 2 from 14 to 15. The new table is shown below (Table 12.19).

As there are only three machine, we can select only three jobs. There is a tie between jobs 3 and 4 when selecting the jobs with the lowest slack. Break this tie, by selecting job 4 with earliest due date and assigning it to machine 2 on which it has the largest processing time. Then, select job 3. It cannot be assigned to machine 2 on which it has the largest processing time since that machine is not free. Assign it on machine 1 on which it has the next largest processing time. Already 2 jobs have been selected, and only one more selection can be made. Now, there are two jobs (jobs 1 and 2) with the same slack and due date. In such a case, both the tied jobs have to be checked. Applying rule 3 or step 2, select any tied job randomly. Suppose we select job 1. This job cannot be assigned to either machine 1 or machine 2 since both the machines are occupied. In a normal case, we should stop here since as many jobs as the number of machines have been checked. But in this case, since the third assignment was based on random selection between the tied jobs, and since the selected job could not be assigned; we could select the other tied job, in this problem, job 2. This job can be assigned to machine 3 on which it has the largest processing time. Thus, the initial assignment is complete.

The subsequent assignments are made in the same way as explained in the illustrative example (Section 12.3). The final solution to this problem is given below.

Total makespan = 20, and total penalty = 6.

Sequence on machine 1: 3 - 1 - 4 - 2 with idle time = 0.
Sequence on machine 2: 4 - 1 - 3 with idle time = 1.
Sequence on machine 3: 2 - 3 with idle time = 0.
Job 1 is on time.
Job 2 is early by 3.

TABLE 12.20
Job-Machine Table

		Machines				Job Total	Due Date	Slack	Weights	Weight Slack	Time When Job Is Available for Next Assignment
		1	**2**	**3**	**4**						
Jobs	1	4	3	1	**7**	15	15	0	3	0	7
	2	5	7	—	6	18	25	7	0	M	0
	3	**8**	3	2	*v*	13	21	8	6	1.33	8
	4	—	9	1	3	13	18	5	0	M	0
	5	4	**6**	3	—	13	24	11	2	5.50	6
Machine Total		21	28	7	16						
Time when machine is free for next assignment		8	6	0	7						

Job 3 is late by 6.
Job 4 is early by 2.

12.4 MINIMIZATION OF TOTAL WEIGHTED TARDINESS PENALTIES–OPEN SHOP (MTWT-OP)

Heuristic in Section 12.3 (MTT-OP) can be extended to accommodate weighted tardiness penalties, that is, each job has a different penalty for finishing late. An example has been illustrated for such a problem.

The steps of the procedure are as follows:

12.4.1 Initial Assignment

Step 1: Following the same steps as in MTT-OP (Section 12.3), construct a job-machine table as shown in Table 12.20. The table shows the processing time for each job on the required machine, the due date, and the tardiness penalty for each. Calculate the total processing time for each job and the total processing time on each machine for the jobs that require that machine. Find the slack for each job by subtracting the total processing time for that job from its due date. Then find the weighted slack for each job by dividing the slack for each by its tardiness penalty. For jobs with zero tardiness penalties, assign a large value M as their weighted slack.

Step 2: Follow the same procedure explained in step 2 of MTT-OP. The only difference is that, in this case, jobs are selected on the basis of lowest weighted slack first instead of the lowest slack first, as was done in MTT-OP. (Note: Jobs with zero tardiness penalty are not assigned in the initial assignment.)

TABLE 12.21
Work Table I (Iteration 2)

	First Half											Second Half			
	Machines				Job Total	Due Date	Slack	Weights	Weight Slack	TJA-Next	Minimum Time When Jobs Can be Placed on Machines			
	1	2	3	4							M1	M2	M3	M4
Jobs 1	4	**3**	1	—	8	15	7	3	2.33	10	8	7	7	—
2	5	7	–	6	18	25	7	0	M	0	8(13)	6(13)	—	7(13)
3	—	3	2	—	5	21	16	6	2.67	8	—	8	8	—
4	—	9	**1**	3	13	18	5	0	M	1	—	6(15)	**0(1)**	7(10)
5	4	—	3	—	7	24	17	2	8.5	6	8	—	6	—
Machine total	13	22	7	9										
TMA-next	8	10	1	7										

12.4.2 Subsequent Assignments

Step 3: Develop a new work table as per step 3 of MTT-OP. The only difference between this table and the one in for MTT-OP is the addition of the weighted slack column, as shown in Table 12.21. The present slack for each job is calculated by subtracting the present job total time for each job from its due date. This value is then recorded in the slack column for that particular job. The weighted slack is calculated by dividing the slack by the tardiness penalty, and the value is placed under the weighted slack column. For the jobs that have a zero tardiness penalty, find the time when the jobs can be placed on the machines and also find the time when the jobs will be free for the next assignment if they are to be placed on these machines in the current iteration. The latter time is the sum of the time when the job can be placed on the machine and the processing time of the job on that machine. The value is placed in parentheses.

Step 4: Select the job with the minimum weighted slack. For that job, find the minimum time when it can be placed on the machines. Check if this time is less than the parentheses value of the jobs (with zero tardiness penalty) on that machine. If so, assign the selected job to that machine. If not, the selected job cannot be assigned in that iteration. If there are more than two machines on which the selected job can be assigned, place the job on the machine on which it has the largest processing time. If that machine is not available for assigning the selected job, place the job on the machine on which it has the next largest processing time. If no machines are available, move to the next job by selecting the job with the next lowest slack and repeating step 4. Once all the jobs, with tardiness penalty greater than zero are assigned, select the jobs with zero tardiness penalty. If more than one job has zero tardiness penalty, select the job with earliest due date. Find the minimum time when the job can be placed on a machine. For that machine, check if the parentheses value of that job is less than the time when jobs (with penalty value greater than zero) can be placed on that machine. If so, assign the job to that machine. If not, the job (with zero penalty) cannot be assigned in this iteration.

Step 5: Once all the possible assignments are made, find the time when the jobs will be available for the next assignment by following the same procedure as in step 5 of the MTT-OP.

Step 6: Then, find the time when the machine will be free to accommodate the next job by following the same procedure as in step 6 of MTT-OP.

Step 7: Repeat steps 3–6 till all the jobs have been assigned.

12.4.3 Illustration Example

The example problem in Section 12.3 is modified to include the weighted penalty for each job and is shown in the job-machine table, Table 12.20. Our aim is to develop a sequence of operation on each machine that will minimize the weighted total tardiness penalties. The sequencing procedure is illustrated next, as we follow the iterations.

Iteration 1 (Initial Assignment): Find the slack for each job by subtracting its due date from its job total. The values are entered under the slack column of Table 12.20. Find the weighted slack for each job by dividing its slack by its weighted penalty. Jobs having a zero tardiness penalty value are assigned an extremely high weighted slack of M. The resulting values are entered under the weighted slack column of Table 12.20.

Since there are only four machines, only four jobs can be selected. Select the job with the lowest weighted slack. In this case, select job 1. Assign it to machine 4, on which it has the largest processing time of 7. Select the job with the next lowest weighted slack. Job 3 is selected and assigned to machine 1, on which it has the largest processing time. Job 5 is selected next and assigned to machine 2. Until now, three jobs are selected and assigned. One more job remains to be selected. But there are no more jobs with a penalty value associated with them. So, we stop our initial assignment here. (Note: Jobs 2 and 4 should not be assigned in the initial assignment since there is no tardiness penalty value associated with them.)

As done in MTT-OP, we calculate the time when the machine is free and the time when the job is free for the next assignment.

Iteration 2: Similar to the MTT-OP, in this iteration, first eliminate the processing times for the jobs on the machines on which they have already been assigned and develop the work table as shown in Table 12.21. Then calculate the job total, machine total, slac,k and weighted slack for each job, and place it under the corresponding column/row. Next, find the minimum time when jobs can be placed on machines. This is also similar to the procedure followed in the previous method. The only difference arises in case of jobs with weighted slack of M. Besides calculating the minimum time when these jobs can be placed on machines, also find the time when these jobs will get free for the next assignment in case they are assigned on the machines in this iteration. The values in the parentheses depict the times. For example, for job 2, the minimum time when job 2 can be placed on machine 1 is 8. In case this job is assigned to machine 1 in this iteration, the time when job 2 will be free for the next assignment will be the sum of the minimum time when job 2 can be placed on machine 1 and the processing time of job 2 on machine 1, that is, 8 + 5 = 13. This value is placed in parentheses.

Select job 1, which has the lowest weighted slack. The minimum time when job 1 can be placed on machines is 7, the time on machines 2 and 3. Try to assign job 1 on machine 2 since it has a larger processing time on that machine. Before assigning job 1 on machine 2, check whether the minimum time when job 1 can be placed on machine 2, that is, 7 is less than the parentheses value on machine 2 for jobs with weighted slack of M. In this case, 7 is less than both 13 for job 2 and 15 for job 4. So assign job 1 to machine 2. Next, select job 3 with the next lowest weighted slack. The minimum time when three can be placed on a machine is 8; the time on machines 2 and 3. Machine 2 has already been assigned job 1. So, select machine 3. Check whether the minimum time when job 3 can be placed on machine 3, that is, 8 is less than the parentheses value on machine 3 for jobs with weighted slack of M. In this case, 8 is greater than 1 for job 4. So we cannot assign job 3 in this iteration. Select the next job. The minimum time when the selected job 5 can be placed on machine is 6, the time on machine 3. But this 6 is also less than 1, the parentheses value of job 4 on machine 3; so job 5 cannot be assigned. Since there are no more jobs with a tardiness penalty value, check for jobs with weighted slack of M. There are two jobs with weighted slack of M, jobs 2 and 4. Select the one with the earlier due date. Job 4 is selected. The minimum time when job 4 can be placed on machine is 0, the time on machine 3. Also, the parentheses value of job 4 on machine 3 is less than the time when jobs with weighted penalty can be placed on machine 3. So we assign job 4 on machine 3. Now, select job 2. The minimum time when this job can be placed on machine is 6, the time on machine 2. But machine 2 is occupied; so job 2 cannot be assigned in this iteration. Thus, in this iteration, job 1 is assigned on machine 2 and job 4 on machine 3.

Once the assignment is done, find the time when the jobs are free for next assignment and when the machines get free. This is done by following the same procedure described in MTT-OP (Table 12.22 through 12.25).

Iteration 3: Following the same procedure as in iteration 2, in this iteration, we assign job 1 to machine 1, job 3 to machine 3 and job 4 to machine 4.

Iteration 4: In this iteration, we assign job 1 to machine 3, job 2 to machine 4, and job 3 to machine 2.

Iteration 5: In this iteration, we assign job 4 to machine 2 and job 5 to machine 1.

Iteration 6: In this iteration, we assign job 2 to machine 1 and job 5 to machine 3.

Iteration 7: In this iteration, we assign the last job remaining, job 2, to machine 2.

Results: The final solution to this problem is given in the following.

Total makespan = 30 and total penalty = 0.

TABLE 12.22
Work Table II (Iteration 3)

		First Half										Second Half			
		Machines				Job Total	Due Date	Slack	Weights	Weight Slack	TJA-Next	Minimum Time When Jobs Can be Placed on Machines			
		1	2	3	4							M1	M2	M3	M4
Jobs	1	**4**	—	1	—	5	15	10	3	3.33	14	**10**	—	10	—
	2	5	7	—	6	18	25	7	0	M	0	8(13)	10(17)	—	7(13)
	3	—	3	**2**	—	5	21	16	6	2.67	10	—	10	**8**	—
	4	—	9	—	**3**	12	18	6	0	M	10	—	10(19)	—	**7(10)**
	5	4	—	3	—	7	24	17	2	8.5	6	8	—	6	—
Machine total		13	19	6	9										
TMA-next		14	10	10	10										

TABLE 12.23
Work Table III (Iteration 4)

		First Half										Second Half			
		Machines				Job Total	Due Date	Slack	Weights	Weight Slack	TJA-Next	Minimum Time When Jobs Can be Placed on Machines			
		1	2	3	4							M1	M2	M3	M4
Jobs	1	—	—	**1**	—	1	15	14	3	4.67	15	—	—	**14**	—
	2	5	7	—	**6**	18	25	7	0	M	16	14(19)	10(17)	—	**10(16)**
	3	—	**3**	—	—	3	21	18	6	3.0	13	—	**10**	—	—
	4	—	9	—	—	9	18	9	0	M	10	—	10(19)	—	—
	5	4	—	3	—	7	24	17	2	8.5	6	14	—	10	—
Machine total		9	19	4	6										
TMA-next		14	13	15	16										

Sequence on machine 1: 3 - 1 - 5 - 2 with idle time = 2.
Sequence on machine 2: 5 - 1 - 3 - 4 - 2 with idle time = 2.
Sequence on machine 3: 4 - 3 - 1 - 5 with idle time = 14.
Sequence on machine 4: 1 - 4 - 2 with idle time = 0.
Job 1 is on time.

TABLE 12.24
Work Table IV (Iteration 5)

	First Half										Second Half			
	Machines				Job Total	Due Date	Slack	Weights	Weight Slack	TJA-Next	Minimum Time When Jobs Can be Placed on Machines			
	1	**2**	**3**	**4**							**M1**	**M2**	**M3**	**M4**
Jobs 1	—	—	—	—	0	15	—	3	—	**15**	—	—	—	—
2	5	7	—	—	12	25	13	0	M	16	16(21)	16(23)	—	—
3	—	—	—	—	0	21	—	6	—	**13**	—	—	—	—
4	—	**9**	—	—	9	18	9	0	M	22	—	**13(22)**	—	—
5	**4**	—	3	—	7	24	17	2	8.5	18	**14**	—	15	—
Machine total	9	16	3	—										
TMA-next	23	22	15	16										

TABLE 12.25
Work Table V (Iteration 6)

	First Half										Second Half			
	Machines				Job Total	Due Date	Slack	Weights	Weight Slack	TJA-Next	Minimum Time When Jobs Can be Placed on Machines			
	1	**2**	**3**	**4**							**M1**	**M2**	**M3**	**M4**
Jobs 1	—	—	—	—	0	15	—	3	—	**15**	—	—	—	—
2	**5**	7	—	—	12	25	13	0	M	23	**18(23)**	22(29)	—	—
3	—	—	—	—	0	21	—	6	—	**13**	—	—	—	—
4	—	—	—	—	0	18	—	0	—	**22**	—	—	—	—
5	—	—	**3**	—	3	24	21	2	10.5	21	—	—	**18**	—
Machine total	5	7	3	—										
TMA-next	23	22	21	16										

Job 2 is late by 5.
Job 3 is early by 8.
Job 4 is late by 4.
Job 5 is early by 3.

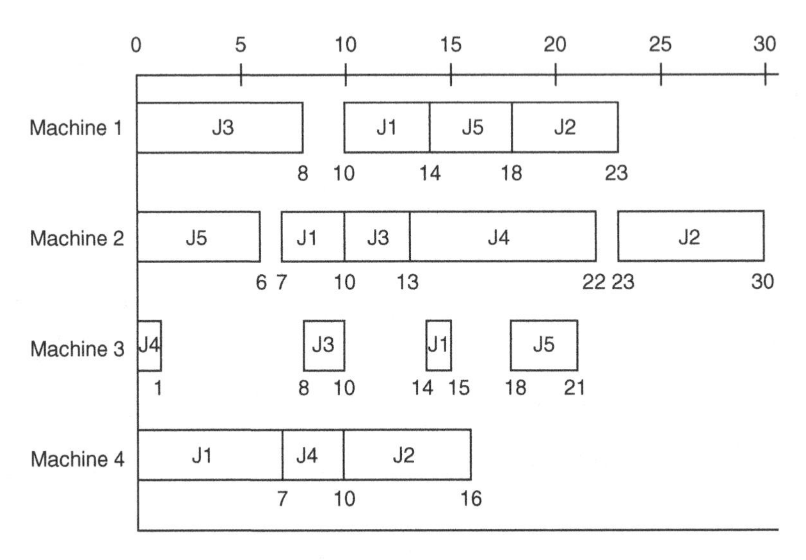

FIGURE 12.5 Gantt chart for the problem illustrated in section 8.4

12.5 SUMMARY

Scheduling in an open-shop problem presents a great challenge. There is much flexibility in job-machine assignment, and therefore, with proper planning, the idle or nonproductive times on the machines, could be reduced if not eliminated. However, because there are a large number of job-machine combinations that can be a part of possible solutions, it is almost impossible to evaluate all alternatives. The chapter presents Pinendo's rule, which gives an optimum makespan when there are only two machines in the open shop. An heuristic to minimize the makespan when we have multiple machines is also presented. The heuristic was tested with Pinendo's rule, and it produced optimum solution for every two-machine problem that was tested.

There are two additional heuristics that are presented in the chapter. The first minimizes the total tardiness, and the second minimizes the total of weighted tardiness penalties. Both are simple to apply and provide good results. The computer programs for the methods are included on the diskette provided.

12.6 PROBLEMS

12.1 Develop a schedule for the following two-machine open job shop job data using the LAPT rule.

	Job 1	Job 2	Job 3	Job 4
Machine 1	1	4	9	3
Machine 2	5	7	1	2

12.2 In an automobile manufacturing plant, six parts are to be processed on four machines. All six parts are independent of each other and can be processed on any machine required by them at any time. In other words, this is a job shop scheduling problem. Find the minimum makespan for processing all the jobs on all the machines required by them. The data are as follows:

	Machine A	Machine B	Machine C	Machine D
Job 1	7	6	9	7
Job 2	4	3	1	—
Job 3	5	7	—	6
Job 4	8	3	2	—
Job 5	—	9	1	3
Job 6	4	6	3	—

12.3 Consider the open-shop problem with six jobs having the following processing times:

	Job 1	Job 2	Job 3	Job 4	Job 5	Job 6
Machine 1	10	7	3	1	12	6
Machine 2	6	9	8	2	7	6

Find the minimum makespan for processing the jobs on the machines by using

a. The LAPT method

b. The LTPT method

12.4 The six jobs in Problem 12.2 are scheduled to be completed by certain due dates. The due dates for each job are as follows:

Job	1	2	3	4	5	6
Due date	40	15	25	21	18	24

Each job will bear a tardiness penalty equal to the number of days by which it is late. Find a schedule for all jobs on all the machines required by them such that the total tardiness penalty is minimized.

12.5 In an open-shop, five jobs are to be processed on three machines. Find a schedule that minimizes the total tardiness. Consider unit late penalties. The data are as follows:

	Machine A	**Machine B**	**Machine C**	**Due Dates**
Job 1	0	8	6	18
Job 2	4	1	8	20
Job 3	6	5	1	14
Job 4	4	7	0	18
Job 5	2	9	8	27

12.6 In Problem 12.5, jobs are assigned different tardiness penalty values:

Job	1	2	3	4	5
Late penalty value	2	3	7	5	1

Find a schedule to minimize the total weighted tardiness.

12.7 Using the same data as in Problem 12.6, except that jobs 2 and 4 carry no penalty for finishing late, find a schedule to minimize the total weighted tardiness.

REFERENCES AND SUGGESTED READINGS

Adiri, I. and N. Amit. September 1983. "Route Dependent Open Shop Scheduling" *IIE Transactions*, 15(3): 231–234.

Atwater, F.S., L.L. Bethel, G. Smith, and H.A. Stackman. 1959. *Essentials of Industrial Management*, 2d ed., New York: McGraw-Hill.

Choo, Y. and S. Sahni. May-June, 1981. "Preemptive Scheduling of Independent Jobs with Release and Due Times on Open, Flow and Job Shops" *Operations Research Journal*, 29(3): 511–522.

Conway, R.W., W.L. Maxwell, and L.W. Miller. 1967. *Theory of Scheduling*, Reading, MA: Addison-Wesley.

Gonzalez, T. and S. Sahni. 1976. "Open Shop Scheduling to Minimize Finish Times" *Journal of the Association for Computing Machinery*, 23: 665–679.

Liu, C.-Y. and R.L. Bulfin. July-August, 1988. "Scheduling Open Shops with Unit Execution Times to Minimize Functions of Due Dates" *Operations Research Journal*, 14(3): 257–264.

Lin, H.-F., C.-Y. Liu, and P.-Y. Liu. 1955. "A Heuristic Approach to the Total Tardiness in Non-Preemptive Open-Shop Scheduling" *International Journal of Industrial Engineering*, 2: 25–33.

Pinedo, M. 1955. *Scheduling: Theory, Algorithms, and Systems*, Englewood Cliffs, NJ: Prentice Hall.

13 Manpower Scheduling

Every organization needs manpower to function, regardless of how automated the operations are. The manpower requirement may vary considerably, however, depending on the type of service provided and the demand for such services. Some production facilities, for example, have highly seasonal demands (e.g., toys at Christmas, lawn mowers in summer) and may employ workers, which is also very seasonal. Other organizations have fairly uniform demand throughout the year and may have more or less a constant workforce. In a few organizations, the demand may vary from day to day (in a hospital, for example, the number of nurses required on a given day of the week may depend on days the operations are planned during the week) or even from hour to hour within a day (the number of employees on checkout counters in a supermarket may vary by time of the day and by the day of the week).

Even the employees have variations among them and may be further divided into classes. Some class of employees, such as purchasing agents, are required only on a day shift, while others, such as production workers, are required on all three shifts in a 24-hr-per-day operation (steel mills, refineries, and hospitals). We concentrate here on manpower scheduling problems associated with businesses that work 7 days a week. The examples include organizations such as telephone and cable TV stations, restaurants, fast-food places, grocery stores, and supermarkets. The employees work a standard 40 hr or 5 days a week and get two consecutive (if possible) days off. The objective is to develop a schedule that satisfies the daily requirements and, at the same time, employ a minimum number of workers.

The manpower scheduling problem has become more prominent with changing lifestyles. With the aid of an information highway, a growing portion of the population is preferring to work flexible hours. Organizations, especially service institutions, must respond to demands that are no longer restricted to the business hours of 8:00 AM to 5:00 PM. Thus, the purpose of employee scheduling is to ensure that employees are working at the time when they are needed and are meeting customer demands for goods and/or services. Employee scheduling is necessary to utilize manpower effectively and efficiently. Some simple but effective algorithms have been developed to produce optimal manpower schedules. The generally preferred linear programming models for such problems, rather than the alternatives, are effective and give near-optimal solutions with considerable less effort and time.

In the following sections, we present a number of algorithms that work for specific scheduling problems. These may be classified according to the following criteria:

- On off-day pattern-examples: consecutive days off, weekdays off, nonconsecutive days off
- The workforce, that is, homogeneous or heterogeneous
- Varying daily workforce requirement

A few terms commonly used in employee scheduling may need further clarification. These are

1. Homogeneous workforce: The workforce is considered to be homogeneous when the number of available hours of each worker is the same and workforce requirement remains constant throughout the shift. Examples: most full-time employees in manufacturing industries fall in this category.
2. Heterogeneous workforce: The workforce is considered to be heterogeneous when the available hours vary from one employee to another, and/or the workforce requirement varies within a shift. Examples: fast-food restaurants. Scheduling with a heterogeneous workforce is referred to as tour scheduling.
3. Work stretch: It is defined as the number of days an employee is scheduled to work without any off-days in-between. Normally, the work stretch is 4 or 5 days in duration.

In every employee schedule, legal requirements and labor/management agreements prescribe permissible days-on and days-off. Usually, 40 hr of work is translated into a permissible work stretch of 5 days, so that every employee is given 2 days off in a week. The scheduling problem involves planning days-on and days-off for each worker, and still meeting the daily minimum manpower requirement of the organization.

If the total number of workers employed is W and if each person works 5 days a week, then we have $5 \times W$ workdays available in a week. If the number of workers required on day i is r_i, $i = 1, 2, \ldots, 7$, then the weekly requirement is Σr_i. For a schedule to be feasible, one condition is that the weekly manpower available be at least equal to the total manpower required, or $5 \times W >= \Sigma r_i$. If $5 \times W$ is strictly greater than Σr_i, then we have excess of manpower available, and the difference between what is available and what is required is called slack. The second condition for the schedule to be feasible is that for any given day the manpower available must be equal to or greater than the manpower required on that particular day. The efficiency of the schedule is defined as: (the total employee workdays required in a week)/(the total employee workdays available in a week).

13.1 CONSECUTIVE DAYS-OFF SCHEDULING

Tibrewala, Philippe, and Browne (1972) (TPB) developed an efficient algorithm for generating a schedule that gives every employee 2 consecutive days off per week. In addition, the requirement may change from day to day. We describe here their algorithm with a few changes using following notations.

Notations

r_i = requirements for the i-th day in the cycle, $i = 1, 2, 3, 4, \ldots, 7$.

s_k = number of men idle for the pair of days $(k, k+1)$. Initially all s_k's are set to zero.

TABLE 13.1
Number of Workers off Each Day of the Week

Monday	Tuesday	Wednesday	Thursday	Friday	Saturday	Sunday
$s_7 + s_1$	$s_1 + s_2$	$s_2 + s_3$	$s_3 + s_4$	$s_4 + s_5$	$s_5 + s_6$	$s_6 + s_7$

Procedure

Step 1

1. Choose 2 consecutive days $(k, k + 1)$ for each value of k, $k = 1, 2, \ldots, 7$. Since the same demand pattern repeats itself each week, the demand for $k = 8$ is also the demand for $k = 1$. Determine the maximum demand in each pair. We get seven values, one from each pair. Determine the minimum of these seven values, and call it m. Determine the pair and associated k, from which m is selected.
2. If $(k, k + 1)$ is not unique, choose, from among the tied ones, the pair that has a minimum sum of the requirements.
3. If $(k, k + 1)$ is still not uniquely determined, select any pair among the tied ones, such that $(k - 1, k)$ is not member of the set of tied pairs (illustrated in example).
4. If there is no such pair, then all r_i must be equal, in which case any pair $(k, k + 1)$ can be chosen.

Step 2: Increase by 1 the value of s_k for k chosen in step 1, and decrease by 1, all r_i's, except for $i = k$ and $k + 1$.

Step 3: Repeat steps 1 and 2 until all requirements are satisfied, that is, until all r_i values as modified in step 2 are reduced to zero.

The value of each slack, s_i resulting from the procedure indicates the workers off on 2 consecutive days i and $i + 1$. The total number of workers off on a given day i, is given by the sum of two slack values, as shown in Table 13.1. Also, the total workforce W, is equal to the sum of all slack values, that is,

$$W = s_1 + s_2 + s_3 + s_4 + s_5 + s_6 + s_7$$

13.1.1 Illustrative Example

The requirement for each day of the week is as given in Table 13.2. We wish to develop an efficient schedule that gives two consecutive days off to each worker.

The procedure is displayed in Table 13.3. The daily requirements r_i ($i = 1, 2, 3, \ldots, 7$) are shown in column 2 (iteration 1). The maximum value of the worker requirement in the first pair, that is, with $k = 1$, is Max (6, 8) = 8. Similarly, the maximum value in the next pair, that is, $k = 2$, is Max (8, 10) = 10. Proceeding in the same manner, maximum value in the last pair is the evaluation between Sunday

TABLE 13.2
Daily Worker Requirement

Days	Monday	Tuesday	Wednesday	Thursday	Friday	Saturday	Sunday
Requirements	6	8	10	7	12	4	2

TABLE 13.3
Application of TPB Algorithm

Iteration	1	2	3	4	5	6	7	8	9	10	11	12	13
r_1	6	5	4	3	3	2	2	1	1	0	0	0	0
r_2	8	7	6	5	4	3	2	1	1	0	0	0	0
r_3	10	9	8	7	6	5	4	3	2	1	0	0	0
r_4	7	6	5	4	3	2	1	0	0	0	0	0	0
r_5	12	11	10	9	8	7	6	5	4	3	2	1	0
r_6	4	4	4	4	3	3	2	2	1	1	0	0	0
r_7	2	2	2	2	2	2	2	2	1	1	0	0	0
k	6	6	6	7	6	7	6	1	6	1	6	6	
Case	*a*	*a*	*c*	a	c	a	c	a	c	*a*	*c*	*c*	

$s_1 = 2$; $s_2 = 0$; $s_3 = 0$; $s_4 = 0$; $s_5 = 0$; $s_6 = 8$; $s_7 = 2$; $W = 12$.

and Monday requirements, that is, between 2 and 6, the maximum value is 6. If we list these seven values, they are 8, 10, 10, 12, 12, 4, and 6. The minimum among these is 4; therefore $m = 4$. The pair where this value is associated has $k = 6$, which is noted in the row denoted by k. The condition from step 1 that made us choose $k = 6$ value is condition a which is noted in the row marked Case. All r_i's are decreased by one except for the pair with $k = 6$, and the values are displayed in the next column, which form the data for iteration 2. The procedure continues in each iteration. Next, we describe one sample iteration of interest.

Iteration 3 shows how step 1.3 may be applicable. The maximum value in each pair is 6, 8, 8, 10, 10, 4, and 4, respectively. The minimum among these is four associated with two pairs, one with $k = 6$, and other with $k = 7$. To break the tie, we apply step 1.2. The sums of the current requirements (data from iteration 3) for the associated pairs are $4 + 2 = 6$ and $2 + 4 = 6$, respectively. Again, we have a tie, so we go to step 1.3. The elements $(k - 1, k)$ for each pair are (10, 4) for $k = 6$ and (4, 2) for $k = 7$. The pair (10, 4) is not the member of tied pair as against (4, 2) is, (again a pair that was tied in step 1.1), and hence $k = 6$. The requirements for all days except 6 and 7 are reduced by 1 to form the next cycle data.

The s_k's can be easily obtained from Table 13.3 by adding the number of times k has a value that is equal to i. For example, k has value of 6 in iterations 1, 2, 3, 5, 7, 9, 11, and 12; hence $s_6 = 8$. The total workers required is the sum of all s_i's, which

TABLE 13.4
Daily Availability and Requirements Based on the Schedule

Days	Monday	Tuesday	Wednesday	Thursday	Friday	Saturday	Sunday	Total
Workers off	4	2	0	0	0	8	10	
Workers on	8	10	12	12	12	4	2	60
Workers required	6	8	10	7	12	4	2	49

is 12. The scheduling efficiency is $49/60 = 0.817$ or 81.7%. Table 13.4 shows the summary describing how many workers are working and are off on each day.

13.2 ROTATING DAYS (WEEKENDS) OFF

In the previous section, we presented an algorithm that gives each worker two consecutive days-off during a week. In this section, we present a method that gives two specific days off, namely weekends, to as many workers as possible, but then gives some workers nonconsecutive days off to keep the total workforce, W, in a reasonable limit. Such a policy maintains high efficiency of the schedule and satisfies the common need to have weekends off.

The arrangement can be made fair to all workers by rotating each employee through the program rather than assigning him/her with a fixed time slot. In the schedule, where some workers are assigned weekends off while others are assigned nonconsecutive days off, a fixed schedule for an employee indicates preferential treatment while the rotating schedule creates equality. On a rotating schedule, all employees eventually work through each time slot with associated on and off days.

The algorithm presented here is developed by Burns and Carter (1985) (BC). It has the following characteristics:

1. The number of employees required each day of the week can vary.
2. The algorithm solution meets the following constraints:
   a. Each employee works exactly 5 days per week.
   b. Each employee has every other weekend off (or each worker is given at least A out of every B weekends off).
   c. No one works more than 6 consecutive days.

The algorithm starts by determining the minimum number of employees, W, needed to satisfy the weekly requirements. W is calculated based on three lower bounds set as follows:

1. Weekend constraint ($L1$): the number of employees available each weekend must be sufficient to meet the maximum weekend demand. Also, an employee who works on one weekend must have following weekend off.

Hence,

$$W >= 2n$$

where, n = maximum weekend demand.

2. Total demand constraint ($L2$): the total number of employees must be sufficient to meet the total weekly demand.
   Total weekly demand

$$\text{Workforce } W \geq \left[\frac{1}{5} \sum_{l=1}^{7} n_i \right]$$

   when [ ] is the higher integer value of the resultant.
3. Maximum daily demand constraint ($L3$): The number of employees must be sufficient to meet the maximum demand on any day.

$$W = \text{Max}\{n_i\}, i = 1, 2, 3, \ldots, 7.$$

Procedure

The Burns and Carter algorithm considers a span of B weeks (2 weeks, in our case), to design A (one, in our case) weekends off for each employee. More often than not, we are looking at scheduling multiple weeks. The steps of the procedure are as follows:

Step 1: *Compute the minimum workforce.* Calculate the three bounds $L1$, $L2$, and $L3$, as shown earlier. The minimum workforce (W) required is the maximum of these three values.

Step 2: *Schedule the weekends off.* We need at least n (a maximum requirement on any 1 day during the weekend) workers to work on any day during the weekend, so the remaining ($W - n$) workers can be given the weekend off. Assign the first weekend off to the first ($W - n$) workers, assign the second weekend off to the next ($W - n$) workers, and continue this process until repetition occurs.

Step 3: *Determine the additional off-day pairs.* For any week, each employee must have exactly 2 days off. Since we have already assigned ($W - n$) Sundays and ($W - n$) Saturdays off to some employees, we must assign an additional $2n$ days off to other employees, so that all W workers will have exactly 2 days off in a week. To do this, calculate the surplus of employees during the weekday by subtracting the daily requirement from W, chosen in step 1. That is,

Surplus $Sj = W - n_j$, where n_j = daily requirements Monday through Friday ($j = 2, 3, \ldots, 5$).

Since we have n workers available on the weekend, the weekend surplus is obtained by subtracting the daily requirements on Saturday and Sunday from the maximum requirement(s) on any day during the weekend. This daily surplus is used to determine which days some workers can be off ($2n$ days, as stated in the earlier step) as required.

TABLE 13.5
Employee Categories for BC Algorithm

Category	Weekend 1	Week 1	Weekend 2
Type T1	Off	(No off days needed)	Off
Type T2	Off	(1 off days needed)	On
Type T3	On	(1 off days needed)	Off
Type T4	On	(2 off days needed)	On

TABLE 13.6
Daily Requirements

Days	Monday	Tuesday	Wednesday	Thursday	Friday	Saturday	Sunday
Requirements	2	6	8	10	7	12	4

Next, iteratively construct a list of n pairs of off days (numbered from 1 to n) as follows:

1. Choose day k such that $S_k = \text{Max}\ \{S_j\}$
2. Choose any $i! = k$ such that $S_i > 0$. If $S_i = 0$ for all $i! = k$, set $i = k$.
3. Add the pair $\{k, i\}$ to the list and decrease S_i and S_k by 1.
4. Repeat this procedure n times.

Step 4: *Assigning off-day pairs in week 1.* In week 1, employees will fall into one of four categories, depending on the weekends off at the beginning and end of the week, as shown in Table 13.5.

Each weekend, we have n employees working, therefore |T2|ggT3|. This identity allows us to pair a T2-type employee with a T3-type employee. If the T2 employee gets the "earliest" day off, then the associated T3 employee gets the "latest" day off. In this way, each employee of type T4 gets 2 days off, and each one with T2 and T3 gets 1 day off during Monday through Friday, as required.

The following numerical example illustrates this algorithm.

13.2.1 Illustrative Example

We illustrate the application of Burns and Carter algorithm by applying it to the same data as used in TDK algorithm. For convenience, the data from Table 13.2 are reproduced in Table 13.6, with the week starting on Sunday. Our objective is to give each employee one weekend off in the work stretch of 2 weeks.

TABLE 13.7
Assigning Weekends and Weekdays Off

Empl.	S	S	M	T	W	T	F	S	S	M	T	W	T	F	S	S
1	0	0						0	0							
2	0	0						0	0							
3	0	0						0	0							
4	0	0						0	0							
5	0	0													0	0
6	0	0													0	0
7	0	0													0	0
8	0	0													0	0
9								0	0						0	0
10								0	0						0	0
11								0	0						0	0
12								0	0						0	0

The first step is to calculate the minimum workforce (W) required within the three constraints (L_1, L_2, L_3):

Weekend constraint: $W >= 2n = 2 \times 4 = 8$.

Total demand constraint

$$\text{Workforce } W \geq \left[\frac{1}{5} \sum_{1}^{7} n_i \right] = \lceil 49/5 \rceil = 10$$

Maximum daily demand constraint: $W = 12$

We choose W as the maximum value of the three values, and hence $W = 12$.

Next, we apply step 2, and start to create an employee schedule for each of 12 employees over a 2-week period as shown in Table 13.7. The table has 12 rows; one for each employee and 14 columns; one for each day in a 2-week period (we have shown 16 columns to show following weekend assignment). Since we require a maximum of four workers on the weekend (maximum demand on weekend), another $12 - 4 = 8$ can be given weekends off. Thus we assign employees 1 through 8, the first weekend off (shown by 0), employees 9 through 4 (next 8 employees, in rotation), second weekend off, 5 through 12, the third weekend off, and so on.

Next, we must decide how to give an additional $4 \times 2 = 8$ days off during the week. For this, we must decide on four "pairs." The calculations are shown in Table 13.8. We start by calculating the surplus of workforce for each day. For Monday through Friday, it is obtained by subtracting the daily requirement for each day from the theoretical minimum workforce, that is, W, which is 12, and for the weekend by subtracting the weekend requirement from the maximum requirement of 4, during the weekend. These values are shown in Table 13.8 for the first four iterations.

TABLE 13.8
Surplus Calculation and Pair Determination

Iteration	Requirements	Sunday (2)	Monday (6)	Tuesday (8)	Wednesday (10)	Thursday (7)	Friday (12)	Saturday (4)
1	Surplus	2	6	4	2	5	0	0
	Pair 1		1			1		
2	Surplus	2	5	4	2	4	0	0
	Pair 2			1		1		
3	Surplus	2	5	3	2	3	0	0
	Pair 3		1	1				
4	Surplus	2	4	2	2	3	0	0
	Pair 4		1		1			
	Surplus	2	3	2	1	3	0	0

Select the 2 days that have maximum surplus, and form it as the first pair. Here, Monday and Thursday are selected as the elements of the first pair since they have the maximum surplus. In order to decide the next pair, reduce the surplus on this pair (Monday and Thursday) by 1 employee and again (in the second iteration) select a pair of days with maximum surplus but *not the same pair if possible*, that was chosen before. So, we have chosen 4 and 4 pair (Tuesday, Thursday) rather than 5 and 4 pairs (if we had chosen 5 and 4 pairs, we still get a solution to the problem; it will just be different from what is illustrated). Continue this process until we have four (since $n = 4$) pairs.

The following step shows assigning additional days off.

Since employees 1 through 4 have both weekends off, they belong to type T1. Employees 5 through 8 belong to type T2 since they have the first weekend off and work during the second weekend. Employees 9 through 12 are type T3 since they work the first weekend and have the second weekend off. We do not have any type T4 employees.

Since each employee must have 2 days off in a week, we have to assign additional days off to employees 5 through 12, that is, to type T2 and type T3 employees (type T1 employees already have 2 days off in a week, so they do not need any additional days off). In the previous step, we have developed four pairs. In pair 1, we have Monday and Thursday as the pair. The decision is as follows: if a type T2 (T3) employee is given Monday off, then a type T3 (T2) employee should be given Thursday off. Following the same rule, their assignments are made and displayed in Table 13.9 and also in Table 13.10. The arrows in Table 13.9 indicate the pair of days off matched with the employees, number 9 employee with number 5, number 10 with number 6, and so on. It should be noted that in week 2, type 1, 2, and 3 employees are not the same as those in week 1, and the assignment changes. The assignments are shifted by four, the maximum weekend requirement. Employee number 5 is matched with number 1, number 6 with number 2, and so on. After a number of weeks, the assignments will repeat, forming a cycle. The scheduling efficiency is $98/120 = 0.8167$ or 81.67%.

TABLE 13.9
Assigning Weekends and Weekdays Off

Employee	*S*	*S*	*M*	*T*	*M*	*T*	*F*	*S*	*S*	*M*	*T*	*W*	*T*	*F*	*S*	*S*
1	0	0						0	0				0			
2	0	0						0	0				0			
3	0	0						0	0		0					
4	0	0						0	0			0				
5	0	0				0				0					0	0
6	0	0				0					0				0	0
7	0	0		0						0					0	0
8	0	0			0					0					0	0
9			0					0	0						0	0
10				0				0	0						0	0
11			0					0	0						0	0
12			0					0	0						0	0

TABLE 13.10
Non-Weekend Days Off

Employee Number	Days Off	Employee Number in T2 and His/Her Day Off in Week 1	Employee Number in T3 and His/Her Day Off in Week 1
1	M, Th	9, Monday	5, Thursday
2	T, Th	10, Tuesday	6, Thursday
3	M, T	11, Monday	7, Tuesday
4	M, W	12, Monday	8, Wednesday

13.3 MONROE'S ALGORITHM

Monroe (1972) developed a simple algorithm in 1972 that is still quite popular. The algorithm tries to give 2 consecutive days off (not necessarily the weekends as in BC) with least number of workers, W. If it is not possible, then we have three choices:

1. See if the daily requirements can be modified by shifting the portion of requirements from 1 day to another, so that W workers can be given two consecutive days off.
2. Increase the number of workers in excess of W, so that we are able to give two consecutive days off.
3. Give nonconsecutive days off to some workers.

Procedure

The basic procedure is described in the following steps.

Step 1: Calculate the theoretical minimum number of employees (W) needed as

$W = \text{Max}\ [T/5, n_j]$

where T = sum of the daily requirements.

n_j = Maximum number of employees required on any day.

W must be an integer. If $T/5$ is not an integer, round it up to the higher integer. The additional capacity thus created may be used to increase the worker assignments for the days having highest workload per man.

Step 2: For each day, calculate the required days off (RDO) by subtracting each day's requirement from W.

Step 3: Construct consecutive day pairs such as MT (Monday–Tuesday), TW (Tuesday–Wednesday), WT (Wednesday–Thursday), ..., SS (Saturday–Sunday), SM' (Sunday and Monday of the next week), and M'T' (Monday–Tuesday of the next week). Assign about half of the Tuesday's RDO to an MT pair. Assign the remaining RDOs to consecutive day pairs in the manner shown in step 4.

Step 4: Subtract the MT pair value from Tuesday's RDO and assign the remaining value to the TW pair. Subtract TW assignment from Wednesday's RDO, and assign the remaining value to WT pair; continue this procedure until we have assigned a value to M'T' pair.

Step 5: If M'T' value is equal to MT value, and all assignments are positive, go to step 6. If they are not equal, or if some trial values are negative, a second iteration is needed (if the second iteration was performed, go to step 7). The initial assignment for the second iteration to MT pair is the average value of MT and M'T' values in the first iteration. Apply the procedure from step 4 to get the assignments for the remaining pair of days. At this point, if there is a solution to the problem without further modification, then the MT value is equal to the M'T' value, and the values for the all day pairs are positive go to step 6. If this is not the case, go to step 7.

Step 6: The numbers assigned in the first trial, if feasible (from top of step 5) or of needed the second trial, indicate the number of workers off on two consecutive days as indicated by the pair.

Step 7: We come to this step only because the solution is not feasible at this point. Let n be the largest negative number in the second trial. There are three ways that may make the solution feasible:

1. Shift the RDOs between days so that when steps 1 through 6 are applied again there is a feasible solution. It may be accomplished if we can increase by n, RDO for a day in the pair where there is the maximum negative value in the second iteration and decrease a positive RDO by equal amount without making it negative.
2. Increase the number of workers by n.
3. Develop a sequence where there are some nonconsecutive days off. This number should, however, be minimum, and the minimum is n. Assign n

TABLE 13.11
Daily Requirements for Example II

Days	Monday	Tuesday	Wednesday	Thursday	Friday	Saturday	Sunday
Requirements	5	4	3	5	5	4	3

TABLE 13.12
Application of Monroe's Algorithm to Example II

Days	Monday	Tuesday	Wednesday	Thursday	Friday	Saturday	Sunday	Next Mond M′
Requirements	5	4	3	5	5	4	3	5
Excess			1					
Total	5	4	4	5	5	4	3	5
RDOs required	1	2	2	1	1	2	3	1
Consecutive day pairs	MT	TW	WTh	ThF	FS	SS	SM′	M′T′
First trial	1	1	1	0	1	1	2	−1
Second trial	0	2	0	1	0	2	1	0

nonconsecutive pairs of RDOs, where n is the largest negative number in the final pairs. One day of each nonconsecutive pair must be assigned to the day of the week with the largest number of RDOs.

13.3.1 Illustrative Example II

A repair facility that works 7 days a week has the daily worker requirements as shown in Table 13.11. Wednesday has been typically a slow day, and only three workers are scheduled on that day. We wish to develop the worker schedule that gives two consecutive days off to each worker.

The application of Monroe's algorithm is displayed in Table 13.12. The total of the daily requirements for a week is 29. Since each person is available to work 5 days per week, we need a minimum of $[29/5] = 6$ workers (the notation $[a/b]$ indicates the next higher integer value of the division a/b, if the division results in a noninteger). All of the daily requirements are less than 6; therefore, set $W = 6$. The total workdays available with six workers are $6 \times 5 = 30$. Hence, there is one extra working day available. We assign that to the day with minimum manpower on the assumption that work load per person on that day would be the highest and assignment of an additional worker will alleviate this overload.

Table 13.12 displays all days of the week plus Monday of the next week. The Monday of the next week is essential since, in developing consecutive day pairs, we go from MT of the present week to M'T' of the next week, giving eight pairs of days in total. The daily requirements are noted, and the excess man–day available is assigned to Wednesday, since that is the day where the requirement is minimum. The total number of workers to be allocated per day is noted in the row called total. Each day's RDOs are determined by subtracting the daily total from W. For example, Monday's RDO is $(6 - 5) = 1$. The next row denotes the pair of consecutive days.

All allocations in the first and second trial must be integers. The allocation to MT in the first trial is half of Tuesday's RDO (if the Tuesday's RDO is odd, assign only the integer portion of the division). The successive allocations to the pairs are found by subtracting from the next day's RDO the allocation made to the preceding pair. For example, TW allocation is RDO for Tuesday—allocation to MT, which is equal to $2 - 1 = 1$. Similarly, Wednesday's allocation is (RDO for W – TW allocation) = (2 – 1) = 1. The allocations in the first trial are complete when an allocation to M'T' is made.

Again, the second trial allocation starts with MT allocation. However, this value now should be half of the assignments for MT and M'T' in trial 1 (if the division is not an integer, take the integer portion of the division). Other values for the pairs are calculated applying the same rule that we have used in trial 1. For example, the TW assignment is: (RDO for Tuesday – MT assignment in the second trial) = (2 – 0) = 2.

Complete the allocation to all pairs including M'T'. If there are no negative values and if the MT allocation is the same as the M'T' allocation, then we have a solution. The value associated with each pair in the second trial indicates the number of workers off on those 2 consecutive days. For example, two workers are off on Tuesday and Wednesday, one worker is off on Thursday and Friday, and so on. It is interesting to note that in total three workers are off on Sunday, two from the Saturday–Sunday pair and one from the Sunday–Monday pair.

13.3.2 Illustrative Example III

Suppose now the daily requirements are slightly modified, as shown in Table 13.13. The application of Monroe's algorithm (shown in Table 13.13) results in an infeasible solution since the second trial assignments are negative. There are three ways to obtain a feasible solution, as shown in Tables 13.14, 13.15, and 13.16.

13.3.3 Realignment of a worker

Table 13.14 shows the analysis when one worker is realigned. Only one worker is chosen for realignment since the maximum negative value in the second trial in Table 13.13 is -1. Monday requirement is decreased by one, since Monday (MT pair under M) was one of the days in Table 13.13 that shows -1 assignment in the second trial. To compensate for decreasing the requirement of Monday, we increase the requirement for Sunday, since Sunday has the least requirement. The Table 13.14 shows that we do have a feasible solution. Thus, if it is possible to realign the requirements, then solution may be feasible.

TABLE 13.13
Data and Solution of Example III

Days	Monday	Tuesday	Wednesday	Thursday	Friday	Saturday	Sunday	Next Mond M′	Weekly Total
Requirements	5	4	3	5	4	5	3	5	29
Excess			1						
Total	5	4	4	5	4	5	3	5	30
RDOs	1	2	2	1	2	1	3	1	
Required Consecutive day pairs	MT	TW	WTh	ThF	FS	SS	SM′	M′T′	
First trial	1	1	1	0	2	−1	4	−3	
Second trial	−1	3	−1	2	0	1	2	−1	

TABLE 13.14
Realign Working/RDO Distribution

Days	Monday	Tuesday	Wednesday	Thursday	Friday	Saturday	Sunday	Next Mond M′	Weekly Total
Requirements	4	4	3	5	4	5	4	4	29
Excess			1						
Total	4	4	4	5	4	5	4	4	30
RDOs	2	2	2	1	2	1	2	2	
Required Consecutive day pairs	MT	TW	WTh	ThF	FS	SS	SM′	M′T′	
First trial	1	1	1	0	2	−1	3	−1	
Second trial	0	2	0	1	1	0	2	0	

13.3.4 Increasing the Workforce

Another way to obtain feasibility is by increasing the workforce by an amount that equals the maximum negative value in the second trial. In our case, it is -1, so we increase W from 6 to 7. But, by doing so, we have created five additional man–workdays. Assign them, one per day, to the days with the least initial requirements, as shown in Table 13.15. RDOs per day are now $W = 7$ minus the daily requirement. All the calculations of Monroe's algorithm are shown in Table 13.15. The final feasible solution is again given by the second trial numbers.

TABLE 13.15
Add Extra Worker

Days	Monday	Tuesday	Wednesday	Thursday	Friday	Saturday	Sunday	Next Mond M′	Weekly Total
Requirements	5	4	3	5	4	5	3	5	29
Excess			1						
Extra worker			1	1	1	1		1	
Total	5	5	5	6	5	5	4	5	35
RDOs	2	2	2	1	2	2	3	2	
Required Consecutive day pairs	MT	TW	WTh	ThF	FS	SS	SM′	M′T′	
First trial	1	1	1	0	2	0	3	−1	
Second trial	0	2	0	1	1	1	2	0	

TABLE 13.16
Assigning Nonconsecutive Days Off

Days	Monday	Tuesday	Wednesday	Thursday	Friday	Saturday	Sunday	Next Mond M′	Weekly Total
Requirements	5	4	3	5	4	5	3	5	29
Excess			1						
Total	5	4	4	5	4	5	3	5	30
RDOs	1	2	2	1	2	1	3	1	
Required Nonconsecutive					1		1		
Remaining RDO Consecutive day pairs	MT	TW	WTh	ThF	FS	SS	SM′	M′T′	
First trial	1	1	1	0	1	0	2	−1	
Second trial	0	2	0	1	0	1	1	0	

13.3.5 Assigning Nonconsecutive Days Off

We can resolve the non-feasibility in Table 13.13 by giving one employee (maximum negative number in second trial in Table 13.13) nonconsecutive days off. Table 13.16 shows the results. Nonconsecutive days are given (as far as possible) to the days where RDOs are maximum. The feasible solution is obtained.

13.4 TOUR SCHEDULING

So far, we have seen manpower scheduling, where the workforce has been a homogeneous group. In this section, we discuss the scheduling of workers that form a heterogeneous group. This is especially true in service industries such as fast-food places where the daily working hours are typically longer than 8, but not necessarily long enough to accommodate two or three full shifts. In addition, manpower requirements may vary from day to day and even from hour to hour.

We discuss here a simple algorithm developed by Bechtold and Showalter. We assume that each worker works for 8 hr with a 1 hr break for lunch, which is unpaid (thus, the shift is 9 hr long). We may have multiple shifts, each starting at a different time of day, to cover longer than 8 hr of operation.

Define for each day

r_t = The number of employees required during the tth time period
x_t = The number of employees assigned to the shift beginning in hour t
s = The number of possible shifts in each operating day
h = The number of hours in each operating day
n = The number of hours worked in each shift prior to the 1-hr meal break

13.4.1 Procedure

Phase I: Daily Schedule

In phase I, we develop the daily schedule for each shift. The procedure examines each day independently. For each day, we determine the number of workers to start on each shift, by applying the following steps.

Step 0: Initialize the number of employees starting in any shift by setting all shift assignments to zero, that is, set $x_i = 0$ for $t = (1 \text{ to } s)$.

Step 1: Set the first shift assignment to the first hour by requirement. Since these workers are working till their meal time and then taking a 1-hr break and working again, for the total of 8 hr, reduce the corresponding hour's requirements by the first hour's requirement (minimum being 0) for the hours where the workers are working.

That is,

$x_t = r_1$ and $r_t = \text{Max}\ (r_t - r_1, 0)$ for $t = (1 \text{ to } n)$ and for $t = n + 2 \ldots., 9$.

Step 2: Set the workers starting on the last shift equal to the requirement of the last hour of operation. This is because, for the schedule to be feasible, at least the last shift must cover the last hour of the workday. Reduce the manpower requirement for the hours where these people will be working, by the last hour's requirement (the minimum value for the result is zero).

That is,

$x_s = r_h$ and $r_t = \max(r_t - r_h,\ 0)$ for $t = s \ldots, s + n - 1$ and for $t = s + n + 1 \ldots h$.

Step 3: If the requirements for all hours are reduced to zero, stop; we have a schedule. If not go to step 2.

TABLE 13.17
Weekly Demand Requirements (in Number of Employees)

Hours of the day	Days of the Week						
	Monday	Tuesday	Wednesday	Thursday	Friday	Saturday	Sunday
1	3	4	4	4	2	4	1
2	5	3	3	5	5	4	3
3	4	2	5	3	4	5	4
4	5	4	2	2	5	3	5
5	2	1	1	5	4	2	3
6	5	5	4	3	5	4	2
7	1	3	3	4	3	5	4
8	4	2	5	3	2	1	1
9	5	5	2	2	4	4	5
10	3	1	1	4	2	3	2
11	5	4	4	1	5	2	4
12	1	3	3	1	1	4	3

That is,
if $r_t = 0$ for $1 \leq t \leq h$, stop. Otherwise continue.

Phase II: Weekly Schedule

Phase II requires application of an algorithm to develop a weekly schedule. We will apply Monroe's algorithm to get the weekly days on and off routine for each shift.

13.4.2 Illustrative Example

The daily requirements on an hourly basis for a restaurant that is open 12 hr a day is shown in Table 13.17. The restaurant is situated in a business district and attracts customers mainly on weekdays, and night times on weekends. There are four possible shifts A, B, C, and D, starting on hour 1, 2, 3, and 4, respectively, to which a full-time employee can be assigned. Each employee works 4 hr, takes 1 hr break, and then works the remaining 4 hr. We wish to develop the hiring policies for the restaurant.

We illustrate the application of the algorithm by applying it to Monday's requirements. They are replicated in row 1 of Table 13.18. By following step 1, the three employees required in the first hour of the day (shown in bold letters in row 1) are assigned to the earliest daily work schedule A (shown in row 2 column A). These workers are working for first 4 hr, taking 1 hr off, and then working again for the next 4 hr. This fact is also noted in row 2. The remaining demand for each hour (row 3) is calculated by subtracting from row 1 the assigned number of employees in the hours that are worked as shown in row 2.

TABLE 13.18
Phase I Application to Monday Daily Shift Schedule Requirements

	Hours of Day												Shift Schedule			
	1	**2**	**3**	**4**	**5**	**6**	**7**	**8**	**9**	**10**	**11**	**12**	**A**	**B**	**C**	**D**
1. Number required	**3**	5	4	5	2	5	1	4	5	3	5	1				
Number assigned	**3**	3	3	3	0	3	3	3	3	0	0	0	3			
2. Number required	0	2	1	2	2	3	0	1	2	3	5	**1**				
Number assigned	0	0	0	1	1	1	1	0	1	1	1	**1**				1
3. Number required	0	**2**	1	1	1	2	0	1	1	2	4	0				
Number assigned	0	**2**	2	2	2	0	2	2	2	2	0	0		2		
4. Number required	0	0	0	0	0	2	0	0	0	0	**4**	0				
Number assigned			4	4	4	4	0	4	4	4	**4**				4	
5. Number required	0	0	0	0	0	0	0	0	0	0	0					
Employees assigned to shift schedule													3	2	4	1

TABLE 13.19
Employee Daily Shift Schedule Assignment

	Day of Week						
Shift Schedule	**Monday**	**Tuesday**	**Wednesday**	**Thursday**	**Friday**	**Saturday**	**Sunday**
A	3	4	4	4	3	5	4
B	2	0	0	4	2	0	0
C	4	0	1	0	0	0	0
D	1	4	3	1	5	4	4

Next is the application of step 2. One employee is required in the last hour of the day (shown in bold letters in row 3), and is assigned to the last daily work schedule, *D*. The employees still required (shown in row 3) each hour are calculated and shown in row 5.

Continual application of step 4 gives the results in row 9, at which point the remaining requirements are zero. The total number of employees scheduled to each shift for Monday is shown at the right-hand corner of Table 13.19. Similarly, the daily shift schedules for Tuesday through Sunday are developed, and the results are shown in Table 13.20. The numbers in the table show the number of workers that will start each shift on each day of the week.

TABLE 13.20
Phase II Application On-Off Days for Shift Schedule A

Days	Monday	Tuesday	Wednesday	Thursday	Friday	Saturday	Sunday	Next Mond M′
Requirements	3	4	4	4	3	5	4	5
Excess	2				1			
Total	5	4	4	4	4	5	4	5
RDOs	1	2	2	2	2	1	2	1
Required Consecutive day pairs	MT	TW	WTh	ThF	FS	SS	SM′	M′T′
First trial	1	1	1	1	1	0	2	−1
Second trial	0	2	0	2	0	1	1	0

TABLE 13.21
Phase II Application On-Off Days for Shift Schedule A

Hour	1	2	3	4	5	6	7	8	9	10	11	12
Monday to schedule	3	2	5	4	2	5	3	4	4	2	1	1
Full time	1	1	1	1	1	1	1	1	1			
Remain	2	1	4	3	2	4	2	3	3			
Part time	2	2	2	2								
Remain	0	0	2	1	2	4	2	3	3	2	1	1
Full time				1	1	1	1	0	1	1	1	1
Remain	0	0	2	0	1	3	1	3	2	1	0	0
Part time							1	1	1	1		
Remain			2	0	1	3	0	2	1	0	0	0
Full time			1	1	1	1	0	1	1	1	1	
Remain			1	0	0	2	0	1	0	0	0	
Part time			1	1	1	1						
Remain			0	0	0	1	0	1	0	0	0	0
Part time						1	1	1	1			
Remain			0	0	0	0	0	0	0	0	0	0

Phase II

The next step is to determine for each shift the worker schedule over the 7-day period so that each worker works 5 days and is off for 2 days. We can apply Monroe's algorithm to obtain a workers on-off schedule. Table 13.21 shows one such schedule; this one is specific to the shift schedule A. Here, we have six workers being assigned to shift A. Similar calculations with other shifts will determine the workers needed for each shift, and thus we can determine the total of all workers needed to satisfy the present requirements shown in Table 13.17.

13.4.3 Other Variations

In the previous section, it is presumed that either there is only one type of task, and every employee is capable of performing this task or if there are multiple tasks necessary for the job, all the workers are capable of performing any assigned task. We also assumed that all workers are full-time workers. In some service organizations, the workers must be treated individually for two reasons. First, the workers may be full-time or part-time workers, varying the hours of the day and days of the week that they are available. Second, workers differ in the tasks they are qualified to perform. In a fast-food restaurant, for example, we may have cooks, cashiers, servers, and cleaners as the major classification of the tasks. Different workers can perform different combinations of the tasks. Some workers can be cooks and cleaners, while others can be cashiers, servers, and cleaners. The scheduling of such a problem is also referred to as tour scheduling.

Consider a problem where we can employ part-time workers who can work 4 hr at a time as long as there is one full-time worker to supervise two part time employees. Part time employees of course cost less and are therefore preferred, plus they are available any time we need them as long as they work for 4 hr in a stretch. The example illustrates the application of tour scheduling algorithm.

In some organizations, especially in the health care industry where critical shortages in a specialized area may exist, it is common to have a shift of 12 hr with the employee working 3 days on and 4 days off in one week and 4 days on and 3 days off in the second week. Many algorithms exist to assign days on and off for such a situation.

13.5 THREE CONSECUTIVE DAYS OFF

Now consider a problem where we want to give 3 consecutive days off using as few workers as possible. Again, list the daily requirements in a table. The remaining steps of the procedure are as follow:

Step 1: Calculate the theoretical minimum number of employees W needed as:

$$\mathrm{W} = [\mathrm{Max}(T/4, n_j)]$$

where T is the sum of the daily requirements for a week, and n_j is the maximum number of employees required on the day. We divide T by 4 here because an employee is only working 4 days a week. W must be an integer. If $T/4$ is not an integer, round it to the next higher integer.

Step 2: Calculate the RDO by subtracting each day's requirement from W.

Step 3: Construct three consecutive day triplets such as SMT (Sunday–Monday–Tuesday), MTW (Monday–Tuesday–Wednesday), ..., SSM (Saturday–Sunday–Monday), and SSM' (Saturday–Sunday–next Monday).

Step 4: Let a_k = the number of workers who are off on three consecutive days, $k-1$, k, and $k+1$. For example, for $k=1$ workers are off on SMT, for $k=2$, they are off on MTW.

Assign four consecutive members of a_k as a group. For example, the first group is a_1, a_2, a_3, and a_4, and the second group is a_2, a_3, a_4, and a_5, and so on. If the value of k exceeds 7, then go back in the loop to start from 1. For example,

a_6, a_7, a_1, and a_2

For each group, set the difference between first and last element a's equal to difference between RDO for second and third element.

Thus, we have,

In the group 1, $g_1 = \{a_1, a_2, a_3, a_4\}$, set $a_1 - a_4 = RDO_2 - RDO_3$
In the group 2, $g_2 = \{a_2, a_3, a_4, a_5\}$, set $a_2 - a_5 = RDO_3 - RDO_4$
In the group 3, $g_3 = \{a_3, a_4, a_5, a_6\}$, set $a_3 - a_6 = RDO_4 - RDO_5$
In the group 4, $g_4 = \{a_4, a_5, a_6, a_7\}$, set $a_4 - a_7 = RDO_5 - RDO_6$
In the group 5, $g_5 = \{a_5, a_6, a_7, a_1\}$, set $a_5 - a_1 = RDO_6 - RDO_7$
In the group 6, $g_6 = \{a_6, a_7, a_1, a_2\}$, set $a_6 - a_2 = RDO_7 - RDO_1$
In the group 7, $g_7 = \{a_7, a_1, a_2, a_3\}$, set $a_7 - a_3 = RDO_1 - RDO_2$

From the preceding, it follows that:

$$a_1 = a_4 + (RDO_2 - RDO_3)$$
$$a_2 = a_5 + (RDO_3 - RDO_4)$$
$$a_3 = a_6 + (RDO_4 - RDO_5)$$
$$a_4 = a_7 + (RDO_5 - RDO_6)$$
$$a_5 = a_1 + (RDO_6 - RDO_7)$$
$$a_6 = a_2 + (RDO_7 - RDO_1)$$
$$a_7 = a_3 + (RDO_1 - RDO_2)$$

From these relationships, we can then develop equations such as

$$a_1 = a_2 + x_2 = a_3 + x_3 = a_4 + x_4 = a_5 + x_5 = a_6 + \underline{x}_6 = a_7 + x_7$$

where x_i is some constant based on the preceding relationships.

Set min $\{a_k\}$ equal to an integer number starting with 0 and increasing by 1, and solve for remaining a_i's, so that the sum of $a_i = W$, if possible.

If $\sum a_k = W$ then the solution is optimal. Stop.

Step 5.

If $\sum a_k > W$, then answer is infeasible. There are two ways to resolve this infeasibility.

1. Add an extra worker.
2. Assign nonconsecutive days to one worker.

Both approaches are illustrated by an example later.

TABLE 13.22
Initial Table

	Monday	Tuesday	Wednesday	Thursday	Friday	Saturday	Sunday	Total
Required	5	4	3	5	5	4	3	29
Excess			1			1	1	32
Total	5	4	4	5	5	5	4	
RDO	3	4	4	3	3	3	4	
Day pair	SMT	MTW	TWT	WTF	TFS	FSS	SSM′	
a_k	a_1	a_2	a_3	a_4	a_5	a_6	a_7	
Counter	1	2	3	4	5	6	7	

Example

Consider the following data. In Table 13.22, based on the daily requirements, we can calculate W as

$W = \max(29/4, 5) = 8$ and, therefore, workdays available with 8 workers $= 8 \times 4 = 32$. The remaining table is completed by first assigning excess to days with the smallest requirements.

Applying step 4 (work with counter if it is more convenient)

In the group 1, $g_1 = \{a_1, a_2, a_3, a_4\}$, set $a_1 - a_4 = RDO_2 - RDO_3 = 0$
In the group 2, $g_2 = \{a_2, a_3, a_4, a_5\}$, set $a_2 - a_5 = RDO_3 - RDO_4 = 1$
In the group 3, $g_3 = \{a_3, a_4, a_5, a_6\}$, set $a_3 - a_6 = RDO_4 - RDO_5 = 0$
In the group 4, $g_4 = \{a_4, a_5, a_6, a_7\}$, set $a_4 - a_7 = RDO_5 - RDO_6 = 0$
In the group 5, $g_5 = \{a_5, a_6, a_7, a_1\}$, set $a_5 - a_1 = RDO_6 - RDO_7 = -1$
In the group 6, $g_6 = \{a_6, a_7, a_1, a_2\}$, set $a_6 - a_2 = RDO_7 - RDO_1 = 1$
In the group 7, $g_7 = \{a_7, a_1, a_2, a_3\}$, set $a_7 - a_3 = RDO_1 - RDO_2 = -1$

Hence,

$a_1 = a_4$
$a_2 = a_5 + 1$
$a_3 = a_6$
$a_4 = a_7$
$a_5 = a_1 - 1$
$a_6 = a_2 + 1$
$a_7 = a_3 - 1$

Therefore,

$a_1 = a_4 = a_7 = a_3 - 1 = a_6 - 1 = a_2 = a_5 + 1$
Or $a_1 = a_2 = a_3 - 1 = a_4 = a_5 + 1 = a_6 - 1 = a_7$

TABLE 13.23
Solution to Example 1

	Monday	Tuesday	Wednesday	Thursday	Friday	Saturday	Sunday	Total
Required	5	4	3	5	5	4	3	29
Excess			1			1	1	32
Total	5	4	4	5	5	5	4	
RDO	3	4	4	3	3	3	4	
Day triplet	SMT	MTW	TWT	WTF	TFS	FSS	SSM′	
Workers off								
$a_1 = 1$	1	1					1	
$a_2 = 1$	1	1	1					
$a_3 = 2$		2	2	2				
$a_4 = 1$			1	1	1			
$a_5 = 0$				0	0	0		
$a_6 = 2$					2	2	2	
$a_7 = 1$	1					1	1	
Total workers off	3	4	4	3	3	3	4	

TABLE 13.24
Data for Example 2

	Monday	Tuesday	Wednesday	Thursday	Friday	Saturday	Sunday	Total
Required	5	7	8	6	10	10	4	50
Excess	1						1	2
Total	6	7	8	6	10	10	5	
RDO	7	6	5	7	3	3	8	
Day pair	SMT	MTW	TWT	WTF	TFS	FSS	SSM	
a_k	a_1	a_2	a_3	a_4	a_5	a_6	a_7	

Since no assignment can be negative, set Min $\{a_k\} = 0$. By observation, it is clear that a_5 should have the least value. Hence, set $a_5 = 0$, and solving for others, we get $a_1 = 1$, $a_2 = 1$, $a_3 = 2$, $a_4 = 1$, $a_5 = 0$, $a_6 = 2$, and $a_7 = 1$. Sum of a_i, which is the number of workers required, is 1 + 1 + 2 + 1 + 0+ 2 + 1 = 8, and matches the initial value of W, and hence we have a feasible solution (Table 13.23).

Note that the total number of workers off on any day matches with the RDO for that day, and hence it is an optimum solution.

Example 2

Consider now another example with the requirements as shown in the Table 13.24.

$$T = \text{Max}(50/4, 10) = 13$$

Iteration 2

	Monday	Tuesday	Wednesday	Thursday	Friday	Saturday	Sunday	Total
Required	5	7	8	6	10	10	4	50
Excess	2			1			3	6
Total	7	7	8	7	10	10	7	
RDO	7	7	6	7	4	4	7	
Day pair	SMT	MTW	TWT	WTF	TFS	FSS	SSM	
a_k	a_1	a_2	a_3	a_4	a_5	a_6	a_7	

Workdays available with 13 workers = 13 × 4 = 52. Excess = 52 – 50 = 2. Applying step 4,

In the group 1, $g_1 = \{a_1, a_2, a_3, a_4\}$, set $a_1 - a_4 = RDO_2 - RDO_3 = 1$
In the group 2, $g_2 = \{a_2, a_3, a_4, a_5\}$, set $a_2 - a_5 = RDO_3 - RDO_4 = -2$
In the group 3, $g_3 = \{a_3, a_4, a_5, a_6\}$, set $a_3 - a_6 = RDO_4 - RDO_5 = 4$
In the group 4, $g_4 = \{a_4, a_5, a_6, a_7\}$, set $a_4 - a_7 = RDO_5 - RDO_6 = 0$
In the group 5, $g_5 = \{a_5, a_6, a_7, a_1\}$, set $a_5 - a_1 = RDO_6 - RDO_7 = -5$
In the group 6, $g_6 = \{a_6, a_7, a_1, a_2\}$, set $a_6 - a_2 = RDO_7 - RDO_1 = 1$
In the group 7, $g_7 = \{a_7, a_1, a_2, a_3\}$, set $a_7 - a_3 = RDO_1 - RDO_2 = 1$

Furthermore,

$a_1 = a_4 + 1$
$a_2 = a_5 - 2$
$a_3 = a_6 + 4$
$a_4 = a_7$
$a_5 = a_1 - 5$
$a_6 = a_2 + 1$
$a_7 = a_3 + 1$

Thus,

$a_1 = a_4 + 1 = a_7 + 1 = a_3 + 2 = a_6 + 6 = a_2 + 7 = a_5 + 5$
$a_1 = a_2 + 7 = a_3 + 2 = a_4 + 1 = a_5 + 5 = a_6 + 6 = a_7 + 1$

Therefore,
$a_k = \{7, 0, 5, 6, 2, 1, 6\}$; $\sum a_k = 27 >= 13$, and the solution is infeasible.
a. Add an extra worker(s)
One alternative is to add an extra worker. Start a new iteration with $W = 14$ and $T = 14 \times 4 = 56$. Excess $= 6$. Distribute the excess to the days where requirements are in the smallest amount, so that the overall total from day to day is as uniform as possible.

Iteration 2

From the table we get,

Nonconsecutive days: Iteration 2

	Monday	Tuesday	Wednesday	Thursday	Friday	Saturday	Sunday	Total
Required	5	7	8	6	10	10	4	50
Excess	1						1	2
Total	6	7	8	6	10	10	5	
RDO	7	6	5	7	3	3	8	
Nonconse	1			1			1	
	6	6	5	6	3	3	7	
Day pair	SMT	MTW	TWT	WTF	TFS	FSS	SSM	
a_k	a_1	a_2	a_3	a_4	a_5	a_6	a_7	

$a_1 = a_4 + 1, a_2 = a_5 - 1, a_3 = a_6 + 3, a_4 = a_7, a_5 = a_1 - 3, a_6 = a_2$, and $a_7 = a_3$
And further,
$a_1 = a_4 + 1 = a_7 + 1 = a_3 + 1 = a_6 + 4 = a_2 + 4 = a_5 + 3$
$a_1 = a_2 + 4 = a_3 + 1 = a_4 + 1 = a_5 + 3 = a_6 + 4 = a_7 + 1$
Therefore,
$a_k = \{4, 0, 3, 3, 1, 0, 3\}; \sum a_k = 14 = W$, and the solution is feasible.

13.5.1 Assign Nonconsecutive Days Off

The second alternative is to give at least one employee nonconsecutive days off. In our table, nonconsecutive days off are given to the days where RDOs are maximum.

Iteration 2

Since one of the workers is assigned to have nonconsecutive days off, the number of the workers left is $W = 12$.

From the table, we get,

$a_1 = a_4 + 1, a_2 = a_5 - 1, a_3 = a_6 + 3, a_4 = a_7, a_5 = a_1 - 4, a_6 = a_2 + 1$ and $a_7 = a_3$

So, $a_1 = a_4 + 1 = a_7 + 1 = a_3 + 1 = a_6 + 4 = a_2 + 5 = a_5 + 4$

Or $a_1 = a_2 + 5 = a_3 + 1 = a_4 + 1 = a_5 + 4 = a_6 + 4 = a_7 + 1$

Therefore, $a_k = \{5, 0, 4, 4, 1, 1, 4\}$. $\sum a_k = 19 \geq W$, and the solution is infeasible. Assign one more worker non consecutive day off.

Iteration 3

Since two workers are assigned nonconsecutive days off, the remaining workers are $W = 11$.

From the table, we get,

$a_1 = a_4, a_2 = a_5, a_3 = a_6 + 2, a_4 = a_7, a_5 = a_1 - 3, a_6 = a_2$, and $a_7 = a_3 + 1$

So, $a_1 = a_4 = a_7 = a_3 + 1 = a_6 + 3 = a_2 + 3 = a_5 + 3$

Or $a_1 = a_2 + 3 = a_3 + 1 = a_4 = a_5 + 3 = a_6 + 3 = a_7$

Therefore, $a_k = \{3, 0, 2, 3, 0, 0, 3\}$, $\sum a_k = 11 = W$, and the solution is feasible.

Nonconsecutive days: Iteration 3

	Monday	Tuesday	Wednesday	Thursday	Friday	Saturday	Sunday	Total
From iteration 2:	6	6	5	6	3	3	7	
		1		1			1	
	6	5	5	5	3	3	6	
Day pair	SMT	MTW	TWT	WTF	TFS	FSS	SSM	
a_k	a_1	a_2	a_3	a_4	a_5	a_6	a_7	

13.5.2 Special Case and Further Discussion

What happens if the number of workers required for each day does not change. Let us consider the following cases and get a further understanding of the definition of the problem.

	Monday	Tuesday	Wednesday	Thursday	Friday	Saturday	Sunday	Total
Required	5	5	5	5	5	5	5	35
Excess		1						1
Total	5	6	5	5	5	5	5	
RDO	4	3	4	4	4	4	4	
Day pair	SMT	MTW	TWT	WTF	TFS	FSS	SSM	

The solution is $a_k = \{1, 1, 1{,}2 , 1, 1, 2\}$; $\sum a_k = 9 = W$, and the solution is feasible.

Similarly, for other constant requirements, the following table summarizes the results.

For MOD $(R/4) = 0$.

Required Workers/day	Total Requirement W	a_1	a_2	a_3	a_4	a_5	a_6	a_7
$R_1 = 4$	$W_1 = 7$	1	1	1	1	1	1	1
$R_2 = 8$	$W_1 = 14$	2	2	2	2	2	2	2
$R_3 = 12$	$W_3 = 21$	3	3	3	3	3	3	3
...	...	...	...	...	...	...	...	...
$R_n = 4 + 4n$	$W_n = 7 + 7n$ For $n = 0, 1{,}2, \ldots$.	$n+1$	$n+1$	$n+1$	$n+1$	$n+1$	$n+1$	$n+1$

For MOD $(R/4) = 1$.

Required Workers/day	Total Requirement W	a_1	a_2	a_3	a_4	a_5	a_6	a_7
$R_1 = 1$	$W_1 = 2$	0	0	0	1	0	0	1
$R_2 = 5$	$W_2 = 9$	1	1	1	2	1	1	2
$R_3 = 9$	$W_3 = 16$	2	2	2	3	2	2	3
...	...	...	...	...	...	...	...	...
$R_n = 1 + 4n$	$W_n = 2 + 7n$ For $n = 0, 1,2, \ldots$	n	n	n	$n+1$	n	n	$n+1$

MOD $(R/4) = 2$.

Required Workers/day	Total Requirement W	a_1	a_2	a_3	a_4	a_5	a_6	a_7
$R_1 = 2$	$W_1 = 4$	1	0	0	0	1	1	1
$R_2 = 6$	$W_1 = 11$	2	1	1	1	2	2	2
$R_3 = 10$	$W_3 = 18$	3	2	2	2	3	3	3
...	...	...	...	...	...	...	...	...
$R_n = 2 + 4n$	$W_n = 4 + 7n$ For $n = 0, 1,2, \ldots$	$n+1$	n	n	n	$n+1$	$n+1$	$n+1$

MOD $(R/4) = 3$.

Required Workers/day	Total Requirement W	a_1	a_2	a_3	a_4	a_5	a_6	a_7
$R_1 = 3$	$W_1 = 6$	1	0	1	1	1	1	1
$R_2 = 7$	$W_1 = 13$	2	1	2	2	2	2	2
$R_3 = 11$	$W_3 = 20$	3	2	3	3	3	3	3
...	...	...	...	...	...	...	...	...
$R_n = 3 + 4n$	$W_n = 6 + 7n$ For $n = 0,1,2, \ldots$	$n+1$	n	$n+1$	$n+1$	$n+1$	$n+1$	$n+1$

From the previous tables, we can notice that when the daily requirement, R, goes up by 4, the number of workers needed, W, goes up 7.

13.6 SUMMARY

Employee scheduling is an important aspect of industrial scheduling. For industries working 5 days a week, with one, two, or three shifts and no employee rotations, employee scheduling may not be a challenge. Each shift may be viewed independently with constant employee requirements, and a fixed schedule may be developed. The weekends are naturally off for all employees.

This is not the case when an industry works 7 days a week and the requirements change from day to day from hour to hour. It is somewhat difficult to develop a schedule that employs minimum personnel and still meets the legal and contractual terms. These may include situations where employees work 40 hr a week, or must have 2 consecutive days off in a week or at least have 1 weekend off in the work span of 2 weeks. This chapter illustrates some of the popular and effective methods that are available to achieve these objectives.

13.7 PROBLEMS

13.1 As the maintenance supervisor, your responsibilities include developing the employee schedule for the maintenance department. When you became the maintenance supervisor, the department was already using a scheduling system whereby each maintenance technician worked 5 days per week, had every other weekend off, and did not work more than 6 consecutive days. You are using the same system and must generate the next schedule. Generate this schedule given the following daily requirements:

	Sunday	Monday	Tuesday	Wednesday	Thursday	Friday	Saturday
Requirements	6	7	5	7	8	4	3

13.2 Using the TPB algorithm, develop an efficient schedule that gives 2 consecutive days off to each worker, if the daily worker requirements are as follows:

	Sunday	Monday	Tuesday	Wednesday	Thursday	Friday	Saturday
Requirements	6	7	5	7	8	4	3

13.3 Using the same method as in Problem 13.2, develop a schedule for the following data:

	Sunday	Monday	Tuesday	Wednesday	Thursday	Friday	Saturday
Requirements	2	8	9	7	11	10	3

13.4 Using BC algorithm, develop a schedule for the following data with the objective of giving each employee one weekend off in a period of 2 weeks:

	Sunday	Monday	Tuesday	Wednesday	Thursday	Friday	Saturday
Requirements	6	8	10	12	4	5	3

13.5 Develop a schedule for the following data to give each employee a minimum of one weekend off every 2 weeks using the BC algorithm.

	Sunday	Monday	Tuesday	Wednesday	Thursday	Friday	Saturday
Requirements	2	8	9	7	11	10	3

13.6 Develop a schedule for the following data that gives two consecutive days off to each employee. Use Monroe's algorithm; Also develop schedule when we need 3 consecutive days off.

	Sunday	Monday	Tuesday	Wednesday	Thursday	Friday	Saturday
Requirements	5	7	9	11	6	8	4

13.7 Using Monroe's algorithm, develop a schedule for the following data:Also develop schedule when 3 consecutive days off are required.

	Sunday	Monday	Tuesday	Wednesday	Thursday	Friday	Saturday
Requirements	3	6	8	10	12	4	5

13.8 The data given in the following table indicate the hourly requirements on a daily basis for a fast-food restaurant open 12 hr a day, 7 days a week. There are four possible shifts: A, B, C, and D, starting at hours 1, 2, 3, and 4, respectively. Develop the hiring policies for the restaurant.

Hours of Day	Monday	Tuesday	Wednesday	Thursday	Friday	Saturday	Sunday
1	3	3	4	3	2	4	2
2	3	3	3	4	4	4	2
3	4	5	4	4	5	4	3
4	5	5	6	5	6	6	4
5	5	4	4	4	6	6	4
6	4	4	5	5	5	4	4
7	2	2	3	3	3	4	2
8	2	1	2	1	3	3	1
9	3	4	3	2	4	4	2
10	6	5	4	4	6	5	3
11	6	6	5	5	6	6	5
12	4	5	5	5	6	6	4

13.9 Given the following daily requirements for ABC Company, develop the schedule for a weekly pattern of 5 days with rotating weekends off. The company wants its maximum workforce requirement in any week to be $W = 14$.

a. Find the manpower requirement for Monday.
b. Using the BC algorithm, develop a schedule.

c. Develop a schedule using Monroe's algorithm.
d. Calculate the scheduling efficiency of both methods.

Monday	Tuesday	Wednesday	Thursday	Friday	Saturday	Sunday
x	10	14	11	13	7	6

13.10 City sanitation department needs the following number of employees each week:

Monday	Tuesday	Wednesday	Thursday	Friday	Saturday	Sunday
23	30	22	25	26	21	0

There is no waste collection on Sunday, so no workers work on Sundays. Develop an employee schedule using the TPB algorithm, and calculate your scheduling efficiency. Note that each employee should get at least 2 days off per week.

13.11 Employee requirement for a fast-food restaurant is based on sales volume, which varies from day to day during a week. Past records show that one employee can handle sales of $150 in any one shift. Assuming that the sales volume is same for all the shifts, determine the minimum number of employees required on each day based on the following sales volume:

	Monday	Tuesday	Wednesday	Thursday	Friday	Saturday	Sunday
Sales volume ($)	1500	1800	1350	1175	1475	1640	1050

Develop the schedule for a work pattern of 5 days on, with consecutive days off. Use (1) TPB, (2) BC, and (3) Monroe's algorithms.

13.12 The working hours for a restaurant are 8 AM to 5 PM. There are two possible shifts during these working hours; the hourly employee requirement is shown in the following table. Each employee works for 8 hr with a 1-hr lunch break. Develop a schedule using Bechtold and Showalter MSSH algorithm.

Hours of Day	Monday	Tuesday	Wednesday	Thursday	Friday	Saturday	Sunday
1	3	2	4	1	3	2	1
2	3	2	4	1	3	2	1
3	5	2	7	5	3	4	5
4	5	6	7	5	3	4	5
5	5	6	7	5	5	6	5
6	4	6	6	7	6	6	7
7	4	4	6	7	6	8	7
8	5	4	6	5	6	8	6
9	5	4	5	5	7	8	6

13.13 The daily requirements on an hourly basis for a restaurant that is open 12 hr a day are shown in the following table. There are four shifts A, B, C, and D. Develop a schedule using Bechtold and Showalter MSSH algorithm.

Hours of Day	Monday	Tuesday	Wednesday	Thursday	Friday	Saturday	Sunday
1	2	4	2	3	1	2	2
2	2	4	2	3	1	2	2
3	5	4	2	3	4	4	3
4	5	6	6	5	4	4	3
5	7	6	6	5	5	3	6
6	7	6	6	5	5	3	6
7	7	5	5	6	7	4	6
8	4	5	5	6	7	4	4
9	4	5	5	6	3	5	4
10	4	4	7	4	3	5	5
11	5	4	7	4	2	3	5
12	5	4	7	4	2	3	5

REFERENCES AND SUGGESTED READINGS

Alfares H.K 2003 "Four-Day Workweek Scheduling with two or three Consecutive Days off", Journal of mathematical modeling and Algorithms, Vol 2, Number 1, 2003 pp 67–80

Bailey, J. 1985. "Integrated Days-Off and Shift Personnel Scheduling" *Journal of Computers and Industrial Engineering*, 9(4): 395–404.

Baker, K.R. 1974. "Scheduling a Full-Time Workforce to Meet Cyclic Staffing Requirements" *Management Science*, 20(12): 1561–1568.

Baker, K.R., R.N. Burns, and M.W. Carter. 1979. "Staff Scheduling with Day-Off and Workstretch Constraints" *AIIE Transactions*, 11(4): 286–292.

Baker, K.R. and M.J. Magazine. 1977. "Workforce Scheduling with Cyclic Demands and Day-Off Constraints" *Management Science*, 24(2): 161–167.

Bechtold, S.E. 1981. "Workforce Scheduling for Arbitrary Cyclic Demands" *Journal of Operations Management*, 1(4): 155–167.

Bechtold, S.E. and M.J. Showalter. 1987. "A Methodology for Labor Scheduling in a Service Operating System" *Decision Sciences*, 18(1): 89–107.

Burns, R.N. and M.W. Carter. 1985. "Workforce Size and Single Shift Schedules with Variable Demands" *Management Sciences*, 31(5): 599–607.

Emmons, H. 1985. "Workforce Scheduling with Cyclic Requirements and Constraints on Days Off, Weekends Off, and Workstretch" *IIE Transactions*, 17(1): 8–16.

Wang James 2005, "On 4 days on 3 days off schedule", Louisiana Teen term paper.

14 Industrial Sequencing I: Scheduling on NC Machines

So far, we have been studying scheduling in somewhat generalized terms. In the next two chapters, we present heuristics that are much more application specific. In some instances, we may be faced with a problem for which no "ready made" heuristic is available. In next few chapters present cases of how we might think through the problems to get good sequencing rules, even though at the first glance the problem may seem to be overwhelming. In developing such procedures, we may have to depend on the techniques that we learned somewhere else. Our ability to transfer knowledge from one field to another is very helpful in constructing sequencing rules.

One such area is the group technology. Group technology techniques have been used in manufacturing to develop cohesive manufacturing cells. Parts with common characteristics are grouped together and produced in a single cell, so that the same machines, jigs, and fixtures can be used to produce the parts. For example, for 1000 parts produced in the plant, perhaps 20 groups can be formed, each group needing machines specific to produce the products assigned to the group. There are many algorithms to form machine groupings and assign components, but we shall use a simple procedure called tabular approach (Sule -89) in Figure 14.1.

14.1 TABULAR APPROACH IN GROUP FORMING

The method starts with a 0-1 table called machine component matrix table. This table shows the machine that each component (e.g., part or job) needs in its production; 1 indicating the use of the machine and 0 or blank indicating nonuse of the machine. For example, in Table 14.1, there are 20 jobs produced using seven machines. Job 1 needs machines 3 and 4 in its production, whereas job 5 needs machines 1, 2, 5, and 7, and so on.

There are a number of steps in the group forming procedure. They are stated next and are illustrated later by their application.

Step 1: Develop a machine-to-machine relationship table. A machine-to-machine relationship table indicates the number of jobs that require both machines (note the jobs may not necessarily be processed in sequence on these machines). The table is developed by observing the machine-components matrix in which "1" indicates the use of the machine and a blank or "0" indicates nonuse of the machine in processing a specific job. Comparing the columns for any two machines and counting

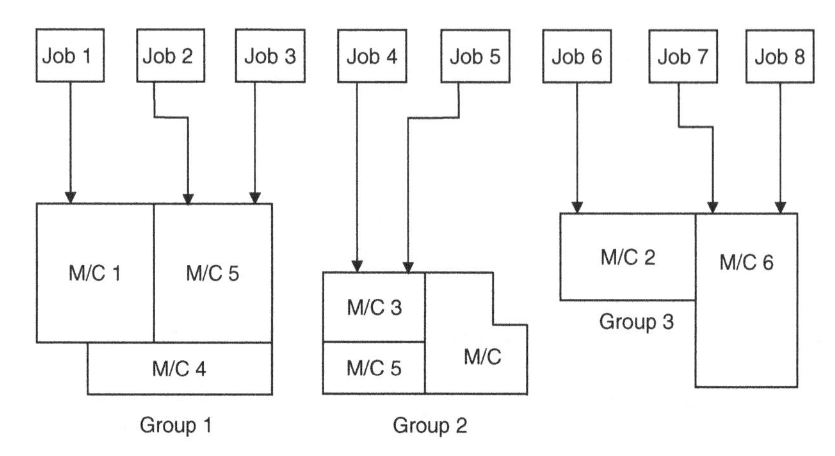

FIGURE 14.1 Group technology in scheduling.

TABLE 14.1
Machine-Component Matrix

Job (Component)	Machine 1	2	3	4	5	6	7
1			1	1			
2			1	1			
3			1	1			
4					1		1
5	1	1			1		1
6		1				1	1
7				1			
8	1						
9					1	1	1
10							1
11	1		1				
12	1	1			1		1
13		1					1
14						1	1
15			1	1			
16			1				
17						1	1
18						1	1
19	1	1			1		1
20	1		1	1			

the number of times "1" appears common in these two columns indicates the number of jobs requiring the corresponding two machines. Note that the machine-to-machine relationship table is symmetrical about the diagonal, and therefore, only the elements above or below the diagonals are needed.

Step 2: Pick the largest element in machine-to-machine table and designate it as the present value of relationship counter (RC).

Step 3: Define a value of minimum percentage. A measure of effectiveness of joining a component to a group consisting of other components is defined by the analyzer at the beginning of the problem; for example, 50% (or $P = 0.5$). It states that, in the group forming process, the closeness an entering machine must have with all the existing machines within a group in order for the entering machine to join that group. Based on the percentage value an analyzer chooses, the solution may have different number of cells. As we shall see later, some machines may be duplicated, that is, they may be part of more than one group. How many times the machines are duplicated may also be influenced by the value of the percentage. Note that this percentage, designated as P, is only defined once during the process of finding a solution to the problem.

There is one distinct advantage in this procedure. We might subject the same machine to machine data to different values of P and perhaps obtain some alternate solutions that could be evaluated further.

Step 4: Starting with the first row, examine each row for an elemental value that equals RC.

Step 5: If none of the associated machines in the row and column are already in a group, then form a group consisting of these two machines and go to step 7. If both machines are already assigned to the same group, ignore the observation and go to step 7. If one of the machines in the pair is in a group and the other one has not been assigned yet, go to step 6a. If both machines are assigned, but to different groups, go to step 6b.

Step 6a: Calculate the "closeness ratio" (CR) of the entering machine with each group that has already been formed. A CR is defined as the ratio of the total of all relationships the entering machine has with the machines that are currently in the group to the total number of machines that are presently assigned to that group.

The entering machine is placed in a group that has the maximum closeness ratio (MCR), as long as this maximum is greater than or equal to the "minimum threshold value" (MTV), calculated as P multiplied by the present value of RC. If the value of MCR is less than MTV, then a new group is formed, consisting of two machines having the relationship value that equals the present value of RC. Go to step 7.

Step 6b: Duplication of one or more machines is suggested. There are two possible alternatives, and they are checked sequentially in the order of importance. The first alternative is to duplicate one additional machine of either type (for illustrative purposes, designate the machines in the pair as machine A and machine B), and place it in the appropriate cell, and the second alternative is to duplicate both machines, one of each type, and form a new group or place the appropriate one in each of the existing groups. The following rules are suggested:

1. Calculate the effect of duplicating one machine. Check machine A as the entering machine for the groups where machine B exists and B as the entering machine for the groups where A exists. Determine the MCR, from all

the groups that are checked and note the associated group and entering machine.

2. If MCR > RC × *P*, the noted machine is duplicated and assigned to the associated group. Go to step 7.
3. If the MCR in the previous calculation was less than RC × *P*, a check must be made to see if both machines should be duplicated. From the previous calculations determine the MCR for the groups where A is the entering machine (MCRA) and the MCR for the groups where B is the entering machine (MCRB). Calculate the index value as maximum RC × *P*/2. If both MCRA and MCRB are greater than the index value and |MCRA − MCRB| < *P* × RC/2, duplicate both machines and place each in an appropriate group. If either MCRA or MCRB is greater than the index value, regardless of the value of |MCRA − MCRB|, form a new group consisting of machines A and B. Go to step 7.
4. If none of the preceding conditions exist, ignore this observation and go to step 7 since the contribution of any duplicating component in improving the efficiency of grouping is very limited.

Step 7: Check to see if all machines are assigned to groups. If they are, go to step 9; otherwise continue.

Continue the check of the machine-to-machine table with the present value of RC proceeding sequentially in rows. If an element is found that is equal to the present value of RC, go to step 5. If no such element is found, go to step 8.

Step 8: Reduce the value of RC to the next value in its descending order of magnitude and return to step 5.

Step 9: Assign each job to an appropriate group. This is accomplished by examining each job and assigning it to a group that has the most machines it needs.

Example

We shall illustrate the procedure by applying it to the data in Table 14.1. The machine-to-machine matrix for the data is shown in Table 14.2. For example, the entry in row 1

TABLE 14.2
Machine-to-Machine Table

Mac/Mac	1	2	3	4	5	6	7
1	—	3	2	1	3	0	3
2		—	0	0	3	1	5
3			—	5	0	0	0
4				—	0	0	0
5					—	1	5
6						—	5
7							—

column 3 is obtained by comparing columns 1 and 3 of Table 14.1 and counting the number of jobs requiring both machines, namely 2, machines 11, and 20.

To begin the grouping procedure, let the value of P be set to 0.5, a percentage chosen arbitrarily (step 3).

Iteration 1: The maximum value among the elements in Table 14.2, that is, RC is 5. The first value of 5 is associated with machines 2 and 7, and, therefore, they are joined together to form the first group, G1 (step 5).

Iteration 2: Machines 3 and 4 also have five relationships. Since none of these machines are already in a group, a new group is formed consisting of machines 3 and 4. Thus, we have in G1: 2, 7, and in G2: 3, 4.

Iteration 3: With RC = 5, there exists a relationship between machines 5 and 7 (from now on, relation is simply expressed as –, e.g., 5–7). Machine 7 already belongs in group 1, and therefore machine 5 becomes the entering variable, and step 6a is applied. The associated calculations are shown in Table 14.3. Note the relationship of the entering machine 5 is checked with the machines in all groups that are formed at this time. The total for each group indicates the total number of machines in that group and the total of associated relationship numbers. The critical ratio for group 1 is $8/2 = 4$, while that for group 2 is $0/2 = 0$. The maximum of these numbers is the value of the maximum critical ratio (MCR). The MTV is calculated by multiplying the present value of the RC, which is 5 by the percentage value (P) defined initially by us.

$$\text{RC} = 5\text{: Relationship: } 5\text{–}7$$

Since MCR > MTV (4 > 2.5), join machine 5 to G1. Thus, we have G1: 2, 7, 5, and G2: 3, 4.

Iteration 4: RC = 5, 6–7.

1. Machine 7 is in group 1, and machine 6 is the entering machine. The corresponding calculations are shown in Table 14.4.

TABLE 14.3
Check for Entering Machine 5

Entering	Existing Groups				
Machine	G1: Machines	Relation	G2: Machines	Relation	MTV
5	2	3	3	0	$5 \times 0.5 = 2.5$
	7	5	4	0	
Total	2	8	2	0	
CR	$8/2 = 4$		$0/2 = 0$		
MCR		4			

TABLE 14.4
Check for Entering Machine 6

Entering Machine	Existing Groups				MTV
	G1: Machines	Relation	G2: Machines	Relation	
6	2	1	3	0	5 × 0.5 = 2.5
	7	5	4	0	
	5	1			
Total	3	7	2	0	
CR	7/3 = 2.3		0/2 = 0		
MCR		2.3			

TABLE 14.5
Check for Component 1

Entering Machine	Existing Groups						MTV
	G1: Machine	Relation	G2: Machine	Relations	G3: Machine	Relations	
1	2	3	3	2	6	0	3 × 0.5 = 1.5
	7	3	4	1	7	3	
	5	3					
Total	3	9	2	3	2	3	
CR	9/3 = 3		3/2 = 1.5		3/2 = 1.5		
MCR			3				

Since MCR < MTV (2 < 2.5), machine 6 cannot be joined to G1. Therefore, we form a new group consisting of machines 6 and 7 (step 6a). We now have three groups, namely G1: 2, 7, 5; G2: 3, 4, and G3: 6, 7. Since all the machines are not yet assigned to groups, we continue the process.

Iteration 5. RC = 3, 1–2.

2. There are three groups to check now. The calculations are similar to the ones shown in Table 14.4. The details are shown in Table 14.5.

Since MCR > MTV (3 > 1.5), machine 1 is joined to group 1. The present assignments are as follows:

G1: 2, 7, 5, 1; G2: 3, 4, and G3: 6, 7.

All the machines are now assigned to groups. We, therefore, proceed to the job assignment phase, step 9 of the procedure.

TABLE 14.6
Job Assignments

Group	Machines	Jobs
1	2, 7, 5, 1	4, 5, 6, 8, 9, 10, 11, 12, 13, 19
2	3, 4	1, 2, 3, 7, 15, 16, 20
3	6, 7	14, 17, 18

Step 9 of the procedure calls for assigning jobs to groups. Assign a job to a group based on the common machines. A group that has the most number of common machines with the machines needed by the job is the group to which the job is assigned. For example, job 6 needs machines 2, 6, and 7. Group 1 has the most machines common with it, namely, machines 2 and 7, and therefore, job 6 is assigned to group 1. The results are displayed in Table 14.6.

A computer program for an IBM-PC compatible machine is provided on the web site. The program can solve problems of any reasonable dimension.

14.2 JOB SEQUENCING TO MINIMIZE TOOL CHANGEOVERS IN FLEXIBLE MANUFACTURING SYSTEMS

A combination of group technology and heuristic is used for scheduling N jobs on a single machine with automatic tool changer (ATC). The problem is prominent in flexible manufacturing systems where the efficiency in operation is in part, obtained by the use of ATC. However, the complications associated with an ATC makes the existing single-machine scheduling algorithms unworkable.

To process a job on a machine may require multiple tools. An ATC, like chuck on a lathe, has the ability to change tools quickly and automatically. A chuck, for example, may have six faces, each holding a tool needed for the job. It can rotate quickly, contacting the necessary tool on the job when required.

An ACT allows for processing of jobs/parts in random sequence, to adjust for in-process and/or finished inventory and to respond, if necessary, to demand changes. ATC allows the job to be processed in any sequence by loading on the machine the required tools for the job. Yet, it may require about 4 min to make a tool change and associated adjustments, if a tool is to be replaced on the tool changer itself. If the jobs are available at the beginning of the processing, it is possible to reduce the makespan by arranging these jobs in a sequence that would minimize the total tool changes.

14.2.1 Assumptions and Problem Statement

The assumptions of the problem are as follows:

1. All jobs are available at the beginning of the planning period with known tool requirements.

2. The maximum number of tools required by any job is at the most equal to the capacity of the tool magazine.
3. Any tool can be placed in any position on the tool magazine (Figure 14.2).

Consider, for example, the same problem that is illustrated in Bard. The data is shown as a job-tool matrix in Table 14.7. Here, "1" indicates the use of a tool in production of a job, while a blank indicates the non use of the tool. The tool magazine

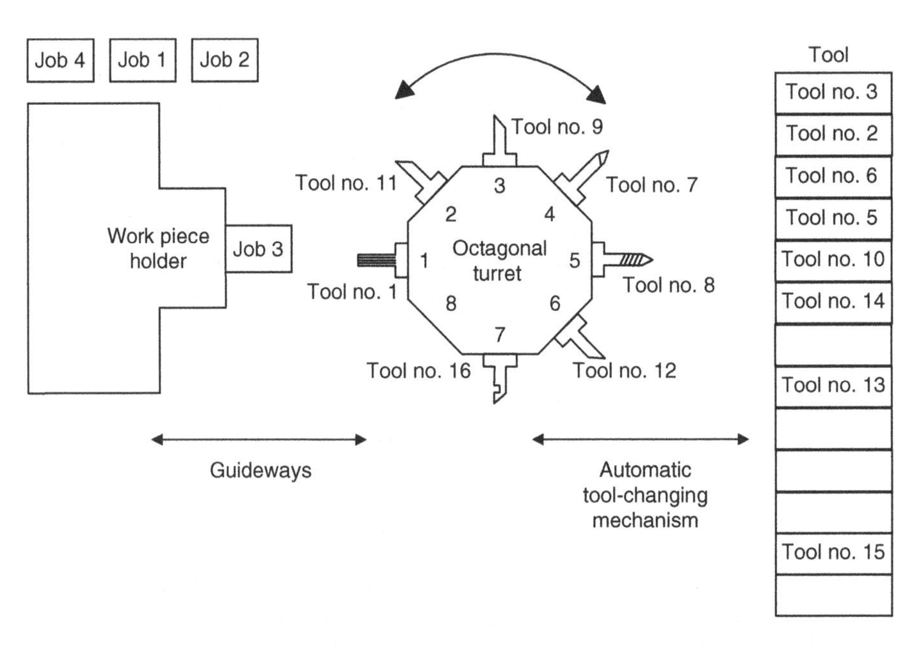

FIGURE 14.2 Turret lathe with automatic tool-changing mechanism.

TABLE 14.7
Job-Tool Matrix

	Tools								
Job	**1**	**2**	**3**	**4**	**5**	**6**	**7**	**8**	**9**
1		1			1		1	1	
2						1		1	
3	1		1		1				
4				1					
5				1		1	1		1
6			1		1	1		1	
7	1		1				1		
8		1			1				1
9				1		1	1		1
10			1			1			

has capacity to set four tools at a time. The objective is to develop a job sequence that would minimize the total tool changeovers in processing all jobs.

14.2.2 Solution Procedure

The solution procedure has 13 steps. We use group technology concepts first to develop cohesive groups of jobs and tools and then to develop another heuristic procedure to sequence these groups and jobs so that we have minimum tool changeovers. The details are as follows:

Step 1: Pre-sort job-tool matrix by planning a job j2 immediately after a job j1 if the tools required for job j2 are the same or subset of tools required for job j1. Such scheduling requires no tool change and we can reduce the dimension of the problem by ignoring job j2 in the remaining procedure by assuming that when j1 is planned, all its sub-jobs, such as job j2, would immediately follow.

Step 2: Using the job-tool matrix, apply the modified tabular method for tool grouping (here, tools are the machines). The method described in Section 14.1 is modified in one aspect. A group is fathomed, that is, not allowed any more tools, when it contains the number of tools equal to the maximum number of tools allowed on the tool changer. The tools forming a group are called the group tools for that group.

Step 3: Assign a job to a group based on number of common tools between group tools and tools required for the job. A group that has the most number of tools common with the job is the group to which the job is assigned.

The first three steps form the tool grouping, so that in a group there are no more tools than the maximum capacity of the tool magazine. The next steps are to develop further sequencing.

Step 4: Find the total number of times each tool is used in processing all the jobs that need scheduling.

Step 5: Find the total number of times each tool is used in each group for processing jobs assigned to that group.

Step 6: Construct a planning table and a scheduling table as described next. The planning table is used to determine which tool to replace, if any, as the jobs are scheduled in the scheduling table.

Step 7: Develop the planning table (as illustrated in Table 14.10 for the example problem). Construct, for each tool, three columns. The first column, marked as U, indicates if the tool is used by the job. The second column, designated as G, indicates the remaining times the tool will be used in production of the remaining jobs from the presently scheduled group. The third column, denoted as T, indicates the total number of times the tool will be used in processing all the remaining jobs (from the group that is presently being scheduled as well as all the groups that still remained to be scheduled).

When a job is assigned in the scheduling table, it is also recorded in the planning table. For each tool, the three columns are adjusted as necessary to reflect its use or non use in the production of the job. If the tool is used, the corresponding tool frequencies in columns G and T are reduced by one, indicating one less remaining job that would use the tool in the presently scheduled group as well as in the jobs that are still remained to be scheduled.

Step 8: Select a start group and mark it in the planning table by creating a row for the group. Enter in that row, in columns G and T for each tool, the cumulative times each tool is needed for processing the jobs in that group as well as for processing all jobs from all groups. Pick from the group a job that has the maximum tools common to the group tools. If there are multiple jobs that satisfy this condition, choose one of them at random. Place it in the scheduling table. This is done by recording job and associated tools in the table (Table 14.11 illustrates the scheduling table for the example problem).

Step 9: Find a new job to schedule from the group that is being worked on (if no job is available, go to step 11). This is done by comparing the tools required for the new job with the ones presently available in the tool magazine (i.e., tools required for the previous job) and selecting a job that has most tools in common. If there is a tie, select the job that requires least tool changes. Enter it in the scheduling table, and mark the common tools. If the new job requires tools that are not presently in the magazine, a question arises as to which tools from the magazine must be replaced. This is where the information from the planning table is used. The following two rules are applied.

1. Replace only the number of tools that must be replaced to allow the production of the new job.
2. Replace a tool that is least frequently used in production of the remaining jobs from that group (the data given in G column for each tool). If there is a tie, break it based on the least used frequency for the remaining jobs (data in T for each tool). If there is still a tie, break it randomly.

Step 10: Assign new tools in appropriate positions for the new job in the scheduling table. Modify the planning table data to reflect the completion of scheduling for the job by modifying, as necessary, the columns for each tool. Also cross out the scheduled job from the group to reflect its assignment in the scheduling table. Go back to step 9.

Step 11: Since all the jobs in the previous group have been scheduled, remove the group from further consideration. If all the groups are scheduled, go to step 12. If a new group must be selected for scheduling, compare the tool requirements for the new group (group tools) with the tools that are presently in the magazine (last line of the scheduling table) and select a group that has most common tools with it. If there are more groups than one that meet this condition, break the tie based on jobs within these groups. Select a group that has a job that requires the most tools from the tools presently in the magazine. Create a row in the planning table for the selected group, and enter, for each tool, the associated cumulative tool use for the group and remaining cumulative tool use for all jobs, in columns G and T, respectively. Go to step 9.

Step 12: After all the groups and jobs have been scheduled, compare the successive rows of the scheduling table to find out the tool changes required in this schedule.

Step 13: Repeat steps 7 through 12 by making another group as the starting group for the procedure. If each group has been used as the start group, stop and choose the schedule that gives minimum tool changes.

14.2.3 Illustrative Example

Consider the data in Table 14.7. The tool magazine has the maximum capacity of 4. Applying step 1 reveals that jobs 2 and 10 are subsets of job 6 and jobs 4 and 5 are subsets of job 9. Therefore, jobs 2 and 10 are scheduled immediately after job 6 and 4, and job 5 immediately after job 9. Take off the jobs 2, 10, 4, and 5 from the data set for the remaining steps of the procedure.

Applying the steps 2 and 3 results in three groups with tool-job assignments, as shown in Table 14.8.

The application of step 4 gives Table 14.9, showing frequency of each tool used by all jobs and also by the jobs assigned in each group. Note that the total tool use for all jobs must equal the sum of tool use in all groups, since the jobs are distributed in groups.

Select one group, say group 1, to start the process. Start the planning table, Table 14.10, by drawing 9 columns for 9 tools and 3 subcolumns within each column to indicate use, cumulative group frequency, and cumulative job frequency for each tool. The columns are designated as U, G, and T, respectively.

TABLE 14.8
Tool Groupings and Job Assignments

Group	Tools Assigned to the Group	Jobs Assigned to the Group
1	1, 3, 5, 7	3, 6, 7
2	2, 5, 8, 9	1, 8
3	4, 6	9

TABLE 14.9
Frequency of Tool Use

	Tools								
	1	2	3	4	5	6	7	8	9
All jobs	2	2	3	1	4	2	3	2	2
G1	2	0	3	0	2	1	1	1	0
G2	0	2	0	0	2	0	1	1	1
G3	0	0	0	1	0	1	1	0	1

TABLE 14.10
Planning Table

	Tools																										
	1			2			3			4			5			6			7			8			9		
Job	U	G	T	U	G	T	U	G	T	U	G	T	U	G	T	U	G	T	U	G	T	U	G	T	U	G	T
GR1		2	2		0	2		3	3		0	1		2	4		1	2		1	3		1	2		0	2
3	x	1	1				x	2	2				x	1	3												
7	x	0	0				x	1	1										x	0	2						
6							x	0	0				x	0	2	x	0	1				x	0	1			
GR2		0	0		2	2		0	0		0	1		2	2		0	1		1	2		1	1		1	2
1				x	1	1							x	1	1				x	0	1	x	0	0			
8				x	0	0							x	0	0										x	0	1
GR3		0	0		0	0		0	0		1	1		0	0		1	1		1	1		0	0		1	1
9										x	0	0				x	0	0	x	0	0				x	0	0

TABLE 14.11
Scheduling Table

Group	Job	Tools in the Tool Magazine Positions				Tool Changes
		1	2	3	4	
GR1	3	1	3	5		
	7	1	3	5	7	
	6	6	3	5	8	2
GR2	1	2	7	5	8	2
	8	2	7	5	9	1
GR3	9	6	7	4	9	2
Total tool changes						7

Since group 1 is chosen for scheduling, the first row of the planning table, GR1, is filled with the appropriate cumulative tool use data for the group and for all jobs. The tools assigned to the group, that is, group tools, are 1, 3, 5, and 7. The jobs that have the most common tools to the group tools are jobs 3 and 7, each having three tools in common. Choose one of these two jobs at random, say job 3, to schedule as the first job. Enter the job in the scheduling, Table 14.11, and in planning table and modify the columns U, G, and T for each tool as appropriate.

To pick the next (and subsequent) job to schedule, select a job from the group that requires most common tools with the present arrangement. Both jobs 6 and 7 have two tools common. Break the tie by choosing the job that requires the least tool changes. Here, it is job 7 that requires one tool change as against job 6, which requires two tool changes. Job 7 is, therefore, scheduled in the second place, and the planning and scheduling table modified accordingly. By the way, since the first job, job 3 requires only three tools (with magazine capacity of 4), and the following job, job 7, requires a tool (tool 7) that is not required by job 3; tool 7 would be placed in the forth position of the tool magazine when job 3 tools are set to avoid an unnecessary changeover.

After scheduling the last job from group 1, that is, job 6, the tool magazine has in it tool numbers 6, 3, 5, and 8. Group 2 listing has the most common components with it, namely 5 and 8, and therefore group 2 is scheduled next. In Planning Table 14.10, for the associated row marked by GR2, the group 2 frequencies for all tools are inserted in their respective G columns. Also, the last modified entries for T columns from the previous group are rewritten in GR2 row for associated T columns.

Jobs 8 and 1 from group 2 have, respectively, 2 and 1 tools common with the present arrangement. Therefore, job 1, with the largest common components, is selected next for scheduling. The tools required for job 1 are 2, 7, 5, and 8, of which 5 and 8 are already in the magazine. Tools 6 and 3 are replaced by tools 2 and 7, respectively.

Job 8 requires tools 2, 5, and 9, of which 2 and 5 are already in the tool magazine. To determine whether tool 7 or 8 should be replaced by tool 9, check in

TABLE 14.12
Number of Tool Changes Using Different Starting Group

Start Group 2						Start Group 3					
Job	Tools				Changes	Job	Tools				Changes
1	2	5	7	8		9	4	6	7	9	
8	2	5	9	8	1	6	3	6	5	8	3
6	3	5	6	8	2	3	3	1	5	8	1
3	3	5	6	1	1	7	3	1	5	7	1
7	3	7	6	1	1	1	2	8	5	7	2
9	4	7	6	9	2	8	2	8	5	9	1
Total tool changes					7	Total tool changes					8

the planning table, the remaining group frequencies for tool 7 and 8. Both G values are zero. Next, check the remaining frequencies for the entire job, that is, T values. Tool 7 has $T = 1$, while tool 8 has $T = 0$. Hence, tool 8 is replaced by tool 9.

The procedure is continued until all groups and jobs are scheduled. The final schedule is displayed in Table 14.11. It requires seven tool changes.

Similarly, starting the procedure with group 2 results in the following job sequence: 1, 8, 6, 3, 7, and 9, with seven tool changes. Starting with group 3 results in 9, 6, 3, 7, 1, and 8 as the job sequence with eight tool changes. The tools in the tool magazine are shown in Table 14.12. A minimum of seven tool changes are necessary to process all jobs.

Here, we studied a method for sequencing N jobs on a machine equipped with an ATC, the objective being to minimize the makespan for the jobs by minimizing the tool changes. The procedure divides the problem into three phases following three different trends of thoughts. In the first phase (i.e., step 1), the dimension of the problem is reduced by permitting only those jobs that are not the subset of any other job to proceed further for planning. The second phase (steps 2 and 3) applies the principles of group technology to develop groups with similar jobs. The third phase (steps 4 through 13) sequences jobs and identifies the position for tool change that minimizes the total tool changeovers. The illustrative example shows the simplicity of the procedure, making it appealing to the end user. The procedure has been computerized and is included on the website provided.

14.3 HEURISTIC TO MINIMIZE THROUGHPUT TIME ON AN NC MACHINE

The problem discussed here is that of scheduling n jobs on an automated machine, such as a numerically controlled milling machine, equipped with a turret that can hold a number of tools, one on each of its faces (quite often, the chuck has eight faces). The objective is to minimize makespan to complete all jobs. The machine is capable of changing cutting tools rapidly by rotating the turret and bringing the right tool in contact with the surface of the work piece. Such a set-up may be viewed as similar to a machine equipped with an ATC. Similar to an automatic chuck, an ATC can hold a number of tools (based on the capacity of the tool magazine), and therefore we can perform multiple sequential operations without stopping the machine for a tool change.

At first glance, above problem seems to be that of scheduling n jobs on a single machine, and we may try to utilize one of the numerous single-machine sequencing algorithms [2] to optimally (or near optimally) sequence the jobs. But single-machine scheduling methods do not apply to a machine equipped with an ATC. Since the job holding fixture on an NC machine is generally universal, that is, same fixture can hold different jobs, the majority of time associated with a job setup is due to the time taken in changing the tool(s). Therefore, to minimize makespan for a fixed set of jobs, one must schedule jobs in such a manner as to minimize the total tool changes. We shall categorize such tool change as an external tool change since a new tool(s) is fetched and installed on the chuck. This problem then is similar to the ATC problem discussed in Section 14.2.

However, it is possible to further improve the performance on throughput time if attention is also given to the positions of the tools on the turret. Rotation of a turret to the "right" position takes time. If the tools demanded by two consecutive operation for a job are in two sequential positions on the turret, it would take less time to process a unit than if the tools are further apart. This difference could quickly add up, depending on the number of units (demand) processed for each job. We shall designate changing of the tool positions of the tools that are presently on the turret as the internal tool changes.

Before a new job begins, a decision must be made as to which tools should be installed in which positions on the turret. Furthermore, if the tools from the previous job could also be used for the new job then the decision may also include whether to adjust the positions of the tools. The tools required for any two consecutive operations in the new job may not be presently adjacent to each other.

There are two time factors that must be considered: one associated with tools and the other with jobs. For tools, it is necessary to know how long it takes to exchange the tools externally (removing an old tool and replacing it with a new tool that is presently not on the turret) and internally (changing the positions of two tools that are presently on the turret).When the tools are internally interchanged, there are two time elements that influence the time for the exchange: a fixed time associated with the procedure itself and a time that depends on the number of tools and positions to be interchanged. In a double swap, two tools presently on the turret are exchanged, while in a single swap, a tool is moved to an empty position.

For each job, we may have to establish the amount of additional time that would be required, if the tools necessary for sequential operations are not adjacent to each other. To process a job, each unit within the job must be subjected to the necessary tools. This may mean rotating the turret through a number of positions where the appropriate tool is located. In a job where a large number of units are processed, the total rotation time due to nonoptimum positioning of the tools could be more than the time it would take to interchange tools in adjacent positions.

14.3.1 Problem Definition and Assumptions

The objective is to develop a job sequence and tool arrangement on the turret for each job, so as to minimize the overheads associated with processing all jobs. Since the process times (metal-cutting times) are constant, this objective would also lead to minimization of the throughput time or makespan for the fixed set of jobs. The following assumptions are made in developing the procedure:

1. All *n* jobs are available at the beginning of planning period, with known tool requirements.
2. The number of units to be processed in each job i (i.e., demand for job i), d_i, is known.
3. The maximum number of tools, e_i, required for a job i is less than or equal to the capacity Z of the turret.
4. A fixed time of *F* is required to transport a tool from the magazine and place it in any position on the turret or remove a tool from the turret and place it in the magazine. T_i are the number of such transfers for job i. This is also noted as an external tool change.
5. The time necessary to change positions of any two tools that are presently on the turret in locations m and k is a linear function: $A + Bt^1_{mk}$, where t^1_{mk} is the number of turret turns that these two locations are apart in l-th double-tool swap. A and B are two constants.

If a tool that is presently on the turret in position "u" is moved to an empty position "v" on the turret, then the time required is, $G + Ht^q_{uv}$, where G and H are constants. t^q_{uv} is the number of turret turns that locations u and v are apart in q-th single-tool swap. Both of these are noted as internal tool changes, and generally $G < A$ and $H \leq B$.

6. The time necessary to turn the turret between tool use is KC_{ij}, where C_{ij} is the number of turret turns required to go from tool position in operation $j - 1$ to tool position in operation j, for job i. K is a constant.
7. It takes a fixed time of "R" units to go from last tooling position of the previous job to the first tooling position of the next job. Also it takes L_i units to go from the last tool position to the first tool position of the same job.
8. If some tools are to be removed from the turret and new tools are to be placed, the sequence of operation is: remove the tools first, make turret tool interchanges if necessary and then place the new tools.
9. Job setup time is shorter than the tool change time.

Mathematically, the problem could be stated as follows: Develop a job sequence and tool arrangement on the turret for each job so as to minimize.

$$R(n-1) + \sum_{\{i=1\}}^{n} FT_i + \sum_{\{i=1\}}^{n} \sum_{\{m=0\}}^{m_i} (A + Bt_{lk}^{m}) + \sum_{\{i=1\}}^{n} \sum_{\{q=0\}}^{q_i} (G + Ht_{lk}^{q})$$

$$+ \sum_{\{i=1\}}^{n} \sum_{\{j=1\}}^{e_i} KC_{ij}d_i + \sum_{\{i=1\}}^{n} L_i(d_i - 1)$$

where m_i and q_i are the number of double and single swaps for job i, respectively.

Each term in the preceding expression represents a time factor if the appropriate action (tooling position change, external tool change, double swap, single swap, number of turns of turret, return to tool position after each unit) is applicable for the job, if not the value of the term is zero. The number of alternatives available is $n!({}^{Z}C_{ei})$, which makes the problem NP complete, and hence the following heuristic method is suggested for its solution.

14.3.2 Heuristic Procedure and Its Application

The heuristic procedure suggested here requires evaluation of two alternative approaches and selecting the best between them. In the first alternative, the initial tool arrangement on the turret is optimized with respect to a job with the largest demand, and other jobs are chosen to minimize the internal tool changes. This approach may result in more than minimum external tool changes. In the second alternative, the initial job sequence is developed to minimize the external tool changes by first applying the procedure described in Section 14.2, and internal tool changes are made to further optimize the results. Here, the time for the internal tool changes may be more than what is possible with alternative one. Depending on the times associated with external and internal tool changes, one alternative may provide better results than the other for a specific set of data.

The steps of the heuristic procedure are illustrated by simultaneously applying them to a numerical example. The turret can hold a maximum of Z number of tools. For illustration purposes, we will assume that Z is 8 in our example. There are six jobs, and Table 14.13 shows their tool requirements and demands.

The following data is in relative time units:

F = Time required to transport a tool between magazine and turret (remove or place a tool) = 500 units

Time necessary to change tools in two positions (double tool swap) = $280 + 10t_{mk}$

Time necessary to move a tool an empty position (single-tool swap) = $100 + 10t_{mk}$

Time for one turret (turn to go from one tool position to next sequential position) = 1

TABLE 14.13
Tool Use Data for Jobs

Job #	Tools Used (Tool Numbers) in Sequence								Demand
1	8	9	1	6	7	5	10	11	30
2	8	5	4						100
3	6	9	7	10	2	5			500
4	8	6	1	7	11	3			500
5	8	1	4	10	5	6			10
6	6	9	7	5					500

TABLE 14.14
Distances (Turns) between Turret Positions

Turret Position	1	2	3	4	5	6	7	8
1	—	1	2	3	4	3	2	1
2	1	—	1	2	3	4	3	2
3	2	1	—	1	2	3	4	3
4	3	2	1	—	1	2	3	4
5	4	3	2	1	—	1	2	3
6	3	4	3	2	1	—	1	2
7	2	3	4	3	2	1	—	1
8	1	2	3	4	3	2	1	—

Since the turret is octagonal, the minimum turns require to go from one position to any other position, assuming the turret can move either clockwise or anticlockwise directions, are given in the Table 14.14.

Mathematically, such distance between location I and j could be expressed as $d_{ij} = d_{ji} = \min(|I - j|, |I - j + n|)$. Installation of new tools may result in a few rotations of the turret. Since the location of the final position after these rotations is not known in advance, it is assumed it takes four units, that is, the maximum possible, to go from the resultant position to the initial tool position of the next job. It is important to realize that this condition occurs only once per job.

14.3.3 Alternative 1: Start with a Job Having the Largest Demand

14.3.3.1 Discussion of the Procedure

There are five steps in the procedure. They are explained in conjunction with their applications to the illustrative example.

Step 1: Similar to the procedure in Section 14.2, start a planning table. The planning table (Table 14.15 shows the completed planning table for our example) consists of

TABLE 14.15
Planning Table

	1		2		3		4		5		6		7		8		9		10		11	
Tool	**U**	**F**	**U**	**F**	**U**	**F**	**U**	**F**	**U**	**F**	**U**	**F**	**U**	**F**	**U**	**F**	**U**	**F**	**U**	**F**	**U**	**F**
Initial freq.		3	1			1		2		5	5		4			4		3		3		2
Jobs																						
3		3	x	0		1		2	x	4	x	4	x	3		4	x	2	x	2		2
6		3		0		1		2	x	3	x	3	x	2		4	x	1		2		2
2		3		0		1	x	1	x	2		3		2	x	3		1		2		2
5	x	2		0		1	x	0	x	1	x	2		2	x	2		1	x	1		2
1	x	1		0		1		0	x	0	x	1	x	1	x	1	x	0	x	0	x	1
4	x	0		0	x	0		0		0	x	0	x	0	x	0		0		0	x	0

Note: U = Tool used; F = Remaining job frequencies.

a log of all tools and their use and non-use frequencies. For each tool, note the total number of times the tool is used in processing all jobs in the initial frequency row. As a job is scheduled, for each tool utilized in the job, the corresponding tool use frequency is reduced by one. The resultant (the value under F) for each tool therefore reflects the number of yet-to-be-scheduled jobs that would require the associated tool. If, to process a job, a tool needs to be removed and another placed on the turret, the tool with minimum current tool use frequency is chosen for removal. If there is a tie, it is broken in a random manner. Also necessary is a tool line for each job. It indicates the tool placement in each position (jaw) of the turret while processing the job. As a job is scheduled, the tool arrangement may change to reflect the optimization. This is explained in the following steps.

Step 2: To start scheduling, select a job with maximum demand. In case of a tie, select, from the tied jobs, the job that requires the most tools. If there is still a tie, obtain from planning table the total frequency for the tools used by the job, and determine the sum of these frequencies. Select the job that had the maximum frequency. If there is still a tie, break it randomly.

Load the tools necessary for the job in the consecutive positions on the turret in the same order as they are required for the job. Record the tool use in the planning table and in the tool line for the job. Calculate the time required for the job.

In our example, the maximum demand is associated with jobs 3, 4, and 6. Since there are ties, determine the job that needs the maximum tools. Tool requirements for jobs 3, 4, and 6 are 6, 6, and 4, respectively. Since there again is a tie between jobs 3 and 4, break the tie with regard to the total use of the tools. For job 3, tools 6, 9, 7, 10, 2, and 5 are used. Their total frequency from the planning table (Table 14.15 initial frequency row) is: tool 6 is used five times, tool 9 is used three times, tool 7 is used four times . . . , giving $5+3+4+3+1+5=21$. Similarly, for job 4, the total tool use frequency is $4+5+3+4+2+1=19$. Since job 3 has the maximum total tools used frequency, it is scheduled first.

Job 3 is noted in the planning table, and tool-use frequencies for tools 6, 9, 7, 10, 2, and 5 are reduced by one. The tools required for job 3 are arranged in the sequential order on the turret as shown by the tool line for job 3 in the first row of Table 14.17.

The total time in processing a job consists of three factors. First is the time necessary to place all new tools on the turret. The second factor is associated with the time for processing a unit. For each unit processed, the time to change the position of the turret to get to the next tool (rotate the turret) is calculated. The third factor is associated with initialization of the tool position for processing the next unit. In some cases, when a setup for a new unit is required, this time may be considered as a part of the setup time. In our example, we shall include the rotation time from last tool use of the previous unit to the first tool use of the next unit as a part of total idle time.

Hence, we have the total processing time for job 3 $=500\times 6+500(1+1+1+1+1)+499\times 3=6997$. The values in the preceding equation are derived as follows: There are six tools placed on the turret, each requiring 500 time units. The total units produced in job 3 are 500, the distances between successive tool positions for the job is 1 time unit each. To turn the chuck back from position 6 to position 1 after the completion of a unit takes 3 time units (from Table 14.14) for 499 units.

For the sake of keeping the procedural steps together, in the following discussion, the entire procedure is first explained before continuing its application to the example problem.

Step 3: Compare tool requirements for each job with the tools that are presently on the turret. Determine the number of common tools and also the number of new tools (outside tools) that a job needs. Select for loading, the job requiring the least number of outside tools. If there is a tie, break it by selecting the job that has the most number of common tools. If there still is a tie, select the job having the maximum demand. If there is still a tie, break it randomly.

Step 4: For the job selected, determine the parameter "P" as explained next. Suppose the operation with a tool located in position $j - 1$ has just been completed. However, the tool required for the next sequential operation is located in position k on the turret. The turret-turning time would be minimum if this tool could be placed in position j (the minimum of one turret turn to go from position $j - 1$ to j cannot be avoided). But that may mean taking out the tool that presently is in location j and interchanging it with the tool in location k. Recall that to interchange any two tools the time required is given by $A + Bt_{jk}$. If the tools are to be interchanged, there must be associated savings when the job is processed. This savings is due to the reduction in turret turns when processing a unit. The additional cost (over and above the time necessary to go from position $j - 1$ to position j) for turret turns is given by KC_{jk}, where C_{jk} is the distance (in terms of turns) between positions j and k on the chuck and K is constant. For a job with demand d_i, the additional cost of turret turns is $(KC_{jk})d_i$. If by making tool interchange the turret turn time can be reduced, then the tool change may be desirable. At minimum, the increase in the time due to tool interchange should be equal to decrease in time in turret turns, that is, $A + Bt_{jk} \leq (KC_{jk})d_i$.

Substituting the data from the example, this becomes: $280 + 10t_{jk} \leq 1 \times C_{jk}d_i$.

For each job, the demand is known. In determining whether to interchange tools or not, we can evaluate the preceding inequality for the known positions j and k. If the left-hand side is less than the right-hand side, tools should be switched.

It is more convenient, however, to approximate the value for minimum allowable tool separation index "P" for the job. The minimum cost, that is, value of the LHS is when $t_{jk} = 1$, that is, the next required tool is in location $j + 1$ rather than j. Thus, the minimum value of the LHS is $280 + 10 = 290$. Using actual demand, determine the minimum number of positions P; the two subsequent tools should be apart to justify the tool change. This is obtained by solving the following inequality for the smallest value of P.

$$290 \leq 1 \times P \times d_i$$

If tool separation is less than P, do not interchange, if more than P, interchange.

Step 5: Calculate the processing time for the job that is loaded. Remove it from yet-to-be-scheduled list. Go back to step 3 if there exists any job yet to be scheduled. If all jobs are scheduled, calculate the total tool change, tool turn and other processing times.

14.3.3.2 Applications of Steps 3 through 5 to the Illustrative Example

Job 3 has been loaded (end of step 2). To select the next job in the sequence, determine tools that are the common ones and tools that are presently not on the turret (outside tools) for the jobs yet to be scheduled. Table 14.16 shows the details.

Job 6 has the least number of outside tools and hence it is scheduled next, and the planning table is modified. With a demand of 500 units, the P value is obtained by, $290 \leq 500P$, or $P = 1$. That is, all tools required for job 6 must be in sequential order. The current tool arrangement and the required tool arrangement for job 6 are shown in Table 14.7.

Hence, interchange tools 10 and 5 from positions 4 and 6, respectively. The tool line for job 6 is shown in the second line of Table 14.17. The time necessary to complete the job is, $4 + (280 + 10 \times 2) + 500 \times (1 + 1 + 1) + (499 \times 3) = 3301$.

The terms in the above equation are for times necessary for tool initialization for the first unit after new tool arrangements, tool exchange, sequential processing, and reinitialization of the tool position from fourth position to the first after completion of each unit.

To determine the next job to sequence, develop Table 14.18, indicating the tools common and outside to the tools presently on the turret.

Since job 2 needs the least outside tools, load job 2 next. Enter the data in the planning table, and modify the tool frequencies for the tools used by job 2.

To determine P for the job that has a demand of 100 units, we have: $290 \leq 100P$, or $P = 3$.

TABLE 14.16
Job Selection for Second Position

Jobs yet to be scheduled	1	2	4	5	6
Common tools	5	1	2	3	4
Outside tools	3	2	4	3	0

TABLE 14.17
The Current Tool Arrangement: After Job 3

	Turret Position							
	1	**2**	**3**	**4**	**5**	**6**	**7**	**8**
Current tools	6	9	7	10	2	5		
Job 6 requirement	6	9	7	5				

TABLE 14.18
Common and Outside Tools for Jobs after Job 6

Jobs yet to be scheduled	1	2	4	5
Common tools	5	1	2	3
Outside tools	3	2	4	3

TABLE 14.19
Current Tool Arrangement after Job 6

Turret position	1	2	3	4	5	6	7	8
Tool	6	9	7	5	2	10		

TABLE 14.20
Final Job Sequence and Tool Arrangement

Job Sequence	Turret Positions and Tool Placement							
	1	2	3	4	5	6	7	8
3	6	9	7	10	2	5		
6	6	9	7	5	2	10		
2	6	9	8	5	4	10	7	
5	6	9	8	5	4	10	7	1
1	6	9	8	5	11	10	7	1
4	6	8	9	5	3	11	7	1

Tools required for jobs 2 are 8, 5, and 4, and the present tool arrangement is shown in Table 14.19.

Since only tool 5 is presently on the turret, the other two tools would have to be brought in. Tool 8 could be placed in location 7; however, the distance between locations 4 and 7 is 3 units, which is equal to the present value of P, and therefore a tool switch would have to be made.

According to the planning table, tool 7 still has use, while tool 2 is not used for any of the remaining jobs. Therefore, the least time-consuming alternative is to move tool 7 from position 3, to an empty position 7, bring in tool 8 in position 3 and remove tool 2 and replace it with tool 4. The corresponding tool arrangement is displayed in the third row of Table 14.20.

The corresponding time requirement is $4 + 3 \times 500 + 100 + 10 \times 4 + 100(1 + 1) + 99 \times 2 = 2042$.

The process is repeated until all jobs are scheduled. For the sake of completeness, the results of the steps are tabulated below in Table 14.21.

TABLE 14.21
Summary of Remaining Scheduling

Next Job	*P*	Action Taken	Time Cost
5	30	None	$4 + 500 \times 1 + 10(3 + 3 + 1 + 2 + 3) + 9 \times 2 = 642$
1	10	Remove tool 4, replace tool 11	$4 + 2 \times 500 + 30(1 + 2 + 1 + 2 + 3 + 2 + 1) + 29 \times 2 = 1422$
4	1	Remove tool 10, move 11 to pos 6, bring tool 3 in position 5, interchange tool 8 and 9	$4 + 2 \times 500 + (280 + 10) + (100 + 10) + 500(1 + 1 + 1 + 1 + 1) + 1499 \times 3 = 5401$

Since all jobs are done, remove all tools from the chuck. The time required is $8 \times 500 = 4000$.

The total time to process all jobs is $6997 + 3301 + 2042 + 642 + 1422 + 5401 + 4000 = 23805$.

14.3.4 Alternative 2: Minimize the External Tool Changes

The steps involved in the second alternative are as follows:

Step 1: Determine the minimum tool change sequence using Section 14.2 method (this would have been the solution when only the external tool changes are minimized).

Step 2: Apply steps 4 and 5 from alternative 1 for each job in the sequence obtained in step 1.

Using step 1 (we leave the calculations to the reader), the minimum external tool change sequence is jobs 3, 6, 2, 5, 1, and 4. Since the job sequence is same as alternative 1, the resulting time is also equal in this case. It is interesting to note that without the internal tool changes, that is, applying only step 1 of the alternative 2, the preceding job sequence would have resulted in 26,721 time units.

14.3.5 Select the Best Alternative

In this example, the time and sequence associated with alternatives 1 and 2 are the same. Hence, select the associated job sequence and tool arrangement as the best.

14.3.5.1 Further Comments

As mentioned earlier, the first alternative tries to minimize internal tool changes (tool changes on the turret), while the second alternative minimizes the external tool changes (tool changes from turret to tool magazine). Depending on the time values, one alternative may be better than the other for a given example. For instance, suppose the time factors are as follows: time for external tool change 300, time for two tool

TABLE 14.22
New Demands

Job	1	2	3	4	5	6
Demand	500	50	100	30	10	400

interchange $200 + 75t_{mk}$, and time for one tool exchange $150 + 75t_{mk}$. Furthermore, suppose the demands for different jobs are given in Table 14.22.

With the same remaining data as in the previous example, alternative 1 results in job sequence of 1, 6, 3, 5, 2, and 4, with a time requirement of 15,842 units, while the alternative two results in the sequence 3, 6, 2, 5, 1, and 4 with 16,165 units of time. This is despite the fact that alternative 1 has four external tool changes while alternative 2 has only three external tool changes. We shall leave it to the reader to check these calculations in detail.

This section illustrates the need for developing not only the job sequence but also the tool arrangement on the turret when the jobs are scheduled to minimize the makespan.

14.4 SUMMARY

This is the first chapter to discuss specific situations where good solutions can be developed by using application-specific heuristic or a combination of heuristics and decision rules. In the first section, the tabular approach to group technology is explained for instances where parts with common characteristics are grouped together for production in machine cells. Section two presents a method of scheduling jobs on a single machine with an ATC. This 13-step method uses a combination of group technology and heuristic for schedule development. In the last section, a heuristic is developed to schedule jobs to minimize throughput time on an NC machine. All these methods lead to optimization of plant operations.

14.5 PROBLEMS

14.1 Using the tabular approach, develop the machine groups for the following data. Also make the job assignments to the machine groups.

	Machine					
Job	**1**	**2**	**3**	**4**	**5**	**6**
1	1		1			
2		1	1			
3		1	1	1		
4			1	1		
5		1			1	1

(Continued)

Continued

	Machine					
Job	**1**	**2**	**3**	**4**	**5**	**6**
6		1			1	1
7		1	1		1	
8				1		1
9			1		1	
10	1		1			
11	1	1				
12	1	1				
13	1					1
14		1				1
15		1				1
16		1		1		

14.2 The data for 20 jobs that are to be produced on seven machines are given in the following table. Develop machine groups, and indicate job assignments to groups. Also, indicate any machine duplications.

	Machine						
Job	**1**	**2**	**3**	**4**	**5**	**6**	**7**
1					1	1	
2					1	1	
3					1	1	
4	1			1			
5	1			1			
6	1			1			
7		1			1	1	
8		1			1	1	
9	1	1	1	1			
10			1	1	1		
11			1			1	1
12	1	1	1				1
13		1	1				1
14		1					1
15	1		1	1			
16				1	1		
17						1	1
18		1					1
19			1			1	
20				1	1		

14.3 A CNC machine center uses an ATC to speed up the time for setup due to tool changes. The following ten jobs are to be processed on the machine center, which has an ATC capacity of four tools. Develop the sequence that will minimize the number of tool changes and thus the makespan as well.

	Tools						
Job	**1**	**2**	**3**	**4**	**5**	**6**	**7**
1	1			1			
2	1		1				
3		1			1		
4		1		1			
5			1		1		
6						1	1
7	1		1				1
8		1		1		1	
9	1		1				
10		1	1				

14.4 Given the following data on the number of jobs to be processed and their requirements for tools using an automatic tool charger with four heads, develop a schedule to minimize tool changeover and makespan.

	Machine						
Job	**1**	**2**	**3**	**4**	**5**	**6**	**7**
1			1				1
2				1	1	1	
3		1				1	
4		1	1				
5			1				1
6	1					1	
7	1	1					
8	1	1					
9			1	1			
10					1	1	
11						1	1

14.5 Determine the schedule and minimum throughout time for the data in Table 10.11 if the demands are changed as indicated in the following:

Job	Demand
1	100
2	100
3	400
4	500
5	400
6	50

REFERENCES AND SUGGESTED READINGS

Bard, J.F. 1988. "A Heuristic for Minimizing the Number of Tool Switches on a Flexible Machine" *IIE Transactions*, 20: 382–391.

Potts, C.N. and L.N. Van Wassenhove. 1991. "Single Machine Tardiness Sequencing Heuristic" *IIE Transactions*, 23: 346–354.

Sule, D.R. 1989. "A Systematic Approach for Machine Cell Formation in Cellular Manufacturing" *Proceedings, International Industrial Engineering Conference*, Toronto, pp. 619–624.

Sule, D.R. 1993. "Job Sequencing to Minimize Tool Changeover in Flexible Manufacturing" *International Journal of Production Planning and Control*, 4: 46–53.

Sule, D.R. 1994. *Manufacturing Facilities: Location, Planning and Design*, 2d ed., Boston: PWS Publishing.

15 Industrial Sequencing II: Electronic Assemblies: Component Tape Assemblies on a Sequencer

Products made with electronic components play a major role in our everyday life. Most of the products use a printed circuit board (PCB) or printed circuit pack (PCP) to store and carry out instructions. This and the next chapter discuss examples of sequencing in electronic manufacturing, first in forming component tapes for robotics insertion machines, and second in producing PCBs using a feeder-type machine.

There are two ways of securing components on the board in PCB production, using surface mount technology or using insertion through-hole technology. In both cases, an appropriate component must be placed in the right position on the board and then either soldered or otherwise connected to the board. Typically, PCBs have hundreds of different components. They are installed on the board by either a robotic insertion machine or by a feeder-type machine. In the feeder-type machine, a head picks a component in its slot as it appears on the input tape. It then inserts the component in the appropriate position on the board as programmed in its computer. In the feeder-type machine, each input tape is made up of only one type of component. The head of the machine may have more than one slot (similar to our hand and fingers), each slot picking up a component from separate tapes. The head then can make a stroke on the board, moving in X–Y coordinates and placing each component from each slot in appropriate position before returning back to load for the next stroke. We will study the feeder-type machines in the next chapter. In this chapter, we will concentrate on the robotic insertion process.

In a robotic insertion machine, the robot picks a component from its input tape and places it in the position specified by the computer and then return and picks up the next component in the sequence from the tape and continues on the next cycle. The input tape has many different components on it as the robotic arm can pick and place each type of component in the right place. Since there may be hundreds of different components installed on each PCB, it is important that each input tape supplied to each robotic arm, have components in the right sequence. The assembling of an input tape is done on a machine called the sequencer. In this chapter, we shall develop a sequencing procedure for a sequencer. We present a real-life example to illustrate the dimension of the problem.

15.1 A HEURISTIC PROCEDURE FOR TAPE ASSEMBLING ON A SEQUENCER

In an article, Fathi and Taheri (1989) describe the operation of a sequencer (or sequencing machine) in PCP (board) manufacturing. A sequencer is used in the preliminary step to arrange the components on an output tape (which is the input tape on the insertion machine). Each output tape contains the proper set of components in a sequence to assemble a certain type of PCP. Since component types, their number, and their sequence may change from one PCP to another, a unique tape is required for each slot of the insertion machine in each PCP assembly. As a result, at times, such an output tape is simply referred to as a "PCP" for convenience.

Similar to an insertion machine, a sequencer also has a feeding mechanism called a dispenser, which has slots known as dispensing heads. Each dispensing head may be loaded with an input tape consisting of the *same components*. The sequencer takes the required components from the dispensing heads and assembles them to form an output tape (PCP).

Figure 15.1 shows a four-head sequencer. Input tapes consist of components 1, 2, 3, and 4. The output tape (PCP) consists of these four components in the sequence that the robot will use in making a PCB that has components 1, 2, 3, and 4. To produce another PCB with, for example, components 1, 2, and 5, a new PCP tape will have to be made on the sequencer. To produce such a tape will need component tapes 1, 2, and 5 as the input tapes on the sequencer. This necessitates changing of either input tape 3 or 4 to input tape 5, thus requiring one changeover. We want to minimize such changeovers in production of a set of PCP.

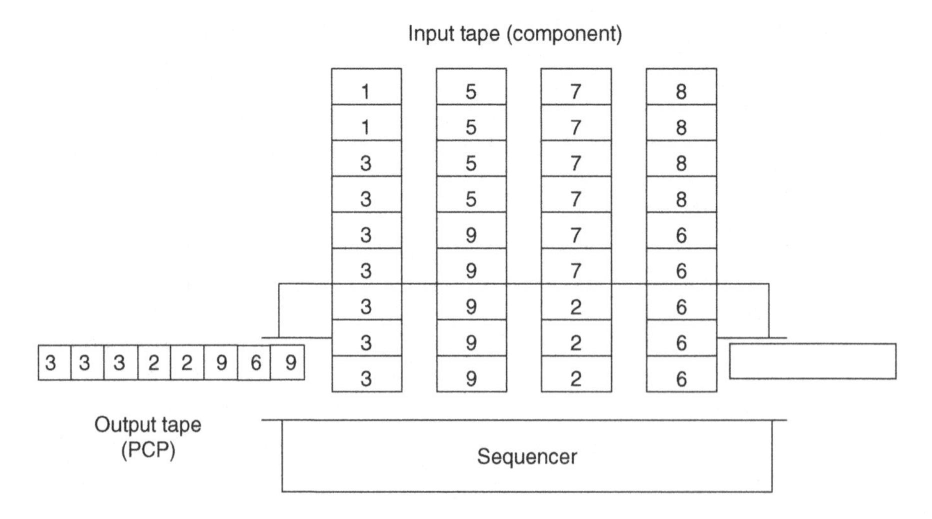

FIGURE 15.1 Scheduling a sequencer to produce PCPs.

Every change in an input tape on a sequencer requires stopping and adjusting the machine, resulting in the associated downtime. The problem then may be stated as follows.

15.1.1 Problem Statement

There are M types of output tapes to be produced, each with a different demand. There are N sets of distinct components available, each on a tape called an input tape. To avoid the confusion between output tape and input tape, an input tape, since it holds a *specific type* of component, is referred to as a "component," and an output tape as a PCP from now on. The components are loaded on a dispensing head of a sequencer during the production of a PCP. There are C dispensing heads on a sequencer on which the components can be installed (generally $C \ll N$). There may be as many as L sequencers, independent of each other, that are available to produce the necessary PCPs in the quantities required. The problem is to develop the component loading schedule for each sequencer that minimizes the total changeovers and also results in a balanced load on each sequencer.

Consider an illustrative example. Table 15.1 shows the data for 45 PCPs that are to be produced using 20 components. Here, "1" represents the use of a component in the production of a PCP and a "blank" the nonuse of the component. Table 15.1 also shows the demand for each PCP. There are two sequencers, each with seven heads. We might notice that each PCP requires no more than seven components. The problem is to determine which PCPs should be produced on which sequencer, and then to develop the detailed production schedule such that it would minimize the number of component changeovers and also allow the work to be as evenly distributed as possible, between the two sequencers.

15.1.2 Solution Procedure

The solution procedure is divided into two phases. In the first phase, using the concept of group technology, similar components are grouped together. PCPs assigned to the group can then be produced mainly by using the components from the group. We shall call this phase I of the solution procedure. Since the concepts and procedure for group forming was illustrated in Chapter 14, we shall assume the reader is familiar with the procedure.

Once the components and PCPs are grouped, the next step is to develop a procedure for sequencing. There are two questions that need to be answered. First, which components should be placed on which head of the sequencer to produce a scheduled PCP, and second, which of the remaining PCPs should follow the present PCP that has just completed its production. To answer these questions, phase II procedure is suggested.

As we proceed through the procedure, the reader will observe that there are some similarities between the approach taken here and the one we studied in Chapter 14 in NC machine sequencing problem. We first start by application of group technology to make coherent groups of components, and then we develop planning and scheduling tables to direct us in scheduling. However, due to the large number of

TABLE 15.1
Components Used in Each PCP and the PCP Demand

	Components																				
PCP	**1**	**2**	**3**	**4**	**5**	**6**	**7**	**8**	**9**	**10**	**11**	**12**	**13**	**14**	**15**	**16**	**17**	**18**	**19**	**20**	**Demand**
1	1	1	1		1																100
2		1	1	1	1	1															50
3							1	1	1		1										200
4	1	1			1		1	1													150
5				1		1	1	1	1	1											75
6								1		1	1			1	1	1					80
7					1	1		1		1		1		1	1						150
8			1		1				1	1					1			1			200
9	1			1	1						1		1								55
10														1	1	1		1	1		125
11	1				1						1	1	1				1		1		90
12	1	1	1						1	1											65
13											1			1		1		1			120
14		1		1	1		1								1						150
15								1		1		1		1		1					170
16				1		1	1	1				1									65
17				1	1			1				1		1							105
18																1	1	1	1	1	120
19						1	1	1	1	1											85
20		1			1	1			1		1										50

21							1		1			1					1	1			100
22	1		1	1	1		1														110
23									1		1			1	1	1					185
24					1	1								1			1	1			160
25		1						1	1		1		1								75
26	1			1						1					1		1				75
27						1	1				1			1							90
28	1				1			1			1		1								100
29											1				1		1	1		1	85
30								1			1			1			1			1	95
31	1		1	1				1	1												100
32			1			1	1				1	1									50
33									1			1	1			1	1				150
34		1			1				1		1	1				1					145
35						1			1						1			1			95
36					1		1			1				1		1					150
37				1		1					1	1	1								200
38	1	1			1		1	1													180
39						1			1			1	1		1						170
40			1				1	1			1			1							140
41	1														1		1		1	1	80
42					1				1							1	1	1			100
43				1		1				1		1		1					1		120
44			1	1									1	1					1		150
45	1	1						1	1	1								1			75

components involved, the planning table is constructed slightly differently. Also, since we may have more than one sequencer, the decisions as to which group of components should be scheduled on which sequencer become critical.

15.1.3 Phase II: Scheduling

There are a number of steps in phase II of the procedure. These steps are explained next.

Step 1: Calculate the ideal load or capacity for each sequencer. This is done by dividing the total load (i.e., sum of demand for each PCP) by the number of sequencers available. When a PCP is scheduled on the sequencer, the available capacity of the sequencer is decreased by the demand associated with the PCP. The problem could be examined by following either of the two approaches stated in the following:

The first approach or approach A: once a group is assigned to a sequencer, load all the PCPs from the group onto that sequencer. The next group (phase I may result in multiple groups), however, is only loaded on the sequencer if the capacity still exists. This alternative gives minimum component changeovers.

The second approach or approach B: load a new PCP only if the capacity is still available on the sequencer. This alternative gives a more balanced load on each sequencer.

Step 2: Assign component groups that are most diverse from each other (with the least number of common components) to different sequencers.

Development of loading pattern for PCPs on a sequencer is carried out by using two tables—a planning table and a loading table. The information from the planning table is used to decide how to change the components on the sequencer, while the loading table indicates the sequence of loading of the PCPs and actual component positions on the sequencer. The details are explained in the following steps.

Step 3: Develop a planning table and a scheduling table for a sequencer as follows:

Planning table. This table consists of component listings for the PCPs associated with the group that is presently assigned to the sequencer. The table is divided into two sections. The first contains components within the group, while the second section contains components required by PCPs that are external to the group.

Recall that there are only a finite number of heads on each sequencer and, because of the group-fathoming process in the grouping procedure, no group has a number of components that exceed this number. In each sequencer's planning table, therefore, the first section consists of components that belong to the group.

The second part of the table is associated with the components that are external to the group and is constructed gradually. For each PCP that is in the group, a recording is made of the components that are already in the table. The components that are not presently in the table but are required for the scheduled PCP are inserted in the second part.

For clarity of presentation, these steps are explained with a small example. Suppose the group components consists of components 11, 2, 13, 8, 5, 9, and 6

TABLE 15.2
A PCP Group and its Components

PCP	Components
1	11, 2, 13, 8, 5, 9, 17
2	11, 5, 6
3	13, 8, 9, 6, 15, 3
4	11, 2, 8, 9, 17
5	11, 2, 8, 5, 6, 15

TABLE 15.3
Illustrative Planning Table for the Data in Table 15.2

	Components Within Group							Components Outside Group			
PCP	**11**	**2**	**13**	**8**	**5**	**9**	**6**	**17**	**15**	**3**	**4**
1	x	x	x	x	x	x		x			
2	x				x		x				
3			x	x		x	x		x	x	
4	x	x		x		x		x			x
5	x	x		x	x		x		x		
Imp 1	4	3	2	4	3	3	3	2	2	1	1
Imp 2	3	2	1	3	2	2	3	1	2	1	1
Imp 3	2	1	1	2	2	1	3	0	2	1	1

for a seven-head sequencer, and there are five PCPs that belong to this group, each requiring components as shown in Table 15.2.

In the planning table (Table 15.3), the group components are noted in the first part of the table. Components outside the group are then listed one at a time, as explained next.

For the first PCP, the required components are 11, 2, 13, 8, 5, and 9, of which 11, 2, 13, 8, 5, and 9 are the same as the components within the group. They are recorded in the first part of Table 15.3 by marking the corresponding "x"s. The component 17 is not from within group components. It also has not been recorded yet as needed outside group component. Therefore, it is entered into the second part of the table, components outside group, and marked as being used by PCP 1. The other four PCPs in our example are similarly recorded. Note in recording PCP 4, that component 17 is already in the table and, therefore, is merely marked by "x" as also being used by PCP 4.

Next, we calculate the number of times each component is used in loading on the sequencer all PCPs from this group. This is simply done by adding x's in each column of the planning table. The sum is shown as Imp 1 row (importance row) in Table 15.3 (other Imp rows are explained later). These numbers give the present relative importance of each component in loading the remaining PCPs from that group. The higher the number, the more often that component will be used in assembling the PCPs that are yet to be built. This information is used in developing the loading table, specifically, to determine which of the presently assigned components on the sequencer should be replaced, if need be. The details are shown in the loading table discussion.

The Imp row is continuously updated as a PCP is taken out from the planning table and scheduled into the loading table. Thus, we have Imp 1, Imp 2, Imp 3, etc.

For example, the present relative importance of each component in scheduling the five PCPs is given by Imp 1. If PCP 1 is scheduled in the loading table, it is no longer in the planning stage. The components associated with PCP 1 are then taken off from the total, giving the new values shown as Imp 2 row. Thus, in this instance, the relative importance of component 11 is reduced from 4 to 3, that of component 2 from 3 to 2, etc.

Loading table. This table is used to determine how to load the components on the sequencer. In particular, it determines which component should be placed on which head of the sequencer. This is a complex process since loading of each PCP may require modification in one or more of the dispensing head's component assignment. To decide the position of the changeover, the information obtained from the planning table is very useful.

To start, load the sequencer with a PCP having the largest number of common components with the initial component listing of the group. If there is a tie, load the component that requires the least number of changes from the present component arrangement. Select and load the PCP and corresponding components on the sequencer and modify the importance row of planning table accordingly.

In our example PCP 1 has six components common to the group, and this is the largest common component number for any PCPs in this group. Load PCP 1 on the loading table as displayed in Table 15.4a. The table shows the components assigned to each head of the sequencer.

The modified importance row is shown as Imp 2 in the planning table.

Step 4: Before loading the next PCP, check if approach A or B from step 1 is to be followed. For approach A, if all the PCPs from the present group are not yet loaded, go to step 5. If all the PCPs from the present group are assigned, the next group is only loaded if the capacity is still available on the sequencer. If such is the case, go to step 6. Otherwise, the present sequencer is loaded to its maximum allowable capacity. Start the scheduling process on the next sequencer, if there is one, by going back to

TABLE 15.4a
Illustrative Loading Table: First PCP Loading

Head	1	2	3	4	5	6	7
PCP 1	11	2	13	8	5	9	17

step 3. If all the sequencers are loaded, stop; we have the optimum solution to the problem. Go to step 7.

If the second approach is to be followed, check to see if the sequencer capacity has been exceeded; if it has, stop loading on the present sequencer. Remove from the group the PCPs that are already loaded so far, thus retaining the group with only non-assigned PCPs from that group, and go to the next sequencer, that is, go to the beginning of step 3. Otherwise, go to step 5 to continue loading the present group. If the present group is completely loaded and there still exists a capacity on the sequencer, go to step 6 to load a new group. Again, if all sequencers are loaded, we have the optimum solution. Stop, and go to step 7.

Step 5: Compare the components necessary for all unloaded PCPs in the group, with the components *presently* on the sequencer (last line of the loading table). Select a PCP that has the most common components with it as the next PCP for loading. If the selected PCP needs a few components that are not presently on the sequencer, then the sequencer also has a few components that are not used by the scheduled PCP. This is the case because the initial problem statement indicates that no PCP requires the number of components that are more than the number of heads on the sequencer. A few of these immediately nonusable components would have to be removed from the sequencing head to make the room for the required components. To that end, select for replacement from the components that are not used, those components that are of least importance, as suggested by the latest importance row in the planning table. Replace only the minimum necessary that would permit the production of the new PCP.

Once the PCP is scheduled and entered into the loading table, remove it from the planning table. Modify the importance row, and select the next PCP to load, following the procedure described earlier, returning to step 4.

In our example, suppose capacity is still available on the sequencer. PCP 4 has the most number of common components with the present state of the sequencer, which has components 11, 2, 13, 8, 5, 9, and 17 on it. Hence, PCP 4 is loaded next. It requires no outside component; therefore, no replacement is needed. The result is shown in Table 15.4b.

The next PCP to load is number 5 (assuming capacity exists), since it has the most, that is, four components, common to the present sequencer assignment. It does have a requirement for two components, 6 and 15, that are presently not on the sequencer. To determine where the components 6 and 15 should be loaded, examine the current Imp row, that is, Imp 3. Component 17 has the least value that is, 0, and therefore it would be replaced by component 6 or 15. We choose 6 randomly. The next

TABLE 15.4b
Illustrative Loading Table: Second PCP Loading

Head	1	2	3	4	5	6	7
PCP 1	11	2	13	8	5	9	17
4	x	x		x		x	x

TABLE 15.4c
Illustrative Loading Table: Third PCP Loading

Head	1	2	3	4	5	6	7
PCP 1	11	2	13	8	5	9	17
4	x	x		x		x	x
5	x	x	15	x	x	6	

TABLE 15.4d
Equivalent Illustrative Loading Table: Third PCP Loading

Head	1	2	3	4	5	6	7
PCP 1	11	2	13	8	5	9	17
4	11	2	13	8	5	9	17
5	11	2	15	8	5	9	6

least value is associated with component 13, which would be replaced by 15. The associated results are shown in Table 15.4c.

Thus, the sequencer now has components 11, 2, 15, 8, 5, 9, and 6 on it. It might be more convenient to display the loading table by merely indicating the component assignment on each head as each of the PCPs is loaded. Thus, Table 15.4d is an equivalent form of the Table 15.4c.

A new Imp row would be constructed and the procedure continued.

Step 6: When all the PCPs in the group are loaded, and there is still some capacity left in the sequencer, choose the next group to load, such that the initial component listing of the new group has the highest number of common components with the latest arrangement on the sequencer (the last line of the loading table). Go to step 3. If all the groups are completely loaded, stop. We have the desired loading pattern on all sequencers. Go to step 7.

Step 7: Check the successive rows of the loading table(s) to determine the total changeovers required for the schedule.

15.1.4 Solution to the Example Problem

We shall now illustrate the entire procedure by applying it to the data shown in Table 15.1. Utilizing the first phase of the procedure results in six groups. Each group's components and associated PCPs are shown in Table 15.5.

Applying step 1 of phase II results in the determination of the ideal load for each sequencer. This is a total load of 5184 being distributed between two sequencers, or 2592 per sequencer. The least number of common components between any two groups is 1, and that occurs between groups 2 and 6. Therefore, group 2 could be

TABLE 15.5
Component Groups and PCPs Assigned to Each Group

Group	Components	PCP
1	1, 2, 4, 5, 8, 9, 11	1, 2, 4, 9, 12, 14, 20, 22, 25, 26, 28, 31, 34, 38, 45
2	7, 8, 9, 10, 11, 14, 16	3, 5, 6, 7, 13, 15, 19, 23, 27, 30, 36, 40
3	4, 5, 6, 7, 9, 11, 12	16, 21, 32, 37, 39, 43
4	4, 5, 9, 11, 12, 13, 14	17, 33, 44
5	5, 9, 14, 15, 16, 17, 18	8, 10, 18, 24, 29, 35, 42
6	1, 11, 12, 13, 17, 19, 20	11, 41

TABLE 15.6
Planning Table

	Components Within Group							Components Outside Group								
PCP	7	8	9	10	11	14	16	4	6	15	5	12	18	17	20	3
3	X	X	X		X											
5	X	X	X	X				X	X							
6		X		X	X	X	X			X						
7		X		X		X			X	X	X	X				
13					X	X	X						X			
15		X		X		X	X					X				
19	X	X	X	X					X							
23			X		X	X	X			X						
27	X				X	X			X							
30		X			X	X								X		
36	X			X		X	X				X					
40	X	X			X	X										X
Imp 1	6	8	4	6	7	9	5	1	4	3	2	2	1	1	1	1
Imp 2	6	7	4	5	6	8	4	1	4	2	2	2	1	1	1	1
Imp 3	6	6	4	4	6	7	3	1	4	2	2	1	1	1	1	1
Imp 4	6	6	3	4	5	6	2	1	4	1	2	1	1	1	1	1

loaded on the first sequencer, and group 6 on the second. For the first sequencer with group 2 PCPs, we can construct a planning table, shown here as Table 15.6. PCP number 6 has the highest number of common components with the initial group listing; therefore, it is shifted to the loading table. Table 15.7a shows part of the loading table for discussion, while the complete table is illustrated as Table 15.7b.

The next step is to recalculate the totals for group 2 after removing PCP 6 from the planning table. The results is shown as Imp 2 row in Table 15.6. There are 2512 units of capacity still available on the sequencer, hence another PCP is scheduled. Look for the PCP that has the highest number of common components with the latest arrangement, that is, PCP 6. The planning table points to PCP 7 and PCP 15 having the same

TABLE 15.7a
Partial Loading (Scheduling) Table

	Sequencer I: Head						
PCP	1	2	3	4	5	6	7
6	8	10	11	14	15	16	—
15	8	10	11	14	15	16	12
23	8	10	11	14	15	16	9

TABLE 15.7b
Load-Based Scheduling for Sequencer 1

			Components on Slot							
PCP	Load	Cumulative Load	1	2	3	4	5	6	7	Changeovers
6	80	80	8	10	11	14	15	16	—	0
15	170	250	8	10	11	14	15	16	12	0
23	185	435	8	10	11	14	15	16	9	1
3	200	635	8	10	11	14	7	16	9	1
19	85	720	8	10	11	14	7	6	9	1
5	75	795	8	10	4	14	7	6	9	1
27	90	885	8	10	11	14	7	6	9	1
40	140	1025	8	10	11	14	7	6	3	1
30	95	1120	8	10	11	14	17	6	20	2
13	120	1240	8	10	11	14	16	6	18	2
36	150	1390	8	10	5	14	16	6	7	2
7	150	1540	8	10	5	14	12	6	15	2
16	65	1605	8	4	7	14	12	6	15	2
32	50	1655	11	4	7	3	12	6	15	2
37	200	1855	11	4	7	13	12	6	15	1
39	170	2025	9	4	7	13	12	6	15	1
21	100	2125	9	4	7	18	12	6	17	2
43	120	2245	10	4	14	19	12	6	17	3
24	160	2405	10	18	14	19	5	6	17	2
10	125	2530	16	18	14	19	5	15	17	2
18	120	2650	16	18	20	19	5	15	17	1

number (4) of common components. As PCP 15 requires fewer changes in the present arrangement than PCP 7, it is selected next. This requires 0 changeovers. Recalculate the importance row, removing PCP 15 to the loading table (Imp 3 row). There is still capacity available on the sequencer. Following the procedure, the next PCP to load is PCP 23, which requires just one changeover. The candidate for the component replacement is a component presently on the sequencer that is of least importance, in this case component 12. Continue applying the procedure in a similar manner until

TABLE 15.7c
Load-Based Scheduling for Sequencer 2

			Components on Slot							
PCP	Load	Cumulative Load	1	2	3	4	5	6	7	Changeovers
11	90	90	17	5	1	19	11	12	13	0
41	80	170	17	20	1	19	11	12	15	2
33	150	320	17	9	16	19	11	12	13	3
44	150	470	14	4	3	19	11	12	13	3
17	105	575	14	4	5	8	11	12	13	2
9	55	630	1	4	5	8	11	12	13	1
28	100	730	1	4	5	8	11	12	13	0
4	150	880	1	4	5	8	11	2	7	2
38	180	1060	1	4	5	8	11	2	7	0
14	150	1210	1	4	5	8	15	2	7	1
22	110	1320	1	4	5	8	3	2	7	1
1	100	1420	1	4	5	8	3	2	7	0
2	50	1470	1	4	5	8	3	2	6	1
31	100	1570	1	4	5	8	3	2	9	1
12	65	1635	1	10	5	8	3	2	9	1
45	75	1710	1	10	5	8	18	2	9	1
20	50	1760	6	10	5	8	11	2	9	2
25	75	1835	13	10	5	8	11	2	9	1
34	145	1980	12	10	5	16	11	2	9	2
26	75	2055	1	10	15	17	4	2	9	4
8	200	2255	18	10	15	17	3	5	9	3
42	100	2355	18	16	15	17	3	5	9	1
35	95	2450	18	16	15	17	6	5	9	1
29	85	2535	18	10	15	17	20	11	9	2

the sequencer is filled to close to 2592 tape capacity. In this case, the closest possible capacity is either 2530 or 2650. Here, we choose to load the sequencer to the capacity of 2650. Counting the number of changeovers from Table 15.7b, we note the schedule has a total of 30 changeovers.

The loading on the second sequencer starts with group 6 PCPs. Since the details of the procedure are the same, the particulars are not described here. The results are shown in Table 15.7c.

The total load on the second sequencer is 2535, with 34 changeovers for the schedule. Thus, the number of changeovers necessary to complete the production of all PCPs is $30 + 34 = 64$.

When the objective is to minimize the changeovers while maintaining as even a distribution of load as possible, that is, following alternative 2 from step 3, the resulting sequences and load distributions are obtained as shown in Tables 15.8a and 15.8b. The load on sequencer 1 is 3130 with 35 changeovers and for sequencer 2 is 2055 with 28 changeovers for the total of 63 changeovers. As expected, this policy has reduced the total changeovers by 1 from the previous policy; however,

TABLE 15.8a
Group Base Scheduling on Sequencer 1

PCP	Load	Cumulative Load	Components on Slot 1	2	3	4	5	6	7	Changeovers
6	80	80	8	10	11	14	15	16	—	0
15	170	250	8	10	11	14	15	16	12	0
23	185	435	8	10	11	14	15	16	9	1
3	200	635	8	10	11	14	7	16	9	1
19	85	720	8	10	11	14	7	6	9	1
5	75	795	8	10	4	14	7	6	9	1
27	90	885	8	10	11	14	7	6	9	1
40	140	1025	8	10	11	14	7	6	3	1
30	95	1120	8	10	11	14	17	6	20	2
13	120	1240	8	10	11	14	16	6	18	2
36	150	1390	8	10	5	14	16	6	7	2
7	150	1540	8	10	5	14	12	6	15	2
16	65	1605	8	4	7	14	12	6	15	2
32	50	1655	11	4	7	3	12	6	15	2
37	200	1855	11	4	7	13	12	6	15	1
39	170	2025	9	4	7	13	12	6	15	1
21	100	2125	9	4	7	18	12	6	17	2
43	120	2245	10	4	14	19	12	6	17	3
24	160	2405	10	18	14	19	5	6	17	2
10	125	2530	16	18	14	19	5	15	17	2
18	120	2650	16	18	20	19	5	15	17	1
29	85	2735	16	18	20	11	5	15	17	1
42	100	2835	16	18	20	9	5	15	17	1
8	200	3035	10	18	3	9	5	15	17	2
35	95	3130	6	18	3	9	5	15	17	1

TABLE 15.8b
Group-Based Scheduling for Sequencer 2

PCP	Load	Cumulative Load	Components on Slot 1	2	3	4	5	6	7	Changeovers
11	90	90	17	5	1	19	11	12	13	0
41	80	170	17	20	1	19	11	12	15	2
33	150	320	17	9	16	19	11	12	13	3
44	150	470	14	4	3	19	11	12	13	3
17	105	575	14	4	5	8	11	12	13	2
9	55	630	1	4	5	8	11	12	13	1
28	100	730	1	4	5	8	11	12	13	0
4	150	880	1	4	5	8	11	2	7	2

TABLE 15.8b
Continued

PCP	Load	Cumulative Load	Components on Slot 1	2	3	4	5	6	7	Changeovers
38	180	1060	1	4	5	8	11	2	7	0
14	150	1210	1	4	5	8	15	2	7	1
22	110	1320	1	4	5	8	3	2	7	1
1	100	1420	1	4	5	8	3	2	7	0
2	50	1470	1	4	5	8	3	2	6	1
31	100	1570	1	4	5	8	3	2	9	1
12	65	1635	1	10	5	8	3	2	9	1
45	75	1710	1	10	5	8	18	2	9	1
20	50	1760	6	10	5	8	11	2	9	1
25	75	1835	13	10	5	8	11	2	9	1
34	145	1980	12	10	5	16	11	2	9	2
26	75	2055	1	10	15	17	4	2	9	4

this has happened at the expense of some unevenness in the load distribution on the sequencers.

15.2 SUMMARY

An efficient method for PCP distributing and processing on sequencers is presented in this chapter. The procedure is important since changeover times can contribute considerably to the overall operation times of the sequencers, and any reductions in these times provide improvement in the productivity. The new method is shown to be more effective than the one presently available in public literature. This methodology should help in cost reduction and more efficient management in the electronic industry.

15.3 PROBLEMS

15.1 Determine which PCPs should be produced on which sequencer and develop the production schedule for the data given in Table 15.1 if the PCP demands are changed according to the following:

PCB	Demand	PCB	Demand	PCB	Demand
1	20	16	115	31	180
2	40	17	130	32	185
3	60	18	195	33	205
4	210	19	155	34	70
5	80	20	105	35	80

(Continued)

Continued

PCB	Demand	PCB	Demand	PCB	Demand
6	65	21	100	36	170
7	45	22	50	37	190
8	125	23	35	38	100
9	180	24	165	39	200
10	170	25	115	40	60
11	110	26	35	41	130
12	50	27	95	42	105
13	190	28	130	43	95
14	70	29	145	44	205
15	75	30	135	45	95

15.2 Given the following data on components and PCPs, determine which PCPs should be produced on which sequencer, and develop the schedule knowing two sequencers are available, each with five heads.

	Components												
PCP	1	2	3	4	5	6	7	8	9	10	11	12	Demand
1		1				1							75
2	1	1	1										90
3	1			1	1		1	1	1				60
4	1						1	1	1				150
5		1	1	1						1			135
6		1			1	1							40
7		1								1	1	1	90
8			1		1	1			1				80
9			1	1	1								145
10			1							1	1	1	100
11				1				1	1				55
12	1			1			1			1			65
13		1		1				1				1	135
14					1	1	1	1	1				175
15				1		1		1		1		1	80
16							1		1	1	1		45
17								1	1	1	1	1	85
18		1		1		1		1					115
19			1	1	1	1	1						130
20								1		1		1	95

15.3 Make the PCP determination and develop a schedule using data from Problem 15.2 if the sequencers each have six heads.

15.4 Change the demand in Problem 15.2 to those following and make the new PCP determinations and schedule.

PCP	Demand	PCP	Demand
1	250	11	35
2	70	12	90
3	150	13	165
4	50	14	205
5	195	15	155
6	275	16	120
7	135	17	220
8	65	18	190
9	115	19	245
10	215	20	140

REFERENCES AND SUGGESTED READINGS

Bridges, R.O. 1977. "Sequencing of Components for Automatic Assembly on Printed Wiring Boards" *Western Electric Engineer*, 21(14).

Fathi, Y. and J. Taheri. 1989. "A Mathematical Model for Loading the Sequencers in a Printed Circuit Pack Manufacturing Environment" *International Journal of Production Research*, 27(8): 1305–1316.

Richerson, F.K. 1976. "Computer Program Controls Component Sequencer Costs" *Assembly Engineering*, 24(24).

Sabah, U.R., Edward, D.M., and S.-D. Faruqui. 1985. "An Integer Programming Application to Solve Sequencer Mix Problems in Printed Circuit Board Production" *International Journal of Production Research*, 23(3): 543–561.

Sule, D.R. "A Systematic Approach for Machine Grouping in Cellular Manufacturing" *Proceedings of the International Industrial Engineering Conference*, 1989, Toronto, pp. 619–624.

16 Industrial Sequencing III: Sequencing Feeder for Component Tape Assembly

The production of printed circuit board (PCB) requires the assembling of various electrical and electronic components at specific locations on the circuit board. The components are discrete devices such as resistors, capacitors, diodes, transistors, and/or connectors.

Modern PCB production utilizes computerized machines to insert components using one of two processes. The first is called the automated insertion process, in which the machines have insertion heads, each of which can be loaded with one type of component. Typically, motions in the x and y directions are obtained by either allowing heads to move or the table with the PCB to move. These automatic insertion machines are also referred to as feeders. Figure 16.1 displays a sketch of a feeder used in our discussion.

The other process is called the robotic insertion process. Robots are more flexible because they can insert any component at any location on the board. In this process, the components arrive on a sequenced tape, so that the robotic arm has only to pick and place each component in its place. The sequenced tape, which is also called an output tape, is assembled by a machine called a sequencer, as we have seen in Chapter 15.

The procedure illustrated here is for a feeder-type machine. An example problem consisting of a machine with six slots (heads) is used to illustrate the procedure. In this machine, each slot can hold one component. A number of units of a single component type are packed on one tape. Thus, once a tape is attached to a slot, it essentially determines the component type that is assigned to the slot. All slots are mounted on an arm called the dispenser that moves as a single unit. Initially, the dispenser is at the home position (i.e., 0-0 coordinates). From here, the dispenser moves forward so that an appropriate slot is over the first location on the board where a component is to be inserted, and infuses the component from the slot. It then moves directly to the next location under the new slot and inserts another component there. Movements in the horizontal and vertical directions are made simultaneously by the arm. This procedure continues, and when all the assigned slots have placed their components, the arm (and therefore all slots) moves back to the home position. This signifies the end of a stroke. Each slot may then pick up one more component from the tape and

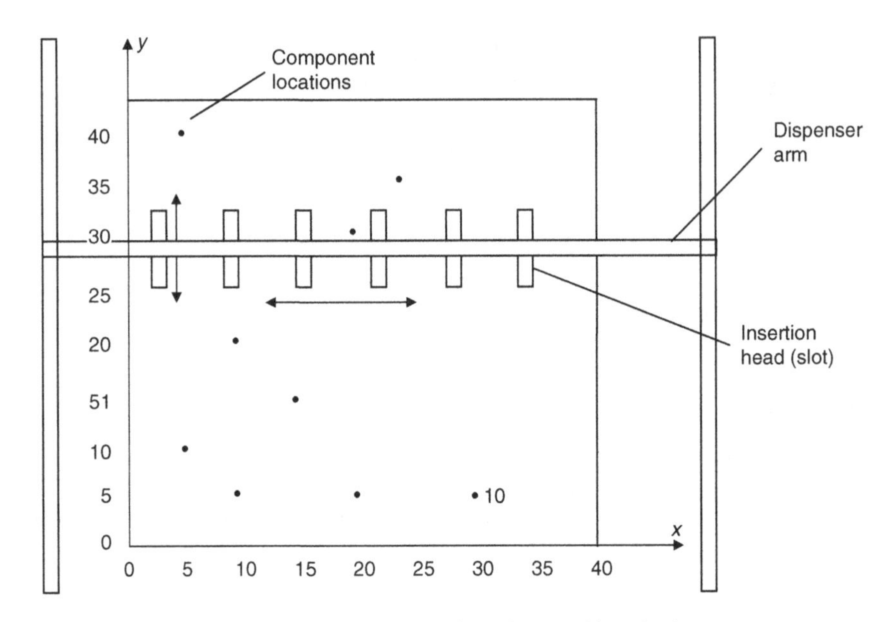

FIGURE 16.1 Automatic insertion machine (feeder).

set out on the next stroke. The procedure continues until all the locations designated to the feeder have been assigned the appropriate components.

Multiple boards are processed as a batch, one after another. However, each board may require only a subset of all the components that have been placed on the feeder. This is because the setup time for the feeder can be very large compared to the processing time. One common practice, therefore, is to group similar boards in a batch and setup the feeder so that there is no changeover required in processing these boards.

The insertion machine may have further restriction on the movement of the slots. The slots can access the locations anywhere along the y (vertical) direction, but are restricted along the x (horizontal) direction, as illustrated in Figure 16.2. This complicates the scheduling process.

16.1 PROBLEM DESCRIPTION

The total time taken for insertion consists of two controllable factors. The first is the number of strokes utilized, and the second is the inter-board movement of the arm in each stroke. In insertion machines, the stroke length, that is, going from the home position to the board and then coming back to the home position, is a dominating factor in the total distance traveled by the arm. Any reduction in the number of strokes used to insert all components contributes much more to the reduction in the total time than optimizing board travel pattern. Therefore, emphasis is placed on reducing the number of strokes.

The objective then is to minimize the number of strokes required to complete the placement of all components on a batch of PCBs. In order to minimize the number of

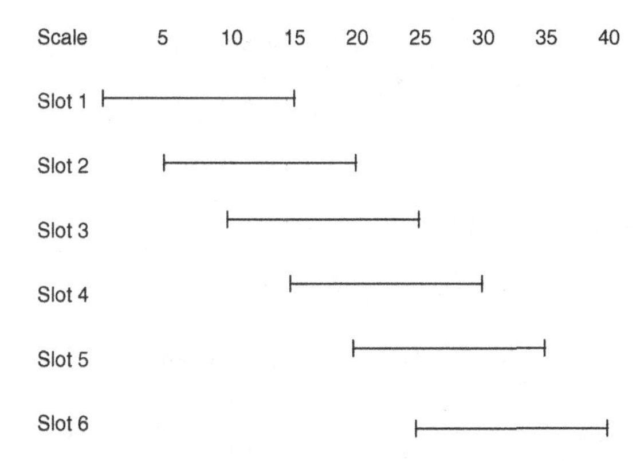

FIGURE 16.2 Reach of the slots along the x direction.

strokes, each slot must be loaded with the correct component. It is also necessary to assign the location that each slot is required to serve in each stroke.

Exhaustive enumeration of the component assignment problem is almost impossible. For example, in a feeder with 20 slots, and 5 distinct components, the number of ways the components can be assigned to slots is 5^{20} (9.537×10^{13}). Only a small percentage of these are feasible solutions. On a Sparc Station (Unix operating system), it takes days to solve such a problem exhaustively. Therefore, a means must be found to reduce the computational time and provide good results. A heuristic solution presented in the following gives good solution (usually an optimum solution) with dramatic savings in computational times.

The input to the problem consists of information about the components on the board(s) and information on the feeder. For each component on the board, the data consists of:

1. The component type.
2. The x and y coordinates on the board where the insertion should be made.

Information about the feeder includes:

1. The number of slots on the feeder (in our example, 6).
2. The reach of each slot in the x direction (y direction reach is not necessary since all the slots can reach all locations in the y direction).

The boards are grouped as a batch, and the entire batch is processed on the same setup. Therefore, if the total number of distinct components exceeds the number of slots available, the result is an infeasible solution.

In brief, the problem has the following settings:

1. Every location is accessible to at least one slot on the feeder.
2. There is no change in the setup while processing a group of boards. Therefore, the total number of distinct component types is always less than the total number of slots available.

3. There is no restriction on the reach of the slots in the y direction. However, there are restrictions on slots travel in the x direction.
4. Each slot can carry only one component type.

16.2 HEURISTIC PROCEDURE

The heuristic has three stages. The details of each stage are described in associated sections:

First, it may be useful to get an overview of each stage.

1. *Develop basic component assignments to slots*: We make an allocation of component tapes to the slots so that all locations on all the boards can be assembled using a minimum number of slots. The purpose is to have as many empty slots as possible and still get a feasible solution so that there is an opportunity to reduce in the number of strokes in the improvement stage.
2. *Assignment stage*: The assignment of locations from the boards to the slots with component tapes is done in this stage. It generates the stroke sequence, that is, the order in which the locations are visited, and the appropriate components are inserted. First, the priorities for all the locations are calculated, and then each is assigned to a slot based on the priority level and distance from the slot.
3. *Improvement stage*: With application of the previous two stages, the assignment of locations is complete, but the solution may not be optimal. The empty slots may be assigned specific components to further reduce the strokes. To maximize the improvement, a measure is developed that determines a savings in the number of strokes that may result by adding an additional slot to a particular component. This measure is used to find the most efficient way to utilize the empty slots.

Next, each stage of the heuristic procedure is described in detail. An example is solved simultaneously to explain the procedure. The batch consists of two boards. Table 16.1 shows the x position of each location on each board as well as the component type needed in that position. Table 16.2 shows data concerning the feeder, in particular, the data on reach of the slots. The home positions indicate the positions of the slots when the dispenser is at rest. L-reach and R-reach indicate possible travel in the left and right directions of the x axis (Figure 16.3).

16.2.1 Developing Basic Component Assignments to the Slots

At the start of the allocation procedure, all slots on the feeder are empty, which means that no component tape has been assigned to any of the slots. At this stage, all the slots are also unblocked. Blocking is done to reserve a slot for critical location, such as a location that can be reached by only a particular slot.

TABLE 16.1
***x* and *y* Locations and Component Types on Each Board**

	Board 1			Board 2		
Location	*x*	*y*	Type	*x*	*y*	Type
1	5	40	1	10	40	3
2	25	35	3	25	40	1
3	20	30	2	10	35	2
4	10	20	1	15	30	2
5	35	20	1	20	30	3
6	15	15	2	5	25	2
7	5	10	1	35	25	1
8	10	5	1	15	20	2
9	20	5	3	30	15	1
10	30	5	1	5	10	1

TABLE 16.2
Slots' Home Positions and Left and Right Reach Coordinates

Slot Number	L-Reach	R-reach	Home *x*	Home *y*
1	0	15	7.5	0
2	5	20	12.5	0
3	10	25	17.5	0
4	15	30	22.5	0
5	20	35	27.5	0
6	25	40	32.5	0

Step 1: Prepare a table that shows the maximum number of locations of each component type (all boards included) that every slot can reach. Table 16.3 is the result for the data.

Start the analysis in step 2 with the first component location on the first board.

Step 2: For each location on the board, check if any of the slots presently carry a component type that is the same as required in this location (in the first case, there is none) and if the associated slot can reach the location.

a. If the conditions in step 2 are satisfied, it means the present location can be assembled by the existing slot/component allocation. Make the assignment to the existing slot that has the maximum frequency for the associated component type. Go to step 3.

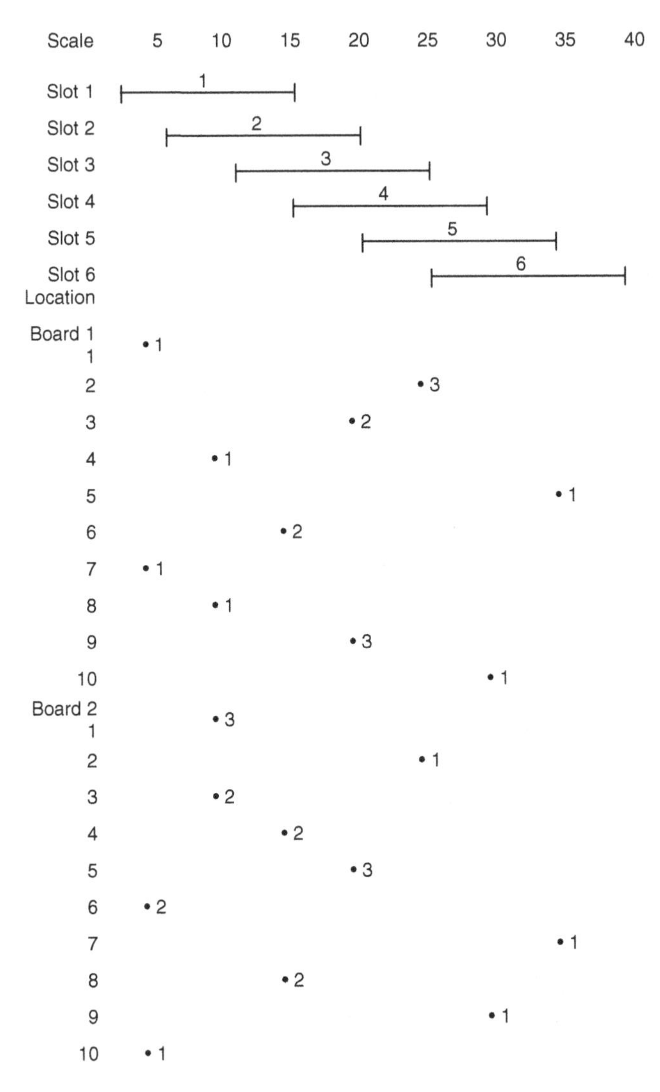

FIGURE 16.3 Locations on the boards with respect to slot reach and component assignment on the slots.

For the first location, all the slots are unassigned. Slot numbers 1 and 2 can reach component location 1 and can cover 5 and 5 locations, respectively, of component type 1 (refer Tables 16.1, 16.2, and 16.3). Therefore, assign component type 1 to slot 1 (shown in Table 16.4). In case of a tie, the heuristic chooses the first of these slots. The column under final assignment indicates the component types that are allocated to each slot. For example, the first row has an arrangement of 100,000 (0 is left blank in Table 16.4 for ease of reading). This means amongst all slots from 1 to 6, only slot 1 has component 1 assigned and others are blank or have no component assigned.

TABLE 16.3
Number of Locations that a Slot Can Reach

Component Type	Slots 1	2	3	4	5	6
1	5	5	3	3	5	5
2	5	6	5	4	1	0
3	1	3	4	3	3	1

TABLE 16.4
Slot Allocations

	Component		Assignments to Slots		Initial Assignment of Component Type to a Slot					
Board	Location	Type	Old	New	Slot 1	Slot 2	Slot 3	Slot 4	Slot 5	Slot 6
1	1	1		1	1					
	2	3		3			3			
	3	2		4				2		
	4	1	1		1				1	
	5	1		5						
	6	2	4					2		
	7	1	1		1					
	8	1	1		1					
	9	3	3				3			
	10	5	5						1	
2	1	3	3				3			
	2	1	5						1	
	3	2		2		2				
	4	2	2			2				
	5	3	3				3			
	6	2	2			2				
	7	1	5						1	
	8	2	2			2				
	9	1	5						1	
	10	1	1		1					

b. If similar components are on the feeder but associated slots cannot reach the current location, then search among the empty slots, and allocate the component type associated with the location to the slot that can reach a maximum number of locations of the same component type.

For example, although location 5 on board 1 needs component type 1, it cannot be reached by slot 1 (the only slot with type 1 assigned so far).

Thus, assign a new slot. Slot numbers 5 and 6 can reach component location 5 and can cover 5 and 5 component locations of component type 1, respectively. Thus, assign component type 1 to slot 5, since slot 5 is the first of the slots.

c. If the component type is not on the feeder, select the slot that can reach the maximum number of locations of the same component type. For example, location 2 on board 1 needs component type 3. Component type 3 is not currently on the feeder. Slots 1, 2, 3, and 4 can reach location 2 and can cover 1, 3, 4, and three locations of type 3, respectively. Thus, assign location 2 to slot 3 by assigning component type 3 to slot 3.
d. If none of the empty slots can reach component location and all slots that are filled with some other component type, this is a critical location. Then, from the slots that have already been allocated but unblocked, select the slot that is within reach of the component location under consideration and has the highest density for the component type under consideration (Table 16.3) and make the assignment. Block this slot. Reset all the unblocked slots to empty slots and go to step 2.

Step 3: Check if all locations under consideration have been examined. If yes, go to the next board; start with component location 1 and return to step 2. If not, go to the next component location on the present board and repeat step 2.

Step 4: Check if all locations on all boards have been checked and found feasible. If yes, then a feasible solution has been reached. If not, then the problem is infeasible with the present number of slots. An alternative is to make the problem feasible by reducing the variety of components in the batch (by perhaps choosing a different board combination). If it is not possible, then the remaining components from the batch would need additional setup. Applying steps 1 through 4 results in the solution displayed in the first part of the Table 16.4 for board 1.

In a similar manner, work out initial feasible assignments for other boards by repeating steps 2–4.

The results for two boards are shown in Table 16.4.

The final component assignment on the feeder can also be tabulated as shown in Table 16.5. This is the basic solution. A zero indicates that no component is assigned to the associated slot.

TABLE 16.5
A Feasible Solution

Slot	1	2	3	4	5	6
Component type	1	2	3	2	1	0
Board 1: accessible locations	1, 4, 7, 8		2, 9	3, 6	5, 10	
Board 2: accessible locations	10	3, 4, 6, 8	1, 5		2, 7, 9	

16.2.2 Developing the Initial Assignment

This phase is divided into three parts: calculations of initial priority, location assignment, and smoothing of the assignments.

16.2.2.1 Calculation of Priorities for Locations

The priorities are calculated in order to give preference to the critical locations, that is, those locations that are accessible by only one slot.

Step 1: Start with the first location on the first board.

Step 2: Calculate the priority level of the location as the number of slots allocated with a similar component type that can reach the present location. For example, component location 1 on board 1 has a priority of 1. This means that only one slot can reach that location.

Step 3: Go to the next location and repeat step 2 until priorities for all locations on the present board are determined. Table 16.6 shows the priorities (P) for all the locations (L) on the boards.

Step 4: Repeat step 2 and 3 until all the boards in the batch have been analyzed.

16.2.2.2 Location Assignments

In this phase, the locations on the board are assigned to the slots. It may take more than one stroke to place a component in all locations. We start with the stroke value of 1. For a new board, all the locations are initially unassigned. Set the location priority to 1 (the location with priority 1 must be assigned to the first accessible slot, while a location with higher priority can be accessed by more than one slot, and is hence checked later).

Step 1: Start with the first allocated slot.

Step 2: Check if the component assigned to the slot is required on the board. If yes, proceed to step 3; if not, select the next assigned slot and go back to the top of step 2. If all the slots are examined, select a new board and go to step 1.

Step 3: List all the locations with present priority level that can be reached by the slot and require the same component as is in the slot. If there are such locations, go to step 4. If there are no locations with present priority level, but there is a location(s) that requires the component but has a higher priority level(s), then increase the priority

TABLE 16.6
Priority Levels

Location	1	2	3	4	5	6	7	8	9	10
Board 1: AI	1	1	2	1	2	2	1	1	1	1
Board 2: AI	1	2	2	2	1	2	2	2	2	2

level by 1 and go back to the top of step 3. If there are no locations requiring the component, discontinue examining the slot, go to next slot in the sequence and go back to step 2. If all the slots are examined, select a new board and go to step 1. If all the boards are examined go to the step called smoothing the assignment.

Step 4: The location assignment is made on the basis of the shortest distance between location and the present position of the slot. The distance is calculated using the distance formula (even though its not shown in our tables, the y coordinates of the positions are known).

$$d = \sqrt{(x - x1)^2 + (y - y1)^2}$$

Calculate the distance(s) between location(s) listed in step 3 and the present position of the slot. Assign location to the slot based on shortest distance. Go to the next slot and back to step 2. If all the slots are examined and the board still has unassigned locations, increase stroke value by 1 and go to step 1.

In our example, start with the first board. The board requires components 1, 2, and 3 only.

Start with the first slot that has been allocated a component type required for the board. The initial value of the stroke is 1, and the location priority is 1. Slot 1 has component 1 as required for board 1. There are no locations with component type 1 and priority 1 within reach of slot 1.

Increment the priority level by one to two. There are four locations of priority 2 and component type 1 that can be covered by slot 1. Using equation 1, calculate the distances of locations from the home position of slot 1. The values are displayed in the Table 16.7, and the calculations are displayed in Table 16.7a.

Select the location nearest to the slot's present position. In our example, the locations 1, 4, 7, and 8 are at distances of 40.08, 20.16, 10.31, and 5.59, respectively. Thus, assign component location 8 to slot 1 on stroke 1.

Since the dispenser moves as one unit to any location, the slots are all at the new position, and the next location will be accessed from this position. Hence, it is necessary to calculate the new relative positions for the next slot (calculate positions for the next slot that has been assigned a component type, and repeat steps 6 and 7). In the example problem, the next slot after slot 1, which carries a component type that is required by board 1, is slot 2. The new position for slot 2 is the location of slot $1 +$ the distance between slot 1 and 2, that is, x coordinate of component location $8 + 5$ (distance between neighboring heads). Thus, the new x position of slot 2 is $10 + 5 = 15$. y coordinate is the same as the y coordinate of the location 8. Slot 2 can reach locations 3 and 6 (locations with priority 2) at a distances of 25.49 and 10, respectively. Assign the board location 6 to slot 2.

Repeating the procedure for all boards results in Table 16.8, which shows the initial assignment with the basic feasible solution. Row 1 indicates the slots on the dispenser. Row 2 indicates the component types that are assigned to the corresponding slots. Below row 2 are the two matrices corresponding to the two boards. The elements of the matrices are the location numbers (from Table 16.1).

TABLE 16.7
Location Assignment

Board	Stroke	Slot	Component Type	Component Location	Distance from Slot	Priority	Location Allotted
1	1	1	1	1	40.08	2	
				4	20.16	2	
				7	10.31	2	
				8	5.59	2	8
		2	2	3	25.49	2	
				6	10	2	6
		3	3	2	20.62	1	
				9			
		4	2	3		2	
		5	1	5	14.14	2	5
				10	25.50	2	
	2	1	1	1	40.08	2	
				4	20.16	2	
				7	10.31	2	7
		3	3	2			2
		5	1	10			10
	3	1	1	1	40.08	2	
				4	20.16	2	4
	4	1	1	1			1
2	1	1	1	10			10
		2	2	3	25.49	2	
				4	20	2	
				6	18.03	2	
				8	10	2	8
		3	3	1	22.36	1	
				5	10	1	5
		4	2	4			4
		5	1	2	11.18	2	2
				7	15.81	2	
				9	18.02	2	
	2	2	2	3	35.09	2	
				6	26.10	2	6
		3	3	1			1
		5	1	7	18.02	2	7
				9	26.93	2	
	3	2	2			3	3
		5	1	9			9

TABLE 16.7a
Location Distance Calculation

		Present Location			Component Location		
Board	**Slot**	$x1$	$y1$	**Component**	$x2$	$y2$	$d = \sqrt{(x1 - x2)^2 + (y1 - y2)^2}$
1	1	7.5	0	1	5	40	40.08
				4	10	20	20.16
				7	5	10	10.31
				8	10	5	5.59
	2	15	5	3	20	30	25.49
				6	15	15	10
	3	20	15	2	25	35	20.62
				9	20	5	10
	5	25	30	5	35	20	14.14
				10	30	5	25.5
	1	7.5	0	1	5	40	40.08
				4	10	20	20.16
				7	5	10	10.31
2	2	15	10	3	10	35	25.49
				4	15	30	20
				6	5	25	18.03
				8	15	20	10
	3	20	20	1	10	40	22.36
				5	20	30	10
	5	20	30	2	25	40	11.18
				7	35	25	15.81
				9	30	15	18.02
	2	12.5	0	3	10	35	35.09
				6	5	25	26.10
	5	20	40	7	35	25	18.02
				9	30	15	26.93

TABLE 16.8
Initial Location Assignments to Slots for Each Stroke

	Slot	**1**	**2**	**3**	**4**	**5**	**6**
	Component	**1**	**2**	**3**	**2**	**1**	—
Board	Stroke						
1	1	8	6	9	3	5	—
	2	7	—	2	—	10	—
	3	4	—	—	—	—	—
	4	1	—	—	—	—	—
2	1	10	8	5	4	2	—
	2	—	6	1	—	7	—
	3	—	3	—	—	9	—

16.2.3 Smoothing the Assignment

In this stage, if possible, the locations are distributed evenly among the slots carrying the same component type. This may reduce the number of strokes necessary, using the present slot allocation.

Step 1: Start with board 1.

Step 2: Note the component types on the board and arrange them in ascending order. Start with the first component type on the present board. Board 1 has component types 1, 2, and 3. So, start with component type 1.

Step 3: Find a, the number of component locations of the above component type. In this case, there are six locations on board 1 requiring component type 1. This number may be determined by counting all the locations in board 1 assigned to the slots 1 and 5.

Step 4: Find b, the number of slots that carry the same component type.
The number of slots having component type 1 is 2 (i.e., slots 1 and 5).

Step 5: Divide a by b and round to the next highest integer. This gives the theoretical number of strokes required by the component type, based on the present slot assignments.
In our case, the number of strokes required for component type 1 on board 1 is $6/2 = 3$.

Step 6: Check if the number of strokes for all slots with the component type under consideration is less than or equal to the theoretical number from step 5.

a. If yes, then consider the next component type and go on to step 7.
b. If not, then arrange the slots of the component type under consideration in the ascending of number of locations assigned. The objective is to shift locations from slots having more strokes than the theoretical value to slots having less strokes than the theoretical value. For example, for component type 2 (on board 2), the theoretical number of strokes is 2, and slots 2 and 4 have three and one strokes assigned, respectively. Thus, smoothing may be possible. Arranging the slots in the ascending order of the number of locations assigned results in the sequence 4, 2.
   i. Start with slot with minimum location assignment (Mincl), This is the slot to which other locations may be transferred. In this case, start with slot 4.
   ii. Select the slot having maximum component locations (Maxcl), This is the slot from which component locations are to be transferred. In this case, slot 2.
   iii. Check if slot Mincl can take up any component locations from Maxcl. If possible, transfer components till the number of component locations on Mincl reach the theoretical stroke value or no further transfers are possible. In our example, component location 8 is transferred from slot 2 to slot 4.

TABLE 16.9
Smoothing Technique for Board 1

	Original Solution			Smooth Solution		
Board	Component Type	Slot(s)	Stroke Assigned	Reallocate to Slot	Reallocate from Slot	Component Location
1	1	1	4	Not required		
		5	2			
	2	2	1			
		4	1			
	3	3	2			
2	1	1	1			
		5	3			
	2	2	3	4	2	8
		4	1			
	3	3	2			

iv. Select the slot with the next highest number of strokes (continuing until a level above the present Mincl is reached), as new Maxcl. Repeat step 3.

v. Select the slot having next lowest number of strokes as the new Mincl, and repeat steps 2, 3, and 4. Continue until all the slots have been considered.

Step 7: Repeat steps 3–6 until all component types have been analyzed. The results of steps 3–6 are shown in Table 16.9.

Step 8: Repeat steps 3–7 for all boards.

The solution at the end of this section is presented in Table 16.10. The table shows the location assignments with the feasible solution. Row 1 indicates the slots on the dispenser. Row 2 indicates the component types that are assigned to the corresponding slots. Below row 2 are the two matrices corresponding to the two boards. The elements of the matrices are the location numbers from Table 16.1.

16.2.4 Improving the Assignment

In this section, the empty slots are utilized to optimize the solution obtained from the previous section, that is, the smooth solution. The slots are dedicated to those component types that result in the maximum stroke reduction. This is achieved in two stages.

In the first stage, a check is made to determine the number of strokes that may be reduced from each board, along with the number of slots that must be dedicated to achieve this reduction. In the second stage, a check is made to see if it is physically possible to implement the improvement suggested in the first stage. The two stages

TABLE 16.10
Smooth Location Assignment to Slots for Each Stroke

	Slot	**1**	**2**	**3**	**4**	**5**	**6**
	Component		**1**	**2**	**3**	**2**	**1**
Board	Stroke						
1	1	8	6	9	3	5	—
	2	7	—	2	—	10	—
	3	4	—	—	—	—	—
	4	1	—	—	—	—	—
2	1	10	6	5	4	2	—
	2	—	3	1	8	7	—
	3	—		—	—	9	—

are repeated until no more empty slots are available or no further improvements can be made.

Step 1: Start with the first board.

Step 2

Calculate (B), the number of empty slots available. If B = 0, go to step 4.

Determine the distinct component (DC) types that appear in the last stroke (LS) of the board.

For example, only component type 1 on board 1 is in the last stroke, that is, stroke 4 (LS = 4). So DC = {1}.

Keep decreasing the number of strokes by one until there exist a component type that does not belong to DC, in the last stroke for the board. This is defined as the next level of strokes (NS). For example, on board 1, component types 3 and 5 are present at stroke 2, and do not belong to set DC. Therefore, the NS = 2.

a. If the number of distinct components in the last stroke LS is more than the number of empty slots available, no improvement is possible on this board. Theoretically, there can be no improvement unless we allocate at least one slot more for each distinct element of DC. Go to step 4.
b. If the number of elements in DC is less than or equal to the number of empty slots, then there is a possibility of improvement.

Step 3

1. Start with the first component of set DC. Assume an allocation of an additional head for this component type. Let Hi represent the number of additional slots for component type i, in set DC. In the present example, component 1 will get an additional slot hence, H1 = 1 for board 1.
2. Calculate t, the total number of locations assigned to all the slots carrying this component type, and which the number of strokes required is presently

greater than NS. In our example, NS is 2 (step 2. iii) and slot 1 has four locations assigned and, therefore, require four strokes. Hence, $t = 4$.

3. Calculate d, the total number of slots that meet the criteria in [2]. So, $d = 1$ (slot 1).
4. Divide t by (d + Hi) (rounding off to the next higher integer) to find the new level of strokes for the component type i, Ti. In our case, the new level of strokes T1 for the component type 1 is 2 (i.e., $4/2 = 2$; round off to the next higher integer if required).
5. Repeat 1–4 for all components in DC.
6. Select the maximum value of Ti among all the elements in DC, and call it MTi. There is only one value of Ti, which is 2, and hence MT1 = 2.
   - If MTi is equal to the LS:
     a. If additional empty slots are available, that is, B > 0, then tentatively allocate one more slot to the component type associated with Mti, that is, Hi = Hi + 1, recalculate the number of empty slots, and repeat steps 2 through 6.
     b. If no additional slots are available, then no further improvement is possible on this board. Go to step 4.
   - If MTi is less than LS:
     a. Also, if MTi is less than or equal to NS, then the possible improvement is the difference between LS and NS. Go to step 4.

        In our example, MT1 is 2 and NS is 2, and since they are equal, the possible improvement is LS − NS = 4 − 2 = 2 strokes. The calculations are shown in the following. The values of t and d are indicated and also MTi, which is the number of strokes for the component type indicated, after the additional slots have been tentatively added.

        Board 1. Comp type 1, $t = 4$, $d = 1$, additional slots required = 1, Ti = 2. MTi = 2.

        Repeating the steps 2 through 6, no further improvement is possible at this stage for board 1.
     b. If MTi is greater than NS, then the improvement is the difference between the LS and the MTi. Repeat steps 2 through 6 by tentatively adding one additional slot for the component type associated with MTi.

Step 4: Repeat steps 2 and 3 for all the boards.

Steps are applied to the boards. For board 2, MTi is 2 and LS is 3. Therefore, the number of strokes that can possibly be reduced is 1 (3 − 2). Calculations are shown in Table 16.11 for both boards.

Step 5: By now, all the boards have been examined for possible improvement. Determine an index as follows:

Improvement per slot (IPS) per board = improvement possible on the board/total number of additional slots dedicated to the board. Therefore, the IPS for board j

TABLE 16.11
Application of Steps 1 and 2

	Comp. Type	*t*	*d*	Additional Slots Required	T_i	MT_i
Board 1	1	4	1	1	2	2
Board 2	1	3	1	1	2	2

TABLE 16.12
Establishment of Board Priorities

Board	Distinct Component(s) Type in the Last Stroke, DC	Priority
1	1	1
2	1	1

by adding k slots is defined as impro[j][k] $= x$, where x is the number of strokes saved. In the example, for board 1, the improvement possible by allocating one additional slot is 1. The following IPS values are obtained for all the boards.

impro[1][1] $= 2/1 = 2$.
impro[2][1] $= 1/1 = 1$.

Step 6: Calculate the priority for each board as the number of distinct components that are common between the DC sets of all the boards. In this case, the DC sets and priority rating of all boards are displayed in Table 16.12.

Board 1 has component type 1 in the final stroke, which also appears in the final stroke of board 2; therefore, it has a priority level of 1. Board 2 has component type 1 in common with board; therefore, it has a priority of 1.

Step 7

Arrange the IPS in the decreasing order.

1. If two boards have the same IPS, then:
   (i) If the requirement on the additional slots is the same, preference is given to the higher priority level. If there is a tie at the priority level also, then select the first board from the list.
   (ii) If the additional slots requirement is different, then give preference to the board with fewer additional slots requirement.

Step 8: Start with the highest level of improvement possible (IPS). In this case, board 1 has the highest IPS of 2. The associated additional slot requirement is 1, and priority level is also 1. Therefore, we start with the first board, board 1.

Step 9

1. Begin with the first component type in Hi for the board associated with the highest value of IPS. In our example, it is component type 1 on board 1.
2. Considering only the locations that have this component type on those slots that are present in the last stroke (LS), find an empty slot that can reach the maximum number of these locations.
   a. If it is not possible to find a slot that can reach the component locations under consideration, then improvement is not possible on this board. Go to step 2 of the improvement stage. In this example, the only location (location 1) in the last stroke cannot be reached by the only empty slot (slot 6). Therefore, go to step 2 of the improvement stage.
   b. If it is possible to do so, then improvement is possible. In the example, for board 2, empty slot 6 can reach a total of three locations of component type 1. Therefore, select slot 6 for allocation.
3. Check if the number of additional slots allocated equals the number of additional slots required (Hi)
   a. If this matches, then allocate this component type on the selected slot. As H1 for board 1 is 2 and 1 slot has already been allocated, no further slots are required.
   b. If it does not match, then repeat step 2, excluding the slots selected up till now.
4. Take the next distinct component type from DC and repeat substeps 2 and 3 in step 9. In the example, for both boards 1 and 2, there is only one component type present in the last stroke. The final allocation at this stage is

   Board 1, number of additional slots selected = 0.

   Board 2, number of additional slots selected = 1, which is slot 6.

Step 10: At this stage, although the additional slots have been assigned for allocation, it is not known whether it will actually result in the savings predicted. This is because, the additional slots allocated may not be able to place enough locations of the component type to result in the stroke reduction. Therefore, reassign only component locations for the board under consideration using the assignment and smoothing routine, while retaining the original component-slot allocation with the addition of new slots as determined in the previous steps for the selected component assignments. Check if the total number of strokes reduces. In our example for board 2, one additional slot has been assigned, which is slot 6 for component type 1. The new solution is displayed in Table 16.13.

16.2.5 Check for Improvement

1. If there is no reduction in the total number of strokes, then improvement is not possible.

TABLE 16.13
Improved Solution to Board 1

Slots	1	2	3	4	5	6
Component	1	2	3	2	1	1
Stroke 1	10	6	5	4	2	7
Stroke 2	—	3	1	8	9	—

2. If there is a reduction in the total number of strokes, then compare the reduction in the number of strokes with the savings predicted in the improvement stage.
   a. If they are equal, make the solution permanent. In the example, the new assignment for board 2 required only 2 strokes, and the previous assignment required 3 strokes, resulting in a saving of 1 stroke. Also, this equals the savings predicted, so make the solution permanent.
   b. If they are not equal, then check if the saving is higher than the next value of IPS calculated in step 7, and if so, make the present solution permanent. For example if the reassignment results in the saving of 2 strokes when the IPS value predicted a saving of 4 strokes. So, condition a is not satisfied, and we go to condition b. Now, if the next value of IPS is 1 stroke, then it would be better to maintain the reassignment, as savings of 2 strokes is higher than the next level of IPS which is 1 stroke.
   c. If (b) is not valid, then discard the allocation of the additional slots and repeat step 9 for the next level of IPS.

As all the required number of slots have been assigned, a solution may be possible.

Step 1: In this step, the values of IPS are adjusted since some component types may have been allocated in the previous step. If an assignment at step 10 has been made permanent, then from the remaining values of IPS

1. Choose the next level of IPS, which is the next board to be selected for improvement.
2. Search for the same component types as those allocated in step 10. If a same component type is found, then compare the number of additional slots that would be required for the present value of IPS (determined in step with the slots allocated for the associated component type in step 10.
   a. If the additional slots required are less than or equal to the slots assigned for the component type in step 10, then discard this particular component type from further consideration for IPS and reduce the additional slot requirement proportionally. This would mean adjusting the k value in impro[j][k].
   b. If not, then reduce the additional slot requirement for the component type in IPS by the number of slots allocated.

TABLE 16.14
The Final Solution

	Slot	**1**	**2**	**3**	**4**	**5**	**6**
	Component	**1**	**2**	**3**	**2**	**1**	**1**
Board	Stroke						
1	1	8	3	9	6	10	—
	2	7	—	2	—	5	—
	3	4	—	—	—	—	—
	4	1	—	—	—	—	—
2	1	10	6	5	4	2	7
	2	—	3	1	8	9	—

3. Repeat step 12.2, taking the next level of improvements IPS, till all levels of improvement have been evaluated .
4. Also, delete those values of improvements from consideration that require more number of slots than presently available.

Step 2: Repeat steps 9 and 10 and step 1 of the improvement routine till all levels of improvements are evaluated.

In the example, only board 2 could have an improvement because the additional slot could not reach any components for board 1, which would have resulted in a reduction of the number of strokes.

Step 3

Check if the there are any empty slots available.

1. If yes, and if there was at least one new assignment made at step 10, then go to step 1 (improvement routine). If there was no new assignment made, then no further improvement is possible.
2. If no, then the assignment is over.

The final improved solution is shown in Table 16.14. The solution results in 6 strokes as compared to 7 in the smoothing assignment of Table 16.10.

16.3 PROBLEMS

16.1 Solve the problem illustrated in this chapter if the component types are changed as indicated:

Location	1	2	3	4	5	6	7	8	9	10
Board 1	2	3	2	3	1	2	2	1	3	1
Board 2	1	1	2	2	3	3	2	3	1	2

16.2 Using the method presented in the chapter, solve the sequencing problem with the following data, assuming that a feeder having six slots is to be used.

	Board 1			Board 2		
Location	x	y	Type	x	y	Type
1	5	30	3	20	30	1
2	25	30	2	30	30	2
3	10	25	1	5	25	3
4	15	25	1	20	25	3
5	30	25	2	25	25	2
6	20	20	2	15	20	1
7	35	20	3	30	20	2
8	5	15	1	35	20	1
9	25	15	3	5	15	1
10	15	10	2	15	15	1
11	25	10	3	35	15	3
12	30	10	1	20	10	2

Slot Number	L-Reach	R-reach	Home x	Home y
1	0	15	7.5	0
2	5	20	12.5	0
3	10	25	17.5	0
4	15	30	22.5	0
5	20	35	27.5	0
6	25	40	32.5	0

16.3 Solve Problem 16.2 if the component types are changed as follows:

Location	1	2	3	4	5	6	7	8	9	10	11	12
Board 1	3	1	2	2	1	3	1	1	3	2	2	3
Board 2	2	1	3	2	3	1	1	2	2	3	1	3

REFERENCES AND SUGGESTED READINGS

Ahmadi, J., S. Grotzinger, and D. Johnson. 1988. "Component Allocation and Partitioning for a Dual Delivery Placement Machine" *Operations Research*, 36(2): 176–191.

Ball, M.O. and J.M. Magazine. 1988. "Sequencing of Insertions in Printed Circuit Assembly" *Operations Research*, 36(2): 192–201.

Carmon, T., O.Z. Maimon, and M.E. Dar-el. 1989. "Group Setup for PCB Assembly" *International Journal of Production Research*, 27(10): 1795–1810.

Cunningham, P. and J. Brown. 1986. "A Lisp Based Heuristic Scheduler for Automatic Insertion in Electronics Assembly" *International Journal of Production Research*, 24(6): 1395–1408.

Fathi, Y. and J. Taheri. 1989. "A Mathematical Model for Loading the Sequencers in a Printed Circuit Pack Manufacturing Environment" *International Journal of Production Research*, 27(8): 1305–1316.

Maimon, O.Z. 1991. "Grouping Methods for PCB Assembly" *International Journal of Production Research*, 29(7): 1379–1390.

Randhwa, U.S., D.E. McDowell, and S. Faruqui. 1985. "An Integer Programming Application to Solve Sequencer Mix Problems in PCB Production" *International Journal of Production Research*, 23(3): 543–552.

Sule, D.R. 1992. "A Heuristic Procedure for Component Scheduling in Printed Circuit Pack Sequencer" *International Journal of Production Research*, 30(5): 111–118.

17 Industrial Sequencing IV: Scheduling in Flexible Manufacturing

Two topics are discussed in this chapter. The first is, how to schedule groups for processing in flexible flow shop manufacturing when there are group-dependent setup times and the objective is to minimize the make span. The second topic discussed is how to assign tasks to robots in automated robot assembly operation.

17.1 A CDS-BASED, TWO-PHASE ALGORITHM FOR GROUP SCHEDULING WITH GROUP-DEPENDENT SETUP TIMES

17.1.1 INTRODUCTION

A flow shop consists of n jobs to be processed on m machines where all jobs have the same processing sequence on the machines. The processing time for the ith job on the jth machine is given by P_{ij}. The setup time for the ith job on the jth machine is given by S_{ij}.

Here, we deal with a generalized flow shop where the jobs are grouped into g groups such that all jobs in the group have a common setup on each machine. Once a group is setup on a machine, all jobs in the group must be processed before the next group is set up. Jobs may or may not utilize all machines on the shop floor. However, a job follows the same sequence as all other jobs on the machines that it does utilize.

A Campbell Dudek Smith (CDS)-based algorithm is presented to solve the group scheduling problem in this generalized flow shop. The objective is the minimization of makespan.

17.1.2 HEURISTIC

The procedure comprises of two phases and is outlined in the following:

17.1.2.1 Phase I: Preliminary Sequence

1. Apply the CDS algorithm to sequence jobs within each group independently to obtain the internal job sequences. Setup times are disregarded in this step. The shortest processing time (SPT) rule is used to break ties in generating the sequences.
2. Select and designate the minimum makespan for each group as M'_g.

3. Plot the machine timelines for each group independently on a Gantt chart. Machine timelines for each group start at time 0 in this step. Setup for each machine is performed at time 0, and jobs are then loaded as per the selected sequence.
4. From the machine timelines plotted in step 3, mark the total time required by each group g on each machine j. Designate this as T_{gj}. T_{gj} includes the idle times for the machines and the wait times between machines for the jobs.
5. Apply the CDS algorithm to obtain the external group sequence. In this step, the P_{ij} values in the conventional CDS table are replaced by the T_{gj} values.
6. Plot the overall machine timelines on a Gantt chart for each machine, using external group sequence obtained in step 5. Designate this as the current solution and designate this as the current makespan, M_c.

17.1.2.2 Phase II: Makespan Reduction by Minimizing the Machine Idle Time

1. Identify the machine on which the makespan is realized. Designate this as the critical machine.
2. Examine the sequence for idle time on the critical machine. Designate all the jobs on the critical machine occurring after the first idle period on that machine as the critical jobs.
3. For each job that has been sequenced after the first idle period on the critical machine, trace back the job schedule on the preceding machines and mark any slack, if available.
4. Continue tracing back the job schedule on preceding machines until you reach a machine on which there is no slack available.
5. Identify bottleneck jobs on this machine. It may be possible to safely delay these jobs and schedule them after the critical jobs to try and reduce the makespan. If bottleneck jobs exist, go to step 6. If no such jobs are available, then the current makespan is the minimum makespan and the current sequence is the optimum sequence for the problem.
6. This step requires a minimum number of passes that equals the number of bottleneck jobs. In each pass, one of the bottleneck jobs is sequenced after the critical jobs on the earliest machine that has zero slack. The total makespan is calculated. If the makespan is less than M_c, the new sequence is designated as the current solution and the value of M_c is updated to the new value.

17.1.3 Illustrative Example

Consider the following example with three groups and five machines, the data for which is provided in Table 17.1.

Since it is a flow shop, all the jobs follow the machine routing: 1-2-3-4-5. Where the jobs do not utilize all the machines, the rest of the routing is still preserved.

TABLE 17.1
Example Data: Time required by each Job on each Machine

	Job *j*	Machine 1	2	3	4	5
Group I	1	0	0	0	24	0
	2	51	70	64	69	0
	3	0	88	9	0	0
Group II	1	66	42	0	92	33
	2	1	98	95	0	0
	3	66	94	98	0	0
	4	0	10	0	45	42
	5	0	0	0	88	32
Group III	1	0	86	52	35	76
	2	0	0	0	43	0
Setup time per group	Group I	2	20	2	14	27
	Group II	45	74	82	80	55
	Group III	18	23	38	59	91

TABLE 17.2
Data for Step 1: Group I Jobs without Setup Times

	Job *j*	Machine 1	2	3	4	5
Group I	1	0	0	0	24	0
	2	51	70	64	69	0
	3	0	88	9	0	0

17.1.3.1 Phase I: Preliminary Sequence

The algorithm starts by considering each group separately and applying the CDS algorithm to obtain the internal sequence for each group.

Consider group I separately as shown in Table 17.2. The processing times are listed, and the setup times are ignored in this step. We apply the CDS algorithm to sequence the jobs in this group.

The four subproblems generated in the CDS algorithm are listed in Table 17.3. Here, the processing times are calculated as follows:

$$A_{1j} = P_{1j} \qquad B_{1j} = P_{5j}$$
$$A_{2j} = P_{1j} + P_{2j} \qquad B_{2j} = P_{5j} + P_{4j}$$
$$A_{3j} = P_{1j} + P_{2j} + P_{3j} \qquad B_{3j} = P_{5j} + P_{4j} + P_{3j}$$
$$A_{4j} = P_{1j} + P_{2j} + P_{3j} + P_{4j} \qquad B_{4j} = P_{5j} + P_{4j} + P_{3j} + P_{2j}$$

TABLE 17.3
CDS Subproblems for Group I

	Machine							
Job j	A_{1j}	B_{1j}	A_{2j}	B_{2j}	A_{3j}	B_{3j}	A_{4j}	B_{4j}
1	0	0	0	24	0	24	24	24
2	51	0	121	69	185	133	254	203
3	0	0	88	0	97	9	97	97

In subproblem 1, S1, with machines A1 and B1, both job 1 and 3 have the lowest processing time on machine A1. To break the tie, we examine their processing times on machine B1, and we shall use the SPT rule to break the tie. However, they also have the same processing time on machine B1. The tie is then broken arbitrarily. Job 1 is placed in the first position, and job 3 is placed in the last position. Job 2 is placed in the middle. The complete sequence generated is 1-2-3.

The sequences yielded by the other three sub-problems are as follows:

S2: 1-2-3
S3: 3-2-1
S4: 1-2-3

The makespan is calculated for each sequence, and the sequence with the minimum makespan is marked as selected. In this case, the selected sequence, with the minimum makespan, is sequence 1-2-3. The corresponding minimum makespan is designated as M′1, for group I, and is M′1 = 254.

We now plot the machine timelines for the selected sequence for group I. The setup times are included in the timelines. Setup for each machine starts at time 0. The Gantt chart is shown in Figure 17.1.

Similarly, the group internal sequences and machine timelines for groups II and III are shown on Gantt charts in Figures 17.2 and 17.3, respectively.

Generation of external group sequence
From the Gantt charts, we note the total time required by each group, g, on each machine, j, as T_{gj}. The T_{gj} values include the setups and the idle times on each machine. These T_{gj} values are tabulated in Table 17.4.

We are now ready to obtain the external group sequence. The T_{gj} values listed in Table 17.4 are treated as if they were processing times for individual jobs on the machines. The CDS algorithm is applied directly to the values in Table 17.4. The sequence with the minimum makespan is chosen as the external group sequence.

Table 17.5 lists the CDS subproblems and the sequences obtained for each subproblem.

We see that all the subproblems yield the same sequence, 3-2-1. Hence, this is accepted as the external group sequence, and the Gantt chart is plotted. This is designated as the current solution and the makespan, $M_c = 657$. The Gantt chart is shown in Figure 17.4.

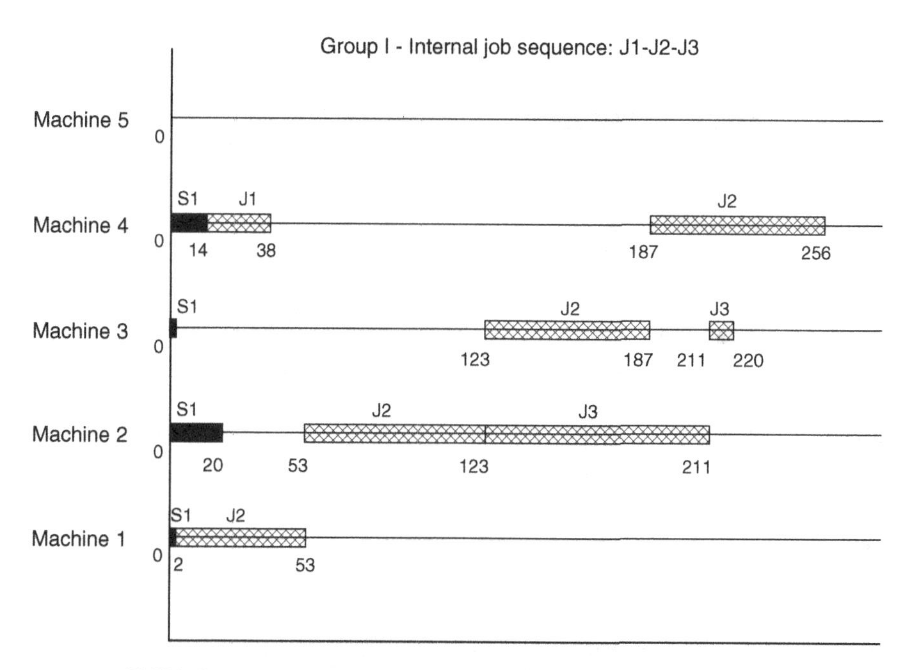

FIGURE 17.1 Group I; internal job sequence and machine timelines.

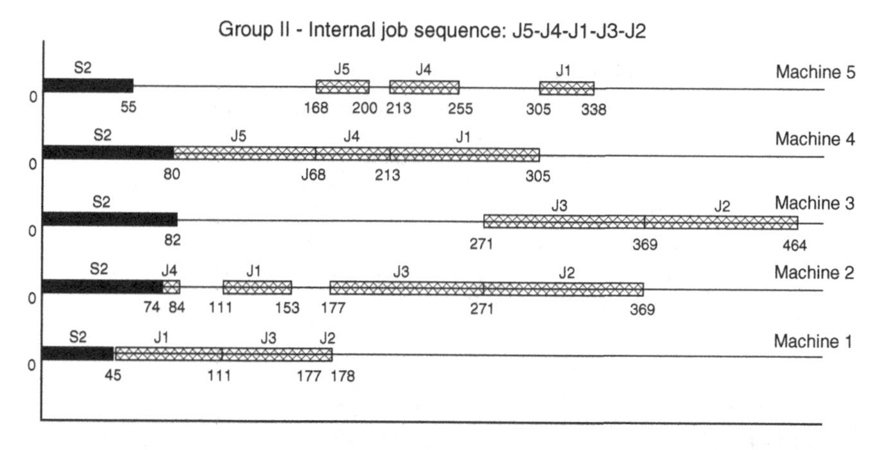

FIGURE 17.2 Group II; internal job sequence and machine timelines.

17.1.3.2 Phase II: Makespan Reduction by Minimizing the Machine Idle Time

The machine on which the makespan is realized is machine 4. This is designated as the critical machine. The first idle period on the machine occurs from time $T = 102$ to $T = 161$. All the jobs that have been sequenced after this idle period are designated as critical jobs. Thus, the critical jobs are G3/J1, G2/J5, G2/J4, G2/J1, G1/J1, and G1/J2.

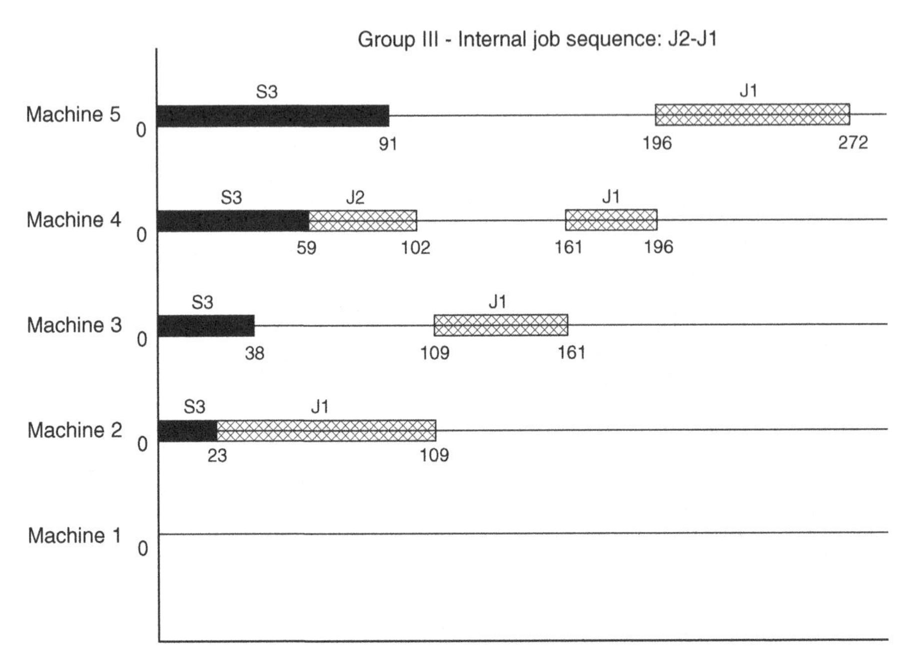

FIGURE 17.3 Group III; internal job sequence and machine timelines.

TABLE 17.4
Processing Times for Groups on Machines

	Machine				
Group	**1**	**2**	**3**	**4**	**5**
I	53	211	220	256	0
II	178	369	464	305	338
III	0	109	161	196	272

TABLE 17.5
CDS Subproblems to Find External Group Sequence

	CDS Equivalent Machine							
Group	A_1	B_1	A_2	B_2	A_3	B_3	A_4	B_4
I	53	0	264	256	484	476	740	687
II	178	338	547	643	1011	1107	1316	1476
III	0	272	109	468	270	629	466	738
Sequence	3-2-1		3-2-1		3-2-1		3-2-1	

We start with the first critical job in the sequence: G3/J1. This job is processed from $T = 161$ to $T = 196$ on the critical machine. We now trace the schedule for this job on the preceding machines. This is given by:

Machine 3: $T = 109 - 161$
Machine 2: $T = 23 - 109$

Since the period $T = 0$ to $T = 23$ on machine 2 is used for setup for group 3, it is obvious that there is no slack available on this job and G3/J1 cannot be moved any earlier on machine 4. Similarly, we find that G2/J5, G2/J4, G2/J1, and G1/J1 also cannot be moved.

The precedence schedule for G1/J2 is as follows:

Machine 1: $T = 180 - 231$
Machine 2: $T = 447 - 517$
Machine 3: $T = 524 - 588$
Machine 4: $T = 588 - 657$

We see that slack is available between machines 1 and 2 and between machines 2 and 3. G1/J2 is delayed because of G2/J3 and G2/J2 on machine 3. These jobs are in turn delayed because of G2/J1 on machines 1 and 2. If G2/J1 is sequenced after G2/J3 and G2/J2, the overall makespan is not affected. The idle period from $T = 539$ to $T = 588$ accommodates the delay in G2/J1. Thus, G2/J1 is the bottleneck job. Accordingly, we now change our internal sequence for group 2 from J5-J4-J1-J3-J2 to J5-J4-J3-J2-J1. The new Gantt chart is shown in Figure 17.5. The new makespan is 650.

Once again, the machine on which the makespan is realized is machine 4. This is designated as the critical machine. The first idle period on the machine occurs from time $T = 102$ to $T = 161$. All the jobs that have been sequenced after this idle period are designated as critical jobs. Thus, the critical jobs are G3/J1, G2/J5, G2/J4, G2/J1, G1/J1, and G1/J2.

We repeat the procedure as before to identify any further potential reduction in makespan. We see that we are constrained by the requirement that all jobs within a group have a common setup. Thus, we are allowed to shuffle the sequence of jobs within a group, but we may not have more than one setup for any given group. Further, since we have already verified during the generation of the external sequence that the group sequence 3-2-1 yields the minimum overall makespan, we cannot change the external sequence at this stage.

Hence, the optimum makespan is 650, and the current solution is the optimum solution.

17.1.4 Results

This CDS-based algorithm is compared with a modified Nawaz Enscore and Ham algorithm (NEH) and with a genetic algorithm (GA), all of which solve the same problem.

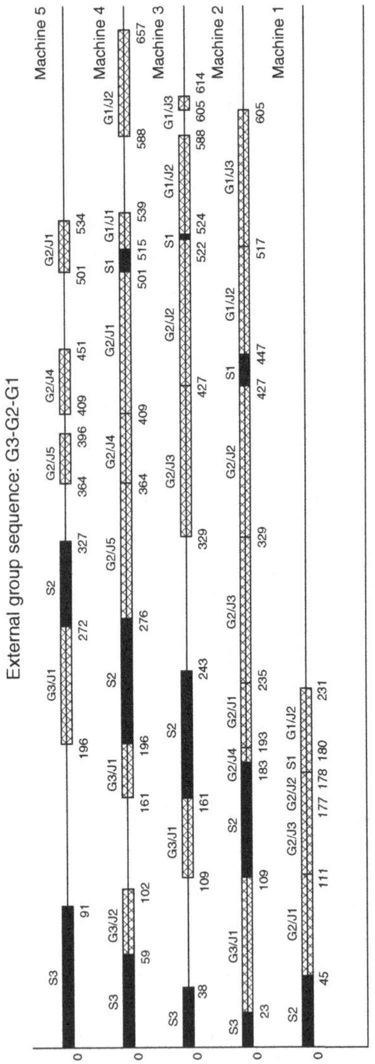

FIGURE 17.4 Preliminary sequence.

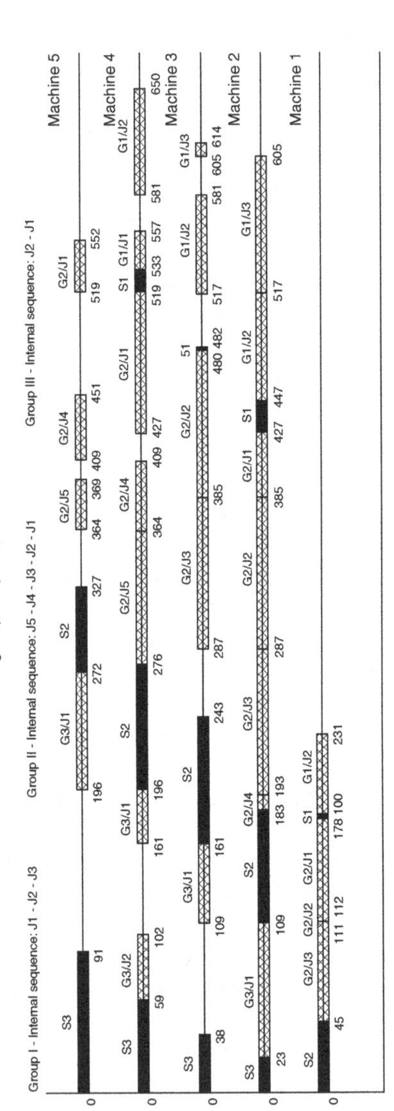

FIGURE 17.5 Final sequence.

TABLE 17.6
Results

Metric	CDS based Algorithm	Genetic Algorithm (GA)	Modified Nawaz, Enscore, Ham (NEH)
Makespan	650	650	650
Buffer capacities			
Machine 1	1	1	
Machine 2	4	4	
Machine 3	1	1	
Machine 4	1	1	
Machine idle time	507	507	

The CDS algorithm yields the same makespan as NEH and GA for the problem instance cited. The GA also measures the buffer capacity required and the job waiting times for the optimum schedule. While the buffer capacity is a valid statistic, it is suggested here that once the overall makespan has been minimized, the job waiting time is only of secondary importance. Hence, the algorithm does not address this statistic and it is not included in the comparison. It is suggested here that the machine idle time is a better indicator of the efficiency of the algorithm since a smaller machine idle time would indicate that the machine is freed earlier. The proposed algorithm yields the same machine idle times as the GA. The proposed algorithm also yields the same buffer capacities as the GA. The results are summarized in Table 17.6.

17.2 MULTIPLE ROBOT IN ASSEMBLY OPERATIONS

In number of automated cells, especially in electronic assemblies, two robots operate simultaneously to assemble a unit. They work in the same work area on the product while collecting parts from their respective part storage areas. The objective is to assign tasks (parts) to each robot such that they do not collide with each other while completing the assembly in the shortest possible time. Generally, tasks are independent of each other and can be performed in any sequence. The assembly is complete when all the tasks are done (Figure 17.6).

There are two observations in the problem. Collision of robots can be avoided if we prevent them from being in the same work area (or work envelope) at the same time, and assembly operation would be completed earliest if the ideal time in the work area is as small as possible.

We can divide the robot time needed for each component into two parts—one outside the work area, and the second, inside the work area. The outside work area time includes traveling from outside the work envelope to reach for the part, grabbing the part, lifting it, and bringing it back to the work area envelope. The inside time includes moving inside the work envelope to the work area from assembling the part to the work piece, releasing the grip and clearing the work piece and traveling back to the edge of the work envelope. So, for each part to be assembled on a product,

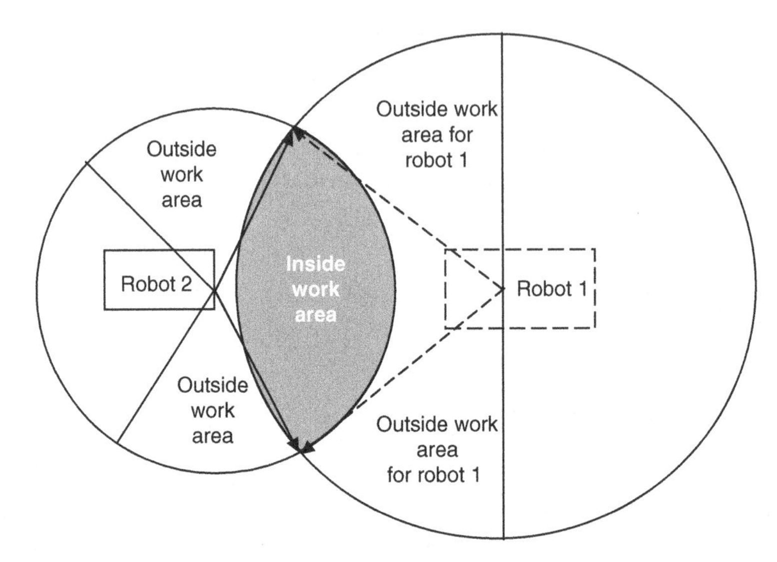

FIGURE 17.6 Two robots in work area.

the time required to perform all the tasks outside the work envelope are denoted as an outside task time, and the sum of the times needed for all the tasks inside the work envelope are called inside task time.

As there are multiple components to be assembled to make the final product, and we want to minimize the idle time in the work area, one observation is obvious. The outside time element for the next planned task should be less than or equal to the inside time value for the present task that is in the work area. In such case, the alternate robot will be waiting to enter the work area while the present robot is completing its assignment, and the work area will have no idle time.

Robotic assembly operations are performed on continuous basis for a long period of time. Once one product is made, the same cycle repeats itself, and therefore, we can start planning with any task in the list and complete all other tasks. The next cycle will repeat itself, starting with the initial task chosen.

The procedure for assigning the tasks between two robots is as follows.

1. List all tasks (elements) in the assembly, listing for each outside time and inside time. The sum of outside and inside times is the assembly time or the time for the robot to perform the task.
2. Select any task as the first task for the sequence to be formed. Place this task at the beginning and again at the of the sequence that we are developing.
3. The subsequent assignments are done alternately in forward and backward directions of the sequence.

In the forward direction, observe the inside time for the task that is presently in the work area. Select the next task in the sequence such that its outside time is equal to or as close as possible but less than the observed inside time. This guarantees no idle time at the work center. Remember that the tasks are assigned to alternate robots and, hence, if outside time of the immediately preceding task is larger than the inside time for the next task, then there is idle time in the work area, leading to a delay in completion of the assembly. If no such element can be found, then select a component that has minimum positive deviation (difference between outside time of the next task and inside time of the present task), which is a part of the total idle time in the system.

In the backward direction, we reverse the procedure. Start with the same initial task (it is the first task for the next assembly). Note the outside time of the present task. Select the next task whose inside time is equal to or minimally larger than the selected outside time. If no such task is found, then choose the one with least negative difference.

When both sequences merge in the center, if we have even number of tasks in the system then we will have two task left that are yet to be placed. We can develop two independent competed sequences. First, when out of two remaining tasks, a task is chosen following the forward sequence, and the last task is assigned and in the second sequence, a task is chosen following backward sequence, and the remaining is assigned. From these two sequences, choose the one that gives minimum work area idle time, as the best sequence.

When total number of tasks are odd, there is only one task that can be placed in the middle of the sequence, and two sequences are not formed.

Sometimes, the choice of the next task is not very clear. We may have to try possible alternatives and select the one that crates a minimum idle time in the entire sequence.

17.2.1 Illustrative Example

Suppose we have nine tasks within the assembly and two robots performing the operations.

Task	1	2	3	4	5	6	7	8	9
Inside time	2.9	3.6	6.3	2.8	1.3	4.2	5.0	4.0	4.2
Outside time	3.0	1.2	4.2	3.8	2.0	4.6	3.2	1.5	2.8

Choose a task, say task 1, and assign it to start and end of the sequence.

	Start of Sequence	End of Sequence
Task	1	1
Inside time	2.9	2.9
Outside time	3.0	3.0
⟶ Forward direction		⟵ Backward direction

The next task selected in forward direction is task 9, with outside time of 2.8, just smaller than the inside time of task 1. The task selected in the backward direction is task 2 with inside time of 3.6, just larger than the outside time of task 1. The idle time for task 9 is = max{0, (outside time for task 9 – inside time of task 1)} = 0. And, for task 1 in backward direction, the idle time = $\max(0, 3 - 3.9) = 0$.

	Start of Sequence		End of Sequence	
Task	1	9	2	1
Inside time	2.9	4.2	3.6	2.9
Outside time	3.0	2.8	1.2	3.0
Idle time	0	0		
⟶ Forward direction			⟵ Backward direction	

Continuing in similar manner, the next task assignments are 3 and 5, respectively.

	Start of Sequence			End of Sequence		
Task	1	9	3	5	2	1
Inside time	2.9	4.2	6.3	1.3	3.6	2.9
Outside time	3.0	2.8	4.2	2.0	1.2	3.0
Idle time	0	0	0	0		
⟶ Forward direction				⟵ Backward direction		

Continuing in similar manner, the final sequence is as follows:

	Start of Sequence						End of Sequence			
Task	1	9	3	6	7	8	4	5	2	1
Inside time	2.9	4.2	6.3	4.2	5.0	4.0	2.8	1.3	3.6	2.9
Outside time	3.0	2.8	4.2	4.6	3.2	1.5	3.8	2.0	1.2	3.0
Idle time	0	0	0	0	0	0	0	0	0	
⟶ Forward direction							⟵ Backward direction			

Since this sequence has the zero total idle time, it is an optimum sequence. At times, we may have to try a few alternatives if it is suspected that one may reduce the idle time. For example, in choosing task 2 before task 1 in the backward direction, with the outside time of 3.0, we had a choice to select task 2 with 3 – 3.6 difference between outside and inside times or task 4 with 2.8 units of inside time. We choose task 3 since 3.6 > 3.0, which crates 0 idle time instead of task 4, which would have crated $3.0 - 2.8 = 0.2$ units of idle time, even though the difference between 3.8 and 3 is larger than 3 and 2.8. In some other problem, we may have to try both alternatives.

Jobs are assigned to robots from the sequence in alternate manner till the cycle for each robot repeats. In our example, the assignments are as follows.

Robot I:	1	3	7	4	2	9	6	8	5	1	Repeat
Robot II:	9	6	8	5	1	3	7	4	2	9	Repeat

17.2.2 More than Two Robots

In case we have n multiple robots to perform the assembly, the same procedure that was illustrated for two robots can be applied with few variations. Apply only the forward phase. For the next task, the outside task time, needs to be less than or equal to sum of $(n - 1)$ inside times of previously assigned sequential tasks.

For example, suppose we have three robots to perform an assembly. There are many sequences can be made, but just to examine if the previous sequence is optimum, check the following. The outside time for task 3, 4.2, should be greater than the sum of inside times of previous two elements, namely 1 and 9. In this case, it is $2.9 + 4.2 > 4.2$. A similar analysis shows the sequence is optimum.

	Start of Sequence						End of Sequence			
Task	1	9	3	6	7	8	4	5	2	1
Inside time	2.9	4.2	6.3	4.2	5.0	4.0	2.8	1.3	3.6	2.9
Outside time	3.0	2.8	4.2	4.6	3.2	1.5	3.8	2.0	1.2	3.0
Idle time		0	0	0	0	0	0	0	0	0

The task assignments are as follows:

Robot I	1	6	4	1	Repeat the sequence
Robot II	9	7	5	9	Repeat the sequence
Robot III	3	8	2	3	Repeat the sequence

EXERCISE

17.1 For a three-machine, two-group problem, develop the optimum schedule.

		Machines		
	Job j	1	2	3
Group I	1	0	0	0
	2	25	70	14
	3	0	58	9
Group II	1	56	32	0
	2	1	98	95
	3	66	74	98
	4	0	10	0
	5	0	0	0
Setup time per group	Group I	2	20	2
	Group II	40	54	58

17.2 For five-machine, three-group problem, develop the optimum sequence.

		Machine				
	Job *j*	**1**	**2**	**3**	**4**	**5**
Group I	1	0	0	0	24	0
	2	50	70	64	69	0
Group II	1	66	42	0	92	33
	2	1	98	95	0	0
	3	66	94	98	0	0
Group III	1	0	86	52	35	76
	2	0	0	0	43	0
Setup time per group	Group I	2	20	2	14	27
	Group II	45	74	82	80	55
	Group III	18	23	38	59	91

17.3 Develop a robot assembly sequence with two robots, and calculate the cycle time.

Task	1	2	3	4	5	6	7	8	9
Inside time	2.6	4.2	5.2	3.8	1.7	5.1	5.7	4.6	5.0
Outside time	4.1	1.9	5.1	4.4	2.7	5.3	4.0	2.2	2.3

17.4 Develop a robot assembly sequence with 3 robots. Also determine the cycle time.

Task	1	2	3	4	5	6	7	8
Inside time	3.5	4.6	3.8	4.7	5.4	6.6	7.3	5.3
Outside time	4.2	3.1	5.6	4.5	5.8	4.2	3.8	6.3
Task	9	10	11	12	13	14	15	
Inside time	7.2	3.8	6.3	2.4	1.5	3.6	7.6	
Outside time	7.2	6.4	6.2	2.4	5.3	4.2	2.8	

REFERENCES AND SUGGESTED READINGS

Nawaz, M., E.E. Enscore Jr., and I. Ham. 1983. "A Heuristic Algorithm for the *m*-Machine, *n*-Job Flow-shop Sequencing Problem" *Omega*, 11(1): 91–95.

Schaller, J. 2000. "A Comparison of Heuristics for Family and Job Scheduling in a Flow-Line Manufacturing Cell" *International Journal of Production Research*, 38(2): 287–308.

Skorin-Kapov, J. and A.J. Vakharia. 1993."Scheduling a Flow-Line Manufacturing Cell: A Tabu Search Approach" *International Journal of Production Research*, 31(7): 1721–1734.

Wemmerlov, U. and A.J. Vakharia. 1991. "Job and Family Scheduling of a Flow-Line Manufacturing Cell: A Simulation Study" *IIE Transactions*, 23(4): 383–393.

Yang, D.-L. and M.-S. Chern. 2000 "Two-Machine Flowshop Group Scheduling Problem" *Computers and Operations Research*, 27: 975–985.

Appendix: Computer Program Description

This chapter gives a brief description about the computer programs that have been distributed along with this book in the diskette. The title of the programs and their executable file names are given, followed by the program description. All text in *italics* are messages displayed by the programs and all text in **bold** are user inputs. All the programs can be run by typing the executable file names followed by the ENTER key at the DOS command prompt. For example, to run single-machine common due date problem, type,

C:> COMMON.EXE

All the single-machine problem programs use the Menu format as shown in Figure A.1. As soon as the single-machine programs are run, the title of the program is displayed in the center, and the following message appears at the bottom,

Press Enter to continue...

The user should respond by pressing the ENTER key. Once the user responds by pressing the ENTER key, the Menu as shown in Figure A.1 appears. Option 1 is used to **Input** a new set of data, option 2, to **Execute** the program, option 3, to **Load** a set of data from a file (already saved), option 4, to **Save** the set of data currently in memory onto a file, option 5, for **Data correction**, that, is, for making changes to the data currently in memory, and option 6, to **Exit** or quit the program. Choosing option 6 will take the user back to the DOS prompt.

The first nine programs deal with single-machine problems. The name of the executable files are shown within parentheses.

Main Menu

1. *Input new data*
2. *Execute program*
3. *Load data from a file*
4. *Save data to a file*
5. *Data correction routine*
6. *Exit*

FIGURE A.1 Main Menu for all single-machine problem programs.

1. Single-machine problem with early and late penalties (SINGLE.EXE)

a. When option 1 is chosen, that is, Input data, the user is prompted for the number of jobs.

Enter the number of jobs:

The user should respond by entering the number of jobs to be processed on the machine. Once the user responds with an integer value, the next screen appears. For example, if the user enters **3**, the following screen appears.

JOB	*PROC. TIME*	*DUE DATE*	*LATE PENALTY*	*EARLY PENALTY*
1				
2				
3				

Here, each row will contain information pertaining to a job. The numbers in the left-most column represent the job numbers. Each job will require the following information to be entered in separate columns. The time taken for the job to be processed on the machine, its due date (date of shipment), the penalty incurred for every time unit of delay, that is, the late penalty, and the penalty incurred for every time unit on early completion (storage cost, etc.), that is, the early penalty.

The cursor is initially positioned at the column for the first job's processing time. The user should respond by entering the processing time followed by the *Enter* key. If the value entered is a proper integer value, it will be accepted and the cursor will move to the next column for this job, that is, the due date and so on. If the value entered is not proper, the cursor stays at the same position for the user to reenter the value. Once all the information for this job is accepted, the cursor moves to the next row for the next job, and the same process is repeated.

If the number of jobs exceed 9 (maximum allowable rows in one screen), the user is prompted whether the information entered on this page is correct (y) or not (n). The user should respond either "y" for yes or "n" for no. If the users' response is "y," the next screen appears with the next set of nine job numbers, and the process repeats until the information for all the jobs are accepted. If the user responds with a "n" for a no, the following message appears,

Enter the job #, p/d/l/e:

The user should respond by entering the job number followed by a comma and either a *p* (for processing time), *d* (for due date), *l* (for late penalty), or *e* (for early penalty). The cursor is then positioned at the corresponding row as given by the job number and the corresponding column as given by either the processing time, due date, and late or early penalty. The user is allowed to make changes here. Once the change has been made, the user is again prompted whether the information entered on this page is correct (y) or not (n). This process is repeated until the user responds with a "y."

Once all the information is entered, the Main Menu is displayed again.

b. When option 2 is chosen, that is, Execute program, the user is prompted with the following message,

Give name of the output file <SINGLE.OUT>:

The user can respond by pressing the *Enter* key, in which case by default the output of the program is stored in the file SINGLE.OUT. The user can also enter a different file name for storing the output of the program. The program is then executed, the output is stored in the corresponding file, and the main menu is displayed again. This option works only when a set of data is already in memory.

c. When option 3 is chosen, that is, Load from file, the user is prompted to enter a file name to load a set of data to the memory. Once the data is loaded, the main menu is displayed again.

d. When option 4 is chosen, that is, Save to a file, the user is prompted to enter a file name to save the set of data currently in memory. Once the data is saved, the main menu is displayed again. This option works only when a set of data is already in memory.

e. When option 5 is chosen, that is, Data correction, the next screen appears with the information about the first nine jobs. This option works only when a set of data is already in memory.

The user is prompted whether the information displayed on this page is correct (y) or not (n). The user should respond with either "y" for yes or "n" for no. If the users' response is "y," the next screen appears with the next set of nine job numbers, and the process repeats until the information for all the jobs are verified. If the user responds with an "n" for a no, the following message appears,

Enter the job #, p/d/l/e:

The user should respond by entering the job number followed by a comma and either a *p* (for processing time), *d* (for due date), *l* (for late penalty), or *e* (for early penalty). The cursor is positioned at the corresponding row as given by the job number and the corresponding column as given by either the processing time, due date, late, or early penalty. The user is allowed to make changes here. Once the change has been made the user is again prompted whether the information entered on this page is correct "y" or not "n." This process is repeated until the user responds with a "y."

Once all the information is entered, the main menu is displayed again.

f. When option 6 is chosen, that is, Exit, the execution of the program is stopped.

2. Common due date problem (COMMON.EXE)

a. When option 1 is chosen, that is, Input data, the user is prompted for the number of jobs.

Enter the number of jobs:

The user should respond by entering the number of jobs to be processed on the machine. Once the user responds with an integer value, the next screen appears.

The following screen contains information pertaining to job number 1. Information for job number 2 is yet to be fed.

JOB	*PROC. TIME*	*DUE DATE*	*LATE PENALTY*	*EARLY PENALTY*
1	**10**	?	**3**	**1**
2				

Since an optimum due date is to be determined the column *DUE DATE* is skipped and a question mark is placed as soon as the user enters the processing time. The cursor then moves to the next column that is, *LATE PENALTY*, and so on.

b. When option 2 is chosen, that is, Execute program, the user is prompted with the following message,

Give name of the output file <COMMON.OUT>:

The user can respond by pressing the *Enter* key in which case by default the output of the program is stored in the file COMMON.OUT. The user can also enter a different file name for storing the output of the program. The program is then executed, and the output stored in the corresponding file and the Main menu is displayed again. This option works only when a set of data is already in memory.

Options **c**, **d**, **e**, and **f** remain the same as in single-machine problem with early and late penalties.

If the number of jobs exceed 9 (maximum allowable rows in one screen), the user is prompted whether the information entered on this page is correct (y) or not (n). The user should respond with either "y" for yes or "n" for no. If the users' response is "y," the next screen appears with the next set of nine job numbers and the process repeats until the information for all the jobs are accepted. If the user responds with an "n" for a no, the following message appears,

Enter the job #, p/l/e:

The user should respond by entering the job number followed by a comma and either a *p* (for processing time), *l* (for late penalty), or *e* (for early penalty). The cursor is then positioned at the corresponding row as given by the job number and the corresponding column. The user is allowed to make changes here. Once the change has been made, the user is again prompted whether the information entered on this page is correct (y) or not (n). This process is repeated until the user responds with a "y."

3. Early and late due dates problem (DUAL.EXE)

a. When option 1 is chosen, that is, Input data, the user is prompted for the number of jobs.

Enter the number of jobs:

The user should respond by entering the number of jobs to be processed on the machine. Once the user responds with an integer value, the next screen appears.

The following screen contains information pertaining to job number 1. Information for job number 2 is yet to be fed.

JOB	*PROC. TIME*	*EARLY DUE*	*LATE DUE*	*LATE PENALTY*	*EARLY PENALTY*
1	**10**	**15**	**20**	**4**	**1**
2					

EARLY DUE is the early due date, that is, if a job is completed before this date it will incur an early penalty and *LATE DUE* is the late due date, that is, if a job is completed after this date then it will incur a late penalty.

b. When option 2 is chosen, that is, Execute program, the user is prompted with the following message,

Give name of the output file <DUAL.OUT>:

The user can respond by pressing the *Enter* key in which case by default the output of the program is stored in the file DUAL.OUT. The user can also enter a different file name for storing the output of the program. The program is then executed and the output stored in the corresponding file and the Main menu is displayed again. This option works only when a set of data is already in memory.

Options **c**, **d**, **e**, and **f** remain the same as in single-machine problem with early and late penalties.

If the number of jobs exceed 9 (maximum allowable rows in one screen), the user is prompted whether the information entered on this page is correct (y) or not (n). The user should respond either "y" for yes or "n" for no. If the users' response is "y," the next screen appears with the next set of nine job numbers, and the process repeats until the information for all the jobs are accepted. If the user responds with an "n" for a no, the following message appears,

Enter the job #, p/d/u/l/e:

The user should respond by entering the job number followed by a comma and either a *p* (for processing time), *d* (for early due date), *u* (for late due date), *l* (for late penalty), or *e* (for early penalty). The cursor is then positioned at the corresponding row as given by the job number and the corresponding column. The user is allowed to make changes here. Once the change has been made the user is again prompted whether the information entered on this page is correct "y" or not "n." This process is repeated until the user responds with a "y."

4. Minimization of average delay (MTARDY.EXE)

a. When option 1 is chosen, that is, Input data, the user is prompted for the number of jobs.

Enter the number of jobs:

The user should respond by entering the number of jobs to be processed on the machine. Once the user responds with an integer value, the next screen appears.

The following screen contains information pertaining to job number 1. Information for job number 2 is yet to be fed.

JOB	*PROC. TIME*	*DUE DATE*
1	**10**	**15**
2		

b. When option 2 is chosen, that is, Execute program, the user is prompted with the following message,

Give name of the output file <MTARDY.OUT>:

The user can respond by pressing the *Enter* key, in which case by default the output of the program is stored in the file MTARDY.OUT. The user can also enter a different file name for storing the output of the program. The program is then executed and the output stored in the corresponding file and the Main menu is displayed again. This option works only when a set of data is already in memory.

Options **c**, **d**, **e**, and **f** remain the same as in single-machine problem with early and late penalties.

If the number of jobs exceed 9 (maximum allowable rows in one screen), the user is prompted whether the information entered on this page is correct (y) or not (n). The user should respond either "y" for yes or "n" for no. If the users' response is "y," the next screen appears with the next set of nine job numbers and the process repeats until the information for all the jobs are accepted. If the user responds with an "n" for a no, the following message appears,

Enter the job #, p/d:

The user should respond by entering the job number followed by a comma and either a *p* (for processing time) or a *d* (for due date). The cursor is then positioned at the corresponding row as given by the job number and the corresponding column. The user is allowed to make changes here. Once the change has been made, the user is again prompted whether the information entered on this page is correct (y) or not (n). This process is repeated until the user responds with a "y."

5. Minimization of maximum delay (TMAX.EXE)

a. When option 1 is chosen, that is, Input data, the user is prompted for the number of jobs.

Enter the number of jobs:

The user should respond by entering the number of jobs to be processed on the machine. Once the user responds with an integer value, the next screen appears. The following screen contains information pertaining to job number 1. Information for job number 2 is yet to be fed.

JOB	*PROC. TIME*	*DUE DATE*	*LATE PENALTY*	*EARLY PENALTY*
1	**10**	**20**	**3**	**1**
2				

b. When option 2 is chosen, that is, Execute program, the user is prompted with the following message,

Give name of the output file <TMAX.OUT>:

The user can respond by pressing the *Enter* key, in which case by default the output of the program is stored in the file TMAX.OUT. The user can also enter a different file name for storing the output of the program. The program is then executed and the output stored in the corresponding file and the Main menu is displayed again. This option works only when a set of data is already in memory.

Options **c**, **d**, **e**, and **f** remain the same as in single-machine problem, with early and late penalties.

If the number of jobs exceed 9 (maximum allowable rows in one screen), the user is prompted whether the information entered on this page is correct (y) or not (n). The user should respond either "y" for yes or "n" for no. If the users' response is "y," the next screen appears with the next set of nine job numbers, and the process repeats until the information for all the jobs are accepted. If the user responds with an "n" for a no, the following message appears,

Enter the job #, p/d/l/e:

The user should respond by entering the job number followed by a comma and either a *p* (for processing time), *d* (for due date), *l* (for late penalty), and *e* (for early penalty). The cursor is then positioned at the corresponding row as given by the job number and the corresponding column. The user is allowed to make changes here. Once the change has been made, the user is again prompted whether the information entered on this page is correct (y) or not (n). This process is repeated until the user responds with a "y."

6. Minimize maximum number of tardy jobs (NTARDY.EXE)

a. When option 1 is chosen, that is, Input data, the user is prompted for the number of jobs.

Enter the number of jobs:

The user should respond by entering the number of jobs to be processed on the machine. Once the user responds with an integer value, the next screen appears. The following screen contains information pertaining to job number 1. Information for job number 2 is yet to be fed.

JOB	*PROC. TIME*	*DUE DATE*
1	**10**	**2**
2		

b. When option 2 is chosen, that is, Execute program, the user is prompted with the following message,

Give name of the output file <NTARDY.OUT>:

The user can respond by pressing the *Enter* key, in which case by default the output of the program is stored in the file NTARDY.OUT. The user can also enter a different file name for storing the output of the program. The program is then executed and the output stored in the corresponding file and the Main menu is displayed again. This option works only when a set of data is already in memory.

Options **c**, **d**, **e**, and **f** remain the same as in single-machine problem with early and late penalties.

If the number of jobs exceed 9 (maximum allowable rows in one screen), the user is prompted whether the information entered on this page is correct (y) or not (n). The user should respond either "y" for yes or "n" for no. If the users' response is "y," the next screen appears with the next set of nine job numbers, and the process repeats until the information for all the jobs are accepted. If the user responds with an "n" for a no, the following message appears,

Enter the job #, p/d:

The user should respond by entering the job number followed by a comma and either a *p* (for processing time), *d* (for due date). The cursor is then positioned at the corresponding row as given by the job number and the corresponding column. The user is allowed to make changes here. Once the change has been made, the user is again prompted whether the information entered on this page is correct "y" or not "n." This process is repeated until the user responds with a "y."

7. Slack Introduction (SLACK.EXE)

a. When option 1 is chosen, that is, Input data, the user is prompted for the number of jobs.

Enter the number of jobs:

The user should respond by entering the number of jobs to be processed on the machine. Once the user responds with an integer value, the next screen appears. The following screen contains information pertaining to job number 1. Information for job number 2 is yet to be fed.

JOB	*PROC. TIME*	*DUE DATE*	*LATE PENALTY*	*EARLY PENALTY*
1	**10**	**20**	**3**	**1**
2				

b. When option 2 is chosen, that is, Execute program, the user is prompted with the following message,

Give name of the output file <SLACK.OUT>:

The user can respond by pressing the *Enter* key, in which case by default the output of the program is stored in the file SLACK.OUT. The user can also enter a different file name for storing the output of the program. The program is then executed and the output stored in the corresponding file and the Main menu is displayed again. This option works only when a set of data is already in memory.

Options **c**, **d**, **e**, and **f** remain the same as in single-machine problem with early and late penalties.

If the number of jobs exceed 9 (maximum allowable rows in one screen), the user is prompted whether the information entered on this page is correct (y) or not (n). The user should respond with either "y" for yes or "n" for no. If the users' response is "y," the next screen appears with the next set of nine job numbers, and the process repeats until the information for all the jobs are accepted. If the user responds with an "n" for a no, the following message appears,

Enter the job #, p/d/l/e:

The user should respond by entering the job number followed by a comma and either a *p* (for processing time), *d* (for due date), *l* (for late penalty), and *e* (for early penalty). The cursor is then positioned at the corresponding row as given by the job number and the corresponding column. The user is allowed to make changes here. Once the change has been made, the user is again prompted whether the information entered on this page is correct (y) or not (n). This process is repeated until the user responds with a "y."

8. Minimizing penalty for Jobs arriving at different times (ARRPENAL.EXE)

a. When option 1 is chosen, that is, Input data, the user is prompted for the number of jobs.

Enter the number of jobs:

The user should respond by entering the number of jobs to be processed on the machine. Once the user responds with an integer value, the next screen appears. The following screen contains information pertaining to job number 1. Information for job number 2 is yet to be fed.

JOB	*PROC. TIME*	*ARR. TIME*	*DUE DATE*	*LATE PENALTY*	*EARLY PENALTY*
1	**10**	**0**	**21**	**3**	**1**
2					

b. When option 2 is chosen, that is, Execute program, the user is prompted with the following message,

Give name of the output file <ARRPENAL.OUT>:

The user can respond by pressing the *Enter* key, in which case by default the output of the program is stored in the file ARRPENAL.OUT. The user can also enter a different file name for storing the output of the program. The program is then executed and the output stored in the corresponding file and the main menu is displayed again. This option works only when a set of data is already in memory.

Options **c**, **d**, **e**, and **f** remain the same as in single-machine problem with early and late penalties.

If the number of jobs exceed 9 (maximum allowable rows in one screen), the user is prompted whether the information entered on this page is correct (y) or not (n). The user should respond with either "y" for yes or "n" for no. If the users' response is "y," the next screen appears with the next set of nine job numbers, and the process repeats until the information for all the jobs are accepted. If the user responds with an "n" for a no, the following message appears,

Enter the job #, p/a/d/l/e:

The user should respond by entering the job number, followed by a comma and either a *p* (for processing time), *a* (for arrival time), *d* (for due date), *l* (for late penalty), and *e* (for early penalty). The cursor is then positioned at the corresponding row as given by the job number and the corresponding column. The user is allowed to make changes here. Once the change has been made, the user is again prompted whether the information entered on this page is correct (y) or not (n). This process is repeated until the user responds with a "y."

9. Batch sequencing with sequence dependent setup times (BATCH.EXE)

a. When option 1 is chosen, that is, Input data, the user is prompted for the number of jobs.

Enter the number of jobs:

The user should respond by entering the number of jobs to be processed on the machine. Once the user responds with an integer value, the next screen appears. The following screen contains information pertaining to job number 1. Information for job number 2 is yet to be fed.

JOB	*PROC. TIME*	*DUE DATE*	*JOB TYPE*	*SETUP TIME*
1	**10**	**20**	**1**	**3**
2				

b. When option 2 is chosen, that is, Execute program, the user is prompted with the following message,

Give name of the output file <BATCH.OUT>:

The user can respond by pressing the *Enter* key, in which case by default the output of the program is stored in the file BATCH.OUT. The user can also enter a different file name for storing the output of the program. The program is then executed and the output stored in the corresponding file, and the Main menu is displayed again. This option works only when a set of data is already in memory.

Options **c**, **d**, **e**, and **f** remain the same as in single-machine problem with early and late penalties.

If the number of jobs exceed 9 (maximum allowable rows in one screen), the user is prompted whether the information entered on this page is correct (y) or not (n). The user should respond either "y" for yes or "n" for no. If the users' response is "y," the next screen appears with the next set of nine job numbers, and the process repeats until the information for all the jobs are accepted. If the user responds with an "n" for a no, the following message appears,

Enter the job #, p/d/j/s:

The user should respond by entering the job number followed by a comma and either a *p* (for processing time), *d* (for due date), *j* (for job type), and *s* (for setup time). The cursor is then positioned at the corresponding row as given by the job number and the corresponding column. The user is allowed to make changes here. Once the change has been made, the user is again prompted whether the information entered on this page is correct (y) or not (n). This process is repeated until the user responds with a "y."

The programs that are discussed from now on do not follow the same Menu format as discussed before.

10. Minimum tool changeover problem (FLEX.EXE)

As soon as this program is run, the following message appears,

Do you want to give an input interactively or from a file?

1. Input Interactively from Keyboard
2. Input from a file

Enter your choice by typing '1' or '2':

If the user selects option '1', the following messages appear. The user should enter appropriate values for each as follows,

Enter the maximum magazine capacity: **2**
Enter the total number of jobs: **2**
Enter the total number of tools: **3**
Enter the tools required for job 1: **1 2 3**
Enter the tools required for job 2: **1 2**
Is this OK? Y/N:

If the user selects "n" for no, then he/she is allowed to make changes to the data. If the user selects "y" for yes the following message appears.

Do you want to save it in a file? y/n:

If the user chooses "y" then the following message appears:

Enter the filename with path for saving input: **/scheduling/flex.inp**

Finally the percent value is prompted. The user should enter a value between 0 and 1.

Enter the percent value: **0.5**

If the user chooses option '2' in the beginning, that is, input from a file, then the following messages appear.

Enter the maximum magazine capacity: **3**
Enter the input file name with path: **/scheduling/flex.inp**

The data is read from the file and displayed. The following message then appears.

Is this OK? Y/N:

If the user chooses "n" for no, then the user is allowed to make changes to the data. The following messages then follow,

Do you want to save it in a file? y/n:

If the user chooses "y," then the following message appears:

Enter the filename with path for saving input: **/scheduling/flex.inp**

Finally the percent value is prompted. The user should enter a value between 0 and 1.

Enter the percent value: **0.5**

The output of this program is stored in the file specified by the user (/scheduling/flex.inp), and the program is terminated.

11. Shifting Bottleneck Job Shop problem for Makespan Minimization (MKBOTTLE.EXE)

As soon as this program is run, the following message appears,

Enter the number of jobs: **3**
Enter the number of m/cs: **2**
Enter the machine sequence for job 1
Example - Machine sequence 3,2,1 should be entered as 3-2-1
Enter here -> **1-2**
Enter proc. time for job 1 on machine 1: **11**
Enter proc. time for job 1 on machine 2: **6**
Enter the machine sequence for job 2
Example - Machine sequence 3,2,1 should be entered as 3-2-1
Enter here -> **2-1**
Enter proc. time for job 2 on machine 2: **7**
Enter proc. time for job 2 on machine 1: **5**
Enter the name of the output file: **/scheduling/job_shop.out**

The output of the program is stored in the file specified by the user (/scheduling/job_shop.out), and the program is terminated.

12. Shifting Bottleneck Job Shop problem for Penalty Minimization (MKBOTTLE.EXE)

As soon as this program is run, the following message appears,

Enter the number of jobs: **2**
Enter the number of m/cs: **2**
Enter Due date for job 1: **32**
Enter Late penalty for job 1: **5**
Enter Early penalty for job 1: 1
Enter the machine sequence for job 1
Example - Machine sequence 3,2,1 should be entered as 3-2-1
Enter here -> **1-2**
Enter proc. time for job 1 on machine 1: **11**
Enter proc. time for job 1 on machine 2: **6**

Enter Due date for job 2: **23**
Enter Late penalty for job 2: **2**
Enter Early penalty for job 2: **0**
Enter the machine sequence for job 2
Example - Machine sequence 3,2,1 should be entered as 3-2-1
Enter here -> **2-1**
Enter proc. time for job 2 on machine 2: **7**
Enter proc. time for job 2 on machine 1: **5**
Enter the name of the output file: **/scheduling/job_shop.out**

The output of the program is stored in the file specified by the user (/scheduling/job_shop.out), and the program is terminated

13. Shifting Bottleneck for Optimizing makespan as well as penalty (OPT_MIX.EXE)

This program has the same user interface as that of Shifting Bottleneck for penalty minimization.

14. CEXSPT Job Shop problem with due date constraints (CEXSPT.EXE)

This program has the same user interface as that of Shifting Bottleneck for penalty minimization. The only difference being that this method does not consider late and early penalties for the jobs, and so these information are not asked for.

15. Open Shop problem for makespan minimization (OPENSHOP.EXE)

The following messages appear when this program is run.

Enter no. of jobs: **2**
Enter no. of machines: **2**
Do you want to make changes in the data you just entered (y/n):

If the user responds with a "y," he/she is allowed to make changes to the number of jobs are machine entered. If the user responds with an "n," the following message appears.

Enter the processing time for job 1 on machine 1: **3**
Enter the processing time for job 1 on machine 2: **4**
Enter the processing time for job 2 on machine 1: **6**
Enter the processing time for job 2 on machine 2: **3**
Do you want make changes in the processing times (y/n):

If the user responds with a "n," then the following message appears.

Enter the name of the output file: **/scheduling/openshop.out**

The output of the program is stored in the user-specified file (/scheduling/openshop.out)

If the user responds with a "y," then the following message appears.

Enter the job# for which processing time has to be changed: **2**
Enter the machine# for which processing time has to be changed: **1**
Enter the new processing time for job 2 on machine 1: **5**
Do you want to make any changes (y/n):

If the user responds with a "y," the entire process is repeated again; otherwise, the output file name is prompted as follows.

Enter the name of the output file: **/scheduling/openshop.out**

The output of the program is stored in the user-specified file (/scheduling/openshop.out), and the program is terminated.

Additional programs under development are listed here.

1. Monroe's algorithm
2. Parallel processing

Index

O

P

T

Printed in the United States
by Baker & Taylor Publisher Services